De Gruyter Series in Nonlinear Analysis and Applications 14

Editor in Chief
Jürgen Appell, Würzburg, Germany

Editors
Catherine Bandle, Basel, Switzerland
Alain Bensoussan, Richardson, Texas, USA
Avner Friedman, Columbus, Ohio, USA
Karl-Heinz Hoffmann, Munich, Germany
Mikio Kato, Kitakyushu, Japan
Umberto Mosco, Worcester, Massachusetts, USA
Louis Nirenberg, New York, USA
Boris N. Sadovsky, Voronezh, Russia
Alfonso Vignoli, Rome, Italy
Katrin Wendland, Freiburg, Germany

Luboš Pick
Alois Kufner
Oldřich John
Svatopluk Fučík

Function Spaces

Volume 1

2nd Revised and Extended Edition

De Gruyter

Mathematics Subject Classification 2010: 46E30, 46E35, 46E05, 47G10, 26D10, 26D15, 46B70, 46B42, 46B10.

ISBN 978-3-11-025041-1
e-ISBN 978-3-11-025042-8
ISSN 0941-813X

Library of Congress Cataloging-in-Publication Data

A CIP catalog record for this book has been applied for at the Library of Congress.

Bibliographic information published by the Deutsche Nationalbibliothek

The Deutsche Nationalbibliothek lists this publication in the Deutsche Nationalbibliografie; detailed bibliographic data are available in the Internet at http://dnb.dnb.de.

© 2013 Walter de Gruyter GmbH, Berlin/Boston

Typesetting: le-tex publishing services GmbH, Leipzig
Printing and binding: Hubert & Co. GmbH & Co. KG, Göttingen
∞ Printed on acid-free paper
Printed in Germany

www.degruyter.com

Preface

The book "Function Spaces" [126], published in 1977 by Academia Publishing House of the Czechoslovak Academy of Sciences in Prague and the Noordhoff International Publishing in Leyden, proved over several decades to be a useful tool for specialists working in many different areas of mathematics and its applications. It has, though, for quite some time been unavailable. Since the 1970s, many other books dedicated to the study of function spaces and related topics have appeared. Nevertheless, we saw signs that a new edition of this book could be useful, which of course would be revised according to the rapid development in the field of function spaces over the past 35 years and upgraded in part by a number of new results.

The current book is an attempt to make a step in this direction. Thanks to the effort spent by the de Gruyter Publishing House, the three authors signed below took upon the task. They used as their point of departure the initial book, thus, the current version now has four authors.

It turned out during the preparation of the material for the new edition that the upgraded text is too long for a single monograph. Consequently, we decided to split the material into two volumes.

The first volume is devoted to the study of function spaces, based on intrinsic properties of a function such as its size, continuity, smoothness, various forms of control over the mean oscillation, and so on. The second volume will be dedicated to the study of function spaces of Sobolev type, in which the key notion is the weak derivative of a function of several variables.

During almost a century of their existence, Lebesgue spaces have constantly played a primary role in analysis. However, it has been known almost from the very beginning that the Lebesgue scale is not sufficiently general to provide a satisfactory description of fine properties of functions required by practical tasks. This was noted during the early 1920s by Kolmogorov, Zygmund, Titchmarsh and others, mostly in connection with research of properties of operators on function spaces. Thus, naturally, during the first half of the twentieth century, new fine scales of function spaces have been introduced. The efforts of Young, Orlicz, Hardy, Littlewood, Zygmund, Halperin, Köthe, Marcinkiewicz, Lorentz, Luxemburg, Morrey, Campanato and many others resulted in the development of a powerful and qualitatively new mathematical discipline of function spaces.

This text is intended to be a motivated introduction to the subject of function spaces. It contains important basic information on various kinds of function spaces such as their functional-analytic or measure-theoretic properties, as well as their important characteristics such as mutual embeddings, duality relations and so on. Hence, it can be considered as a reference book and a pointer to other sources.

Summary of the text

The text opens with Chapter 1, which has a purely preliminary character. It contains basic information about vector spaces, topological, metric, normed and modular spaces as well as important ingredients from classical functional analysis. In comparison with the first edition, this chapter was enlarged considerably. In particular, a thorough treatment of the theory of measure and integral was added. There are of course many possible sources available for this purpose. We mostly use the book of Rana [185], where the interested reader will find detailed proofs and many further details. In addition, some new material was added on topological spaces and modular spaces.

In Chapter 2 we present just a very short elementary discussion about spaces of continuous and smooth functions, including Hölder and Lipschitz spaces.

Chapter 3 contains the material concerning Lebesgue spaces from the first edition, only slightly modified and upgraded. Some parts (for instance Section 3.11, devoted to the study of weighted Hardy inequalities) are completely new. A few basic facts about sequence spaces were added, and the topic of modes of convergence was reworked.

In Chapter 4 we present the study of basic properties of Orlicz spaces. Again, this chapter as compared to the first edition is only slightly modified. For traditional reasons, the functions studied in Chapters 3 and 4 are assumed to act on an open subset of the Euclidean space. The main reason for this restriction is the further use of these spaces in Sobolev-type spaces, which will be dealt with in the second volume. For the study of these spaces themselves, of course, such restriction is not necessary. In fact, it is later reduced in the frame of the study of general Banach function spaces from Chapter 6 onwards, noting that Lebesgue and Orlicz spaces are particular examples of rearrangement-invariant (r.i.) Banach function spaces.

Chapter 5 contains elementary properties of Morrey- and Campanato-type spaces in which the mean oscillation of a function is measured. The functions in this chapter are assumed to act on nice domains in the Euclidean space. This chapter is practically unchanged compared to the first edition.

In Chapter 6 we develop the basic theory of Banach function norms and Banach function spaces, which was in some sense a culmination of efforts to cover Orlicz spaces with other types of spaces under a common theme, performed from the 1930s to the 1950s by Orlicz, Lorentz, Luxemburg, Zaanen, Köthe, Halperin and others. This material gradually appeared mostly in works of the mentioned authors. The systematic treatment of this topic can also be found in the following books: Luxemburg and Zaanen [142] or Bennett and Sharpley [14]. Here (except for some minor additions and changes) we follow the excellent exposition of this subject from [14, Chapter 1] almost verbatim (even though our ultimate goal is slightly different, as we are not so much aimed towards interpolation theory).

In Chapter 7 we turn our attention to function spaces in which the norm is purely determined by the size of a given function. Hence, we study the distribution function and the nonincreasing rearrangement, and develop the resulting structure, the so-called rearrangement-invariant (r.i.) function spaces. This stuff can be in some sense traced back to as far as to the 1880s results of Steiner [214], but its first systematic treatment was done only in the 1930s in the work of Hardy, Littlewood and Pólya [101]. Here we once again mostly follow the book by Bennett and Sharpley [14]. The only significant addition is Section 7.11 in which more recent material concerning the important relation between function spaces called almost-compact embedding is studied in great detail. Our main source in this section is [206].

The knowledge of the notion of the nonincreasing rearrangement of a function now enables us to construct new scales of function spaces that could not exist without it. We have seen that, for example, Lebesgue and Orlicz spaces just happen to be rearrangement-invariant Banach function spaces, but for their initial definition we did not need to know this at all. In subsequent chapters however we study function spaces for whose definition the nonincreasing rearrangement is indispensable. First such class of function spaces is that of two-parameter Lorentz spaces, whose elementary properties are studied in Chapter 8. These spaces gradually emerged during the 1950s and 1960s through the efforts of Lorentz, Calderón, Hunt, Peetre, O'Neil, Weiss, Oaklander and others, mostly in connection with some kind of interpolation. We cover some of their elementary properties, embedding characteristics and duality relations.

In the subsequent two chapters we turn our attention to some of the most interesting generalizations of the two-parameter Lorentz spaces. In Chapter 9, we study the important scale of the so-called Lorentz–Zygmund spaces, invented in the 1980s by Bennett, Rudnick and Sharpley, and their generalization to four-parameter spaces, studied in the 1990s by Edmunds, Gurka, Opic and others in connection with various limiting or critical-state problems concerning the action of operators on function spaces. These spaces turned out to be extremely useful in various extremal problems concerning Sobolev inequalities as well as limiting properties of operators. They cover important previously known function classes such as Lebesgue and Lorentz spaces, Zygmund classes of both logarithmic and exponential type, and also the space $L^{\infty,n;-1}$, which appeared (under various different symbols) in connection with the optimal target space for a limiting Sobolev inequality in works by Maz'ya, Hansson, Brézis–Wainger and others. As we know very well from our own research, these spaces often arise in various practical tasks. We thus study them here in great detail, concentrating on their embedding and duality characteristics and basic functional properties. We mostly follow [172] and [73].

In Chapter 10, we investigate the so-called classical Lorentz spaces. These spaces are currently known to be of three different types (denoted as of type Λ, Γ and S) and have been widely studied by many authors. We first concentrate on their basic functional properties such as nontriviality, linearity, normability, quasi-normability, lattice property and so on, and then we focus on their embedding relations. These questions are highly nontrivial and they require the development of certain new methods some of which we present in detail. This is the case for instance of the inequalities involving suprema, which we just quote, or of the modification of the Hardy inequality which concerns two integral operators rather than just one, which we present in detail. The spaces of type S are known to be of great interest because of the gradient inequalities they govern and their connection to Sobolev and Besov spaces that will be studied in the second volume of this book. In particular, they contain functions with controlled nonincreasing rearrangement of mean oscillation. Here we mix classical results with recent ones scattered in many papers. We follow [33, 34, 35, 56, 89]. At some occasions we do not provide all the details of the proofs because this would increase the length of the text enormously, and restrict ourselves to the hints and references.

Finally, Chapter 11 is devoted to a brief account of the generalized Lebesgue spaces with variable exponent. This topic has become very fashionable in recent years and there exist entire schools of top scientists investigating all kinds of variable exponent spaces and their generalizations. We restrict ourselves to some basic information about these spaces, following mostly [122] and [140], and we refer the reader interested in their deep study to the recent book by Diening, Hästö, Harjulehto and Růžička [63].

Acknowledgments

The main objective of this book is to provide pure mathematicians as well as applied scientists with a handbook containing a summary of results concerning various types of function spaces that might be useful for a broad variety of applications. Therefore, naturally, in most cases we do not claim any originality. We took great effort to give full credit for all the results appearing in the text to its discoverers but, obviously, it is almost an impossible task to trace the origins of every detail.

It would be impossible to list all the authors, colleagues and friends who have influenced us in preparation of the second edition of the text.

The exposition was partly inspired by important books in the field such as those by Bennett and Sharpley [14], Maz'ya [149], Krasnosel'skii and Rutitskii [123], Rana [185], Diening, Hästö, Harjulehto and Růžička [63] and others.

Luboš Pick wishes to express his special deep gratitude to Ms. Lenka Slavíková for many stimulating discussions and suggestions that led to great improvement of some parts of the text.

We thank Mr. Komil Kuliev, Ms. Guli Kulieva and Ms. Eva Ritterová for their help with the preparation of the manuscript in LaTeX.

We would like to thank Ms. Anja Möbius from the publishing house De Gruyter for her collaboration and mostly for her patience.

Finally, the authors would be grateful for critical comments and suggestions for later improvements.

The three authors below would like to dedicate this second edition to the memory of the fourth author, the late Professor Svatopluk Fučík, who passed away prematurely not long after the appearance of the first edition of the book [126].

Prague, September 2012 	Luboš Pick, Alois Kufner and Oldřich John

Contents

Preface		v
1	**Preliminaries**	1
	1.1 Vector space	1
	1.2 Topological spaces	2
	1.3 Metric, metric space	6
	1.4 Norm, normed linear space	6
	1.5 Modular spaces	7
	1.6 Inner product, inner product space	10
	1.7 Convergence, Cauchy sequences	11
	1.8 Density, separability	12
	1.9 Completeness	12
	1.10 Subspaces	13
	1.11 Products of spaces	14
	1.12 Schauder bases	14
	1.13 Compactness	15
	1.14 Operators (mappings)	16
	1.15 Isomorphism, embeddings	18
	1.16 Continuous linear functionals	19
	1.17 Dual space, weak convergence	20
	1.18 The principle of uniform boundedness	21
	1.19 Reflexivity	21
	1.20 Measure spaces: general extension theory	22
	1.21 The Lebesgue measure and integral	29
	1.22 Modes of convergence	34
	1.23 Systems of seminorms, Hahn–Saks theorem	36

2 Spaces of smooth functions — 38
- 2.1 Multiindices and derivatives — 38
- 2.2 Classes of continuous and smooth functions — 39
- 2.3 Completeness — 43
- 2.4 Separability, bases — 45
- 2.5 Compactness — 51
- 2.6 Continuous linear functionals — 55
- 2.7 Extension of functions — 59

3 Lebesgue spaces — 62
- 3.1 $\mathscr{L}^p$-classes — 62
- 3.2 Lebesgue spaces — 66
- 3.3 Mean continuity — 67
- 3.4 Mollifiers — 69
- 3.5 Density of smooth functions — 71
- 3.6 Separability — 71
- 3.7 Completeness — 72
- 3.8 The dual space — 74
- 3.9 Reflexivity — 78
- 3.10 The space L^∞ — 78
- 3.11 Hardy inequalities — 83
- 3.12 Sequence spaces — 92
- 3.13 Modes of convergence — 93
- 3.14 Compact subsets — 94
- 3.15 Weak convergence — 95
- 3.16 Isomorphism of $L^p(\Omega)$ and $L^p(0, \mu(\Omega))$ — 96
- 3.17 Schauder bases — 97
- 3.18 Weak Lebesgue spaces — 101
- 3.19 Remarks — 104

4 Orlicz spaces — 108
- 4.1 Introduction — 108
- 4.2 Young function, Jensen inequality — 109
- 4.3 Complementary functions — 115

	4.4	The Δ_2-condition 119
	4.5	Comparison of Orlicz classes 122
	4.6	Orlicz spaces 126
	4.7	Hölder inequality in Orlicz spaces 131
	4.8	The Luxemburg norm 134
	4.9	Completeness of Orlicz spaces 137
	4.10	Convergence in Orlicz spaces 138
	4.11	Separability 143
	4.12	The space $E^\Phi(\Omega)$ 145
	4.13	Continuous linear functionals 151
	4.14	Compact subsets of Orlicz spaces 155
	4.15	Further properties of Orlicz spaces 161
	4.16	Isomorphism properties, Schauder bases 163
	4.17	Comparison of Orlicz spaces 166
5	**Morrey and Campanato spaces**	**173**
	5.1	Introduction 173
	5.2	Marcinkiewicz spaces 173
	5.3	Morrey and Campanato spaces 176
	5.4	Completeness 178
	5.5	Relations to Lebesgue spaces 178
	5.6	Some lemmas 181
	5.7	Embeddings 185
	5.8	The John–Nirenberg space 187
	5.9	Another definition of the space $JN(Q)$ 194
	5.10	Spaces $N_{p,\lambda}(Q)$ 197
	5.11	Miscellaneous remarks 199
6	**Banach function spaces**	**203**
	6.1	Banach function spaces 203
	6.2	Associate space 209
	6.3	Absolute continuity of the norm 216
	6.4	Reflexivity of Banach function spaces 223
	6.5	Separability in Banach function spaces 228

7 Rearrangement-invariant spaces — 237
  - 7.1 Nonincreasing rearrangements — 237
  - 7.2 Hardy–Littlewood inequality — 241
  - 7.3 Resonant measure spaces — 243
  - 7.4 Maximal nonincreasing rearrangement — 249
  - 7.5 Hardy lemma — 251
  - 7.6 Rearrangement-invariant spaces — 253
  - 7.7 Hardy–Littlewood–Pólya principle — 255
  - 7.8 Luxemburg representation theorem — 256
  - 7.9 Fundamental function — 259
  - 7.10 Endpoint spaces — 264
  - 7.11 Almost-compact embeddings — 275
  - 7.12 Gould space — 292

8 Lorentz spaces — 301
  - 8.1 Definition and basic properties — 301
  - 8.2 Embeddings between Lorentz spaces — 305
  - 8.3 The associate space — 307
  - 8.4 The fundamental function — 309
  - 8.5 Absolute continuity of norm — 309
  - 8.6 Remarks on $\|\cdot\|_{1,\infty}$ — 311

9 Generalized Lorentz–Zygmund spaces — 313
  - 9.1 Measure-preserving transformations — 313
  - 9.2 Basic properties — 314
  - 9.3 Nontriviality — 317
  - 9.4 Fundamental function — 318
  - 9.5 Embeddings between Generalized Lorentz–Zygmund spaces — 320
  - 9.6 The associate space — 332
  - 9.7 When Generalized Lorentz–Zygmund space is Banach function space — 353
  - 9.8 Generalized Lorentz–Zygmund spaces and Orlicz spaces — 356
  - 9.9 Absolute continuity of norm — 367
  - 9.10 Lorentz–Zygmund spaces — 372
  - 9.11 Lorentz–Karamata spaces — 373

Contents

10 Classical Lorentz spaces   375

   10.1 Definition and basic properties 375

   10.2 Functional properties 380

   10.3 Embeddings 388

     10.3.1 Embeddings of type $\Lambda \hookrightarrow \Lambda$ 392

     10.3.2 Embeddings of type $\Lambda \hookrightarrow \Gamma$ 393

     10.3.3 Embeddings of type $\Gamma \hookrightarrow \Lambda$ 396

     10.3.4 Embeddings of type $\Gamma \hookrightarrow \Gamma$ 399

     10.3.5 The Halperin level function 401

     10.3.6 Embeddings of type $\Gamma^{p,\infty}(v) \hookrightarrow \Lambda^q(w)$ 404

     10.3.7 The single-weight case $\Gamma^{1,\infty}(v) \hookrightarrow \Lambda^1(v)$ 406

   10.4 Associate spaces 409

   10.5 Lorentz and Orlicz spaces 411

   10.6 Spaces measuring oscillation 412

   10.7 The missing case 425

   10.8 Embeddings 427

     10.8.1 Embeddings of type $S \hookrightarrow S$ 429

     10.8.2 Embeddings of type $\Gamma \hookrightarrow S$ and $S \hookrightarrow \Gamma$ 431

     10.8.3 Embeddings of type $\Lambda \hookrightarrow S$ and $S \hookrightarrow \Lambda$ 434

11 Variable-exponent Lebesgue spaces   437

   11.1 Introduction 437

   11.2 Basic properties 438

   11.3 Embedding relations 445

   11.4 Density of smooth functions 447

   11.5 Reflexivity and uniform convexity 450

   11.6 Radon–Nikodým property 453

   11.7 Daugavet property 455

Bibliography   459

Index   472

Chapter 1

Preliminaries

In this chapter we give a survey of concepts and results from functional analysis that will be used in the text. All results are stated without proofs, which can be found in standard monographs.

1.1 Vector space

Let X be a set of elements denoted by $u, v, w, \ldots$.

Definition 1.1.1. Let *addition* in X be defined, i.e. to every pair $u \in X$, $v \in X$ there corresponds an element $w \in X$ called the *sum* of u and v and denoted by $u + v$:

$$w = u + v.$$

Definition 1.1.2. Let *multiplication* by scalars be defined in X, i.e. to every real number λ (called a *scalar*) and every $u \in X$ there corresponds an element $w \in X$ called the *λ-multiple* of u and denoted by λu (or $\lambda \cdot u$):

$$w = \lambda u.$$

Definition 1.1.3. The set X with addition and multiplication by scalars defined in it is called a *real vector space* if the following axioms are satisfied:

(i) $u + v = v + u$ (symmetry);

(ii) $u + (v + w) = (u + v) + w$ (associativity);

(iii) in X there exists a uniquely determined element denoted by θ and called the *zero element* such that
$$u + \theta = u$$
for every $u \in X$;

(iv) for each $u \in X$ there exists a uniquely determined element in X denoted by $-u$ such that
$$u + (-u) = \theta;$$

(v) $\lambda(u + v) = \lambda u + \lambda v$, $\lambda \in \mathbb{R}$;

(vi) $(\lambda + \mu)u = \lambda u + \mu u$, $\lambda, \mu \in \mathbb{R}$;

(vii) $\lambda(\mu u) = (\lambda\mu)u$, $\lambda, \mu \in \mathbb{R}$;

(viii) $1u = u$;

(ix) $0u = \theta$.

Definition 1.1.4. In a vector space X, the *difference* (or *subtraction*) $u - v$ of two elements $u, v \in X$ is defined by
$$u - v := u + (-v).$$

Definition 1.1.5. Let M be a subset of a vector space X. Denote
$$[M] := \bigcap Y,$$
where the intersection is taken over all vector spaces $Y \subset X$ containing M. Then $[M]$ is called the *linear hull* of M. Obviously, $[M]$ is again a vector space.

Example 1.1.6. In what follows, the elements of a vector space X will usually be real-valued *functions* defined on a certain set R:
$$u = u(x), \ x \in R.$$
In this case addition and scalar multiplication are defined as usual:
$$(u + v)(x) := u(x) + v(x), \ (\lambda u)(x) := \lambda u(x).$$

Remark 1.1.7. We shall speak frequently of a *linear space* or *linear set* instead of a *vector space*.

Definition 1.1.8. A vector space X is called an *algebra* if for every ordered pair $u \in X$, $v \in X$, a *product* uv is defined as an element of X, which satisfies the following axioms for every $w \in X$ and all scalars λ and μ:

(i) $(uv)w = u(vw)$;

(ii) $u(v + w) = uv + uw$;

(iii) $(u + v)w = uw + vw$;

(iv) $(\lambda u)(\mu v) = (\lambda\mu)(uv)$.

1.2 Topological spaces

Notation 1.2.1. Let X be a nonempty set. Then by $\exp X$ we denote the set of all subsets of set X. If $A \subset X$, we denote by A^c the complement of A with respect to X, that is, $X \setminus A$.

Definition 1.2.2. We say that a couple $(X, \mathcal{T})$ is a *topological space* if X is a nonempty set and $\mathcal{T}$ is a system of subsets of X satisfying the following three conditions:

(i) $\emptyset \in \mathcal{T}$ and $X \in \mathcal{T}$ (where $\emptyset$ denotes the empty set).

(ii) If $G_1 \in \mathcal{T}$ and $G_2 \in \mathcal{T}$, then $G_1 \cap G_2 \in \mathcal{T}$.

(iii) If A is an index set of arbitrary cardinality and $A_\alpha \in \mathcal{T}$ for every $\alpha \in A$, then $\bigcup_{\alpha \in A} A_\alpha \in \mathcal{T}$.

The subsets of X belonging to $\mathcal{T}$ are called *open sets* in the space X, and the family $\mathcal{T}$ is called a *topology* on X.

Definition 1.2.3. Let $(X, \mathcal{T})$ be a topological space. If $x \in X$, $B \in \mathcal{T}$ and $x \in B$, then we say that B is a *neighborhood* of x.

Definition 1.2.4. Let $(X, \mathcal{T})$ be a topological space. A family $\mathcal{B} \subset \exp X$ is called a *base* for topology $\mathcal{T}$ on X if every nonempty open subset of X can be represented as a union over sets from $\mathcal{B}$.

Remark 1.2.5. Let $(X, \mathcal{T})$ be a topological space. Then every base $\mathcal{B}$ of the topology $\mathcal{T}$ has the following properties:

(i) For every $G_1, G_2 \in \mathcal{B}$ and every point $x \in G_1 \cap G_2$ there exists a set $G \in \mathcal{B}$ such that $x \in G \subset G_1 \cap G_2$.

(ii) For every $x \in X$ there exists a set $G \in \mathcal{B}$ such that $x \in G$.

Proposition 1.2.6. *Let X be a nonempty set. Let $\mathcal{B}$ be a collection of subsets of X, which has properties* (i) *and* (ii) *of Remark 1.2.5. We denote by $\mathcal{T}$ the collection of all those subsets of X, which can be represented as unions of sets from some subcollection of $\mathcal{B}$. Then $\mathcal{T}$ is a topology on X and $\mathcal{B}$ is a base for this topology.*

Definition 1.2.7. Let X, $\mathcal{B}$ and $\mathcal{T}$ be as in Proposition 1.2.6. Then we say that the topology $\mathcal{T}$ is *generated by the base* $\mathcal{B}$.

Definition 1.2.8. Let $(X, \mathcal{T})$ be a topological space and let $F \in \exp X$. We say that F is *closed* if $F^c \in \mathcal{T}$.

Remark 1.2.9. Let $(X, \mathcal{T})$ be a topological space. Denote by $\mathcal{F}$ the system of all closed subsets of X. Then it follows from Definition 1.2.2 and De Morgan laws that

(i) $\emptyset \in \mathcal{F}$ and $X \in \mathcal{F}$.

(ii) If $F_1 \in \mathcal{F}$ and $F_2 \in \mathcal{F}$, then $F_1 \cup F_2 \in \mathcal{F}$.

(iii) If A is an index set of arbitrary cardinality and $F_\alpha \in \mathcal{F}$ for every $\alpha \in A$, then $\bigcap_{\alpha \in A} F_\alpha \in \mathcal{F}$.

Definition 1.2.10. Let $(X, \mathcal{T})$ be a topological space and let $A \subset X$. Then the set

$$\overline{A} := \bigcap \{F \in \mathcal{F};\ F \supset A\}$$

is called the *closure* of A in X (with respect to the topology $\mathcal{T}$).

Theorem 1.2.11. *Let $(X, \mathcal{T})$ be a topological space. The closure operator has the following properties:*

(i) $\overline{\emptyset} = \emptyset$

(ii) $A \subset \overline{A}$

(iii) $\overline{A \cup B} = \overline{A} \cup \overline{B}$

(iv) $\overline{(\overline{A})} = \overline{A}$

Proposition 1.2.12. *Let X be a nonempty set. Assume that $\mathrm{cl} : \exp X \to \exp X$ is an operator assigning to every set $A \in \exp X$ some set $\mathrm{cl}(A) = \overline{A} \in \exp X$ such that the properties (i)–(iv) of Theorem 1.2.11 hold. Then the family*

$$\mathcal{T} := \{X \setminus A;\ A = \mathrm{cl}(A)\} \tag{1.2.1}$$

is a topology on X. Moreover, the set $\mathrm{cl}(A) = \overline{A}$ is the closure of A in X with respect to $\mathcal{T}$.

Definition 1.2.13. Let X be a nonempty set and let $\mathcal{T}_1, \mathcal{T}_2$ be two topologies on X. We say that the topology $\mathcal{T}_1$ is *weaker* than $\mathcal{T}_2$ (that is, $\mathcal{T}_2$ is *stronger* than $\mathcal{T}_1$) if $\mathcal{T}_1 \subset \mathcal{T}_2$.

Definition 1.2.14. Let $(X, \mathcal{T})$ and $(Y, \mathcal{T}')$ be two topological spaces. A mapping $f : X \to Y$ is called *continuous* if $f^{-1}(G) \in \mathcal{T}$ for every $G \in \mathcal{T}'$.

Example 1.2.15. Let $X = \mathbb{R}$, the set of all real numbers. Then the system

$$\mathcal{T} := \left\{ \bigcup_{n=1}^{\infty} (a_n, b_n) \right\},$$

where (a_n, b_n) are pairwise disjoint nontrivial open intervals, complemented with $\emptyset$ and $\mathbb{R}$ itself, defines a natural topology on $\mathbb{R}$.

Notation 1.2.16. Let $(X, \mathcal{T})$ and $(Y, \mathcal{T}')$ be two topological spaces. Then by $C(X, Y)$ we denote the set of all continuous mappings from X to Y.

Section 1.2 Topological spaces

Definition 1.2.17. Let $(X, \mathcal{T})$ be a topological space and let $\{f_n\}_{n=1}^{\infty}$ be a sequence of functions from X to $\mathbb{R}$. Let $f : X \to \mathbb{R}$ be a function. We say that $\{f_n\}_{n=1}^{\infty}$ is *uniformly convergent* to f if, for every $\varepsilon > 0$, there exists an $n_0 \in \mathbb{N}$ such that for every $n \in \mathbb{N}$, $n \geq n_0$ and for every $x \in X$, one has $|f_n(x) - f(x)| < \varepsilon$. We write

$$f = \lim_{n \to \infty} f_n.$$

Our next aim is to develop some natural topologies on $C(X, Y)$. We begin with the special case when $Y = \mathbb{R}$.

Definition 1.2.18. Let $(X, \mathcal{T})$ be a topological space. We then define the operator cl on $\exp C(X, \mathbb{R})$ in the following way: given a set $A \subset C(X, \mathbb{R})$, we set

$$\operatorname{cl}(A) := \left\{ f \in X; \ f = \lim_{n \to \infty} f_n \text{ for some sequence } \{f_n\}_{n=1}^{\infty} \subset A \right\}. \quad (1.2.2)$$

Proposition 1.2.19. *Let $(X, \mathcal{T})$ be a topological space and let cl be the operator on $\exp C(X, \mathbb{R})$ defined by (1.2.2). Then cl satisfies the properties (i)–(iv) of Theorem 1.2.11.*

Definition 1.2.20. Let $(X, \mathcal{T})$ be a topological space and let cl be the operator on $\exp C(X, \mathbb{R})$ defined by (1.2.2). Then the topology generated by this closure operator on $C(X, \mathbb{R})$ through the formula (1.2.1), is called the *topology of uniform convergence*.

Now we shall define a reasonable topology on the set $C(X, Y)$ for an arbitrary pair of topological spaces.

Definition 1.2.21. Let X and Y be topological spaces. For every pair of sets $A \in \exp X$ and $B \in \exp Y$, we denote

$$M(A, B) := \{ f \in C(X, Y); \ f(A) \subset B \}.$$

Let $\mathcal{T}_Y$ be the topology of Y. Further, denote by $\mathcal{F}_X$ the family of all finite subsets of X. Define next the system

$$\mathcal{B} := \left\{ \bigcap_{i=1}^{k} M(A_i, G_i), \ A_i \in \mathcal{F}_X, \ G_i \in \mathcal{T}, \ i = 1, \ldots, k \right\}.$$

Then the set $\mathcal{B}$ (according to Proposition 1.2.6) generates a topology on $C(X, Y)$. This topology is called the *topology of pointwise convergence*.

Remark 1.2.22. For every topological space X, the topology of uniform convergence is stronger on $C(X, \mathbb{R})$ than the topology of pointwise convergence.

Proof of the assertions in this section as well as many further details on general topological spaces can be found, e.g., in [71].

1.3 Metric, metric space

Definition 1.3.1. Let X be a nonempty set. A nonnegative function ϱ defined on the Cartesian product $X \times X$ is called a *metric* if it satisfies the following axioms for every $u, v, w \in X$:

(i) $\varrho(u, v) = 0$ if and only if $u = v$;

(ii) $\varrho(u, v) = \varrho(v, u)$ (symmetry);

(iii) $\varrho(u, v) \leq \varrho(u, w) + \varrho(w, v)$ (the triangle inequality).

A set X with a metric ϱ will be called a *metric space* and denoted by (X, ϱ).

Definition 1.3.2. Let (X, ϱ) be a metric space and let $G \subset X$. We say that G is *open* in X with respect to ϱ if for every $x \in G$ there exists an $r > 0$ such that the set

$$B_r(x) := \{y \in X;\ \varrho(x, y) < r\}$$

satisfies $B_r(x) \subset G$. The set $B_r(x)$ is called an *open ball centered at x with radius r*.

Remark 1.3.3. If (X, ϱ) is a metric space, then the metric ϱ automatically generates a topology on X, in which open sets are those that are open with respect to ϱ. It is an easy exercise to verify that the axioms of Definition 1.2.2 are satisfied.

1.4 Norm, normed linear space

Definition 1.4.1. Let X be a vector space. A nonnegative function defined on X whose value at $u \in X$ is denoted by $\|u\|$ is called a *norm* on X if it satisfies the following axioms:

(i) $\|u\| = 0$ if and only if $u = \theta$;

(ii) $\|\lambda u\| = |\lambda| \|u\|$ for every $u \in X$ and all scalars λ (the homogeneity axiom);

(iii) $\|u + v\| \leq \|u\| + \|v\|$ for every $u, v \in X$ (the triangle inequality).

A vector space X endowed with a norm $\|\cdot\|$ is called a *normed linear space*; the number $\|u\|$ is called the norm of $u \in X$.

Remark 1.4.2. If it is necessary to specify the vector space X on which the norm is defined we use $\|u\|_X$ instead of $\|u\|$. Sometimes the normed linear space X will be denoted by $(X, \|\cdot\|_X)$.

Definition 1.4.3. If the function $\|\cdot\|$ satisfies only the axioms (ii), (iii) from Definition 1.4.1 then it is called a *seminorm* (or sometimes a *pseudonorm*).

If the function $\|\cdot\|$ satisfies the axioms (i), (ii) from Definition 1.4.1 and there exists a constant $C > 1$ such that (iii) becomes

(iii′) $\|u + v\| \leq C(\|u\| + \|v\|)$ for every $u, v \in X$,

then it is called a *quasinorm*.

Proposition 1.4.4. *Every normed linear space is a metric space with the metric ϱ defined by*
$$\varrho(u, v) := \|u - v\|, \quad u, v \in X.$$
Hence, owing to Remark 1.3.3, it is also a topological space.

Definition 1.4.5. Let X be a normed linear space, $u_0 \in X$, $r > 0$. The set
$$B(u_0, r) := \{u \in X;\ \|u - u_0\| < r\}$$
is called an *open ball* (with center u_0 and radius r).

A subset $M \subset X$ is called an *open set* in X if for every $u_0 \in M$ there exists an $r = r(u_0) > 0$ such that $B(u_0, r) \subset M$.

A subset $M \subset X$ is called a *closed set* in X if $X \setminus M$ is an open set in X.

1.5 Modular spaces

Definition 1.5.1. Let X be a vector space. A function $\varrho : X \to [0, \infty]$ is called *left-continuous* if the mapping $\lambda \mapsto \varrho(\lambda x)$ is continuous on $[0, \infty)$ in the sense that
$$\lim_{\lambda \to 1-} \varrho(\lambda x) = \varrho(x) \quad \text{for every } x \in X.$$

A convex and left-continuous function $\varrho : X \to [0, \infty]$ is called a *semimodular* on X if

(i) $\varrho(\theta) = 0$;

(ii) $\varrho(-x) = \varrho(x)$ for every $x \in X$;

(iii) if $\varrho(\lambda x) = 0$ for every $\lambda \in \mathbb{R}$, then $x = \theta$.

A semimodular is called a *modular* on X if $\varrho(x) = 0$ if and only if $x = \theta$. A semimodular is called *continuous* if the mapping $\lambda \mapsto \varrho(\lambda x)$ is continuous on $[0, \infty)$ for every fixed $x \in X$.

Definition 1.5.2. Let X be a vector space and let ϱ be a semimodular or a modular on X. Then the space
$$X_\varrho := \left\{ x \in X;\ \lim_{\lambda \to 0} \varrho(\lambda x) = 0 \right\}$$
is called a *semimodular space* or a *modular space*, respectively.

Notation 1.5.3. Let X be a vector space and let ϱ be a semimodular on X. We then denote by $\|\cdot\|_\varrho$ the functional, given for every $x \in X$ by

$$\|x\|_\varrho := \inf\left\{\lambda > 0;\ \varrho\left(\frac{1}{\lambda}x\right) \leq 1\right\}. \tag{1.5.1}$$

Remark 1.5.4. Let X be a vector space and let ϱ be a semimodular on X. Then it can be easily shown that the functional $\|\cdot\|_\varrho$ defined by (1.5.1) is a norm on X_ϱ. It is also known as the *Minkowski functional* of the set $\{x \in X;\ \varrho(x) \leq 1\}$. A detailed proof can be found for example in [63, Theorem 2.1.7].

Definition 1.5.5. Let X be a vector space and let ϱ be a semimodular on X. Then the functional $\|\cdot\|_\varrho$ defined by (1.5.1) is called the *Luxemburg norm* on X_ϱ.

Remark 1.5.6. Let X be a vector space and let ϱ be a semimodular on X. Then, for every fixed $x \in X$, the mapping $\lambda \mapsto \varrho(\lambda x)$ is nondecreasing on $[0, \infty)$. Moreover, by the convexity of ϱ, we have

$$\varrho(\lambda x) \begin{cases} \leq \lambda \varrho(x) & \text{for every } \lambda \in [0, 1], \\ \geq \lambda \varrho(x) & \text{for every } \lambda \in [1, \infty). \end{cases} \tag{1.5.2}$$

Proposition 1.5.7. *Let X be a vector space, let ϱ be a semimodular on X and let $x \in X$. Then*

$$\varrho(x) \leq 1 \quad \text{if and only if} \quad \|x\|_\varrho \leq 1.$$

Proof. Assume that $\varrho(x) \leq 1$. Then the definition of the Luxemburg norm also implies that $\|x\|_\varrho \leq 1$.

Now assume that $\|x\|_\varrho \leq 1$. Then, for every $\lambda > 1$, one has $\varrho(\lambda^{-1}x) \leq 1$. Since ϱ is left-continuous, we obtain $\varrho(x) \leq 1$. The proof is complete. $\square$

Proposition 1.5.8. *Let X be a vector space, let ϱ be a semimodular on X and let $x \in X$.*

(i) *If $\|x\|_\varrho \leq 1$, then $\varrho(x) \leq \|x\|_\varrho$.*

(ii) *If $\|x\|_\varrho > 1$, then $\varrho(x) \geq \|x\|_\varrho$.*

(iii) *For every $x \in X$, $\|X\|_\varrho \leq \varrho(x) + 1$.*

Proof. (i) When $x = 0$, there is nothing to prove. Assume that $0 < \|x\|_\varrho \leq 1$. By Proposition 1.5.7 and by the fact that $\|x\|x\|_\varrho^{-1}\|_\varrho = 1$, we obtain

$$\varrho\left(\frac{x}{\|x\|_\varrho}\right) \leq 1.$$

Because $\|x\|_\varrho \leq 1$, the claim follows from (1.5.2).

(ii) If $\|x\|_\varrho > 1$, then for every $\lambda \in (1, \|x\|_\varrho)$, we have $\varrho(\lambda^{-1}x) > 1$. Thus, by (1.5.2), we obtain $\lambda^{-1}\varrho(x) > 1$. Since this is so for an arbitrary $\lambda < \|x\|_\varrho$, the claim follows.

(iii) This is an immediate consequence of (ii). □

Proposition 1.5.9. *Let X be a vector space, let ϱ be a semimodular on X and let $\{x_n\}_{n=1}^\infty$ be a sequence in X. Then $\lim_{n\to\infty} \|x_n\|_\varrho = 0$ if and only if $\lim_{n\to\infty} \varrho(\lambda x_n) = 0$ for every $\lambda > 0$.*

Proof. Assume first that $\lim_{n\to\infty} \|x_n\|_\varrho = 0$ and let $\lambda > 0$. Then, for every $K > 1$, there exists an $n_0 \in \mathbb{N}$ such that for every $n \in \mathbb{N}, n \geq n_0$, one has

$$\|K\lambda x_n\|_\varrho < 1.$$

Therefore also $\varrho(K\lambda x_n) \leq 1$ and, by (1.5.2),

$$\varrho(\lambda x_n) = \varrho\left(\frac{1}{K}K\lambda x_n\right) \leq \frac{1}{K}\varrho(K\lambda x_n) \leq \frac{1}{K},$$

establishing $\lim_{n\to\infty} \varrho(\lambda x_n) = 0$.

Now assume that $\lim_{n\to\infty} \varrho(\lambda x_n) = 0$. Then there exists an $n_0 \in \mathbb{N}$ such that for every $n \in \mathbb{N}, n \geq n_0$, one has $\varrho(\lambda x_n) \leq 1$. Especially, for such n, we have

$$\|x_n\|_\varrho \leq \frac{1}{\lambda}.$$

Since λ was arbitrary, we obtain $\lim_{n\to\infty} \|x_n\|_\varrho = 0$, as desired. The proof is complete. □

Definition 1.5.10. Let X be a vector space, let ϱ be a modular on X and let $\{x_n\}_{n=1}^\infty$ be a sequence in X. We say that $\{x_n\}_{n=1}^\infty$ is *modular convergent* to some $x \in X$ if

$$\lim_{n\to\infty} \varrho(x_n - x) = 0.$$

Corollary 1.5.11. *It follows from Proposition 1.5.8 that if X is a modular space, then the norm convergence always implies the modular convergence.*

Proposition 1.5.12. *Let X be a modular space. Then the modular convergence on X is equivalent to the norm convergence if and only if $\varrho(x_n) \to 0$ implies $\varrho(2x_n) \to 0$.*

Proof. ⇒ Assume that the modular convergence on X is equivalent to the norm. Let $\{x_n\}_{n=1}^\infty$ be a sequence in X such that $\varrho(x_n) \to 0$. Also, $\|x_n\|_\varrho \to 0$, and it follows immediately from Proposition 1.5.8 applied to $\lambda = 2$ that $\varrho(2x_n) \to 0$.

⇐ Assume conversely that the condition holds. Let $\{x_n\}_{n=1}^\infty$ be a sequence in X satisfying $\varrho(x_n) \to 0$. Given $\lambda > 0$, we find $m \in \mathbb{N}$ such that $\lambda \leq 2^m$. Then,

iterating the condition m times, we obtain $\varrho(2^m x_n) \to 0$ as $n \to \infty$. Thus, by (1.5.2), we obtain
$$0 \leq \lim_{n\to\infty} \varrho(\lambda x_n) \leq \lambda 2^{-m} \lim_{n\to\infty} \varrho(2^m x_n) = 0.$$
Owing to Proposition 1.5.9, this establishes $x_n \to 0$, as desired. The proof is complete. □

Theorem 1.5.13. *Let (X, ϱ_X) and (Y, ϱ_Y) be two modular spaces, $X \subset Y$. Assume that there exists a function $h : (0, \infty) \to (0, \infty)$, which is bounded on some right neighborhood of zero. Suppose that*

$$\varrho_Y(x) \leq h(\varrho_X(x)) \quad \text{for every } x \in X. \tag{1.5.3}$$

Then there exists a positive constant C such that, for every $x \in X$,

$$\|x\|_{\varrho_Y} \leq C \|x\|_{\varrho_X}. \tag{1.5.4}$$

Proof. Suppose that the assertion is not true. Then for every $n \in \mathbb{N}$ there exists some $x_n \in X$ such that $\|x_n\|_{\varrho_Y} > n \|x_n\|_{\varrho_X}$. Set

$$\tilde{x}_n := \frac{x_n}{\|x_n\|_{\sqrt{n}\varrho_X}}, \quad n \in \mathbb{N}.$$

Then $\|\tilde{x}_n\|_{\varrho_X} \to 0$ and $\|\tilde{x}_n\|_{\varrho_Y} \to \infty$. By Proposition 1.5.8, we get $\varrho_X(\tilde{x}_n) \to 0$ and $\varrho_Y(\tilde{x}_n) \to \infty$. That, however, is a contradiction with (1.5.3), since h is bounded on some neighborhood of zero. The proof is complete. □

For more details on modular spaces, see, e.g., [63, 160, 163]. Most of the material in this section can be found in [63]. Theorem 1.5.13 is a special case of a more general result in [125], see also [178].

1.6 Inner product, inner product space

Definition 1.6.1. Let X be a vector space. A real-valued function on $X \times X$, whose value at the ordered pair (u, v), $u, v \in X$, is denoted by $\langle u, v \rangle$, is called an *inner product* on X if it satisfies the following axioms:

(i) $\langle u, u \rangle > 0$ for every $u \in X$, $u \neq \theta$;

(ii) $\langle u, v \rangle = \langle v, u \rangle$ for every $u, v \in X$;

(iii) $\langle u + v, w \rangle = \langle u, w \rangle + \langle v, w \rangle$ for every $u, v, w \in X$;

(iv) $\langle \lambda u, v \rangle = \lambda \langle u, v \rangle$ for every $u, v \in X$ and all scalars λ.

A vector space X endowed with an inner product is called an *inner product space* (or a *unitary space*).

Definition 1.6.2. Let X be a unitary space. For $u \in X$, we define

$$\|u\| = \langle u, u \rangle^{\frac{1}{2}}.$$

The inequalities

$$|\langle u, v \rangle| \leq \|u\| \|v\|, \quad u, v \in X,$$

(the so-called *Cauchy–Schwarz inequality*) and

$$\|u + v\| \leq \|u\| + \|v\|, \quad u, v \in X,$$

holds. In particular, every unitary space is a normed linear space with respect to the norm $\|u\|$ generated by the inner product $\langle u, v \rangle$.

1.7 Convergence, Cauchy sequences

Definition 1.7.1. Let X be a metric space with respect to the metric ϱ and let $\{u_n\}_{n=1}^{\infty}$ be a sequence in X. We say that u_n *converges to* $u \in X$ (and write $u_n \to u$ in X) if $\lim_{n \to \infty} \varrho(u_n, u) = 0$, i.e. if for every $\varepsilon > 0$ there exists an $n_0 = n_0(\varepsilon) \in \mathbb{N}$ such that $\varrho(u_n, u) < \varepsilon$ for all $n > n_0$.

We shall also say that u_n converges to u *strongly* in X or *in the norm of X*.

If $u_n \to u$ in X, then the sequence $\{u_n\}_{n=1}^{\infty}$ is said to be *convergent* in X and u is called the *limit* of $\{u_n\}_{n=1}^{\infty}$.

Definition 1.7.2. Let X be a normed linear space, let $\{u_n\}_{n=1}^{\infty}$ be a sequence in X and let $u \in X$. If

$$\lim_{n \to \infty} \left\| u - \sum_{k=1}^{n} u_k \right\|_X = 0,$$

we say that the series $\sum_{n=1}^{\infty} u_n$ *converges to* u in X and write

$$u = \sum_{n=1}^{\infty} u_n.$$

Definition 1.7.3. A sequence $\{u_n\}_{n=1}^{\infty}$ in a metric space (X, ϱ) is called a *Cauchy sequence* if

$$\lim_{m, n \to \infty} \varrho(u_m, u_n) = 0,$$

i.e. if for every $\varepsilon > 0$ there exists an $n_0 = n_0(\varepsilon) \in \mathbb{N}$ such that $\varrho(u_m, u_n) < \varepsilon$ for all $m, n \in \mathbb{N}, m, n > n_0$.

Remark 1.7.4. In every metric space, each convergent sequence is a Cauchy sequence. The converse is not true in general.

The following simple auxiliary assertion will be useful later.

Lemma 1.7.5. *Let $\{u_n\}_{n=1}^{\infty}$ be a Cauchy sequence in a metric space X and let $\{u_{n_k}\}_{k=1}^{\infty}$ be its subsequence. If $u_{n_k} \to u$ in X then $u_n \to u$ in X.*

Proof. Let $\varepsilon > 0$. Then, by the Cauchy property, there exists an index $n_0 \in \mathbb{N}$ such that for every $m, n \in \mathbb{N}$, $m, n \geq n_0$, one has $\varrho(x_m, x_n) < \varepsilon$. Next, there is an index $k_0 \in \mathbb{N}$ such that $n_{k_0} \geq n_0$ and $\varrho(x_{n_{k_0}}, x) < \varepsilon$. Thus, for every $n \in \mathbb{N}$, $n \geq n_{k_0}$, one has
$$\varrho(x_n, x) \leq \varrho(x_n, x_{n_{k_0}}) + \varrho(x_{n_{k_0}}, x) < \varepsilon + \varepsilon = 2\varepsilon.$$
The proof is complete. □

Definition 1.7.6. Let M be a subset of a metric space X. The *closure* of M in X, denoted by $\overline{M}$ or $\overline{M}^X$, is defined as the set of all elements $u \in X$ such that there exists a sequence $\{u_n\}_{n=1}^{\infty}$ in M for which $u_n \to u$ in X.

Remark 1.7.7. Clearly, if M is a subset of a metric space X, then $M \subset \overline{M}$. A set M is closed in X if and only if $M = \overline{M}$.

1.8 Density, separability

Definition 1.8.1. A subset M of a metric space X is said to be *dense* in X if $\overline{M} = X$.

Definition 1.8.2. A metric space X is called *separable* if it contains a countable dense subset.

Remark 1.8.3. Let (X, ϱ) be a metric space. Assume that there exists an uncountable subset M of X and a $\delta > 0$ such that
$$\varrho(u, v) > \delta \quad \text{for every } u, v \in M, \ u \neq v.$$
Then X is not separable.

1.9 Completeness

Definition 1.9.1. A metric space X is said to be *complete* if every Cauchy sequence in X is convergent in X.

Definition 1.9.2. A complete normed linear space is called a *Banach space*.

Definition 1.9.3. Let X be a unitary space endowed with an inner product $\langle u, v \rangle$. If the normed linear space X is complete with respect to the norm $\|u\| = \langle u, u \rangle^{\frac{1}{2}}$, then X is called a *Hilbert space*.

Definition 1.9.4. We say that a normed linear space $(X, \|\cdot\|_X)$ has the *Riesz–Fischer property* if for each sequence $\{u_n\}_{n=1}^\infty$ such that

$$\sum_{n=1}^\infty \|u_n\|_X < \infty, \tag{1.9.1}$$

there exists an element $u \in X$ such that $\sum_{n=1}^\infty u_n = u$ in X, that is,

$$\lim_{n\to\infty} \left\| \sum_{k=1}^n u_k - u \right\|_X = 0.$$

Theorem 1.9.5. *A normed linear space is complete if and only if it has the Riesz–Fischer property.*

Proof. Assume first that X is a Banach space and let $\{u_n\}_{n=1}^\infty$ be a sequence in X such that (1.9.1) holds. Then

$$\left\{ \sum_{k=1}^n u_k \right\}_{n=1}^\infty$$

is a Cauchy sequence in X, hence it converges in X to some element $u \in X$. Thus,

$$u = \sum_{n=1}^\infty u_n.$$

Conversely, assume X is a normed linear space with the Riesz–Fischer property and let $\{u_n\}_{n=1}^\infty$ be a Cauchy sequence in X. Then a subsequence $\{u_{n_k}\}_{k\in\mathbb{N}}$ can be chosen so that

$$\sum_{k=1}^\infty \|u_{n_k} - u_{n_{k+1}}\|_X < \infty.$$

Then, necessarily, the series $\sum_{k=1}^\infty (u_{n_k} - u_{n_{k+1}})$ converges in X. In particular, there exists an element $u \in X$ such that $u_{n_k} \to u$ in X. But then since $\{u_n\}_{n=1}^\infty$ is a Cauchy sequence, we also have $u_n \to u$ in X, as desired. □

1.10 Subspaces

Definition 1.10.1. A subset of a normed linear space X is called a *subspace* of X if it is a linear set which is closed in X.

Remark 1.10.2. We shall distinguish between *linear subsets* and *subspaces*: a linear subset *need not be closed* in X.

Remarks 1.10.3. (i) A subspace M of a normed linear space $(X, \|\cdot\|_X)$ is again a normed linear space with the norm $\|\cdot\|_M$ defined by

$$\|u\|_M := \|u\|_X \text{ for } u \in M.$$

(ii) A subspace of a separable normed linear space X is itself a separable normed linear space.

(iii) A subspace of a Banach space is also a Banach space.

1.11 Products of spaces

Remarks 1.11.1. (i) Let $n \in \mathbb{N}$, and let $X_1, X_2, \ldots, X_n$ be normed linear spaces, $X = X_1 \times X_2 \times \cdots \times X_n$ the Cartesian product of $X_1, \ldots, X_n$, i.e. the set of all (ordered) n-tuples

$$u = (u_1, \ldots, u_n)$$

such that $u_i \in X_i$, $i = 1, \ldots, n$. Then X is also a normed linear space; the norm in X can be defined in various ways, for example

$$\|u\|_X := \left(\sum_{i=1}^n \|u_i\|_{X_i}^p \right)^{\frac{1}{p}} \quad \text{for some } p \in [1, \infty)$$

or

$$\|u\|_X := \max_{i=1,\ldots,n} \|u_i\|_{X_i}.$$

(ii) Let $X_1, \ldots, X_n$ be separable normed linear spaces. Then the product space $X = X_1 \times \cdots \times X_n$ is also a separable normed linear space.

(iii) Let $X_1, \ldots, X_n$ be Banach spaces. Then the product space $X = X_1 \times \cdots \times X_n$ is also a Banach space.

1.12 Schauder bases

Definition 1.12.1. Let X be a Banach space. A sequence $\{u_n\}_{n=1}^\infty$ in X is called a *Schauder basis* of X if for every $u \in X$ there exists a unique sequence $\{a_n\}_{n=1}^\infty$ of scalars such that

$$u = \sum_{n=1}^\infty a_n u_n.$$

The (uniquely determined) numbers $a_n = a_n(u)$ are called the *coefficients* of u with respect to the Schauder basis $\{u_n\}_{n=1}^\infty$.

Remark 1.12.2. Every Banach space with a Schauder basis is separable. The converse implication does not hold, i.e. there exists a separable Banach space without Schauder basis.

Definition 1.12.3. A Schauder basis $\{u_n\}_{n=1}^{\infty}$ in a Banach space X is called *unconditional* if the convergence of a series of the form

$$\sum_{n=1}^{\infty} a_n u_n$$

implies the convergence of the series

$$\sum_{n=1}^{\infty} a_{\pi(n)} u_{\pi(n)}$$

for every permutation π of the set $\mathbb{N}$.

Remark 1.12.4. A basis $\{u_n\}_{n=1}^{\infty}$ is unconditional if and only if at least one of the following conditions holds:

(i) The convergence of a series $\sum_{n=1}^{\infty} a_n u_n$ implies the convergence of the series $\sum_{n=1}^{\infty} \varepsilon_n a_n u_n$ for any choice of ε_n equal to $+1$ or -1.

(ii) The convergence of a series $\sum_{n=1}^{\infty} a_n u_n$ implies the convergence of the series $\sum_{k=1}^{\infty} a_{n_k} u_{n_k}$ for any subsequence $\{n_k\}_{k=1}^{\infty}$ of $\mathbb{N}$.

1.13 Compactness

Definition 1.13.1. Let (X, ϱ) be a metric space. A set $M \subset X$ is said to be *relatively compact* if every sequence in M contains a convergent subsequence (i.e. a subsequence with a limit in X).

If M is closed and relatively compact then M is said to be *compact*.

Definition 1.13.2. Let (X, ϱ) be a metric space. Let $\varepsilon > 0$ and let M be a subset of X. A set E in X is called an *ε-net* of M if for every $u \in M$ there exists a $u_\varepsilon \in E$ such that

$$\varrho(u, u_\varepsilon) < \varepsilon.$$

A set $M \subset X$ is called *totally bounded* (or *precompact*) in X if for every $\varepsilon > 0$ there exists a finite ε-net of M in X.

Remarks 1.13.3. Let (X, ϱ) be a metric space. Then

(i) every relatively compact set in X is bounded;

(ii) every closed subset of a compact set is itself compact;

(iii) if X is complete, then each its subset is relatively compact if and only if it is totally bounded.

Remark 1.13.4. For $i = 1, \ldots, n$, let M_i be a relatively compact subset of a Banach space X_i. Then $M_1 \times \cdots \times M_n$ is a relatively compact subset of $X_1 \times \cdots \times X_n$.

1.14 Operators (mappings)

Definition 1.14.1. Let X, Y be normed linear spaces, and M a set in X. Suppose that a rule is given by which to every $u \in M$ there corresponds a uniquely determined element in Y. We denote this element by

$$Au \quad \text{or} \quad A(u)$$

and say that the rule defines an *operator* A on M. The set M is called the *domain* of the operator A and denoted by $\text{Dom}(A)$; the set

$$\text{Rng}(A) = \{\tilde{u} \in Y; \tilde{u} = Au \text{ for some } u \in M\}$$

is called the *range* of the operator A.

If for all $u, v \in M$ we have $Au \neq Av$ provided $u \neq v$, then to every $\tilde{u} \in \text{Rng}(A)$ is assigned a uniquely determined element $u \in M$ by the rule $Au = \tilde{u}$. We write this as

$$u = A^{-1}\tilde{u}$$

and call A^{-1} the *inverse operator* to A. We have that $\text{Dom}(A^{-1}) = \text{Rng}(A)$, $\text{Rng}(A^{-1}) = \text{Dom}(A) = M$.

We say that the operator A is an operator *from X into Y* which maps M into Y. If $\text{Rng}(A) = Y$, we say that A maps M *onto* Y.

The expressions *function*, *abstract function* or *mapping* are frequently used instead of *operator*.

Definition 1.14.2. Let X, Y be normed linear spaces. An operator A from X into Y is called *linear* if $\text{Dom}(A)$ is a linear set and if the following two conditions are satisfied:

(i) $A(u + v) = Au + Av$

(ii) $A(\lambda u) = \lambda Au$

for all $u, v \in \text{Dom}(A)$ and for every scalar λ.

Definition 1.14.3. Let X, Y be normed linear spaces. An operator A from X into Y is said to be *continuous* if

$$u_n \to u \text{ in } X \quad \text{implies} \quad Au_n \to Au \text{ in } Y$$

for any sequence $\{u_n\}_{n=1}^{\infty}$ provided $u_n \in \text{Dom}(A)$, $u \in \text{Dom}(A)$.

Section 1.14 Operators (mappings)

Definition 1.14.4. Let X, Y be normed linear spaces. A linear operator A from X into Y is said to be *bounded* if

$$\sup \|Au\|_Y < \infty,$$

where the supremum is taken over all $u \in \text{Dom}(A)$ such that $\|u\|_X \leqq 1$.

Remark 1.14.5. A linear operator between two normed linear spaces is bounded if and only if it is continuous.

We shall now introduce the notion of a *norm* of a continuous linear operator (the so-called *operator norm*).

Definition 1.14.6. Let X, Y be normed linear spaces. Let A be a continuous linear operator from X into Y. Then

$$\|A\|_{X \to Y} := \sup \{\|Au\|_Y; \; u \in \text{Dom}(A), \; \|u\|_X \leqq 1\}.$$

Remark 1.14.7. Let X, Y be normed linear spaces and let A be a continuous linear operator from X into Y. Then the following formulas for the norm $\|A\|_{X \to Y}$ of A will often be useful:

$$\|A\|_{X \to Y} = \sup_{\substack{u \in \text{Dom}(A) \\ \|u\|_X = 1}} \|Au\|_Y = \sup_{\substack{u \in \text{Dom}(A) \\ u \neq \theta}} \frac{\|Au\|_Y}{\|u\|_X}.$$

Theorem 1.14.8 (Banach). *Let X, Y be Banach spaces, and A a linear operator from X onto Y with $\text{Dom}(A) = X$ (and $\text{Rng}(A) = Y$). Suppose that A is continuous and that A^{-1} exists. Then A^{-1} is continuous.*

Definition 1.14.9. Let A be a linear operator from a normed linear space X into a normed linear space Y with $\text{Dom}(A) = X$. The operator A is said to be *compact* (or *completely continuous*) if it maps every bounded set in X onto a relatively compact set in Y.

Remark 1.14.10. Every compact linear operator is continuous.

Theorem 1.14.11 (Banach–Steinhaus). *Let X be a Banach space, Y a normed linear space. The sequence $\{A_n\}_{n=1}^{\infty}$ of bounded linear operators from X into Y with $\text{Dom}(A_n) = X$ satisfies*

$$A_n u \to Au \quad \text{in } Y \text{ for every } u \in X$$

if and only if the following two conditions are satisfied:

(i) *the sequence $\{\|A_n\|\}_{n=1}^{\infty}$ is bounded;*

(ii) *$A_n u \to Au$ in Y for every $u \in M$, where M is a dense subset of X.*

1.15 Isomorphism, embeddings

Definition 1.15.1. Two normed linear spaces X, Y are said to be *isomorphic* if there exists a continuous linear operator A such that $\mathrm{Dom}(A) = X$, $\mathrm{Rng}(A) = Y$, and A^{-1} exists and is continuous. This operator A is called an *isomorphism mapping* or briefly an *isomorphism* between X and Y.

Definition 1.15.2. (i) Two metric spaces (X, ϱ) and (Y, σ) are said to be *isometric* if there exists a mapping T such that $\mathrm{Dom}(T) = X$, $\mathrm{Rng}(T) = Y$ and

$$\varrho(x, y) = \sigma(Tx, Ty)$$

for every pair $x, y \in X$.

(ii) Two normed linear spaces X, Y are said to be *isometrically isomorphic* if there exists a linear operator A such that $\mathrm{Dom}(A) = X$, $\mathrm{Rng}(A) = Y$ and

$$\|u - v\|_X = \|Au - Av\|_Y$$

for every pair $u, v \in X$.

Remark 1.15.3. In particular, two isometrically isomorphic spaces are isomorphic.

Remark 1.15.4. Let X, Y be two isomorphic normed linear spaces. Then

(i) if X is separable then Y is also separable;

(ii) if X is complete then Y is also complete;

(iii) if X has a Schauder basis then Y also has a Schauder basis.

Definition 1.15.5. Let X, Y be two normed linear spaces and let $X \subset Y$. We define the *identity operator* Id from X into Y with $\mathrm{Dom}(\mathrm{Id}) = \mathrm{Rng}(\mathrm{Id}) = X$ as the operator which maps every element $u \in X$ onto itself: $\mathrm{Id}\, u = u$, regarded as an element of Y. If the identity operator is continuous, that is, if there exists a constant $c > 0$ such that

$$\|u\|_Y \leq c \|u\|_X \quad \text{for every } u \in X,$$

then we say that the space X is *embedded into the space* Y and we shall call the operator Id the *embedding operator* from X to Y. Alternatively, we may sometimes say that there exists a continuous (or bounded) embedding of X to Y. By the *operator norm* of Id we call the number

$$\|\mathrm{Id}\|_{X \hookrightarrow Y} := \sup_{f \neq 0} \frac{\|f\|_Y}{\|f\|_X}.$$

We shall also call this quantity an *embedding constant*.

Remarks 1.15.6. (i) The operator Id from Definition 1.15.5 is obviously linear.

(ii) While the elements in X and in $\mathrm{Rng}(\mathrm{Id}) \subset Y$ coincide, their respective norms in X and in Y may be different.

Notation 1.15.7. If X and Y are two normed linear spaces and there exists a continuous embedding from X into Y, we write

$$X \hookrightarrow Y.$$

If simultaneously

$$X \hookrightarrow Y \text{ and } Y \hookrightarrow X,$$

then we shall write

$$X \rightleftarrows Y.$$

If the embedding operator is compact (see Definition 1.14.9), we write

$$X \hookrightarrow\hookrightarrow Y.$$

Definition 1.15.8. Let X be a vector space and suppose that $\|\cdot\|_1$ and $\|\cdot\|_2$ are two norms on X. These norms are said to be *equivalent* if there exist constants $c > 0$, $d > 0$ such that

$$c\|u\|_1 \leq \|u\|_2 \leq d\|u\|_1 \text{ for all } u \in X.$$

In other words, the norms $\|\cdot\|_1$ and $\|\cdot\|_2$ are equivalent if and only if $(X, \|\cdot\|_1) \rightleftarrows (X, \|\cdot\|_2)$. In particular, the embedding from $(X, \|\cdot\|_1)$ into $(X, \|\cdot\|_2)$ is an isomorphism.

1.16 Continuous linear functionals

Definition 1.16.1. Let X be a normed linear space. A linear operator from X into $\mathbb{R}$ is then called a *linear functional*.

Remark 1.16.2. In this section we denote linear functionals by Greek letters: $\varphi, \Phi, \ldots$. The value of the functional φ at $u \in X$ will usually be denoted by

$$\varphi(u);$$

the notation

$$\langle \varphi, u \rangle$$

is also very frequently used in the literature.

Since a functional is an operator, some concepts introduced in connection with operators (see Section 1.14) can be transferred to functionals. We shall deal here only with *continuous linear functionals*, that is, continuous linear operators from X into $\mathbb{R}$.

Definition 1.16.3. A *norm* of a linear functional φ is defined by
$$\|\varphi\| = \sup |\varphi(u)|$$
where the supremum is taken over all $u \in \text{Dom}(\varphi)$ such that $\|u\|_X \leq 1$.

Continuous linear functionals defined on linear subsets of X can be *extended* to the entire space X as stated in the following theorem.

Theorem 1.16.4 (Hahn–Banach). *Let φ be a continuous linear functional defined on a linear subset M of a normed linear space X. Then there exists a continuous linear functional Φ defined on X such that*
$$\Phi(u) = \varphi(u) \text{ for } u \in M \text{ and } \|\Phi\| = \|\varphi\|.$$

1.17 Dual space, weak convergence

Definition 1.17.1. Let us denote by X^* the set of all continuous linear functionals defined on X. Then X^* is a vector space if we define the addition of functionals and the multiplication of a functional by a scalar in the natural way, namely
$$(\varphi + \psi)(u) := \varphi(u) + \psi(u), \quad (\lambda\varphi)(u) := \lambda\varphi(u)$$
where $\varphi, \psi \in X^*$ and λ is a scalar.

Remarks 1.17.2. If X is a normed linear space, then the space X^* from Definition 1.17.1, endowed with the norm from Definition 1.16.3, is itself also a normed linear space. Moreover, the space X^* is a Banach space, that is, it is always complete.

Definition 1.17.3. Let X be a normed linear space, $\{u_n\}_{n=1}^{\infty}$ a sequence in X. We say that u_n *converges weakly* to $u \in X$ (notation $u_n \rightharpoonup u$ or $u_n \overset{w}{\to} u$) if $\lim_{n\to\infty} \varphi(u_n) = \varphi(u)$ for every $\varphi \in X^*$.

Remark 1.17.4. Every weakly convergent sequence is bounded.

Theorem 1.17.5 (Banach–Steinhaus theorem for weak convergence). *Let X be a Banach space. The sequence $\{u_n\}_{n=1}^{\infty}$ converges weakly to $u \in X$ if and only if the following two conditions are satisfied:*

(i) *the sequence $\{\|u_n\|_X\}_{n=1}^{\infty}$ is bounded;*

(ii) $\lim_{n\to\infty} \varphi(u_n) = \varphi(u)$ *for all $\varphi \in \Gamma$, where Γ is a dense subset of the dual space X^*.*

Theorem 1.17.6 (Banach–Alaoglu). *Let X be a Banach space and X^* its dual. Then the unit ball*
$$\{\Lambda \in X^*; \|\Lambda\|_{X^*} \leq 1\}$$
is weakly* compact in X^*.

1.18 The principle of uniform boundedness

Theorem 1.18.1 (Uniform boundedness principle). *Let X be a Banach space and Y a normed linear space. Let $\{T_\alpha\}_{\alpha \in I}$ be a set of linear operators from X to Y, where I is an arbitrary index set (without any restriction on its cardinality). Assume that*

$$\sup_{\alpha \in I} \|T_\alpha x\|_Y < \infty \quad \text{for every } x \in X.$$

Then

$$\sup_{\alpha \in I} \|T_\alpha\|_{X \to Y} < \infty.$$

1.19 Reflexivity

Definition 1.19.1. Let X be a Banach space, X^* its dual space. Then we can define the dual of X^* by setting

$$X^{**} := (X^*)^*.$$

Let us denote the elements of X^{**} by $u^{**}, v^{**}, \ldots$ The operator J from X into X^{**} with $\text{Dom}(J) = X$, defined by the formula

$$(Ju)(\varphi) = \varphi(u) \quad \text{for } \varphi \in X^*,\ u \in X,$$

is called the *canonical mapping* from X into X^{**}. (Ju is the element $u^{**} \in X^{**}$ which satisfies $u^{**}(\varphi) = \varphi(u)$.)

Remark 1.19.2. Let X be a Banach space. Denote by $J(X)$ the image of X in the canonical mapping J. Then J is an isometric isomorphism between X and $J(X)$.

Definition 1.19.3. A Banach space X is said to be *reflexive* if

$$J(X) = X^{**}.$$

Remarks 1.19.4. (i) Every subspace of a reflexive Banach space is reflexive.

(ii) A Cartesian product of a finite number of reflexive Banach spaces is a reflexive Banach space.

(iii) A Banach space isomorphic to a reflexive Banach space is reflexive.

(iv) If a Banach space has separable dual space, then it is itself separable.

(v) The dual space of a separable reflexive Banach space is separable.

1.20 Measure spaces: general extension theory

Definition 1.20.1. Let X be a nonempty set and let $\mathcal{S} \subset \exp X$ be a collection of subsets of X. Then $\mathcal{S}$ is called an *algebra* if the following three conditions are satisfied:

(i) $\emptyset, X \in \mathcal{S}$;

(ii) $A \cap B \in \mathcal{S}$ for every $A, B \in \mathcal{S}$;

(iii) $A^c \in \mathcal{S}$ for every $A \in \mathcal{S}$.

Remark 1.20.2. For every collection $\mathcal{F}$ of subsets of a set X, there exists a unique algebra $\mathcal{S}$ of subsets of X such that $\mathcal{F} \subset \mathcal{S}$ and if $\mathcal{S}_1$ is another algebra containing $\mathcal{F}$, then $\mathcal{S} \subset \mathcal{S}_1$.

Definition 1.20.3. Let X be a nonempty set and let $\mathcal{S} \subset \exp X$ be a collection of subsets of X. Then every function $\mu : \mathcal{S} \to [0, \infty]$ is called a *set function* on $\mathcal{S}$. We say that a set function μ is *monotone* on $\mathcal{S}$ if

$$\mu(A) \leq \mu(B) \quad \text{whenever} \quad A, B \in \mathcal{S}, \ A \subset B.$$

We say that μ is *finitely additive* on $\mathcal{S}$ if

$$\mu\left(\bigcup_{i=1}^{n} A_i\right) = \sum_{i=1}^{n} \mu(A_i),$$

for every $n \in \mathbb{N}$ and pairwise disjoint sets $A_i \in \mathcal{S}$, $i = 1, \ldots, n$, such that $\bigcup_{i=1}^{n} A_i \in \mathcal{S}$. We say that μ is *countably additive* on $\mathcal{S}$ if

$$\mu\left(\bigcup_{n=1}^{\infty} A_n\right) = \sum_{n=1}^{\infty} \mu(A_n)$$

for all pairwise disjoint sets $A_n \in \mathcal{S}$, $n \in \mathbb{N}$, such that $\bigcup_{n=1}^{\infty} A_n \in \mathcal{S}$. We say that μ is *countably subadditive* on $\mathcal{S}$ if

$$\mu(A) \leq \sum_{n=1}^{\infty} \mu(A_n)$$

for every $A \in \mathcal{S}$ such that $A \subset \bigcup_{n=1}^{\infty} A_n$, where $A_n \in \mathcal{S}$ for each $n \in \mathbb{N}$.

Definition 1.20.4. Let X be a nonempty set and let $\mathcal{S} \subset \exp X$ be a collection of subsets of X satisfying $\emptyset \in \mathcal{S}$. A countably additive set function $\mu : \mathcal{S} \to [0, \infty]$ is called a *measure* on $\mathcal{S}$ if $\mu(\emptyset) = 0$.

Section 1.20 Measure spaces: general extension theory

We shall now extend a measure to a set function on the entire $\exp X$. We pay for this extension by a possible loss of nice properties of the original measure. In particular, the extended set function need not be countably additive any more.

Definition 1.20.5. Let X be a nonempty set and let $\mathcal{S} \subset \exp X$ be an algebra of subsets of X satisfying $\emptyset \in \mathcal{S}$. Let $\mu : \mathcal{S} \to [0, \infty]$ be a measure on $\mathcal{S}$. For every $A \subset X$ we define

$$\mu^*(A) := \inf \left\{ \sum_{i=1}^{\infty} \mu(A_i); \ A_i \in \mathcal{S}, \ A \subset \bigcup_{i=1}^{\infty} A_i \right\}.$$

The function μ^* is called the *outer measure induced by* μ.

Remark 1.20.6. An outer measure is well-defined since for every $A \in \exp X$ there exists at least one sequence $\{A_n\}_{n=1}^{\infty}$ such that $A \subset \bigcup_{n=1}^{\infty} A_n$. The set function μ^* can attain infinite value.

In the next proposition we shall collect some properties of an outer measure.

Proposition 1.20.7. *Let X be a nonempty set and let $\mu^* : \exp X \to [0, \infty]$ be an outer measure on X. Assume that μ^* is induced by a measure μ defined on an algebra $\mathcal{S} \subset \exp X$. Then*

(i) $\mu^*(A) \geq 0$ *for every $A \in \exp X$;*

(ii) $\mu^*(\emptyset) = 0$;

(iii) μ^* *is monotone;*

(iv) μ^* *is countably subadditive;*

(v) μ^* *is an extension of μ on $\mathcal{S}$ in the sense that $\mu^*(A) = \mu(A)$ for every $A \in \mathcal{S}$.*

We shall now extend the notion of an outer measure to set functions about which we do not *a priori* know that they were induced by a measure.

Definition 1.20.8. Let X be a nonempty set and let $\nu : \exp X \to [0, \infty]$ be a set function such that the properties (i)–(iv) of Proposition 1.20.7 are satisfied. Then ν is called an *outer measure* on X.

We have seen how an outer measure is defined on $\exp X$ by an extension of a given measure on a subalgebra of $\exp X$. Now we shall take the converse path. We start with an outer measure and our aim will be to build a measure from it. This will be done by choosing an appropriate subclass of $\exp X$ on which the given outer measure behaves like a measure.

Definition 1.20.9. Let X be a nonempty set and let $\nu : \exp X \to [0, \infty]$ be an outer measure on X. We say that a set $A \in \exp X$ is ν-*measurable* if

$$\nu(T) = \nu(T \cap A) + \nu(T \cap A^c) \quad \text{for every "test set" } T \in \exp X. \qquad (1.20.1)$$

We shall denote by $\mathcal{M}$ the set of all ν-measurable subsets of X.

Remark 1.20.10. If $A \in \mathcal{M}$, then also $A^c \in \mathcal{M}$ due to the symmetry in (1.20.1).

Definition 1.20.11. Let X be a nonempty set and let $\mathcal{S} \subset \exp X$ be a collection of subsets of X. Then $\mathcal{S}$ is called a σ-*algebra* if the following three conditions are satisfied:

(i) $\emptyset, X \in \mathcal{S}$;

(ii) $\bigcup_{n=1}^{\infty} A_n \in \mathcal{S}$ for every countable sequence $\{A_n\}_{n=1}^{\infty} \subset \mathcal{S}$;

(iii) $A^c \in \mathcal{S}$ for every $A \in \mathcal{S}$.

In the next proposition we shall collect some properties of the class of measurable sets.

Proposition 1.20.12. *Let X be a nonempty set, let $\nu : \exp X \to [0, \infty]$ be an outer measure on X, and let $\mathcal{M}$ be the set of all ν-measurable subsets of X. Then*

(i) *$\mathcal{M}$ is a σ-algebra of subsets of X;*

(ii) *ν is countably additive when restricted to $\mathcal{M}$;*

(iii) *if ν was induced by some measure defined on an algebra $\mathcal{S}$, then $\mathcal{S} \subset \mathcal{M}$;*

(iv) *the set*
$$\mathcal{N} := \{A \in \exp X; \ \nu(E) = 0\}$$
satisfies $\mathcal{N} \subset \mathcal{M}$.

Remark 1.20.13. For every collection $\mathcal{A}$ of subsets of a set X, there exists a unique σ-algebra $\mathcal{S}$ of subsets of X such that $\mathcal{A} \subset \mathcal{S}$ and if $\mathcal{S}_1$ is another algebra containing $\mathcal{A}$, then $\mathcal{S} \subset \mathcal{S}_1$. In such cases, we can say that the σ-algebra $\mathcal{S}$ is *generated* by $\mathcal{A}$ and write $\mathcal{S} = \mathcal{S}(\mathcal{A})$.

Definition 1.20.14. Let X be a topological space. Let $\mathcal{G}$ and $\mathcal{F}$ denote the set of all open and closed subsets of X, respectively. Then

$$\mathcal{S}(\mathcal{G}) = \mathcal{S}(\mathcal{F}).$$

We call the σ-algebra generated by open (or closed) sets the σ-algebra of *Borel subsets* of X and denote it by $\mathfrak{B}(X)$.

Definition 1.20.15. Let X be a nonempty set, let $\mathcal{S} \subset \exp X$ be a collection of subsets of X and let $\mu : \mathcal{S} \to [0, \infty]$ be a set function on $\mathcal{S}$. We say that μ is *finite* on $\mathcal{S}$ if $\mu(A) < \infty$ for every $A \in \mathcal{S}$. We say that μ is σ-*finite* on $\mathcal{S}$ if there exists a sequence $\{X_n\}_{n=1}^\infty$ of pairwise disjoint subsets of sets $X_n \in \mathcal{S}$, $n \in \mathbb{N}$, such that $\mu(X_n) < \infty$ for every $n \in \mathbb{N}$ and $X = \bigcup_{n=1}^\infty X_n$.

Definition 1.20.16. Let X be a nonempty set, let $\mathcal{S} \subset \exp X$ be a σ-algebra of subsets of X and let $\mu : \mathcal{S} \to [0, \infty]$ be a measure on $\mathcal{S}$. The pair $(X, \mathcal{S})$ is then called a *measurable space* and the triple $(X, \mathcal{S}, \mu)$ is called a *measure space*. The elements of $\mathcal{S}$ are called *measurable sets*.

Definition 1.20.17. Let $(X, \mathcal{S}, \mu)$ be a measure space. Define

$$\mathcal{N} := \{A \in \exp X;\ \text{there exists } N \in \mathcal{S} \text{ such that } A \subset N \text{ and } \mu(N) = 0\}.$$

Then the elements of $\mathcal{N}$ are called the μ-*null subsets* of X. We say that $(X, \mathcal{S}, \mu)$ is a *complete* measure space if $\mathcal{N} \subset \mathcal{S}$.

Remark 1.20.18. It is not difficult to realize that every measure can be completed. Therefore it is not such a restriction to assume that the measure space in question is complete.

Definition 1.20.19. Let $(X, \mathcal{S}, \mu)$ be a measure space. A set $A \in \mathcal{S}$ is called an *atom* if $\mu(A) > 0$ and for every set $B \in \mathcal{S}$, $B \subset A$, either $\mu(B) = 0$ or $\mu(A \setminus B) = 0$. The measure space $(X, \mathcal{S}, \mu)$ is called *completely atomic* (or just *atomic* or *discrete*) if there exists a set $M \subset X$ such that $\mu(X \setminus M) = 0$ and $\mu(\{x\}) \neq 0$ for every $x \in M$. The measure space $(X, \mathcal{S}, \mu)$ is called *nonatomic* if there do not exist any atoms in $\mathcal{S}$. We often for short say that *the measure μ is nonatomic*. Such measure is also called *continuous*.

Example 1.20.20. Let X be a nonempty set and let $\mathcal{S} := \exp X$. For $A \in \mathcal{S}$, define

$$\mu(A) := \begin{cases} \text{the number of elements of } A & \text{if } A \text{ is finite,} \\ \infty & \text{if } A \text{ is infinite.} \end{cases}$$

Then μ is a measure. This measure is called the *counting measure* on X.

In the particular case when $X = \mathbb{N}$, the counting measure is denoted by m and is called the *arithmetic measure* on $\mathbb{N}$. The triple $(\mathbb{N}, \exp \mathbb{N}, m)$ is a typical example of a completely atomic space with all atoms having the same measure.

Definition 1.20.21. Let (X, ϱ) be a metric space and let μ be a measure defined on $\mathfrak{B}(X)$. We say that μ is *outer regular* if, for every $A \in \mathfrak{B}(X)$, one has

$$\mu(A) = \inf\{\mu(G);\ G \text{ open},\ A \subset G\}$$
$$= \sup\{\mu(F);\ F \text{ closed},\ F \subset A\}.$$

We say that μ is *inner regular* if, for every $A \in \mathcal{B}(X)$ satisfying $0 < \mu(A) < \infty$ and every $\varepsilon > 0$, there exists a compact set $K \subset A$ such that $\mu(A \setminus K) < \varepsilon$. If a measure is both inner and outer regular, then we say that it is *regular*.

Definition 1.20.22. Let X be a nonempty set and $E \subset X$. The function $\chi_E : X \to [0, \infty)$, defined by
$$\chi_E(x) := \begin{cases} 1 & \text{if } x \in E, \\ 0 & \text{if } x \notin E, \end{cases}$$
is called the *characteristic function* of E.

Definition 1.20.23. Let $(X, \mathcal{S}, \mu)$ be a complete σ-finite measure space. The function $s : X \to [0, \infty)$ is called a *μ-simple function* (or just *simple function*) if it is a finite linear combination of characteristic functions of μ-measurable sets of finite measure, i.e. if there exist an $m \in \mathbb{N}$, real numbers $\{a_1, \ldots, a_m\}$ and disjoint measurable subsets of X of finite measure $\{E_1, \ldots, E_m\}$ such that
$$s(x) = \begin{cases} a_j, & x \in E_j, \ j = 1, \ldots, m, \\ 0, & x \in X \setminus \bigcup_{j=1}^m E_j. \end{cases}$$

Definition 1.20.24. Let (X, ϱ) be a metric space and let $u : X \to [0, \infty]$. The set
$$\overline{\{x \in X; u(x) \neq 0\}},$$
where the bar denotes the closure in the space (X, ϱ), is called the *support* of the function u and is denoted by $\operatorname{supp} u$.

Remark 1.20.25. Let $(X, \mathcal{S}, \mu)$ be a complete σ-finite measure space endowed further with a metric. Then a function $s : X \to \mathbb{R}$ is simple if and only if its range is a finite set and its support is of finite measure.

Definition 1.20.26. Let $(X, \mathcal{S}, \mu)$ be a complete σ-finite measure space and let $A \in \mathcal{S}$. We say that a certain statement, say $V(x)$, holds *almost everywhere* on A (or *a.e.* on A for short) with respect to μ (we will also say *for almost all* $x \in A$) if the set
$$E := \{x \in A;\ V(x) \text{ does not hold}\}$$
satisfies $E \in \mathcal{S}$ and $\mu(E) = 0$.

Definition 1.20.27. Let $(X, \mathcal{S}, \mu)$ be a complete σ-finite measure space and let s be a simple function on X with representation
$$s(x) = \sum_{i=1}^m a_i \chi_{E_i}, \quad x \in X,$$

Section 1.20 Measure spaces: general extension theory

where $a_i \in \mathbb{R}$ and $E_i \in \mathcal{S}$, $i = 1, \ldots, m$. We then define the *integral* of s by

$$\int_X s(x) \, d\mu(x) := \sum_{i=1}^{m} a_i \mu(E_i).$$

Definition 1.20.28. Let $(X, \mathcal{S}, \mu)$ be a complete σ-finite measure space and let $f : X \to [0, \infty]$. We say that f is $\mathcal{S}$-*measurable* (or just *measurable*) if there exists a nondecreasing sequence of nonnegative simple functions $\{s_n\}_{n=1}^{\infty}$ satisfying

$$f(x) = \lim_{n \to \infty} s_n(x), \quad x \in X.$$

We shall denote by $\mathcal{S}_+(X)$ the set of all nonnegative measurable functions on X.

For a nonnegative measurable function $f : X \to [0, \infty]$, we define its *integral* by

$$\int_X f(x) \, d\mu(x) := \lim_{n \to \infty} \int_X s_n(x) \, d\mu(x).$$

The first important property of nonnegative measurable functions is the following theorem on monotone convergence.

Theorem 1.20.29 (monotone convergence theorem). *Let $(X, \mathcal{S}, \mu)$ be a complete σ-finite measure space and let $\{f_n\}_{n=1}^{\infty}$ be a sequence of nonnegative integrable functions defined on X such that $f_n(x) \leq f_{n+1}(x)$ for every $n \in \mathbb{N}$ and $x \in X$. Let*

$$f(x) := \lim_{n \to \infty} f_n(x), \quad x \in X.$$

Then $f \in \mathcal{S}_+(X)$ and

$$\int_\Omega \lim_{n \to \infty} f_n(x) \, d\mu(x) = \int_\Omega f(x) \, d\mu(x) = \lim_{n \to \infty} \int_\Omega f_n(x) \, d\mu(x).$$

Our aim is now to extend the notion of measurability to functions that are not necessarily nonnegative.

Definition 1.20.30. Let $(X, \mathcal{S}, \mu)$ be a complete σ-finite measure space and let $f : X \to [-\infty, \infty]$. We define the *nonnegative part* f^+ of f by

$$f^+(x) := \begin{cases} f(x) & \text{if } f(x) \geq 0, \\ 0 & \text{if } f(x) < 0. \end{cases}$$

Similarly, we define the *nonpositive part* f^- of f by

$$f^-(x) := \begin{cases} 0 & \text{if } f(x) \geq 0, \\ -f(x) & \text{if } f(x) < 0. \end{cases}$$

Remark 1.20.31. The functions f^+ and f^- are nonnegative and
$$f = f^+ - f^-, \quad |f| = f^+ + f^-.$$

Definition 1.20.32. Let $(X, \mathcal{S}, \mu)$ be a complete σ-finite measure space and let $f : X \to [-\infty, \infty]$. We say that f is $\mathcal{S}$-*measurable* (or just *measurable*) if both f^+ and f^- are measurable in the sense of Definition 1.20.28. We denote by $\mathcal{M}(X)$ the class of all measurable functions on X.

Remark 1.20.33. Let $(X, \mathcal{S}, \mu)$ be a complete σ-finite measure space. Then the characteristic function χ_E of a subset E of X is measurable if and only if the set E is measurable.

Theorem 1.20.34 (Fatou lemma). *Let $(X, \mathcal{S}, \mu)$ be a complete σ-finite measure space and let $\{f_n\}_{n=1}^{\infty}$ be a sequence of nonnegative measurable functions on X. Then*
$$\int_X \liminf_{n \to \infty} f_n(x) \, d\mu(x) \leq \liminf_{n \to \infty} \int_X f_n(x) \, d\mu(x).$$

Definition 1.20.35. Let $(X, \mathcal{S}, \mu)$ be a complete σ-finite measure space and let $f : X \to [-\infty, \infty]$. We say that f is μ-*integrable* (or just *integrable*) if both $\int_X f^+ \, d\mu$ and $\int_X f^- \, d\mu$ are finite. We then define the *integral* of f by
$$\int_X f(x) \, d\mu(x) := \int_X f^+(x) \, d\mu(x) - \int_X f^-(x) \, d\mu(x). \tag{1.20.2}$$

In such cases, we say that the integral of f over X *converges*. We denote by $L^1(X, \mathcal{S}, \mu)$ (or just $L^1(X)$ or $L^1(\mu)$) the set of all μ-integrable functions on X.

In the case when exactly one of the values $\int_X f^+(x) \, d\mu(x)$, $\int_X f^-(x) \, d\mu(x)$ is infinite, we again define the integral of f by the formula (1.20.2), that is, as ∞ or $-\infty$, and we say that the integral of f over X *exists* (although it does not converge). When both values $\int_X f^+(x) \, d\mu(x)$ and $\int_X f^-(x) \, d\mu(x)$ are infinite, then we say that the integral of f over X *does not exist*.

We shall now collect basic properties of integrable functions in the following proposition.

Proposition 1.20.36. *Let $(X, \mathcal{S}, \mu)$ be a complete σ-finite measure space and let $f, g : X \to [-\infty, \infty]$ and $a, b \in \mathbb{R}$. Then*

(i) *$f \in L^1(X)$ holds if and only if $|f| \in L^1(X)$ and*
$$\left| \int_X f(x) \, d\mu(x) \right| \leq \int_X |f(x)| \, d\mu(x);$$

(ii) *if $|f(x)| \leq g(x)$ for a.e. $x \in X$ and $g \in L^1(X)$, then $f \in L^1(X)$;*

(iii) if $f(x) = g(x)$ for a.e. $x \in X$ and $f \in L^1(X)$, then $g \in L^1(X)$ and
$$\int_X f(x)\,d\mu(x) = \int_X g(x)\,d\mu(x);$$

(iv) if $f, g \in L^1(X)$, then $(f+g) \in L^1(X)$ and
$$\int_X (af(x) + bg(x))\,d\mu(x) = a\int_X f(x)\,d\mu(x) + b\int_X g(x)\,d\mu(x);$$

(v) if $E \in \mathcal{S}$ and $f \in L^1(X)$, then $f\chi_E \in L^1(X)$ and
$$\int_X \chi_E(x) f(x)\,d\mu(x) = \int_E f(x)\,d\mu(x).$$

Another important result concerning convergence is the following assertion.

Theorem 1.20.37 (Lebesgue dominated convergence theorem). *Let $(X, \mathcal{S}, \mu)$ be a complete σ-finite measure space and let $\{f_n\}_{n=1}^\infty$ be a sequence of measurable functions on X and let $g \in L^1(X)$ be such that*
$$|f_n(x)| \leq g(x)$$
for all $n \in \mathbb{N}$ and almost all $x \in X$. Assume that
$$\lim_{n\to\infty} f_n(x) = f(x) \quad \text{for a.e. } x \in X.$$
Then the following statements hold:

(i) $f \in L^1(X)$;

(ii) $\int_X f(s)\,d\mu(x) = \lim_{n\to\infty} \int_X f_n(x)\,d\mu(x)$;

(iii) $\lim_{n\to\infty} \int_X |f_n(x) - f(x)|\,d\mu(x) = 0.$

1.21 The Lebesgue measure and integral

In this section we apply the abstract general theory developed in Section 1.20 to the particular case when $X = \Omega$, where Ω is an open set in $\mathbb{R}^N$, $N \in \mathbb{N}$, $\mathcal{S}$ is the σ-algebra generated by all N-dimensional intervals and λ is the length function defined by
$$\lambda((a_1, b_1) \times \cdots \times (a_N, b_N)) := \prod_{i=1}^N (b_i - a_i).$$

We denote the set of all N-dimensional intervals in $\mathbb{R}^N$ by $\mathcal{I}_N$.

The restriction to an open set Ω is temporary and will be abandoned in Chapter 6 and subsequent chapters, where more general measure spaces will be considered.

Definition 1.21.1. The *Lebesgue outer measure* λ^* is defined for every $A \in \exp(\Omega)$ by

$$\lambda^*(A) := \inf \left\{ \sum_{n=1}^{\infty} \lambda(I_n \cap \Omega);\ I_n \in \mathcal{I}_N,\ I_m \cap I_n = \emptyset \text{ for } m \neq n,\ A \subset \bigcup_{n=1}^{\infty} I_n \right\}.$$

The corresponding σ-algebra of λ^*-measurable sets is called the σ-algebra of *Lebesgue measurable sets*. The resulting measure is denoted by μ and is called the *Lebesgue measure*. The measure space $(\Omega, \mathfrak{B}(\Omega), \mu)$ is called the *Lebesgue measure space*.

Proposition 1.21.2. *The Lebesgue measure on $\mathbb{R}^N$, $N \in \mathbb{N}$, is a translation-invariant (that is, if $A \in \mathcal{M}$ and $x \in \mathbb{R}^N$, then the set $A + x := \{y \in \mathbb{R}^N,\ y - x \in A\}$ satisfies $A + x \in \mathcal{M}$ and $\mu(A + x) = \mu(A)$) σ-finite regular measure on $\mathfrak{B}(\mathbb{R}^N)$. In fact, it is unique, up to multiplication by a positive constant.*

Convention 1.21.3. Throughout the text, we shall often write $\int_\Omega |f(x)|\,dx$ instead of $\int_\Omega |f(x)|\,d\mu(x)$ when no confusion can arise.

In the rest of this section we shall collect the most important assertions concerning the Lebesgue integral. Some of them will be the "Lebesgue" versions of their appropriate abstract counterparts from Section 1.20.

Theorem 1.21.4 (Levi monotone convergence theorem)**.** *Let $\{f_n\}_{n=1}^{\infty}$ be a sequence of integrable functions on a measurable set $\Omega \subset \mathbb{R}^N$ such that $f_n(x) \leq f_{n+1}(x)$ for every $n \in \mathbb{N}$ and almost all $x \in \Omega$ and that*

$$\int_\Omega f_1(x)\,dx > -\infty.$$

Then, for almost all $x \in \Omega$, the limit

$$f(x) = \lim_{n \to \infty} f_n(x)$$

exists, the function f is integrable and

$$\int_\Omega \lim_{n \to \infty} f_n(x)\,dx = \int_\Omega f(x)\,dx = \lim_{n \to \infty} \int_\Omega f_n(x)\,dx.$$

Theorem 1.21.5 (Lebesgue dominated convergence theorem)**.** *Let $\{f_n\}_{n=1}^{\infty}$ be a sequence of measurable functions on a measurable set $\Omega \subset \mathbb{R}^N$ which converges for almost all $x \in \Omega$ to $f(x)$. Suppose that there exists a function g with the finite Lebesgue integral over Ω such that*

$$|f_n(x)| \leq g(x)$$

for all $n \in \mathbb{N}$ and almost all $x \in \Omega$.

Then f_n, $n \in \mathbb{N}$, and f have finite integrals and

$$\int_\Omega f(x)\,dx = \lim_{n\to\infty} \int_\Omega f_n(x)\,dx.$$

Theorem 1.21.6 (Fatou lemma). *Let $\{f_n\}_{n=1}^\infty$ be a sequence of measurable functions which are nonnegative almost everywhere on Ω. Then*

$$\liminf_{n\to\infty} f_n(x)$$

is integrable and

$$\int_\Omega \liminf_{n\to\infty} f_n(x)\,dx \leq \liminf_{n\to\infty} \int_\Omega f_n(x)\,dx.$$

A generalization of Theorem 1.21.5 is given by the following result.

Theorem 1.21.7 (Vitali). *Let $\{f_n\}_{n=1}^\infty$ be a sequence of functions with finite integrals over a measurable set $\Omega \subset \mathbb{R}^N$. Suppose that*

$$\lim_{n\to\infty} f_n(x) = f(x)$$

for almost all $x \in \Omega$ and let f be an almost everywhere finite function. Suppose that the following condition is satisfied:

(P) *for every $\varepsilon > 0$ there exists a $\delta > 0$ with the property: if $B \subset \Omega$, $\mu(B) < \delta$, then*

$$\int_B |f_n(x)|\,dx < \varepsilon$$

for all $n \in \mathbb{N}$.

Then the function f has a finite integral over Ω and

$$\lim_{n\to\infty} \int_\Omega f_n(x)\,dx = \int_\Omega f(x)\,dx.$$

Very close to the preceding result is the following theorem.

Theorem 1.21.8 (Vitali–Hahn–Saks). *Let $\{f_n\}_{n=1}^\infty$ be a sequence of functions with finite integrals over a measurable set $\Omega \subset \mathbb{R}^N$. Suppose that, for an arbitrary measurable set $E \subset \Omega$, the limit*

$$\lim_{n\to\infty} \int_E f_n(x)\,dx$$

exists and is finite.
Then the condition (P) in Theorem 1.21.7 is satisfied.

The following assertion is a direct consequence of Theorem 1.21.5.

Theorem 1.21.9 (continuous dependence of the integral on a parameter). *Let $\Omega \subset \mathbb{R}^N$ be measurable and let (X, ϱ) be a metric space. Let $f(x, \alpha)$ be defined on $\Omega \times X$. Suppose that*

(i) *for almost all $x \in \Omega$, $f(x, \cdot)$ is continuous on X;*

(ii) *for every $\alpha \in X$, $f(\cdot, \alpha)$ is measurable on Ω;*

(iii) *there exists a g with a finite integral over Ω such that*

$$|f(x, \alpha)| \leq g(x)$$

for all $\alpha \in X$ and almost all $x \in \Omega$.

Then the function F, defined by

$$F(\alpha) := \int_\Omega f(x, \alpha)\,dx, \quad \alpha \in X,$$

is continuous on X.

The following assertion is of great importance.

Theorem 1.21.10 (derivative of the integral with respect to a parameter). *Let $\Omega \subset \mathbb{R}^N$ be measurable, $-\infty \leq a < b \leq \infty$ and let $f(x, \alpha)$ be defined on $\Omega \times (a, b)$. Define the function*

$$F(\alpha) := \int_\Omega f(x, \alpha)\,dx, \quad \alpha \in (a, b),$$

and suppose that

(i) *$F(\alpha)$ is finite for at least one $\alpha \in (a, b)$;*

(ii) *for every $\alpha \in (a, b)$, $f(\cdot, \alpha)$ is measurable on Ω;*

(iii) *the partial derivative*

$$\frac{\partial f(x, \alpha)}{\partial \alpha}$$

exists and is finite for every $\alpha \in (a, b)$ and almost every $x \in \Omega$;

(iv) *there exists a function g with finite integral over Ω such that*

$$\left|\frac{\partial f(x, \alpha)}{\partial \alpha}\right| \leq g(x)$$

for almost every $x \in \Omega$ and all $\alpha \in (a, b)$.

Then, for all $\alpha \in (a, b)$, the integral

$$F(\alpha) = \int_\Omega f(x, \alpha) \, dx$$

is finite and

$$F'(\alpha) = \int_\Omega \frac{\partial f(x, \alpha)}{\partial \alpha} \, dx.$$

To handle integrals over subsets of $\mathbb{R}^N$ we shall use the Fubini theorem (sometimes in literature called the Fubini–Tonelli theorem).

Theorem 1.21.11 (Fubini). *Let $\Omega_i \subset \mathbb{R}^{N_i}$, $i = 1, 2$, be measurable and set $\Omega = \Omega_1 \times \Omega_2$. Let $f(x, y)$ be integrable over Ω. Then for almost all $x \in \Omega_1$ and $y \in \Omega_2$ the integrals*

$$\int_{\Omega_1} f(x, y) \, dx \quad \text{and} \quad \int_{\Omega_2} f(x, y) \, dy$$

exist. Moreover,

$$\int_\Omega f(x, y) \, dx \, dy = \int_{\Omega_1} \left(\int_{\Omega_2} f(x, y) \, dy \right) dx = \int_{\Omega_2} \left(\int_{\Omega_1} f(x, y) \, dx \right) dy.$$

A characterization of measurable functions is given by the Luzin theorem.

Theorem 1.21.12 (Luzin). *Let $\Omega \subset \mathbb{R}^N$ be measurable; let f be defined almost everywhere on Ω. Then f is measurable on Ω if and only if, for every $\varepsilon > 0$, there exists an open set $M \subset \Omega$, $\mu(M) < \varepsilon$, such that the restriction of f to $\Omega \setminus M$ is continuous on $\Omega \setminus M$.*

We shall now recall a useful result on the absolute continuous dependence of an integral on the integration domain.

Theorem 1.21.13 (absolute continuity of integral). *Let f be a function with a finite Lebesgue integral over $\Omega \subset \mathbb{R}^N$. Then, for every $\varepsilon > 0$, there exists a $\delta = \delta(\varepsilon) > 0$ such that for every measurable subset E of Ω with $\mu(E) < \delta$, we have*

$$\left| \int_E f(x) \, dx \right| < \varepsilon.$$

Definition 1.21.14. Let ν be a σ-additive set function defined on the family of all Lebesgue measurable subsets of Ω. Let $\nu(\emptyset) = 0$. We say that ν is *absolutely continuous* with respect to Lebesgue measure μ and write $\nu \in AC[\mu]$, if

$$\mu(E) = 0 \quad \text{implies} \quad \nu(E) = 0$$

for every μ-measurable sets $E \subset \Omega$.

Theorem 1.21.15 (Radon–Nikodým). *Let $\nu \in AC[\mu]$ be a finite set function. Then there exists exactly one function f with a finite Lebesgue integral over Ω such that*

$$\nu(E) = \int_E f(x)\,dx$$

for every Lebesgue measurable subset $E \subset \Omega$.

1.22 Modes of convergence

Throughout this section we shall assume that $(X, \mathcal{S}, \mu)$ is a σ-finite complete measure space. As already noticed, the requirement of completeness is not too restrictive because every measure space can be easily completed. The main reason why it is reasonable to assume completeness is that in a complete space, the following implication holds: if f and g are functions on X, f is μ-measurable and

$$\mu^*(\{x \in X;\ f(x) \neq g(x)\}) = 0,$$

then g is also μ-measurable.

We shall study several types of convergence of a sequence of functions.

Definition 1.22.1. Let f_n, $n \in \mathbb{N}$, and f be μ-measurable functions defined on a σ-finite complete measure space $(X, \mathcal{S}, \mu)$.

(i) We say that the sequence $\{f_n\}_{n=1}^\infty$ converges to f *pointwise* on X if

$$\lim_{n \to \infty} f_n(x) = f(x)$$

for every $x \in X$. We write $f_n \to f$.

(ii) We say that the sequence $\{f_n\}_{n=1}^\infty$ converges to f *uniformly* on X if for every $\varepsilon > 0$ there exists an $n_0 \in \mathbb{N}$ such that for every $n \geq n_0$ and every $x \in X$ we have

$$|f_n(x) - f(x)| < \varepsilon.$$

We write $f_n \rightrightarrows f$.

(iii) If X is further endowed with a metric ϱ, then we say that the sequence $\{f_n\}_{n=1}^\infty$ converges to f *locally uniformly* on X if $\{f_n\}_{n=1}^\infty$ converges to f uniformly on each compact subset K of X. We write $f_n \overset{\text{loc}}{\rightrightarrows} f$.

(iv) We say that the sequence $\{f_n\}_{n=1}^\infty$ converges to f *uniformly up to small sets* on X if for every $\varepsilon > 0$ there exists an $M \subset X$, $\mu(M) < \varepsilon$, such that $\{f_n\}_{n=1}^\infty$ converges uniformly to f on $X \setminus M$.

(v) We say that the sequence $\{f_n\}_{n=1}^{\infty}$ converges to f *almost everywhere* on X if there exists a set $M \subset X$ of measure zero such that $\{f_n\}_{n=1}^{\infty}$ converges pointwise to f on $X \setminus M$. We write $f_n \xrightarrow{\text{a.e.}} f$.

(vi) We say that the sequence $\{f_n\}_{n=1}^{\infty}$ converges to f *in measure* on X if for every $\varepsilon > 0$,
$$\lim_{n \to \infty} \mu(\{x \in X;\ |f_n(x) - f(x)| \geq \varepsilon\}) = 0.$$
We write $f_n \xrightarrow{\mu} f$.

Remark 1.22.2. Let f_n, $n \in \mathbb{N}$, and f be μ-measurable functions defined on a σ-finite complete measure space $(X, \mathcal{S}, \mu)$. Then the following implications hold:

$$f_n \rightrightarrows f \quad \Rightarrow \quad f_n \overset{\text{loc}}{\rightrightarrows} f \quad \Rightarrow \quad f_n \to f \quad \Rightarrow \quad f_n \xrightarrow{\text{a.e.}} f. \tag{1.22.1}$$

Moreover, it is easy to construct examples showing that none of the implications can be reversed. However, there is a certain substitute result in this direction as the following theorem shows.

Theorem 1.22.3 (Egorov). *Let $(X, \mathcal{S}, \mu)$ be a complete σ-finite measure space, let $\{f_n\}_{n=1}^{\infty}$ be a sequence of measurable functions on X and let f be a measurable function on X such that $\lim f_n(x) = f(x)$ a.e. on X. Let $E \subset X$ be such that $\mu(E) < \infty$. Then, for every $\varepsilon > 0$, there exists a measurable set $M \subset E$ such that $\mu(M) < \varepsilon$ and $\{f_n\}_{n=1}^{\infty}$ converges uniformly to f on $E \setminus M$.*

Remark 1.22.4. The assertion of the Egorov theorem can also be restated as follows: If $f_n \xrightarrow{\text{a.e.}} f$ on a set of finite measure, then $f_n \to f$ uniformly up to small sets. As a corollary, we get the following assertion.

Proposition 1.22.5. *Let $(X, \mathcal{S}, \mu)$ be a complete σ-finite measure space such that $\mu(X) < \infty$, let $\{f_n\}_{n=1}^{\infty}$ be a sequence of measurable functions on X and let f be a measurable function on X. Then $f_n \xrightarrow{\text{a.e.}} f$ on X if and only if $f_n \to f$ on X uniformly up to small sets.*

Now we shall study the relations between convergence in measure and other modes of convergence.

Proposition 1.22.6. *Let $(X, \mathcal{S}, \mu)$ be a complete σ-finite measure space such that $\mu(X) < \infty$, let $\{f_n\}_{n=1}^{\infty}$ be a sequence of measurable functions on X and let f be a measurable function on X. Assume that $f_n \xrightarrow{\text{a.e.}} f$ on X. Then the sequence $\{f_n\}_{n=1}^{\infty}$ converges to f in measure on X.*

The converse implication in Proposition 1.22.6 is not true. Moreover, the assumption $\mu(X) < \infty$ is indispensable. We shall demonstrate these facts with examples.

Examples 1.22.7. (i) Let $X = \mathbb{R}$, let μ be the one-dimensional Lebesgue measure and $\mathcal{S}$ the σ-algebra of Lebesgue measurable sets. Define $f_n := \chi_{[n,n+1]}$ and $f(x) := 0$ for every $x \in \mathbb{R}$. Then $f_n \to f$ on $\mathbb{R}$ but

$$\lim_{n \to \infty} \mu(\{x \in \mathbb{R};\ |f_n(x) - f(x)| \geq 1\}) = 1,$$

so $\{f_n\}_{n=1}^\infty$ does not converge to f in measure.

(ii) Let $X = (0, 1)$, let μ be the one-dimensional Lebesgue measure restricted to $(0, 1)$ and $\mathcal{S}$ the σ-algebra of sets of type $A \cap (0, 1)$, where A is a Lebesgue measurable set on $\mathbb{R}$. For every $n \in \mathbb{N}$ we find the unique integer $m \in \mathbb{N}$ such that $2^m \leq n < 2^{m+1}$. Then $n = 2^m + k$ for some $k = 0, 1, \ldots, 2^m$. Define

$$I_n := \left[\frac{k}{2^m}, \frac{k+1}{2^m}\right]$$

(note that the correspondence $n \mapsto I_n$ is unique) and $f_n := \chi_{I_n}$. Define further $f(x) := 0$ for every $x \in \mathbb{R}$. Then $\{f_n\}_{n=1}^\infty$ converges to f in measure but not a.e.

Again, there is at least a "partial" result in the converse direction, as the following theorem shows.

Theorem 1.22.8 (Riesz). *Let $\{f_n\}_{n=1}^\infty$ be a sequence of measurable functions which converges in measure to f on Ω. Then there exists a subsequence $\{f_{n_k}\}_{k=1}^\infty$ which converges to f almost everywhere on Ω.*

We shall return to the study of modes of convergence in Section 3.13.

1.23 Systems of seminorms, Hahn–Saks theorem

Definition 1.23.1. Let X be a Banach space and let $\mathfrak{P}$ be a system of seminorms on X. We say that $\mathfrak{P}$ is *separating* if for every $x \in X$, $x \neq \theta$, there exist some $p \in \mathfrak{P}$ such that $p(x) \neq 0$.

Definition 1.23.2. Let X be a vector space. We say that a subset B of X is *balanced* if for every $x \in B$ and every $\alpha \in [-1, 1]$, we have $\alpha x \in B$.

Theorem 1.23.3. *Let X be a vector space and let $\mathfrak{P}$ be a separating system of seminorms on X. Define*

$$V(p, n) := \left\{x \in X;\ p(x) < \frac{1}{n}\right\}, \quad p \in \mathfrak{P},\ n \in \mathbb{N},$$

and set

$$\mathcal{B}' := \{V(p_1, n_1) \cap \ldots V(p_k, n_k),\ k \in \mathbb{N},\ p_i \in \mathfrak{P},\ n_i \in \mathbb{N},\ i = 1, \ldots, k\}.$$

Then $\mathcal{B}'$ is a convex balanced basis of topology such that every p is continuous and for every bounded $E \subset X$ and every $p \in \mathfrak{P}$, $p(E)$ is bounded in $\mathbb{R}$.

We shall now quote a classical result from measure theory (see, e.g., [105, Exercise 19.68, p. 339]).

Theorem 1.23.4 (Hahn–Saks). *Let $(X, \mathcal{S}, \mu)$ be a measure space and let $\{v_n\}_{n=1}^{\infty}$ be a sequence of finite measures on $\mathcal{S}$. Assume that for every $n \in \mathbb{N}$, the measure v_n is absolutely continuous with respect to μ. Let v be a functional on $\mathcal{S}$. Assume further that for every $E \in \mathcal{S}$ such that $\mu(E) < \infty$, one has*

$$\lim_{n \to \infty} v_n(E) = v(E).$$

Then

(i) *The measures $\{v_n\}$ are uniformly absolutely continuous with respect to μ, that is,*

$$\forall \varepsilon > 0 \; \exists \delta > 0 \; \forall n \in \mathbb{N}, \; \forall E \subset \mathcal{S} : \mu(E) < \delta \Rightarrow v_n(E) < \varepsilon;$$

(ii) *v is a measure, which is absolutely continuous with respect to μ.*

Chapter 2

Spaces of smooth functions

In this chapter we shall focus on spaces of continuous and smooth functions. Since however the role of such spaces for our purpose is in principal auxiliary, we will just briefly survey some of their very basic properties.

2.1 Multiindices and derivatives

Definition 2.1.1. Let N be a positive integer (i.e. $N \in \mathbb{N}$). A vector

$$\alpha = (\alpha_1, \ldots, \alpha_N)$$

with components $\alpha_i \in \mathbb{N}_0$ ($i = 1, \ldots, N$), where $\mathbb{N}_0 = \mathbb{N} \cup \{0\}$, is said to be a *multiindex of dimension* N. The number

$$|\alpha| = \sum_{i=1}^{N} \alpha_i$$

is called the *length* of the multiindex α. We introduce the following notation ("calculus of multiindices"): given multiindices α, β,

$$\alpha + \beta := (\alpha_1 + \beta_1, \ldots, \alpha_N + \beta_N),$$
$$\alpha! := \alpha_1! \cdots \alpha_N!,$$
$$\binom{\alpha}{\beta} := \frac{\alpha!}{\beta!(\alpha - \beta)!},$$

for $\alpha = (\alpha_1, \ldots, \alpha_N)$ a multiindex and $x = (x_1, \ldots, x_N) \in \mathbb{R}^N$,

$$x^\alpha := x_1^{\alpha_1} x_2^{\alpha_2} \cdots x_N^{\alpha_N}.$$

Notation 2.1.2. Let $k \in \mathbb{N}_0$. We denote by $\mathfrak{M}_{N,k}$ the set of all multiindices of dimension N whose length does not exceed k. For the sake of convenience, denote by θ the multiindex with all components zero (i.e. θ is the multiindex with zero length). Further, we denote by $\kappa = \kappa(k) = \kappa_N(k)$ the number of all elements of the set $\mathfrak{M}_{N,k}$, i.e. the number of all *different* multiindices α such that $|\alpha| \leq k$, and by $\tilde{\kappa} = \tilde{\kappa}(k) = \tilde{\kappa}_N(k)$ the number of all elements of the set $\mathfrak{M}_{N,k} - \mathfrak{M}_{N,k-1}$, i.e. the total number of different multiindices α such that $|\alpha| = k$.

Remark 2.1.3. One can easily prove the following relations:
$$\tilde{\kappa}(k) = \frac{(N+k-1)!}{k!(N-1)!},$$
$$\kappa(k) = \frac{(N+k)!}{N!k!}.$$

Notation 2.1.4. The concept of the "classical" partial derivative of a function of N real variables
$$u = u(x), \ x = (x_1, \ldots, x_N),$$
is well known. Here, we shall use the following notation: for α a multiindex
$$D^\alpha u := \frac{\partial^{|\alpha|} u}{\partial x_1^{\alpha_1} \partial x_2^{\alpha_2} \ldots \partial x_N^{\alpha_N}}.$$

In the following we consider functions all of whose derivatives up to a fixed order k are continuous. Thus the fact that for example, the same symbol $D^\alpha u$ for $N = 2$, $k = 2$, $\alpha = (1, 1)$ means both
$$\frac{\partial^2 u}{\partial x_1 \partial x_2} = \frac{\partial}{\partial x_1}\left(\frac{\partial u}{\partial x_2}\right) \quad \text{and} \quad \frac{\partial^2 u}{\partial x_2 \partial x_1} = \frac{\partial}{\partial x_2}\left(\frac{\partial u}{\partial x_1}\right)$$
causes no trouble since $\frac{\partial^2 u}{\partial x_1 \partial x_2} = \frac{\partial^2 u}{\partial x_2 \partial x_1}$ due to the continuity of the derivatives.

2.2 Classes of continuous and smooth functions

Definition 2.2.1. Let Ω be a domain (i.e. an open connected set) in $\mathbb{R}^N$. We introduce:

(i) $C(\Omega)$ or $C^0(\Omega)$ – the set of all functions defined and continuous on Ω;

(ii) $C^k(\Omega)$ with $k \in \mathbb{N}$ – the set of all functions defined on Ω which have continuous derivatives up to the order k on Ω;

(iii) $C^\infty(\Omega)$ – the set of all functions defined on Ω which have derivatives of any order on Ω, i.e.
$$C^\infty(\Omega) := \bigcap_{k=0}^{\infty} C^k(\Omega).$$

(iv) Let $k \in \mathbb{N}_0$. Then $C_0^k(\Omega)$ denotes the set of all functions $u \in C^k(\Omega)$ whose supports are compact subsets of Ω.

(v) We denote
$$C_0^\infty(\Omega) := \bigcap_{k=1}^{\infty} C_0^k(\Omega).$$

Definition 2.2.2. Let Ω be a bounded domain in $\mathbb{R}^N$.

(i) We denote by $C(\overline{\Omega})$ or $C^0(\overline{\Omega})$ the set of all functions in $C(\Omega)$ which are bounded and uniformly continuous on Ω.

(ii) For $k \in \mathbb{N}$ we denote by $C^k(\overline{\Omega})$ the set of all functions $u \in C^k(\Omega)$ such that $D^\alpha u \in C(\overline{\Omega})$ for all $\alpha \in \mathfrak{M}_{N,k}$.

(iii) We denote
$$C^\infty(\overline{\Omega}) := \bigcap_{k=0}^\infty C^k(\overline{\Omega}).$$

Remark 2.2.3. Let Ω be a bounded domain in $\mathbb{R}^N$. Let $k \in \mathbb{N}_0 \cup \{\infty\}$. Then $u \in C^k(\overline{\Omega})$ means that for an arbitrary multiindex α, $|\alpha| \leq k$, there exists a uniquely determined function $u_\alpha \in C(\overline{\Omega})$ and such that the restriction of u_α to Ω coincides with $D^\alpha u$. In what follows, we shall extend $D^\alpha u$ on $\overline{\Omega}$ as u_α for $u \in C^k(\overline{\Omega})$ and $|\alpha| \leq k$.

Notation 2.2.4. Let Ω be a bounded domain in $\mathbb{R}^N$, and let $\lambda \in (0, 1]$.

(i) For $u \in C^k(\Omega)$ and α a multiindex, $|\alpha| \leq k$, let us denote
$$H_{\alpha,\lambda}(u) = \sup_{\substack{x,y \in \Omega \\ x \neq y}} \frac{|D^\alpha u(x) - D^\alpha u(y)|}{|x-y|^\lambda}. \tag{2.2.1}$$

(ii) By $C^{k,\lambda}(\overline{\Omega})$ we denote the subset of all functions $u \in C^k(\overline{\Omega})$ such that
$$H_{\alpha,\lambda}(u) < \infty \text{ for all } \alpha \text{ with } |\alpha| = k.$$

Definition 2.2.5. A function u defined on Ω is said to satisfy the *Hölder condition* with exponent λ, $0 < \lambda \leq 1$, if there exists a nonnegative constant $c = c(u)$ such that
$$|u(x) - u(y)| \leq c|x-y|^\lambda \tag{2.2.2}$$
for all $x, y \in \Omega$. If $\lambda = 1$, we say that u satisfies the *Lipschitz condition*. Accordingly, the functions in $C^{0,\lambda}(\overline{\Omega})$ are said to be λ-*Hölder continuous* or, when $\lambda = 1$, *Lipschitz continuous* or *Lipschitz functions* for short).

Exercise 2.2.6. (i) Show that
$$H_{\alpha,\nu}(u) \leq H_{\alpha,\lambda}(u) \cdot (\operatorname{diam} \Omega)^{\lambda-\nu}, \tag{2.2.3}$$
$$H_{\alpha,\nu}(u) \leq (2 \max_{x \in \overline{\Omega}} |D^\alpha u(x)|)^{\frac{\lambda-\nu}{\lambda}} (H_{\alpha,\lambda}(u))^{\frac{\nu}{\lambda}} \tag{2.2.4}$$

for $0 < \nu \leq \lambda$.

Section 2.2 Classes of continuous and smooth functions

(ii) Show that a function u satisfying condition (2.2.2) with the exponent $\lambda > 1$ is necessarily constant on the domain Ω.

Definition 2.2.7. Let Ω be a domain in $\mathbb{R}^N$, $k \in \mathbb{N}_0$ and $\lambda \in (0, 1]$. Then $C^{k,\lambda,0}(\overline{\Omega})$ denotes the subset of all functions $u \in C^{k,\lambda}(\overline{\Omega})$ satisfying the following condition: For every $\varepsilon > 0$ there exists a $\delta = \delta(\varepsilon, u)$ such that for all $x, y \in \Omega$ with $0 < |x - y| < \delta$

$$\frac{|D^\alpha u(x) - D^\alpha u(y)|}{|x - y|^\lambda} < \varepsilon$$

for all α with $|\alpha| = k$.

Exercise 2.2.8. (i) Show that u belongs to $C^{k,1,0}(\overline{\Omega})$ if and only if u is a polynomial of degree less than or equal to k.

(ii) Clearly,

$$C^{k,\lambda,0}(\overline{\Omega}) \subset C^{k,\lambda}(\overline{\Omega}). \tag{2.2.5}$$

Show that, for $0 < \nu < \lambda \leq 1$,

$$C^{k,\lambda}(\overline{\Omega}) \subset C^{k,\nu,0}(\overline{\Omega}). \tag{2.2.6}$$

Remark 2.2.9. As a consequence of (2.2.3) and (2.2.4) or of (2.2.5) and (2.2.6), we have

$$C^{k,\lambda}(\overline{\Omega}) \subset C^{k,\nu}(\overline{\Omega})$$

for $0 < \nu < \lambda \leqq 1$. However it is not obvious that

$$C^{k+1,\lambda}(\overline{\Omega}) \subset C^{k,\lambda}(\overline{\Omega})$$

or

$$C^{k+1}(\overline{\Omega}) \subset C^{k,\lambda}(\overline{\Omega})$$

holds.

Definition 2.2.10. A domain $\Omega \subset \mathbb{R}^N$ is said to satisfy the *condition* (S) if there exists a constant $M > 0$ with the following property: for every pair of points $x, y \in \Omega$, there exist points $x = z_0, z_1, \ldots, z_n = y$ such that the segments with endpoints z_i, z_{i+1} ($i = 0, 1, \ldots, n - 1$) are subsets of Ω and

$$\sum_{i=1}^{n-1} |z_i - z_{i+1}| \leqq M|x - y|.$$

(Note that n may depend on x and y.)

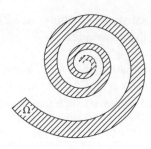

Figure 2.1. A spiraled domain.

Examples 2.2.11. (i) Every convex or star-shaped domain in $\mathbb{R}^N$ satisfies the condition (S).

(ii) Denote by Ω the plane domain lying – roughly speaking – "between two infinite disjoint spirals which converge to the same point" (see Figure 2.1; a precise description of the set Ω can be given using polar coordinates). Then Ω does *not* satisfy the condition (S).

(iii) Another example of a domain which does not satisfy the condition (S) is the open unit circle in $\mathbb{R}^2$ from which the segment $\{[x, y] \in \mathbb{R}^2; x \in (0, 1), y = 0\}$ was removed.

Theorem 2.2.12. *Let $\Omega \subset \mathbb{R}^N$ satisfy condition (S). Then*
$$C^{k+1}(\overline{\Omega}) \subset C^{k,\lambda}(\overline{\Omega}),$$
for $k \in \mathbb{N}_0$ and $\lambda \in (0, 1]$.

The proof of Theorem 2.2.12 is left to the reader. It uses the fact that condition (S) enables one to estimate $|D^\alpha u(x) - D^\alpha u(y)|$ in terms of M from Definition 2.2.10, $\max_{|\beta|=|\alpha|+1} \sup_{z \in \Omega} |D^\beta u(z)|$ and $|x - y|$.

Remarks 2.2.13. All function classes introduced in this section are vector spaces with respect to the sum of two functions and the scalar multiple of a function defined in Example 1.1.6.

Moreover, these spaces are algebras (in the sense of Definition 1.1.8) with respect to the product $w = u \cdot v$ of two functions u, v, defined by
$$w(x) = u(x) \cdot v(x), \quad x \in \Omega.$$

Exercise 2.2.14. Find an estimate for the number $H_{\alpha,\lambda}(u \cdot v)$ if $u, v \in C^{k,\lambda}(\overline{\Omega})$.

2.3 Completeness

Lemma 2.3.1. *Let $k \in \mathbb{N}_0$, $\lambda \in (0, 1]$, and let*

$$\|u\|_k = \sum_{|\alpha| \leq k} \sup_{x \in \Omega} |D^\alpha u(x)| \; (= \|u\|_{C^k(\overline{\Omega})}) \tag{2.3.1}$$

for $u \in C^k(\overline{\Omega})$ and

$$\|u\|_{k,\lambda} = \|u\|_k + \sum_{|\alpha|=k} H_{\alpha,\lambda}(u) \; (= \|u\|_{C^{k,\lambda}(\overline{\Omega})}) \tag{2.3.2}$$

for $u \in C^{k,\lambda}(\overline{\Omega})$.

Then $\|u\|_k$ and $\|u\|_{k,\lambda}$ are norms in the vector spaces $C^k(\overline{\Omega})$ and $C^{k,\lambda}(\overline{\Omega})$, respectively.

Proof. An easy verification of the assertion is left to the reader. □

Remark 2.3.2. For $u_n, u \in C^0(\overline{\Omega})$ ($n = 1, 2, \ldots$) the convergence

$$u_n \to u \text{ in } C^0(\overline{\Omega}),$$

i.e.

$$\lim_{n \to \infty} \|u_n - u\|_0 = 0,$$

is in fact the *uniform convergence* of u_n to u in $\overline{\Omega}$. Analogously, the convergence

$$u_n \to u \text{ in } C^k(\overline{\Omega}),$$

i.e. $\lim_{n \to \infty} \|u_n - u\|_k = 0$, is the uniform convergence of all derivatives $D^\alpha u_n$ (with $|\alpha| \leq k$) to $D^\alpha u$ in $\overline{\Omega}$.

Theorem 2.3.3. *Let Ω be a domain in $\mathbb{R}^n$, $k \in \mathbb{N}$ and $\lambda \in (0, 1]$. Then the spaces $C^k(\overline{\Omega})$ and $C^{k,\lambda}(\overline{\Omega})$ are Banach spaces.*

Proof. (I) The completeness of $C^k(\overline{\Omega})$ follows from the fact stated in Remark 2.3.2 and from the properties of uniform convergence.

(II) Let $\{u_n\}_{n=1}^\infty$ be a fundamental sequence in $C^{k,\lambda}(\overline{\Omega})$. Then it is also a fundamental sequence in the space $C^k(\overline{\Omega})$ and from part (I) of the proof it follows that there exists a function $u \in C^k(\overline{\Omega})$ such that

$$\lim_{n \to \infty} \|u_n - u\|_k = 0.$$

Since $\{u_n\}_{n=1}^\infty$ is a fundamental sequence in $C^{k,\lambda}(\overline{\Omega})$, for every $\varepsilon > 0$ there exists an $n_0 = n_0(\varepsilon) \in \mathbb{N}$ such that if $n > n_0$ and $m > n_0$ then

$$\sup_{\substack{x,y \in \Omega \\ x \neq y}} \frac{|D^\alpha(u_n - u_m)(x) - D^\alpha(u_n - u_m)(y)|}{|x-y|^\lambda} < \varepsilon$$

($|\alpha| = k$). When $m \to \infty$ we obtain

$$\sup_{\substack{x,y \in \Omega \\ x \neq y}} \frac{|D^\alpha(u_n - u)(x) - D^\alpha(u_n - u)(y)|}{|x-y|^\lambda} \leqq \varepsilon,$$

i.e.

$$H_{\alpha,\lambda}(u_n - u) \leqq \varepsilon.$$

This means that $u_n - u \in C^{k,\lambda}(\overline{\Omega})$ for all $n > n_0$, so that $u = u_n - (u_n - u)$ also belongs to $C^{k,\lambda}(\overline{\Omega})$ and, further,

$$\lim_{n \to \infty} H_{\alpha,\lambda}(u_n - u) = 0$$

($|\alpha| = k$), which implies

$$\lim_{n \to \infty} \|u_n - u\|_{k,\lambda} = 0. \qquad \square$$

Exercise 2.3.4. Show that $C^{k,\lambda,0}(\overline{\Omega})$ is a subspace of $C^{k,\lambda}(\overline{\Omega})$ and therefore $C^{k,\lambda,0}(\overline{\Omega})$ is a Banach space.

Remark 2.3.5. The norms introduced in (2.3.1) and (2.3.2) can be replaced by various equivalent norms in which, roughly speaking, we can take differences of higher order instead of $H_{\alpha,\lambda}(u)$ or $\|D^\alpha u\|_0$. Thus, we can define on $C^{k,\lambda}(\overline{\Omega})$ with $k \in \mathbb{N}_0$ and $\lambda \in (0, 1)$

$$|||u|||_{k,\lambda} = \|u\|_k + \sum_{|\alpha|=k} \sup_{\substack{0 \neq h \in \mathbb{R}^N \\ x \in \Omega_{2h}}} \frac{|D^\alpha u(x+2h) - 2D^\alpha u(x+h) + D^\alpha u(x)|}{|h|^\lambda}$$

where

$$\Omega_{2h} = \{x \in \Omega;\ x + h \in \Omega,\ x + 2h \in \Omega\}.$$

A good survey in this direction can be found in the work by Miranda [155] and in Triebel [228]. Let us remark here that these equivalent norms are meaningful and useful mainly for spaces $C^k(\overline{\Omega})$ and $C^{k,\lambda}(\overline{\Omega})$ with $0 < \lambda < 1$. The spaces $C^{k,1}(\overline{\Omega})$ of Lipschitz continuous functions play a somewhat exceptional role in the entire theory.

2.4 Separability, bases

Convention 2.4.1. Unless stated otherwise, we shall suppose in this section that Ω is a *bounded domain* in $\mathbb{R}^n$.

First let us state the well-known (first) Weierstrass theorem proved in 1885.

Theorem 2.4.2 (Weierstrass). *Let $-\infty < a < b < \infty$ and $u \in C^0([a, b])$. Then, for every $\varepsilon > 0$ there exists an $n \in \mathbb{N}$ and real numbers $a_0, \ldots, a_n$ such that*

$$|u(t) - a_0 - a_1 t - \cdots - a_n t^n| < \varepsilon$$

for all $t \in [a, b]$.

There are various proofs of this important assertion. We shall give here a proof based on a lemma by Korovkin from [120].

Definition 2.4.3. Let $L : C([0, 1]) \to C([0, 1])$ be a linear operator. We say that L is *monotone* if for every $u, v \in C([0, 1])$ such that $u(t) \geq v(t)$ for all $t \in [0, 1]$, we have

$$(Lu)(t) \geq (Lv)(t) \quad \text{for all } t \in [0, 1].$$

Lemma 2.4.4 (Korovkin). *Let $\{H_n\}_{n=1}^\infty$ be a sequence of monotone linear operators from $C^0([0, 1])$ into $C^0([0, 1])$. Let*

$$\lim_{n \to \infty} \|H_n e_i - e_i\|_0 = 0 \tag{2.4.1}$$

where $e_i(t) := t^{i-1}$, $i = 1, 2, 3$. Then for every function $u \in C^0([0, 1])$,

$$\lim_{n \to \infty} \|H_n u - u\|_0 = 0.$$

Proof. Let $u \in C^0([0, 1])$ be arbitrary. Since u is uniformly continuous, for every $\varepsilon > 0$ there exists a $\delta > 0$ such that

$$|u(t_1) - u(t_2)| < \tfrac{1}{2}\varepsilon \tag{2.4.2}$$

provided $t_1, t_1 \in [0, 1]$, $|t_1 - t_2| < \delta$. For $s, t \in [0, 1]$ we have

$$|u(t) - u(s)| \leq \tfrac{1}{2}\varepsilon + 2\|u\|_0 \delta^{-2}(t - s)^2, \tag{2.4.3}$$

since if $|t - s| < \delta$, inequality (2.4.2) implies (2.4.3), and if $|t - s| \geq \delta$, then

$$|u(t) - u(s)| \leq 2\|u\|_0 \leq 2\|u\|_0 \delta^{-2}(t - s)^2.$$

Let $s \in [0, 1]$ be fixed for the moment. If we denote

$$v(t) := u(t) - u(s) = u(t) - u(s) e_1(t)$$

and

$$w(t) = \tfrac{1}{2}\varepsilon + 2\|u\|_0 \delta^{-2}(t-s)^2 = \tfrac{1}{2}\varepsilon e_1(t) + 2\|u\|_0\delta^{-2}\{e_3(t) - 2se_2(t) + s^2 e_1(t)\},$$

we can rewrite (2.4.3) as

$$-w(t) \leq v(t) \leq w(t).$$

From the monotonicity and linearity of H_n it follows that

$$-(H_n w)(t) \leq (H_n v)(t) \leq (H_n w)(t)$$

with

$$(H_n v)(t) = (H_n u)(t) - u(s)(H_n e_1)(t)$$

and

$$(H_n w)(t) = \tfrac{1}{2}\varepsilon(H_n e_1)(t) + 2\|u\|_0\delta^{-2}\{(H_n e_3)(t) - 2s(H_n e_2)(t) + s^2(H_n e_1)(t)\},$$

which implies for arbitrary $n \in \mathbb{N}$ and for arbitrary $s, t \in [0,1]$ that

$$\begin{aligned}&|(H_n u)(t) - u(s)(H_n e_1)(t)| \\ &\leq \tfrac{1}{2}\varepsilon(H_n e_1)(t) + 2\|u\|_0\delta^{-2}\{(H_n e_3)(t) - 2s(H_n e_2)(t) + s^2(H_n e_1)(t)\}.\end{aligned} \quad (2.4.4)$$

By making $s = t$ the right-hand side in (2.4.4) becomes

$$\begin{aligned}&\tfrac{1}{2}\varepsilon + \tfrac{1}{2}\varepsilon[(H_n e_1)(t) - e_1(t)] + 2\|u\|_0\delta^{-2}\{[(H_n e_3)(t) - e_3(t)] \\ &\quad - 2t[(H_n e_2)(t) - e_2(t)] + t^2[(H_n e_1)(t) - e_1(t)]\}.\end{aligned}$$

Thus,

$$\begin{aligned}|(H_n u)(t) - u(t)| &\leq |(H_n u)(t) - u(t)(H_n e_1)(t)| + |u(t)(H_n e_1)(t) - u(t)e_1(t)| \\ &\leq \tfrac{1}{2}\varepsilon + \{\tfrac{1}{2}\varepsilon + |u(t)| + e_3(t) \cdot 2\|u\|_0\delta^{-2}\}|(H_n e_1)(t) - e_1(t)| \\ &\quad + 4\|u\|_0\delta^{-2}|e_2(t)| \cdot |(H_n e_2)(t) - e_2(t)| + 2\|u\|_0\delta^{-2}|(H_n e_3)(t) - e_3(t)|\end{aligned}$$

and

$$\begin{aligned}\|H_n u - u\|_0 &\leq \tfrac{1}{2}\varepsilon + \{\tfrac{1}{2}\varepsilon + \|u\|_0 + \|e_3\|_0 \cdot 2\|u\|_0\delta^{-2}\}\|H_n e_1 - e_1\|_0 \\ &\quad + 4\|u\|_0\delta^{-2}\|e_2\|_0\|H_n e_2 - e_2\|_0 + 2\|u\|_0\delta^{-2}\|H_n e_3 - e_3\|_0.\end{aligned}$$

Now assumption (2.4.1) implies the assertion. □

Proof of Theorem 2.4.2. Let us define the operators B_n from $C^0([0,1])$ into $C^0([0,1])$ by

$$(B_n u)(t) = \sum_{i=0}^{n} \binom{n}{i} u\left(\frac{i}{n}\right) t^i (1-t)^{n-i}, \quad n \in \mathbb{N}, \ t \in [0,1].$$

Section 2.4 Separability, bases   47

Obviously,

$$(B_n e_1)(t) = 1 = e_1(t),$$
$$(B_n e_2)(t) = t = e_2(t),$$
$$(B_n e_3)(t) = t^2 + \frac{t - t^2}{n} = e_3(t) + \frac{t - t^2}{n},$$

and thus conditions (2.4.1) are satisfied. The operators B_n are linear and monotone, so Lemma 2.4.4 implies that

$$\lim_{n \to \infty} \|B_n u - u\|_0 = 0.$$

However, $B_n u$ is a polynomial of degree at most n (the so-called *Bernstein polynomial*), so that we have proved the assertion of Theorem 2.4.2 for $a = 0, b = 1$.

Using a suitable transformation we obtain the assertion of Theorem 2.4.2 also for $C^0([a, b])$. □

Theorem 2.4.2 asserts that every function in $C^0([a, b])$ can be approximated in the norm of the space $C^0([a, b])$ by a sequence of polynomials. The following generalization (for the proof see, e.g., [161]) to the case of more than one variable holds (let us recall that, according to the above convention, only *bounded* domains are admitted).

Theorem 2.4.5. *Let $u \in C^0(\overline{\Omega})$ and let $\varepsilon > 0$. Then there exists a polynomial*

$$P_n(x) = \sum_{|\alpha| \leq n} a_\alpha x^\alpha, \quad x \in \overline{\Omega}, \tag{2.4.5}$$

where $n \in \mathbb{N}$ and $a_\alpha \in \mathbb{R}$, $|\alpha| \leq n$, such that

$$\|u - P_n\|_0 < \varepsilon. \tag{2.4.6}$$

Now we can easily prove the following result.

Theorem 2.4.6. *The space $C^0(\overline{\Omega})$ is separable.*

Proof. Let $\varepsilon > 0$. Let P_n be the polynomial from (2.4.5) satisfying (2.4.6) and let $\kappa = \kappa_N(n)$ be the number of its coefficients a_α (cf. Notation 2.1.2). Further, let d be a positive number such that $\overline{\Omega}$ is contained in the N-dimensional cube

$$C_d = [-d, d]^N.$$

For any coefficient a_α of P_n there exists a rational number r_α such that

$$|a_\alpha - r_\alpha| < \frac{\varepsilon}{\kappa d^{|\alpha|}} \quad (|\alpha| \leq n). \tag{2.4.7}$$

Denoting by Q_n the polynomial

$$Q_n(x) = \sum_{|\alpha| \leq n} r_\alpha x^\alpha,$$

we have from (2.4.7) that

$$|P_n(x) - Q_n(x)| = \left| \sum_{|\alpha| \leq n} (a_\alpha - r_\alpha) x^\alpha \right| \leq \sum_{|\alpha| \leq n} |a_\alpha - r_\alpha| \cdot |x^\alpha|$$

$$< \sum_{|\alpha| \leq n} \frac{\varepsilon}{\kappa d^{|\alpha|}} = \varepsilon$$

for all $x \in \overline{\Omega}$. Together with (2.4.6) this yields the estimate

$$\|u - Q_n\|_0 \leq \|u - P_n\|_0 + \|P_n - Q_n\|_0 < \varepsilon + \varepsilon = 2\varepsilon.$$

Since the set of all polynomials Q_n, $n = 0, 1, \ldots$ with rational coefficients is countable, the assertion of Theorem 2.4.6 is proved. □

Remark 2.4.7. The separability of $C^0(\overline{\Omega})$ can also be proved without using the Weierstrass theorem, by direct construction of a countable dense subset in $C^0(\overline{\Omega})$ formed by "piecewise linear" continuous functions on $\overline{\Omega}$ with rational values at the points with rational coordinates. For $N = 1$ see, e.g., [222]; the construction for the case $N > 1$ is based on the same idea.

Remark 2.4.8. We note that the boundedness of Ω is indispensable in Theorem 2.4.2. For example, when $N = 1$ and $\Omega = (0, \infty)$, then the space $C^0(\overline{\Omega})$ is not separable. Indeed, for $\varepsilon \in (0, \frac{1}{2})$ and $\mathbb{A} \subset \mathbb{N}$, define

$$\begin{cases} u_\mathbb{A}(k) = 1 & \text{if } k \in \mathbb{A} \\ u_\mathbb{A}(k - \varepsilon) = u_\mathbb{A}(k + \varepsilon) = 0 & \text{if } k \in \mathbb{A} \\ u_\mathbb{A} \text{ is linear} & \text{on } [k - \varepsilon, k], [k, k + \varepsilon], k \in \mathbb{A} \\ u_\mathbb{A}(t) = 0 & \text{for all other } t \in [0, \infty). \end{cases}$$

Obviously, $u_\mathbb{A} \in C^0(\overline{\Omega})$, but for $\mathbb{A} \neq \mathbb{B}$, $\mathbb{A}, \mathbb{B} \subset \mathbb{N}$, we have

$$\|u_\mathbb{A} - u_\mathbb{B}\|_{C^0(\overline{\Omega})} = 1.$$

Since $\exp \mathbb{N}$ is uncountable, $C^0(\overline{\Omega})$ is not separable.

Theorem 2.4.9. *The space $C^k(\overline{\Omega})$, $k \in \mathbb{N}$, is separable.*

Proof. Let $\kappa = \kappa_N(k)$ be the number introduced in Notation 2.1.2 and let Y be the Cartesian product of κ spaces $C^0(\overline{\Omega})$:

$$Y = [C^0(\overline{\Omega})]^\kappa.$$

For $U = \{u_\alpha;\ |\alpha| \leq k\} \in Y$ with $u_\alpha \in C^0(\overline{\Omega})$ we introduce the norm on Y by

$$\|U\|_Y = \sum_{|\alpha| \leq k} \|u_\alpha\|_0.$$

The space Y is a separable Banach space in view of Remark 1.11.1 (ii) and (iii) and Theorem 2.4.6. Let us denote by Y_1 the following subspace of Y:

$$U = \{u_\alpha;\ |\alpha| \leq k\} \in Y_1 \text{ if and only if } u_\alpha = D^\alpha u_\theta. \tag{2.4.8}$$

The space Y_1 is separable, for it is a subspace of the separable space Y (see Remark 1.10.3 (ii)). Further, (2.4.8) determines a one-to-one mapping from Y_1 onto $C^k(\overline{\Omega})$ which is an isomorphism. Thus $C^k(\overline{\Omega})$ is separable and the assertion is proved. □

For the spaces of Hölder continuous functions we have the following theorem.

Theorem 2.4.10. *Let $k \in \mathbb{N}_0$ and $0 < \lambda \leq 1$. Then the space $C^{k,\lambda}(\overline{\Omega})$ is not separable.*

Proof. It suffices to consider the case $N = 1$, $\Omega = (0, 1)$. If we define for $a \in (0, 1)$ a function u_a by

$$u_a(t) = \begin{cases} 0 & \text{for } t \in [0, a], \\ (t-a)^\lambda & \text{for } t \in (a, 1), \end{cases} \tag{2.4.9}$$

then

$$\|u_a - u_b\|_{0,\lambda} \geq 1$$

for $a, b \in (0, 1)$, $a \neq b$. Indeed, denoting $w = u_a - u_b$ we have (for $a < b$)

$$H_{0,\lambda}(w) \geq \frac{|w(b) - w(a)|}{|b-a|^\lambda} = \frac{(b-a)^\lambda}{|b-a|^\lambda} = 1.$$

The set of all functions u_a, $a \in (0, 1)$, is uncountable, and thus the space $C^{0,\lambda}([0, 1])$ is not separable (see Remark 1.8.3).

For the space $C^{k,\lambda}([0, 1])$ with $k \in \mathbb{N}$ we choose a function v_a such that its derivative $v_a^{(k)}$ is the function u_a from (2.4.9), and use integration. The proof is complete. □

Remark 2.4.11. For the case $0 < \lambda < 1$, the assertion of Theorem 2.4.10 also follows from the more general result from [50], where it has been shown that the space $C^{k,\lambda}(\overline{\Omega})$ is isomorphic with the space ℓ^∞, for the definition of ℓ^∞ see Section 3.12. As ℓ^∞ is not separable (see, e.g., [134]), the same assertion also holds for $C^{k,\lambda}(\overline{\Omega})$.

Theorem 2.4.12. *For $k \in \mathbb{N}_0$ and $0 < \lambda \leq 1$, the space $C^{k,\lambda,0}(\overline{\Omega})$ is separable.*

Proof. For $\lambda = 1$ the assertion follows from Exercise 1.1.10 (i), for $0 < \lambda < 1$ from [82], where it is shown that the space $C^{k,\lambda,0}(\overline{\Omega})$ is isomorphic to the space c_0. Since c_0 is separable (see, e.g., [134]), the space $C^{k,\lambda,0}(\overline{\Omega})$ is also separable. □

Separability of a space is a trivial consequence of the existence of a Schauder basis in this space – naturally, provided such a basis exists (see 0.10.3). The results of [52] and [156] immediately imply the following theorem.

Theorem 2.4.13. *The space $C^k(\overline{\Omega})$, $k \in \mathbb{N}_0$, has a Schauder basis.*

Often not only the *existence* of a Schauder basis but also the knowledge of its *actual form* is desirable. The first result concerning the construction of a Schauder basis can be found in [195] for the space $C^0([0, 1])$. Let us note here only that this basis is constructed by means of the *Haar orthogonal system* $\{h_n\}_{n=1}^\infty$ (see Section 3.17): If we define

$$e_1(t) = 1, \quad e_n(t) = \int_0^t h_{n-1}(s) \, ds \quad \text{for} \quad n = 2, 3, \ldots$$

($t \in [0, 1]$), then it can be shown that the sequence $\{e_n\}_{n=1}^\infty$ forms a Schauder basis in $C^0([0, 1])$.

Remark 2.4.14. Let $\{e_n\}_{n=1}^\infty$ be a Schauder basis in $C^0([0, 1])$ and let $k \in \mathbb{N}$. It is easy to see that the functions $e_{n,k}$ ($n = 1, 2, \ldots$) defined by

$$e_{n,k}(t) = \frac{t^{k-1}}{(k-1)!} \quad \text{for} \quad t \in [0, 1] \quad \text{and} \quad n = 1, 2, \ldots, k,$$

$$e_{n,k}(t) = \int_0^t \int_0^{t_{k-1}} \cdots \int_0^{t_2} \int_0^{t_1} e_{n-k}(s) \, ds \, dt_1 \, dt_2 \ldots dt_{k-1}$$

$$= \int_0^t (t-s)^{k-1} e_{n-k}(s) \, ds$$

for $t \in [0, 1]$, $n = k+1, k+2, \ldots$

form a Schauder basis in $C^k([0, 1])$.

Remark 2.4.15. The actual form of Schauder bases in the spaces $C^k(\overline{\Omega})$ for $N > 1$ is known only for some special domains Ω (see, e.g., [51, 52, 191, 196, 197]). For instance, let $\Omega = (0, 1) \times (0, 1)$ and let $\{e_n\}_{n=1}^{\infty}$ be a Schauder basis in $C^1([0, 1])$. We enumerate the set $\mathbb{N} \times \mathbb{N}$ in the following way:

$$(1, 1), (1, 2), (2, 1), (2, 2), \ldots, (n, n), (1, n+1), (2, n+1), \ldots,$$
$$\ldots, (n, n+1), (n+1, 1), (n+1, 2), \ldots, (n+1, n+1), \ldots \quad (2.4.10)$$

and set $h_p(x, y) = e_i(x) \cdot e_j(y)$ where (i, j) is the pth element in the sequence (2.4.10). The functions h_p, $p = 1, 2, \ldots$, form a basis in $C^1(\overline{\Omega})$ (see [196]).

An idea how to construct Schauder bases in the spaces $C^k(\overline{\Omega})$ for domains Ω with "sufficiently smooth" boundary $\partial\Omega$ is contained in [192].

2.5 Compactness

It is not our intention to investigate more deeply compactness of subsets of the spaces considered in this chapter, and, therefore, we give only the fundamental Arzelà–Ascoli theorem on the characterization of compact sets in $C^0(\overline{\Omega})$ and some results on compact sets in $C^k(\overline{\Omega})$ and $C^{k,\lambda}(\overline{\Omega})$. Let us note that Convention 2.4.1 takes place.

Definition 2.5.1. Let K be a subset of $C^0(\overline{\Omega})$. Then K is said to be *equicontinuous* if to every $\varepsilon > 0$ there exists a $\delta = \delta(\varepsilon) > 0$ such that

$$|u(x) - u(y)| < \varepsilon$$

holds for all $u \in K$ and all $x, y \in \overline{\Omega}$ for which $|x - y| < \delta$.

Exercise 2.5.2. Prove the following assertion: A set $K \subset C^0(\overline{\Omega})$ is equicontinuous if and only if for every $\varepsilon > 0$ and every $x \in \overline{\Omega}$ there exists a neighborhood $V(x)$ of x such that

$$\sup_{u \in K} \sup_{y \in V(x)} |u(x) - u(y)| \leq \varepsilon.$$

(Hint: Use the fact that Ω is bounded and consequently $\overline{\Omega}$ is compact in $\mathbb{R}^N$.) In what follows we shall use this equivalent definition of equicontinuity.

Theorem 2.5.3 (Arzelà–Ascoli). *A subset K of $C^0(\overline{\Omega})$ is relatively compact if and only if it is bounded and equicontinuous.*

Proof. (I) Let K be relatively compact. Then for $\varepsilon > 0$ there exists a finite ε'-net $u_1, \ldots, u_r$ ($\varepsilon' = \frac{1}{3}\varepsilon$), i.e. for every $u \in K$ there exists an index $i = i(u)$, $1 \leq i \leq r$, such that

$$\|u - u_i\|_0 < \tfrac{1}{3}\varepsilon. \quad (2.5.1)$$

For $x \in \overline{\Omega}$ and $i \in \{1, 2, \ldots, r\}$ there exists a neighborhood $V_i(x)$ of x such that
$$|u_i(x) - u_i(y)| < \tfrac{1}{3}\varepsilon$$
for all $y \in V_i(x)$. The set $V(x) = \bigcap_{i=1}^{r} V_i(x)$ is a neighborhood of x and
$$|u_i(x) - u_i(y)| < \tfrac{1}{3}\varepsilon \tag{2.5.2}$$
for all $i \leq r$ and and all $y \in V(x)$.

Now, it follows from (2.5.1) and (2.5.2) that for any $u \in K$,
$$|u(x) - u(y)| \leq |u(x) - u_i(x)| + |u_i(x) - u_i(y)| + |u_i(y) - u(y)| < \varepsilon$$
for every $y \in V(x)$. Consequently, K is equicontinuous.

The boundedness of K follows from (2.5.1):
$$\|u\|_0 = \|u - u_i + u_i\|_0 \leq \|u - u_i\|_0 + \|u_i\| \leq \tfrac{1}{3}\varepsilon + \max_{1 \leq i \leq r} \|u_i\|_0.$$

(II) Let K be bounded and equicontinuous. The equicontinuity of K implies that for $x \in \overline{\Omega}$ there exists a neighborhood $V(x)$ of x such that
$$|u(x) - u(y)| < \tfrac{1}{4}\varepsilon$$
for $y \in V(x)$ and $u \in K$. Since $\bigcup_{x \in \overline{\Omega}} V(x) \supset \overline{\Omega}$, a finite set of neighborhoods $V_i = V(x_i)$ $(i = 1, \ldots, r)$ can be chosen so that
$$\bigcup_{i=1}^{r} V(x_i) \supset \overline{\Omega}.$$

As K is bounded, the set $K(x) = \{u(x); u \in K\}$ is also a bounded subset of $\mathbb{R}$ and, consequently, $K(x)$ is relatively compact in $\mathbb{R}$. Therefore $\widehat{K} = \bigcup_{i=1}^{r} K(x_i)$ is also relatively compact in $\mathbb{R}$. Let $\varepsilon' = \tfrac{1}{4}\varepsilon$. Then there exists some $s \in \mathbb{N}$ and a finite ε'-net $\alpha_1, \alpha_2, \ldots, \alpha_s$, i.e.
$$|\beta - \alpha_i| < \tfrac{1}{4}\varepsilon$$
for every $\beta \in \widehat{K}$ and some α_i (with i depending on β).

Let us denote by Φ the set of mappings
$$\varphi : \{1, 2, \ldots, r\} \to \{1, 2, \ldots, s\}$$
and for $\varphi \in \Phi$ set
$$K_\varphi = \{u \in K; |u(x_i) - \alpha_{\varphi(i)}| \leq \tfrac{1}{4}\varepsilon, \ i = 1, \ldots, r\}.$$

We have that $K \subset \bigcup_{\varphi \in \Phi} K_\varphi$ (the set Φ is obviously finite). Indeed, for $u \in K$ we can construct $\varphi \in \Phi$ such that $u \in K_\varphi$ in the following way: For $u(x_1) \in K(x_1)$ there exists an α_t such that

$$|u(x_1) - \alpha_t| \leq \tfrac{1}{4}\varepsilon.$$

We set $\varphi(1) = t$ and define $\varphi(2), \ldots, \varphi(r)$ analogously. Further

$$\operatorname{diam} K_\varphi \leq \varepsilon.$$

Indeed, for $u, v \in K_\varphi$ and $x \in \overline{\Omega}$ we have that $x \in V(x_i)$ for some $i \leq r$ and

$$|u(x) - v(x)| \leq |u(x) - u(x_i)| + |u(x_i) - \alpha_{\varphi(i)}|$$
$$+ |\alpha_{\varphi(i)} - v(x_i)| + |v(x_i) - v(x)| < \varepsilon$$

which follows from the equicontinuity of K and from the definition of the set K_φ. Therefore,

$$\operatorname{diam} K_\varphi = \sup_{u,v \in K_\varphi} \sup_{x \in \overline{\Omega}} |u(x) - v(x)| \leq \varepsilon.$$

We now take one function u_φ from each set K_φ which yields a finite set of functions $\{u_\varphi\}_{\varphi \in \Phi} \subset K$. If $u \in K$ then $u \in K_\varphi$ for some φ and, consequently,

$$\|u - u_\varphi\|_0 \leq \operatorname{diam} K_\varphi \leq \varepsilon.$$

Hence, K is relatively compact and the theorem is proved. $\square$

By virtue of the Arzelà–Ascoli theorem (Theorem 2.5.3) we can easily prove the following theorem.

Theorem 2.5.4. *A set $K \subset C^k(\overline{\Omega})$ is relatively compact if and only if the following conditions are satisfied:*

(i) *K is bounded in $C^k(\overline{\Omega})$;*

(ii) *the sets $K_\alpha = \{D^\alpha u; u \in K\}$ are equicontinuous for all α such that $|\alpha| \leq k$.*

Proof. (i) Let conditions (i) and (ii) be satisfied and let $\{u_n\}_{n=1}^\infty$ be a sequence in K. Because K is also bounded in $C^0(\overline{\Omega})$ and K_0 is equicontinuous, it follows from Theorem 2.5.3 that there exists a subsequence $\{u_{n1}\}$ of $\{u_n\}$ which converges in $C^0(\overline{\Omega})$. Now we use the assertion of Theorem 2.5.3 for the set $\mathcal{K} = \{D^\alpha u_{n1}\}$ with a fixed α, $|\alpha| = 1$. The set $\mathcal{K}$ is bounded in $C^0(\overline{\Omega})$ and equicontinuous, thus there exists a subsequence $\{u_{n2}\}$ of $\{u_{n1}\}$ such that $\{D^\alpha u_{n2}\}$ converges in $C^0(\overline{\Omega})$. Repeating this process with the other derivatives D^β, we obtain after κ steps (for κ see Section 1.1) a subsequence $\{u_{n\kappa}\}$ of $\{u_n\}$ such that $\{D^\gamma u_{n\kappa}\}$ converges in $C^0(\overline{\Omega})$ for all multiindices γ which $|\gamma| \leq k$, i.e. the sequence $\{u_{n\kappa}\}$ converges in $C^k(\overline{\Omega})$.

(ii) Let K be a relatively compact subset of $C^k(\overline{\Omega})$. Then every set K_α ($|\alpha| \leq k$) is relatively compact in $C^0(\overline{\Omega})$ and thus K_α is bounded and equicontinuous in $C^0(\overline{\Omega})$. Hence, it follows that conditions (i) and (ii) of the theorem are satisfied. □

Definition 2.5.5. Let K be a subset of $C^{k,\lambda}(\overline{\Omega})$. Then K is said to be (k, λ)-equicontinuous if for every $\varepsilon > 0$ there exists a $\delta = \delta(\varepsilon) > 0$ such that

$$\frac{|D^\alpha u(x) - D^\alpha u(y)|}{|x - y|^\lambda} < \varepsilon$$

for all $u \in K$, all multiindices α with $|\alpha| = k$ and for all $x, y \in \Omega$ with $0 < |x - y| < \delta$.

Remark 2.5.6. (i) Obviously, $(0, 0)$-equicontinuity coincides with the equicontinuity introduced in Definition 2.5.1.

(ii) Comparison of Definition 2.5.5 with Definition 2.2.7 shows that if K is a (k, λ)-equicontinuous subset of $C^{k,\lambda}(\overline{\Omega})$ then K is a subset of the space $C^{k,\lambda,0}(\overline{\Omega})$.

Theorem 2.5.7. *Let $K \subset C^{k,\lambda,0}(\overline{\Omega})$ and let K be relatively compact in $C^{k,\lambda}(\overline{\Omega})$. Then K is bounded and (k, λ)-equicontinuous in $C^{k,\lambda}(\overline{\Omega})$.*

Proof. The boundedness of K is obvious. Let us suppose that K is *not* (k, λ)-equicontinuous. Then there exist an $\varepsilon > 0$, a sequence of functions $u_n \in K$ ($n \in \mathbb{N}$), a multiindex α ($|\alpha| = k$) and two sequences of points x_n, y_n ($n \in \mathbb{N}$) belonging to Ω such that

$$0 < |x_n - y_n| < \frac{1}{n} \quad \text{and} \quad \frac{|D^\alpha u_n(x_n) - D^\alpha u_n(y_n)|}{|x_n - y_n|^\lambda} \geq \varepsilon. \tag{2.5.3}$$

Since K is relatively compact we can suppose $u_n \to u$ in $C^{k,\lambda}(\overline{\Omega})$ with $u \in C^{k,\lambda,0}(\overline{\Omega})$ (see Exercise 2.3.4). For arbitrary $n \in \mathbb{N}$ we have

$$\varepsilon \leq \frac{|D^\alpha u_n(x_n) - D^\alpha u_n(y_n)|}{|x_n - y_n|^\lambda}$$
$$\leq \frac{|D^\alpha u_n(x_n) - D^\alpha u_n(y_n) - D^\alpha u(x_n) + D^\alpha u(y_n)|}{|x_n - y_n|^\lambda} + \frac{|D^\alpha u(x_n) - D^\alpha u(y_n)|}{|x_n - y_n|^\lambda}.$$

For $n \to \infty$, the right-hand side in the last relation tends to zero. Indeed, the first term tends to zero since $u_n \to u$ in $C^{k,\lambda}(\overline{\Omega})$ and the other term tends to zero since $u \in C^{k,\lambda,0}(\overline{\Omega})$. This is a contradiction. □

Remark 2.5.8. If the domain Ω has some special properties, e.g., if it satisfies condition (S) introduced in Definition 2.2.10, it can be shown that boundedness and (k, λ)-equicontinuity of the set $K \subset C^{k,\lambda,0}(\overline{\Omega})$ is also a sufficient condition for K to be relatively compact in $C^{k,\lambda}(\overline{\Omega})$.

Exercise 2.5.9. Prove the following assertion: A (k, λ)-equicontinuous set $K \subset C^{k,\lambda,0}(\overline{\Omega})$ is relatively compact in $C^{k,\lambda}(\overline{\Omega})$ if and only if it is relatively compact in $C^k(\overline{\Omega})$.

(Hint: First prove that if a (k, λ)-equicontinuous sequence $\{u_n\}_{n=1}^{\infty}$ of functions in $C^{k,\lambda,0}(\overline{\Omega})$ converges in $C^k(\overline{\Omega})$ to a function $u \in C^k(\overline{\Omega})$ then u belongs to $C^{k,\lambda,0}(\overline{\Omega})$ and $u_n \to u$ in $C^{k,\lambda}(\overline{\Omega})$).

We shall now formulate several inclusions between the spaces of smooth functions. We will use the symbols $\hookrightarrow$ and $\hookrightarrow\hookrightarrow$ introduced in Notation 1.15.7.

Theorem 2.5.10. *Let $k \in \mathbb{N}_0$, $\lambda \in (0, 1)$. Then*

$$C^{k,\lambda}(\overline{\Omega}) \hookrightarrow C^k(\overline{\Omega}). \tag{2.5.4}$$

Moreover,

$$C^{k,\lambda}(\overline{\Omega}) \hookrightarrow\hookrightarrow C^k(\overline{\Omega}). \tag{2.5.5}$$

Proof. (i) Embedding (2.5.4) follows from the definition of the spaces under consideration.

(ii) To prove (2.5.5), let K be a bounded set in $C^{k,\lambda}(\overline{\Omega})$. Then K is obviously bounded in $C^k(\overline{\Omega})$ and further, there exists a constant $c > 0$ such that

$$\frac{|D^\alpha u(x) - D^\alpha u(y)|}{|x - y|^\lambda} \leq H_{\alpha,\lambda}(u) \leq c$$

for all $u \in K$ (where $|\alpha| = k$). This implies that the sets $K_\alpha = \{D^\alpha u; u \in K\}$ are equicontinuous for $|\alpha| = k$. It can be shown by integration that the sets K_α are also equicontinuous for $|\alpha| < k$ and the compactness of K in $C^k(\overline{\Omega})$ follows from Theorem 2.5.4. □

Remark 2.5.11. The assertion of Theorem 2.5.10 can be strengthened: For $k \in \mathbb{N}_0$ and $0 < \nu < \lambda \leq 1$,

$$C^{k,\lambda}(\overline{\Omega}) \hookrightarrow\hookrightarrow C^{k,\nu}(\overline{\Omega})$$

(see, e.g., [242]; cf. also Exercise 2.2.6 (i)).

2.6 Continuous linear functionals

Definition 2.6.1. Let K be a compact subset of $\mathbb{R}^N$. We define the space $C(K)$ of all real-valued functions defined and continuous on K. This space is endowed with the norm

$$\|u\|_{C(K)} := \sup_{x \in K} |u(x)|.$$

Our next aim is to characterize the dual space $[C(K)]^*$ of the space $C(K)$. We recall (cf. Definition 1.20.14) that, for a compact subset K of $\mathbb{R}^N$, the smallest σ-algebra containing all closed subsets of K is denoted by $\mathfrak{B}(K)$ and every $M \in \mathfrak{B}(K)$ is called a *Borel subset* of K. Any measure defined on $\mathfrak{B}(K)$ is called a *Borel measure*. We also introduce the symbol

$$\mathrm{var}(\mu, K) := \sup \sum_{i=1}^{m} |\mu(M_i)|$$

where the supremum is taken over all finite systems $\{M_i\}_{i=1}^m$, $M_i \in \mathfrak{B}(K)$ such that $M_i \cap M_j = \emptyset$ for $i \neq j$. As usual, the norm of a functional Φ in $[C(K)]^*$ is given by

$$\|\Phi\| := \sup_{\|u\|_0 \leq 1} |\Phi(u)|.$$

Now we shall formulate the Riesz representation theorem. For the proof see, e.g., [66] or [222].

Theorem 2.6.2 (Riesz). *Let K be a compact subset of $\mathbb{R}^N$ and let Φ be a continuous linear functional on $C(K)$. Then there exists a uniquely determined real Borel measure μ on K such that*

$$\Phi(u) = \int_K u(x)\,d\mu(x)$$

for every $u \in C(K)$. Moreover,

$$\|\Phi\| = \|\Phi\|_{[C(K)]^*} = \sup \sum_{i=1}^{m} |\mu(M_i)|,$$

where the supremum is extended over all finite pairwise disjoint systems $\{M_i\}_{i=1}^m$, $M_i \in \mathfrak{B}(K)$.

Later we shall use the following important result:

Theorem 2.6.3. *The space $C(K)$ is not reflexive.*

Proof. For $\Phi \in [C(K)]^*$ let μ_Φ be the measure from Theorem 2.6.2, i.e. μ_Φ is a measure such that

$$\Phi(u) = \int_K u(x)\,d\mu_\Phi(x)$$

for all $u \in C(K)$. Let x_0 be a fixed point in K and let us define a functional F_0 on $[C(K)]^*$ by

$$F_0(\Phi) = u_\Phi(\{x_0\})$$

Section 2.6 Continuous linear functionals

for every $\Phi \in [C(K)]^*$. The functional F_0 is obviously linear and

$$\sup_{\|\Phi\| \leq 1} |F_0(\Phi)| = \sup_{\mathrm{var}(\mu,K) \leq 1} |\mu(\{x_0\})| = 1,$$

so that F_0 is also bounded and, consequently, $F_0 \in [C(K)]^{**}$. Let us suppose that $C(K)$ is reflexive. Thus there exists a function $u_0 \in C(K)$ such that

$$F_0(\Phi) = \Phi(u_0), \text{ i.e. } \mu_\Phi(\{x_0\}) = \int_K u_0(x) \, \mathrm{d}\mu_\Phi(x)$$

for any $\Phi \in [C(K)]^*$. If we define the Borel measure μ_0 by

$$\mu_0(M) = \int_M u_0(x) \, \mathrm{d}x$$

for $M \in \mathfrak{B}(K)$, then for every $u \in C(K)$,

$$\int_K u(x) \, \mathrm{d}\mu_0(x) = \int_K u(x) u_0(x) \, \mathrm{d}x.$$

Let us denote by Φ_0 the functional in the space $[C(K)]^*$ which is defined by the measure μ_0. Then

$$F_0(\Phi_0) = \mu_0(\{x_0\}) = \int_{\{x_0\}} u_0(x) \, \mathrm{d}x = 0. \tag{2.6.1}$$

On the other hand, Φ_0 is not the null-functional and, consequently, u_0 is not everywhere zero. Therefore,

$$F_0(\Phi_0) = \Phi_0(u_0) = \int_K u_0(x) \, \mathrm{d}\mu_0(x) = \int_K [u_0(x)]^2 \, \mathrm{d}x$$

is a positive number which contradicts (2.6.1). Consequently, $C(K)$ cannot be reflexive. □

Let us now consider the space $C^{k,\lambda,0}(\overline{\Omega})$ with $k \in \mathbb{N}_0$ and $\lambda \in (0,1)$. For $u \in C^{k,\lambda,0}(\overline{\Omega})$, the function

$$F_\beta(x,y) = \frac{D^\beta u(x) - D^\beta u(y)}{|x-y|^\lambda}, \quad |\beta| = k \tag{2.6.2}$$

is defined for $x, y \in \overline{\Omega}$ such that $x \neq y$. If we set

$$F_\beta(x,x) = 0$$

we obtain a function which is defined and continuous on $\overline{\Omega} \times \overline{\Omega}$.

Let us introduce the product space

$$\Pi(\overline{\Omega}) = \underbrace{C(\overline{\Omega}) \times C(\overline{\Omega}) \times \cdots \times C(\overline{\Omega})}_{\kappa\text{-times}} \times \underbrace{C(\overline{\Omega} \times \overline{\Omega}) \times \cdots \times C(\overline{\Omega} \times \overline{\Omega})}_{\tilde{\kappa}\text{-times}}$$

(for the definition of κ and $\tilde{\kappa}$ see Notation 2.1.2). Then we can identify the space $C^{k,\lambda,0}(\overline{\Omega})$ with a closed subspace of $\Pi(\overline{\Omega})$, namely, with the subspace of elements

$$\{(u_\alpha(x), v_\beta(x, y)); \ |\alpha| \le k, \ |\beta| = k\} \in \Pi(\overline{\Omega})$$

corresponding to an element $u \in C^{k,\lambda,0}(\overline{\Omega})$ by the formulas

$$u_\alpha(x) = D^\alpha u(x), \ v_\beta(x, y) = F_\beta(x, y).$$

On the other hand, $\Pi(\overline{\Omega})$ can be identified with the space $C(K)$ where K is a compact subset of $\mathbb{R}^M$ with $M = N(\kappa(k) + 2\tilde{\kappa}(k))$, and the problem of characterizing the continuous linear functionals on $C^{k,\lambda,0}(\overline{\Omega})$ can by virtue of the Hahn–Banach theorem (Theorem 1.16.4) be reduced to the problem considered in Theorem 2.6.2. Hence we have the following theorem.

Theorem 2.6.4. *Let Φ be a continuous linear functional defined on $C^{k,\lambda,0}(\overline{\Omega})$. Then there exist two uniquely determined families of Borel measures*

$$\mu_\alpha \text{ on } \overline{\Omega} \ (|\alpha| \le k), \ v_\beta \text{ on } \overline{\Omega} \times \overline{\Omega} \ (|\beta| = k)$$

such that

$$\Phi(u) = \sum_{|\alpha| \le k} \int_\Omega D^\alpha u(x) \, d\mu_\alpha(x) + \sum_{|\beta| = k} \int_{\overline{\Omega} \times \overline{\Omega}} \frac{D^\beta u(y) - D^\beta u(z)}{|y - z|^\lambda} \, dv_\beta(y, z)$$

(2.6.3)

for every $u \in C^{k,\lambda,0}(\overline{\Omega})$. Moreover,

$$\|\Phi\| = \sum_{|\alpha| \le k} \mathrm{var}(\mu_\alpha, \overline{\Omega}) + \sum_{|\beta| = k} \mathrm{var}(v_\beta, \overline{\Omega} \times \overline{\Omega}).$$

Remark 2.6.5. (i) For functionals on $C^k(\overline{\Omega})$ we can prove analogously the same assertion (with $v_\beta = 0$ for $|\beta| = k$) as in Theorem 2.6.4, i.e. if $\Phi \in [C^k(\overline{\Omega})]^*$ then

$$\Phi(u) = \sum_{|\alpha| \le k} \int_\Omega D^\alpha u(x) \, d\mu_\alpha(x)$$

for every $u \in C^k(\overline{\Omega})$ and

$$\|\Phi\| = \sum_{|\alpha| \le k} \mathrm{var}(\mu_\alpha, \overline{\Omega}).$$

(ii) For functionals on $C^{k,\lambda}(\overline{\Omega})$ we cannot use the above argument because the functions $F_\beta(x, y)$ from (2.6.2) cannot in general be extended in this case to all of $\overline{\Omega} \times \overline{\Omega}$ so as to be continuous on $\overline{\Omega} \times \overline{\Omega}$. Nonetheless, these functions are bounded on $(\overline{\Omega} \times \overline{\Omega}) - S$, where S has measure zero, and instead of $\Pi(\overline{\Omega})$ from (1.7.4.2) we can introduce another product space

$$\Pi_1(\Omega) = \underbrace{C(\overline{\Omega}) \times \cdots \times C(\overline{\Omega})}_{\kappa\text{-times}} \times \underbrace{L^\infty(\Omega \times \Omega) \times \cdots \times L^\infty(\Omega \times \Omega)}_{\tilde{\kappa}\text{-times}} \quad (2.6.4)$$

(for $L^\infty(\Omega \times \Omega)$ see Section 2.11). Then we obtain that every continuous linear functional Φ on $C^{k,\lambda}(\overline{\Omega})$ can be expressed in the form (2.6.3) but with more restrictive conditions on the measure ν_β, $|\beta| = k$. For details see, e.g., [236].

Remark 2.6.6. To complete the assertion of Theorem 2.6.3 it can be shown that the spaces $C^k(\overline{\Omega})$, $C^{k,1}(\overline{\Omega})$ and $C^{k,\lambda}(\overline{\Omega})$, $C^{k,\lambda,0}(\overline{\Omega})$ for $0 < \lambda < 1$ are not reflexive. The nonreflexivity of the last two spaces follows from the following characterization of the second dual space to $C^{k,\lambda,0}(\overline{\Omega})$ (for the details see [87] or [61]):

Let $0 < \lambda < 1$. Then the spaces $[C^{k,\lambda,0}(\overline{\Omega})]^{**}$ and $C^{k,\lambda}(\overline{\Omega})$ are isometrically isomorphic.

On the other hand, $C^{k,1,0}(\overline{\Omega})$ (as a finite dimensional space, see Exercise 2.2.8 (i)) is reflexive.

2.7 Extension of functions

Let u be a real-valued function defined on a set $M \subset \mathbb{R}^N$. A function U defined on $\mathbb{R}^N$ is said to be an *extension* of u if

$$U(x) = u(x) \quad \text{for} \quad x \in M.$$

We shall now study the problem of extending u in such a way that its extension U has – roughly speaking – the same properties as u itself.

First, let us remark that a necessary condition for the existence of continuous extensions of continuous functions is that u is *defined on a closed set*. Indeed, let $\overline{M} \setminus M \neq \emptyset$. We construct an example of a continuous function u on M which has no continuous extension to $\overline{M}$ and, consequently, cannot be extended continuously to all of $\mathbb{R}^N$. Suppose that $x_0 \in \overline{M} \setminus M$. We choose a sequence $x_n \in M$, $n \in \mathbb{N}$, such that

$$x_0 = \lim_{n \to \infty} x_n \text{ and } |x_{n+1} - x_0| < |x_n - x_0|$$

for $n \in \mathbb{N}$. There exists a continuous real-valued function g defined on $\mathbb{R}$ such that

$$g(|x_n - x_0|) = \frac{1}{n\pi} \text{ for } n \in \mathbb{N},$$

$$g(t) = 0 \text{ if and only if } t \leq 0.$$

(For instance, we can set $g(t) = 0$ for $t \leq 0$, $g(t) = \frac{1}{\pi}$ for $t \geq \frac{1}{\pi}$, g linear on the intervals $[|x_{n+1} - x_0|, |x_n - x_0|]$, $n \in \mathbb{N}$.) The function

$$u(x) = \sin \frac{1}{g(|x - x_0|)}$$

is obviously continuous and bounded on M. Nevertheless, a continuous extension of u to $\overline{M}$ does not exist.

Before we state the fundamental results due to Whitney concerning extensions of functions in $C^k(\overline{\Omega})$ and $C^{k,\lambda}(\overline{\Omega})$ we prove one easy assertion. Let us introduce the necessary notation.

Notation 2.7.1. For u defined and continuous on a closed set $M \subset \mathbb{R}^N$ denote

$$\|u\|_{0,M} := \sup_{x \in M} |u(x)|,$$

$$H_{\lambda,M}(u) := \sup_{\substack{x,y \in M \\ x \neq y}} \frac{|u(x) - u(y)|}{|x - y|^\lambda} \quad (0 < \lambda \leq 1).$$

Further, $C^{0,\lambda}(M)$ denotes the space of continuous functions u on M such that

$$\|u\|_{0,M} + H_{\lambda,M}(u) < \infty.$$

The following result from [16] is an extension of the classical Tietze extension theorem (see [223] or [130]).

Theorem 2.7.2. *Let M be a closed set in $\mathbb{R}^N$, let $u \in C^{0,\lambda}(M)$ and let $0 < \lambda \leq 1$. Then there exists an extension $U \in C^{0,\lambda}(\mathbb{R}^N)$ of u such that*

$$\|U\|_{0,\mathbb{R}^N} = \|u\|_{0,M} \tag{2.7.1}$$

and

$$H_{\lambda,\mathbb{R}^N}(U) = H_{\lambda,M}(u). \tag{2.7.2}$$

Proof. For $x \in \mathbb{R}^N$ set

$$v(x) = \sup_{y \in M} \{u(y) - H_{\lambda,M}(u)|x - y|^\lambda\} \tag{2.7.3}$$

and define U by

$$U(x) = \begin{cases} v(x) & \text{for } x \text{ such that } |v(x)| \leq \|u\|_{0,M}, \\ -\|u\|_{0,M} & \text{for } x \text{ such that } v(x) < -\|u\|_{0,M}. \end{cases} \tag{2.7.4}$$

Let $x \in M$. It follows from (2.7.3) that $v(x) \geq u(x)$. If we suppose that $v(x) > u(x)$ then there exists a point $y \in M$ such that

$$u(y) - H_{\lambda,M}(u)|x - y|^\lambda > u(x),$$

Section 2.7 Extension of functions

which leads to a contradiction with the definition of $H_{\lambda,M}(u)$. Therefore,

$$u(x) = v(x) \quad \text{for} \quad x \in M \tag{2.7.5}$$

and (2.7.4) implies

$$U(x) = u(x) \quad \text{for} \quad x \in M,$$

i.e. U is an extension of u. From (2.7.4) it also follows that $\|U\|_{0,\mathbb{R}^N} = \|u\|_{0,M}$ and we have (2.7.1).

Further, let $x, y \in \mathbb{R}^N$. The case $U(x) = U(y)$ is devoid of interest. If $U(x) \neq U(y)$ we can suppose that $U(x) > U(y)$. Then we have from (2.7.4) that

$$0 < U(x) - U(y) \leq v(x) - v(y)$$

and in view of

$$v(x) - v(y) = \sup_{z \in M} \{u(z) - H_{\lambda,M}(u)|x-z|^\lambda\} - \sup_{z \in M} \{u(z) - H_{\lambda,M}|y-z|^\lambda\}$$

$$\leq H_{\lambda,M}(u) \sup_{z \in M} \{|y-z|^\lambda - |x-z|^\lambda\} \leq H_{\lambda,M}(u)|x-y|^\lambda$$

we have

$$H_{\lambda,\mathbb{R}^N}(U) \leq H_{\lambda,M}(u).$$

We have that $H_{\lambda,\mathbb{R}^N}(U) \geq H_{\lambda,M}(u)$ since U is an extension of u. Thus we have (2.7.2) and the proof is complete. □

Remark 2.7.3. Formula (2.7.5) in the foregoing proof shows that the function v from (2.7.3) is also an extension of u; v is continuous and $H_{\lambda,\mathbb{R}^N}(v) < \infty$, but v is not necessarily bounded.

Chapter 3

Lebesgue spaces

3.1 $\mathcal{L}^p$-classes and some integral inequalities

Notation 3.1.1. Let $p \in (0, \infty)$ (the case $p = \infty$ will be considered in Section 3.10). Let Ω be a (Lebesgue) measurable subset of the Euclidean space $\mathbb{R}^N$. Denote by $\mathcal{L}^p(\Omega)$ the set of all real-valued measurable functions f defined almost everywhere on Ω and such that the Lebesgue integral

$$\int_\Omega |f(x)|^p \, dx \tag{3.1.1}$$

is finite.

For $\lambda \in \mathbb{R}$ and $f, g \in \mathcal{L}^p(\Omega)$, we define λf and $f + g$ by the relations

$$(\lambda f)(x) := \lambda f(x), \quad (f + g)(x) := f(x) + g(x)$$

for $x \in \Omega$. For $f \in \mathcal{L}^p(\Omega)$ we set

$$\mathcal{N}_p(f) := \left(\int_\Omega |f(x)|^p \, dx \right)^{\frac{1}{p}}. \tag{3.1.2}$$

Remark 3.1.2. It follows immediately from (3.1.2) that, given $p, q \in [1, \infty)$, we have

$$\mathcal{N}_q\left(|f|^p\right) = \mathcal{N}_{pq}^p(f). \tag{3.1.3}$$

Definition 3.1.3. For every $p \in (1, \infty)$, we define its *conjugate Lebesgue index* p' by the relation

$$\frac{1}{p} + \frac{1}{p'} = 1.$$

In other words, we have

$$p' = \frac{p}{p-1}. \tag{3.1.4}$$

In what follows, we shall first prove some inequalities which will be useful.

Theorem 3.1.4 (Young inequality). *Let φ be a continuous real-valued strictly increasing function defined on $[0, \infty)$ such that*

$$\lim_{u \to \infty} \varphi(u) = \infty$$

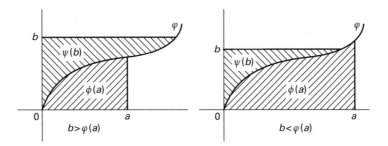

Figure 3.1. Young inequality.

and
$$\varphi(0) = 0.$$

Let $\psi = \varphi^{-1}$. For all $x \in [0, \infty)$, define
$$\Phi(x) := \int_0^x \varphi(u)\, du$$

and
$$\Psi(x) := \int_0^x \psi(v)\, dv.$$

Then, for all $a, b \in [0, \infty)$,
$$ab \leq \Phi(a) + \Psi(b);$$
the equality occurs if and only if $b = \varphi(a)$.

Proof. The above inequality is seen to be highly plausible by observing Figure 3.1. A formal proof can be given by virtue of the fact that
$$\Phi(c) + \Psi(\varphi(c)) = \int_0^c \varphi(u)\, du + \int_0^{\varphi(c)} \psi(v)\, dv = c\varphi(c)$$
for all $c \in [0, \infty)$. □

Corollary 3.1.5. *Given $p \in (1, \infty)$ and nonnegative real numbers a and b, we have*
$$ab \leq \frac{a^p}{p} + \frac{b^{p'}}{p'}, \qquad (3.1.5)$$
where p' is defined by (3.1.4). Equality holds in (3.1.5) if and only if $a^p = b^{p'}$.

Proof. For $u \in [0, \infty)$, define $\varphi(u) = u^{p-1}$; φ is continuous and strictly increasing, $\lim_{u \to \infty} \varphi(u) = \infty$ and $\varphi(0) = 0$. The inverse ψ of φ is given by $\psi(v) = v^{\frac{1}{p-1}}$. The corollary follows immediately from the Young inequality (Theorem 3.1.4). □

Theorem 3.1.6 (Hölder inequality). *Let $p \in (1, \infty)$, $f \in \mathcal{L}^p(\Omega)$ and $g \in \mathcal{L}^{p'}(\Omega)$. Then $fg \in \mathcal{L}^1(\Omega)$ and*

$$\left| \int_\Omega f(x)g(x) \, dx \right| \leq \int_\Omega |f(x)g(x)| \, dx \leq \mathcal{N}_p(f)\mathcal{N}_{p'}(g). \tag{3.1.6}$$

If there exists a positive constant c such that

$$|f(x)|^p = c|g(x)|^{p'} \tag{3.1.7}$$

for almost every $x \in \Omega$, then the equality

$$\int_\Omega |f(x)g(x)| \, dx = \mathcal{N}_p(f)\mathcal{N}_{p'}(g) \tag{3.1.8}$$

holds.

Proof. If either of the functions f or g is zero almost everywhere, then (3.1.6) is trivial. Otherwise, applying inequality (3.1.5) to

$$a = \frac{|f(x)|}{\mathcal{N}_p(f)}, \quad b = \frac{|g(x)|}{\mathcal{N}_{p'}(g)},$$

we have

$$\frac{|f(x)|}{\mathcal{N}_p(f)} \cdot \frac{|g(x)|}{\mathcal{N}_{p'}(g)} \leq \frac{1}{p} \cdot \frac{|f(x)|^p}{\mathcal{N}_p^p(f)} + \frac{1}{p'} \cdot \frac{|g(x)|^{p'}}{\mathcal{N}_{p'}^{p'}(g)} \tag{3.1.9}$$

for all $x \in \Omega$. Integrating (3.1.9) we obtain (3.1.6) since

$$\frac{1}{p} + \frac{1}{p'} = 1.$$

If (3.1.7) is satisfied for some $c \in (0, \infty)$ and almost every $x \in \Omega$, then (3.1.8) follows from Corollary 3.1.5. □

In Section 3.11, we shall need a version of the Hölder inequality for three functions. We shall omit the proof.

Theorem 3.1.7 (Hölder inequality for three functions). *Let $p, q, r \in (1, \infty)$ be such that*

$$\frac{1}{p} + \frac{1}{q} + \frac{1}{r} = 1.$$

Furthermore, let $f \in \mathcal{L}^p(\Omega)$, $g \in \mathcal{L}^q(\Omega)$ and $h \in \mathcal{L}^r(\Omega)$. Then $fgh \in \mathcal{L}^1(\Omega)$ and

$$\left| \int_\Omega f(x)g(x)h(x) \, dx \right| \leq \int_\Omega |f(x)g(x)h(x)| \, dx \leq \mathcal{N}_p(f)\mathcal{N}_q(g)\mathcal{N}_r(h). \tag{3.1.10}$$

Theorem 3.1.8 (Minkowski inequality). *Let $p \in [1, \infty)$ and let $f, g \in \mathcal{L}^p(\Omega)$. Then $f + g \in \mathcal{L}^p(\Omega)$ and*

$$\mathcal{N}_p(f + g) \leq \mathcal{N}_p(f) + \mathcal{N}_p(g). \tag{3.1.11}$$

Proof. For $p = 1$, the assertion follows from the triangle inequality. Suppose that $p \in (1, \infty)$. We first note that $f + g \in \mathcal{L}^p(\Omega)$. Indeed, that follows from integrating over Ω the estimate

$$|f(x) + g(x)|^p \leq (|f(x)| + |g(x)|)^p \leq [2 \max\{|f(x)|, |g(x)|\}]^p$$
$$\leq 2^p (|f(x)|^p + |g(x)|^p),$$

which holds for almost all $x \in \Omega$. In particular, $\mathcal{N}_p(f + g) < \infty$. We can also assume that $\mathcal{N}_p(f + g) > 0$ as otherwise there is nothing to prove. By the Hölder inequality (3.1.6), by (3.1.3) applied to $q = p'$ and by using the relation $p'(p-1) = p$, we get

$$\mathcal{N}_p^p(f + g) = \int_\Omega |f(x) + g(x)|^p \, dx = \int_\Omega |f(x) + g(x)||f(x) + g(x)|^{p-1} \, dx$$
$$\leq \int_\Omega |f(x)||f(x) + g(x)|^{p-1} \, dx + \int_\Omega |g(x)||f(x) + g(x)|^{p-1} \, dx$$
$$\leq \mathcal{N}_p(f) \mathcal{N}_{p'}(|f+g|^{p-1}) + \mathcal{N}_p(g) \mathcal{N}_{p'}(|f+g|^{p-1})$$
$$= (\mathcal{N}_p(f) + \mathcal{N}_p(g)) \mathcal{N}_{p'}(|f+g|^{p-1})$$
$$= (\mathcal{N}_p(f) + \mathcal{N}_p(g)) (\mathcal{N}_p(f+g))^{p-1},$$

and our claim follows on dividing this inequality by $(\mathcal{N}_p(f + g))^{p-1}$. $\square$

Corollary 3.1.9. *The set $\mathcal{L}_p(\Omega)$ is a vector space.*

The Minkowski inequality implies that for $p \in [1, \infty)$ the functional $f \mapsto \mathcal{N}_p(f)$ satisfies the triangle inequality. Furthermore, it obviously satisfies the homogeneity axiom (ii) of Definition 1.4.1.

Remark 3.1.10 (Clarkson inequalities). Let $f, g \in \mathcal{L}^p(\Omega)$. If $p \in [2, \infty)$, then

$$\mathcal{N}_p^p(f + g) + \mathcal{N}_p^p(f - g) \leq 2^{p-1} (\mathcal{N}_p^p(f) + \mathcal{N}_p^p(g)) \tag{3.1.12}$$

while if $p \in [1, 2)$, then

$$\mathcal{N}_p^p(f + g) + \mathcal{N}_p^p(f - g) \leq 2(\mathcal{N}_p^p(f) + \mathcal{N}_p^p(g)). \tag{3.1.13}$$

The proof can be obtained by considering the function

$$\varphi(t) = \frac{(1+t)^p + (1-t)^p}{1 + t^p}, \quad t \in [0, 1].$$

Remark 3.1.11. Let $p \in (0, 1)$ and let $f, g \in \mathcal{L}^p(\Omega)$, $f \geq 0$ and $g \geq 0$. Then, again, $f + g \in \mathcal{L}^p(\Omega)$, but instead of the Minkowski inequality we have the reverse inequality, namely

$$\mathcal{N}_p(f + g) \geq \mathcal{N}_p(f) + \mathcal{N}_p(g). \tag{3.1.14}$$

Remark 3.1.12 (generalized Hölder inequality)**.** Let $n \in \mathbb{N}$ and let $p_i \in (1, \infty)$, $i = 1, \ldots, n$, be such that

$$\sum_{j=1}^{n} \frac{1}{p_j} = 1.$$

Let $f_i \in \mathcal{L}^{p_i}(\Omega)$, $i = 1, \ldots, n$. Then

$$f_1 f_2 \cdots f_n \in \mathcal{L}^1(\Omega)$$

and

$$\int_\Omega |f_1(x) f_2(x) \cdots f_n(x)| \, dx \leq \mathcal{N}_{p_1}(f_1) \mathcal{N}_{p_2}(f_2) \cdots \mathcal{N}_{p_n}(f_n).$$

In particular, if $f_1, f_2, \ldots, f_n \in \mathcal{L}^n(\Omega)$, then

$$f_1 f_2 \cdots f_n \in \mathcal{L}^1(\Omega)$$

and

$$\int_\Omega |f_1(x) f_2(x) \cdots f_n(x)| \, dx \leq \mathcal{N}_n(f_1) \mathcal{N}_n(f_2) \cdots \mathcal{N}_n(f_n).$$

3.2 Lebesgue spaces

Convention 3.2.1. From now on, two measurable functions f, g on Ω will be considered to be equal if $f(x) = g(x)$ for almost all $x \in \Omega$.

Notation 3.2.2. Let $p \in (0, \infty)$. Then we denote by $L^p(\Omega)$ the set of (equivalence classes of) functions from $\mathcal{L}^p(\Omega)$. The elements of $L^p(\Omega)$ will be called "functions." In the case that Ω is an open interval (a, b) in $\mathbb{R}$, we shall simply write $L^p(a, b)$ instead of $L^p((a, b))$.

Lemma 3.2.3. *Let* $p \in [1, \infty)$ *and set*

$$\|f\|_p = \mathcal{N}_p(f). \tag{3.2.1}$$

Then $L^p(\Omega)$ endowed with $\|f\|_p$ is a normed linear space.

Proof. By Corollary 3.1.9, $L^p(\Omega)$ is a linear set. The homogeneity axiom (ii) of Definition 1.4.1 for the norm follows immediately from the properties of the Lebesgue integral. The fact that $\|f\|_p > 0$ if $f \neq 0$ is guaranteed by the equality relation on the set $L^p(\Omega)$. The triangle inequality is a consequence of the Minkowski inequality (3.1.11). □

Exercise 3.2.4. Let $\Omega \subset \mathbb{R}^N$ and let $1 \leq p_2 < p_1 < \infty$.

(i) If $\mu(\Omega) < \infty$ and $f \in L^{p_1}(\Omega)$, then also $f \in L^{p_2}(\Omega)$ and

$$\|f\|_{p_2} \leq (\mu(\Omega))^{\frac{p_1-p_2}{p_1 p_2}} \|f\|_{p_1}. \tag{3.2.2}$$

(Hint: The assertion follows immediately from the Hölder inequality 3.1.6 applied to $g(x) \equiv 1$.)

(ii) If $\mu(\Omega) = \infty$, then there is no inclusion between the sets $L^{p_1}(\Omega)$ and $L^{p_2}(\Omega)$.

3.3 Mean continuity

Convention 3.3.1. Here we will, when convenient, consider a function f defined almost everywhere on Ω to be extended outside of Ω by zero. We thus obtain a function F defined for almost all $x \in \mathbb{R}^N$ by

$$F(x) = \begin{cases} f(x) & \text{if } x \in \Omega, \\ 0 & \text{if } x \notin \Omega. \end{cases}$$

Instead of $F(x)$ we shall often simply write $f(x)$ also for $x \notin \Omega$.

Definition 3.3.2. Let $p \in (0, \infty)$ and $f \in L^p(\Omega)$. The function f is said to be *p-mean continuous* if for every $\varepsilon > 0$ there exists a $\delta = \delta(\varepsilon) > 0$ such that

$$\left(\int_\Omega |f(x+h) - f(x)|^p \, dx \right)^{\frac{1}{p}} < \varepsilon$$

provided $h \in \mathbb{R}^N$, $|h| < \delta$.

We shall now formulate the basic result of this section.

Theorem 3.3.3. *Let $p \in (0, \infty)$ and let Ω be a nonempty open subset of $\mathbb{R}^N$ having finite measure. Then any function $f \in L^p(\Omega)$ is p-mean continuous.*

Proof. Let $\varepsilon > 0$. According to the absolute continuity of the Lebesgue integral (Theorem 1.21.13), there exists an $\eta > 0$ such that for each $E \subset \Omega$ satisfying $\mu(E) < 4\eta$ we have

$$\left(\int_E |f(x)|^p \, dx \right)^{\frac{1}{p}} < \frac{\varepsilon}{4}. \tag{3.3.1}$$

For this η, there exists a $\varrho > 0$ such that $\mu(H_\varrho) < \eta$, where

$$H_\varrho = \{x \in \Omega; \, \text{dist}(x, \partial\Omega) \leq \varrho\}$$

and $\partial\Omega$ is the boundary of Ω defined by $\partial\Omega := \overline{\Omega} \cap \overline{\mathbb{R} \setminus \Omega}$. Set $\Omega_\varrho := \Omega \setminus H_\varrho$. Clearly, f is measurable on Ω_ϱ and thus the Luzin theorem (Theorem 1.21.12) implies the existence of a closed set $F_\eta^1 \subset \Omega_\varrho$ such that the restriction of the function f to F_η^1 is continuous, $\mu(\Omega_\varrho \setminus F_\eta^1) < \eta$ and thus

$$\mu(\Omega \setminus F_\eta^1) < 2\eta.$$

The function f is uniformly continuous on F_η^1 since F_η^1 is compact. Hence, there exists a $\delta \in (0, \varrho)$ such that

$$|f(x+h) - f(x)| < \frac{\varepsilon}{2(\mu(\Omega))^{\frac{1}{p}}} \tag{3.3.2}$$

for all $x, x+h \in F_\eta^1$ and $|h| < \delta$.

Let $|h| < \delta$ and set $F_\eta^2 := \{x \in \Omega; x+h \in F_\eta^1\}$ and $F_\eta := F_\eta^1 \cap F_\eta^2$. The set F_η is closed and $\mu(F_\eta^1) = \mu(F_\eta^2)$ since the Lebesgue measure is invariant with respect to translations. Furthermore, $\mu(\Omega \setminus F_\eta^2) < 2\eta$ and

$$\mu(\Omega \setminus F_\eta) = \mu((\Omega \setminus F_\eta^1) \cup (\Omega \setminus F_\eta^2)) \le \mu(\Omega \setminus F_\eta^1) + \mu(\Omega \setminus F_\eta^2) < 4\eta.$$

Using (3.3.1), we obtain

$$\left(\int_{\Omega \setminus F_\eta} |f(x+h)|^p \, dx\right)^{\frac{1}{p}} + \left(\int_{\Omega \setminus F_\eta} |f(x)|^p \, dx\right)^{\frac{1}{p}} < \frac{\varepsilon}{2},$$

and since (3.3.2) holds for arbitrary $x \in F_\eta$ we have

$$\left(\int_{F_\eta} |f(x+h) - f(x)|^p \, dx\right)^{\frac{1}{p}} < \frac{\varepsilon}{2}.$$

Summarizing, we arrive at

$$\left(\int_\Omega |f(x+h) - f(x)|^p \, dx\right)^{\frac{1}{p}}$$
$$\le \left(\int_{\Omega \setminus F_\eta} |f(x+h)|^p \, dx\right)^{\frac{1}{p}} + \left(\int_{\Omega \setminus F_\eta} |f(x)|^p \, dx\right)^{\frac{1}{p}}$$
$$+ \left(\int_{F_\eta} |f(x+h) - f(x)|^p \, dx\right)^{\frac{1}{p}} < \varepsilon,$$

which completes the proof. □

Exercise 3.3.4. Let Ω be an open subset of $\mathbb{R}^N$ having finite measure, let $p \in (0, \infty)$ and let $f \in L^p(\Omega)$. Then, for every $\varepsilon > 0$, there exists a $\delta > 0$ such that

$$\left(\int_\Omega |f((1+t)x) - f(x)|^p \, dx \right)^{\frac{1}{p}} < \varepsilon$$

provided $t \in \mathbb{R}$, $|t| < \delta$.

3.4 Mollifiers

Notation 3.4.1. We will denote by S the set of all functions φ_0 satisfying

(i) $\varphi_0 \in C_0^\infty(\mathbb{R}^N)$,

(ii) $\varphi_0(x) \geq 0$ for all $x \in \mathbb{R}^N$,

(iii) $\int_{\mathbb{R}^N} \varphi_0(x) \, dx = 1$,

(iv) $\operatorname{supp} \varphi_0 = \{x \in \mathbb{R}^N ; |x| \leq 1\}$.

We shall first note that S is a nonempty set. A classical example of a function belonging to $C_0^\infty(\mathbb{R}^N)$ is

$$\varphi(x) = f(|x|^2 - 1),$$

where

$$f(t) = \begin{cases} e^{\frac{1}{t}} & \text{if } t < 0 \\ 0 & \text{if } t \geq 0. \end{cases}$$

Obviously $\varphi \geq 0$, $\int_{\mathbb{R}^N} \varphi(x) \, dx > 0$ and φ vanishes for $|x| \geq 1$. Multiplying the function φ by the constant

$$\left(\int_{\mathbb{R}^N} \varphi(x) \, dx \right)^{-1},$$

we obtain a function belonging to S. Hence, S is not empty.

Definition 3.4.2. Let $\varepsilon > 0$ and let $\varphi_0 \in S$. For $u \in L^1(\Omega)$, set

$$(R_\varepsilon u)(x) := \varepsilon^{-N} \int_\Omega \varphi_0 \left(\frac{x-y}{\varepsilon} \right) u(y) \, dy, \tag{3.4.1}$$

i.e.

$$(R_\varepsilon u)(x) = \int_{B(0,1)} u(x - \varepsilon y) \varphi_0(y) \, dy, [1] \tag{3.4.2}$$

where $B(0, 1) = \{y \in \mathbb{R}^N ; |y| < 1\}$.

The mapping R_ε defined by (3.4.1) is called a *mollifier*.

---

[1] Note that the Convention 3.3.1 is used here.

Theorem 3.4.3. *Let $p \in [1, \infty)$ and let Ω be a nonempty bounded open subset of $\mathbb{R}^N$. Let $u \in L^p(\Omega)$. Then:*

(i) $R_\varepsilon u \in C^\infty(\mathbb{R}^N)$;

(ii) $\lim_{\varepsilon \to 0+} \|R_\varepsilon u - u\|_p = 0$.

Proof. The continuity of $R_\varepsilon u$ follows from that of φ_0 (see Theorem 1.21.9). Differentiation can be carried under the integral sign so that the differentiability properties of $R_\varepsilon u$ follow from those of φ_0 (see Theorem 1.21.10). Obviously,

$$|(R_\varepsilon u)(x) - u(x)| \leq \int_{B(0,1)} \varphi_0(y) |u(x - \varepsilon y) - u(x)| \, dy \qquad (3.4.3)$$

and

$$|(R_\varepsilon u)(x) - u(x)|^p \leq c \int_{B(0,1)} |u(x - \varepsilon y) - u(x)|^p \, dy, \qquad (3.4.4)$$

where

$$c = \begin{cases} \max_{y \in B(0,1)} \varphi_0(y) & \text{if } p = 1 \\ \left(\int_{B(0,1)} \varphi_0^{p'}(y) \, dy \right)^{\frac{p}{p'}} & \text{if } p > 1. \end{cases}$$

Integrating (3.4.4) and using the Fubini theorem (Theorem 1.21.11), we obtain

$$\int_\Omega |(R_\varepsilon u)(x) - u(x)|^p \, dx \leq c \int_{B(0,1)} \left(\int_\Omega |u(x - \varepsilon y) - u(x)|^p \, dx \right) dy, \qquad (3.4.5)$$

and the assertion (ii) follows from Theorem 3.3.3. $\square$

Theorem 3.4.4. (i) *If the support of $u \in L^1(\Omega)$ is contained in a compact subset K of Ω and $\varepsilon \in (0, \mathrm{dist}(K, \partial\Omega))$, then $R_\varepsilon u \in C_0^\infty(\Omega)$.*

(ii) *For $u \in C_0^0(\Omega)$, we have*

$$\lim_{\varepsilon \to 0+} \max_{x \in \Omega} |(R_\varepsilon u)(x) - u(x)| = 0. \qquad (3.4.6)$$

Proof. (i) Let $\varepsilon \in (0, \mathrm{dist}(K, \partial\Omega))$. Since, clearly, $R_\varepsilon u \in C^\infty(\Omega)$, we only need to show that $R_\varepsilon u$ has a compact support. Given $x \in \Omega$, assume that $(R_\varepsilon u)(x) \neq 0$. Then $x - \varepsilon y \in K$ for some $y \in B(0, 1)$. Thus,

$$\mathrm{dist}(K, \mathbb{R}^N \setminus \mathrm{supp}\, R_\varepsilon u) < \varepsilon.$$

Hence, $\mathrm{supp}\, R_\varepsilon u$ is bounded, and since it is obviously closed, it is a compact subset of Ω.

The assertion (ii) follows immediately from (3.4.2) and the fact that u is uniformly continuous. $\square$

3.5 Density of smooth functions

Theorem 3.5.1. *Let Ω be a nonempty bounded open subset of $\mathbb{R}^N$. Then the set $C_0^\infty(\Omega)$ is dense in $L^p(\Omega)$ for arbitrary $p \in [1, \infty)$.*

Proof. The assertion follows immediately from Theorem 3.4.3, Theorem 3.4.4 and Theorem 1.21.13. □

Exercises 3.5.2. (i) The set of all polynomials in $\mathbb{R}^N$ restricted to a nonempty bounded open set Ω is dense in $L^p(\Omega)$. (Hint: Use the Weierstrass approximation theorem and Theorem 3.5.1.)

(ii) Let $f \in C_0^0(\Omega)$. Then there exists a sequence $\{f_n\}_{n=1}^\infty$ of simple functions such that
$$|f_n(x)| \leq |f(x)| \tag{3.5.1}$$
for every $x \in \Omega$, $n \in \mathbb{N}$, and f is the uniform limit of $\{f_n\}_{n=1}^\infty$.

(iii) The set of all simple functions is dense in $L^p(\Omega)$. (Hint: Use Theorem 3.5.1 and (ii) above.)

(iv) Let $f \in L^p(\Omega)$. Then there exists a sequence $\{f_n\}_{n=1}^\infty$ of simple functions converging to f in $L^p(\Omega)$ and satisfying (3.5.1) for $n \in \mathbb{N}$ and almost all $x \in \Omega$.

3.6 Separability

Theorem 3.6.1. *Let Ω be a nonempty bounded open subset of $\mathbb{R}^N$. Assume that $p \in [1, \infty)$. Then $L^p(\Omega)$ is separable.*

Proof. Every polynomial can be uniformly approximated on $\overline{\Omega}$ by a sequence of polynomials with rational coefficients. It thus suffices to use Exercise 3.5.2 (i). □

Exercise 3.6.2. The previous assertion is also true in case $\Omega = \mathbb{R}^N$. (Hint: For $f \in L^p(\mathbb{R}^N)$, define $\overline{f}$ by
$$\overline{f}(x) = \begin{cases} f(x) & \text{if } |x| < r \\ 0 & \text{if } |x| \geq r, \end{cases} \tag{3.6.1}$$
where r is a sufficiently large rational number. Then use Theorems 1.21.13 and 3.6.1.)

3.7 Completeness

Theorem 3.7.1. *Let Ω be a measurable subset of $\mathbb{R}^N$. Let $p \in [1, \infty)$. Then $L^p(\Omega)$ is a Banach space.*

Proof. Assume that $\{f_n\}_{n=1}^\infty$ is a Cauchy sequence in $L^p(\Omega)$. We want to prove that it is convergent in $L^p(\Omega)$. By Lemma 1.7.5, it is enough to find a subsequence $\{f_{n_k}\}_{k=1}^\infty$ convergent in $L^p(\Omega)$. Let $\{n_k\}_{k=1}^\infty$ be a strictly increasing sequence of positive integers such that

$$\|f_{n_{k+1}} - f_{n_k}\|_p < 2^{-k} \quad \text{for every } k \in \mathbb{N}.$$

For $k \in \mathbb{N}$ and $x \in \Omega$, set

$$g_k(x) := |f_{n_1}(x)| + \sum_{i=1}^{k} |f_{n_{i+1}}(x) - f_{n_i}(x)|.$$

Then $\{g_k(x)\}_{k=1}^\infty$ is a nondecreasing sequence for every $x \in \Omega$. Furthermore, the Minkowski inequality (Theorem 3.1.8) guarantees that, for every $k \in \mathbb{N}$, one has

$$\left(\int_\Omega |g_k(x)|^p \, dx\right)^{\frac{1}{p}} \leq \|f_{n_1}\|_p + \sum_{i=1}^{\infty} \|f_{n_{i+1}} - f_{n_i}\|_p \leq \|f_{n_1}\|_p + 1 < \infty. \quad (3.7.1)$$

Hence, $g_k \in L^p(\Omega)$ for every $k \in \mathbb{N}$ and, moreover, the set $\{g_k\}_{k=1}^\infty$ is uniformly bounded in $L^p(\Omega)$.

Now, define the function

$$g(x) := \lim_{k \to \infty} g_k(x), \quad x \in \Omega.$$

The pointwise limit exists thanks to the monotonicity of the sequence $\{g_k(x)\}_{k=1}^\infty$, so the function g is defined. By (3.7.1) and the Fatou Lemma (Theorem 1.21.6), we have

$$\int_\Omega |g(x)|^p \, dx = \int_\Omega \lim_{k \to \infty} |g_k(x)|^p \, dx \leq \liminf_{k \to \infty} \int_\Omega |g_k(x)|^p \, dx$$

$$\leq \sup_{k \in \mathbb{N}} \int_\Omega |g_k(x)|^p \, dx \leq (\|f_{n_1}\|_p + 1)^p < \infty,$$

hence also $g \in L^p(\Omega)$. Also in particular, the sum

$$\sum_{i=1}^{\infty} \left(f_{n_{i+1}}(x) - f_{n_i}(x)\right)$$

Section 3.7 Completeness

exists for almost every $x \in \Omega$. Consequently, we may define the function

$$f(x) := f_{n_1}(x) - \sum_{i=1}^{\infty} \left(f_{n_{i+1}}(x) - f_{n_i}(x) \right), \quad x \in \Omega.$$

It follows that
$$\lim_{k \to \infty} f_{n_k}(x) = f(x) \quad \text{for a.e. } x \in \Omega. \tag{3.7.2}$$

Next, for every $k \in \mathbb{N}$ and $x \in \Omega$, we have

$$f_{n_{k+1}}(x) = f_{n_1}(x) + \sum_{i=1}^{k} \left(f_{n_{i+1}}(x) - f_{n_i}(x) \right),$$

hence,
$$|f_{n_{k+1}}(x)| \leq g_k(x) \leq g(x),$$

and, in turn,
$$|f_{n_{k+1}}(x)|^p \leq (g(x))^p.$$

Therefore, combining this with (3.7.2) and the Lebesgue dominated convergence theorem (Theorem 1.21.5), we get $f \in L^p(\Omega)$. Finally, we have for every $k \in \mathbb{N}$ and $x \in \Omega$

$$f_{n_{k+1}}(x) - f(x) = \sum_{i=k+1}^{\infty} \left(f_{n_{i+1}}(x) - f_{n_i}(x) \right),$$

hence,
$$|f_{n_{k+1}}(x) - f(x)| \leq \sum_{i=k+1}^{\infty} \left| f_{n_{i+1}}(x) - f_{n_i}(x) \right| \leq g(x),$$

and, in turn,
$$|f_{n_{k+1}}(x) - f(x)|^p \leq (g(x))^p.$$

Thus, using the Lebesgue dominated convergence theorem once again, we end up with
$$\|f_{n_k} - f\|_p \to 0 \quad \text{as } k \to \infty,$$

as desired. The proof is complete. □

Alternative proofs of completeness of the Lebesgue space can be carried out, see, e.g., [185, Exercise 8.4.15, p. 270].

3.8 The dual space of $L^p(\Omega)$ $(1 < p < \infty)$

In this section we shall construct the dual space $[L^p(\Omega)]^*$ of $L^p(\Omega)$ if $1 < p < \infty$. For the case $p = 1$ see Theorem 3.10.11; for the case $p = \infty$ see Theorem 3.10.10.

Theorem 3.8.1. Let $g \in L^{p'}(\Omega)$ and denote

$$\Phi_g(f) := \int_\Omega f(x) g(x)\, dx$$

for $f \in L^p(\Omega)$. Then $\Phi_g \in [L^p(\Omega)]^*$ and

$$\|\Phi_g\| = \|g\|_{p'}.$$

Proof. By the Hölder inequality (Theorem 3.1.6), we have

$$|\Phi_g(f)| \le \|f\|_p \|g\|_{p'}.$$

Thus, Φ_g is a bounded linear functional on $L^p(\Omega)$ (the linearity of Φ_g is obvious), and

$$\|\Phi_g\| \le \|g\|_{p'}.$$

To prove the converse inequality, let $f = |g|^{p'-1} \operatorname{sign} g$; then we have $|f|^p = |g|^{p'}$. Consequently, $f \in L^p(\Omega)$ and $\|f\|_p = \|g\|_{p'}^{p'-1}$. Moreover,

$$\Phi_g(f) = \int_\Omega |g(x)|^{p'}\, dx = \|g\|_{p'}^{p'} = \|f\|_p \cdot \|g\|_{p'},$$

whence $\|\Phi_g\| \ge \|g\|_{p'}$. Altogether, we obtain $\|\Phi_g\| = \|g\|_{p'}$. □

Our aim now will be to prove an assertion in some sense converse to Theorem 3.8.1, namely that an arbitrary bounded linear functional Φ on $L^p(\Omega)$ can be uniquely represented in the form

$$\Phi_g(f) = \int_\Omega f(x) g(x)\, dx, \quad f \in L^p(\Omega),$$

for some $g \in L^{p'}(\Omega)$.

Lemma 3.8.2. Let Ω be a nonempty bounded open subset of $\mathbb{R}^N$. Let g be a measurable function on Ω. Suppose that there exist $p > 1$ and $M > 0$ such that, for arbitrary $f \in L^p(\Omega)$, we have

$$f \cdot g \in L^1(\Omega) \tag{3.8.1}$$

and

$$\left| \int_\Omega f(x) g(x)\, dx \right| \le M \|f\|_p. \tag{3.8.2}$$

Then $g \in L^{p'}(\Omega)$ and $\|g\|_{p'} \le M$.

Section 3.8 The dual space

Proof. The function $f \equiv 1$ is clearly in $L^p(\Omega)$. Therefore, according to (3.8.1), $g \in L^1(\Omega)$. Hence, $|g|^{\frac{1}{p}} \operatorname{sign} g \in L^p(\Omega)$. Using (3.8.1) again, we obtain $g \cdot |g|^{\frac{1}{p}} \operatorname{sign} g \in L^1(\Omega)$, and thus

$$|g|^{\frac{1}{p}+\frac{1}{p^2}} \operatorname{sign} g \in L^p(\Omega).$$

Set

$$f_n = |g|^{\frac{1}{p}+\cdots+\frac{1}{p^n}}, \quad n \in \mathbb{N}.$$

By induction, one can prove $f_n \in L^p(\Omega)$. We have already proved this in the cases when $n=1$ and $n=2$. According to assumption (3.8.2), we have

$$\|g\|_1 \le M \, \|\chi_\Omega\|_p \le M \, (\mu(\Omega))^{\frac{1}{p}}.$$

Further,

$$\int_\Omega |g(x)|^{1+\frac{1}{p}} \, dx \le M^{1+\frac{1}{p}} \, (\mu(\Omega))^{\frac{1}{p^2}}.$$

Set

$$g_n = |g|^{1+\frac{1}{p}+\cdots+\frac{1}{p^n}}, \quad n \in \mathbb{N}.$$

In virtue of $p' = \sum_{n=0}^{\infty} \frac{1}{p^n}$, $g_n \searrow |g|^{p'}$ on the set

$$\Omega_1 = \{x \in \Omega; |g(x)| \le 1\}$$

and $g_n \nearrow |g|^{p'}$ on $\Omega \setminus \Omega_1$. Using induction, we obtain

$$\left| \int_\Omega g_n(x) \, dx \right| \le M^{1+\frac{1}{p}+\cdots+\frac{1}{p^n}} \, (\mu(\Omega))^{\frac{1}{p^{n+1}}}$$

and applying the Levi theorem (Theorem 1.21.4) (first on the set Ω_1, then on $\Omega \setminus \Omega_1$), we obtain

$$\lim_{n \to \infty} \int_\Omega g_n(x) \, dx = \int_\Omega |g(x)|^{p'} \, dx.$$

Thus,

$$\int_\Omega |g(x)|^{p'} \, dx \le M^{p'},$$

and our assertion follows immediately. □

Theorem 3.8.3 (Riesz representation theorem). *Let Ω be a nonempty bounded open subset of $\mathbb{R}^N$. Let $p \in (1, \infty)$ and let Φ be a bounded linear functional on $L^p(\Omega)$. Then there exists exactly one $g \in L^{p'}(\Omega)$ such that*

$$\Phi(f) = \int_\Omega g(x) f(x) \, dx$$

for every $f \in L^p(\Omega)$. Moreover,

$$\|\Phi\| = \|g\|_{p'}.$$

Proof. *Uniqueness.* Let $\Phi \in [L^p(\Omega)]^*$ and let $g_1, g_2 \in L^{p'}(\Omega)$ be such that

$$\Phi(f) = \int_\Omega f(x)g_1(x)\,dx = \int_\Omega f(x)g_2(x)\,dx$$

for arbitrary $f \in L^p(\Omega)$. Set $g = g_1 - g_2$. Then we have

$$\int_\Omega g(x)f(x)\,dx = 0$$

for every $f \in L^p(\Omega)$. Hence,

$$\int_E g(x)\,dx = 0$$

provided $E \subset \Omega$ is measurable, since the characteristic function of a measurable set $E \subset \Omega$ belongs to $L^p(\Omega)$. Put

$$E = \{x \in \Omega; \; g(x) > 0\}.$$

The set E is measurable and it follows from the above reasoning that $\mu(E) = 0$. Thus $g(x) \leq 0$ almost everywhere on Ω. Analogously, we obtain $g(x) \geq 0$ almost everywhere. So g is the zero element in $L^p(\Omega)$.

Existence. Let $\Phi \in [L^p(\Omega)]^*$. For a measurable subset E of Ω, set

$$\nu(E) = \Phi(\chi_E).$$

Then, clearly, $\chi_E \in L^p(\Omega)$ for every measurable subset E of Ω, and

$$|\nu(E)| \leq \|\Phi\| \, (\mu(E))^{\frac{1}{p}}. \tag{3.8.3}$$

Obviously, ν is finite and additive on $\mathfrak{M}$. Let $\{E_n\}_{n=1}^\infty$ be a sequence of measurable subsets of Ω such that $E_i \cap E_j = \emptyset$ if $i \neq j$. Denote

$$E = \bigcup_{n=1}^\infty E_n, \quad F_n = \bigcup_{i=1}^n E_i, \quad G_n = E \setminus F_n.$$

It is easy to see that

$$\nu(E) = \nu(F_n) + \nu(G_n) = \sum_{i=1}^n \nu(E_i) + \nu(G_n). \tag{3.8.4}$$

Moreover,

$$G_1 \supset G_2 \supset \ldots, \quad \mu(G_1) \leq \mu(E) < \infty$$

and
$$\bigcap_{n=1}^{\infty} G_n = \emptyset.$$
Thus,
$$\lim_{n\to\infty} \mu(G_n) = 0$$
and, according to (3.8.3), we also have
$$\lim_{n\to\infty} \nu(G_n) = 0.$$
This together with (3.8.4) implies that ν is σ-additive.

According to the Radon–Nikodým representation theorem (Theorem 1.21.15), there exists a $g \in L^1(\Omega)$ such that
$$\nu(E) = \int_E g(x)\,dx$$
for every $E \in \mathfrak{M}(\Omega)$.

Let $E_1, \ldots, E_k$ be measurable subsets of Ω, $E_i \cap E_j = \emptyset$ for $i \neq j$, let $c_1, \ldots, c_k$ be real numbers. For
$$f = \sum_{i=1}^{k} c_i \chi_{E_i}$$
we have the desired representation of $\Phi(f)$:
$$\Phi(f) = \sum_{i=1}^{k} c_i \nu(E_i) = \int_\Omega g(x) f(x)\,dx.$$

Now, let $f \in L^p(\Omega)$. Then (see Exercise 3.5.2 (iv)) there exists a sequence $\{f_n\}_{n=1}^\infty$ of simple functions such that
$$|f_n(x)| \leq |f(x)|$$
for $n \in \mathbb{N}$ and almost all $x \in \Omega$, and
$$\lim_{n\to\infty} \|f_n - f\|_p = 0.$$
We can suppose that $f_n(x) \to f(x)$ almost everywhere on Ω (if this is not the case, we can consider a suitable subsequence of $\{f_n\}_{n=1}^\infty$ with such properties, the existence of which follows from the proof of Theorem 3.7.1). Thus,
$$f_n(x)g(x) \to f(x)g(x)$$
almost everywhere on Ω and since $\Phi(f_n \chi_E) \to \Phi(f \chi_E)$, it follows from the Vitali–Hahn–Saks theorem (Theorem 1.21.8) that the condition (P) from the Vitali theorem (Theorem 1.21.7) is satisfied. Thus, the Vitali theorem implies $f \cdot g \in L^1(\Omega)$ and
$$\Phi(f) = \lim_{n\to\infty} \Phi(f_n) = \lim_{n\to\infty} \int_\Omega f_n(x)g(x)\,dx = \int_\Omega f(x)g(x)\,dx.$$

Now it suffices to prove $g \in L^{p'}(\Omega)$. This follows from

$$\left| \int_\Omega f(x)g(x)\,dx \right| = |\Phi(f)| \le \|\Phi\| \cdot \|f\|_p,$$

and from Lemma 3.8.2.

Theorem 3.8.1 now implies that $\|\Phi\| = \|g\|_{p'}$. □

3.9 Reflexivity

Theorem 3.9.1. *Let $p \in (1, \infty)$. Then the space $L^p(\Omega)$ is reflexive.*

Proof. Let $x^{**} \in [L^p(\Omega)]^{**}$. By Theorem 3.8.3, there exists a $y^* \in [L^p(\Omega)]^*$ such that, for arbitrary $g \in L^{p'}(\Omega)$, $x^{**}(x^*) = y^*(g)$, where g and x^* satisfy

$$x^*(f) = \int_\Omega f(t)g(t)\,dt$$

for every $f \in L^p(\Omega)$. Applying Theorem 3.8.3 once more we obtain $h \in L^p(\Omega)$ such that

$$y^*(\varphi) = \int_\Omega h(t)\varphi(t)\,dt$$

for every $\varphi \in L^{p'}(\Omega)$. Hence,

$$x^{**}(x^*) = y^*(g) = \int_\Omega g(t)h(t)\,dt = x^*(h).$$

We have thus just proved that for arbitrary $x^{**} \in [L^p(\Omega)]^{**}$ there exists a unique $h \in L^p(\Omega)$ such that for every $x^* \in [L^p(\Omega)]^*$

$$x^{**}(x^*) = x^*(h), \quad \text{i.e.} \quad J(h) = x^{**},$$

where J is the canonical mapping. Hence, $J(L^p(\Omega)) = [L^p(\Omega)]^{**}$. □

Exercise 3.9.2. Show that, for $p \in (1, \infty)$, one has

$$\|u\|_p = \sup_{\|v\|_{p'}=1} \left| \int_\Omega u(x)v(x)\,dx \right| = \sup_{\|v\|_{p'}=1} \int_\Omega |u(x)v(x)|\,dx.$$

3.10 The space L^∞

The space $L^p(\Omega)$ with $p = \infty$ has been so far singled out of the L^p-scale because of its intrinsic properties that are in many ways different from those of the spaces $L^p(\Omega)$ with $p < \infty$. We shall deal with this space in this section.

Section 3.10 The space L^∞

Definition 3.10.1. By $L^\infty(\Omega)$ we denote the set of all measurable functions f defined almost everywhere on Ω such that there exists a constant $K > 0$ and a set $E \subset \Omega$ (both depending on f), $\mu(E) = 0$, with the property

$$|f(x)| \leq K$$

for all $x \in \Omega \setminus E$.

The operations of a sum $f + g$ and of a multiplication by a scalar λf (where $f, g \in L^\infty(\Omega)$ and λ is a scalar) are defined on $L^\infty(\Omega)$ in the same way as in Section 3.2. We obtain immediately that $L^\infty(\Omega)$ is a vector space.

Definition 3.10.2. For $f \in L^\infty(\Omega)$ set

$$\mathcal{N}_\infty(f) := \inf_{\substack{E \in \mathfrak{M}(\Omega) \\ \mu(E)=0}} \{ \sup_{x \in \Omega \setminus E} |f(x)| \}.$$

The number $\mathcal{N}_\infty(f)$ is called the *essential supremum* of f over Ω and will be denoted by

$$\operatorname*{ess\,sup}_{x \in \Omega} |f(x)|.$$

If no confusion can arise, we shall just write

$$\operatorname{ess\,sup} |f(x)|.$$

Remarks 3.10.3. (i) Using the triangle inequality for real numbers and properties of supremum and infimum we obtain immediately the following results for every $f, g \in L^\infty(\Omega)$ and every scalar λ:

(a) $\mathcal{N}_\infty(f + g) \leq \mathcal{N}_\infty(f) + \mathcal{N}_\infty(g)$;

(b) $\mathcal{N}_\infty(\lambda f) = |\lambda| \mathcal{N}_\infty(f)$;

(c) $\mathcal{N}_\infty(f) = 0$ if and only if $f = 0$ almost everywhere on Ω.

Thus, the function $\mathcal{N}_\infty(f)$ is a seminorm on the space $L^\infty(\Omega)$.

(ii) We introduce on the set $L^\infty(\Omega)$ the same equality relation as in Section 3.2, i.e. we shall say that $f = g$ if and only if $f(x) = g(x)$ for almost all $x \in \Omega$. The elements (equivalence classes) of $L^\infty(\Omega)$ will be called functions. For $f \in L^\infty(\Omega)$, set

$$\|f\|_\infty = \mathcal{N}_\infty(f).$$

Then $\|\cdot\|_\infty$ is a norm on the space $L^\infty(\Omega)$.

(iii) It is easy to see that if Ω is a nonempty open bounded subset of $\mathbb{R}^N$ then the set of all continuous functions on $\overline{\Omega}$ is a closed subspace of $L^\infty(\Omega)$ and

$$\|f\|_\infty = \max_{x \in \overline{\Omega}} |f(x)|$$

for all $f \in C(\overline{\Omega})$.

Theorem 3.10.4. *Let Ω be a nonempty open subset of $\mathbb{R}^N$. Then $L^\infty(\Omega)$ is not a separable space.*

Proof. For $r \in (0, \infty)$ and $n \in \mathbb{N}$, set

$$\Omega_{rn} = \{x \in \Omega; \ \mathrm{dist}(x, \partial\Omega) \geq r\} \cap \{x \in \mathbb{R}^N; \ |x| < n\}.$$

The collection $\{\Omega_{rn}\}$ is uncountable. Let χ_{rn} be the characteristic function of the set Ω_{rn}. Clearly, $\chi_{rn} \in L^\infty(\Omega)$ and, for $s \in (0, \infty)$ with $s \neq r$, we have, for some $n \in \mathbb{N}$,

$$\|\chi_{rn} - \chi_{sn}\|_\infty = 1.$$

Thus the space $L^\infty(\Omega)$ is not separable. $\square$

Remark 3.10.5. We note that the Hölder inequality, formulated in Theorem 3.1.6 for $p \in (1, \infty)$, can be very easily extended to the case when $p = 1$ (or, which is the same, $p = \infty$). Indeed, for $f \in \mathcal{L}^1(\Omega)$ and $g \in L^\infty(\Omega)$, one trivially has $fg \in \mathcal{L}^1(\Omega)$ and

$$\left| \int_\Omega f(x)g(x)\, dx \right| \leq \int_\Omega |f(x)g(x)|\, dx \leq \|f\|_1 \|g\|_\infty. \tag{3.10.1}$$

Remark 3.10.6. Let $\Omega \subset \mathbb{R}^N$ have finite measure and let $1 < p < q < \infty$. Then

$$L^\infty(\Omega) \subset L^q(\Omega) \subset L^p(\Omega) \subset L^1(\Omega). \tag{3.10.2}$$

This follows immediately from the basic properties of the Lebesgue integral and from the Hölder inequality (Theorem 3.1.6).

The following two theorems give a characterization of functions in the space $L^\infty(\Omega)$ by means of their norms in the spaces $L^p(\Omega)$ where p is finite.

Theorem 3.10.7. *Let $\mu(\Omega) < \infty$. Let $f \in L^\infty(\Omega)$. Then*

$$\|f\|_\infty = \lim_{p \to \infty} \|f\|_p.$$

Proof. The assertion of the theorem is trivial in the case of $\|f\|_\infty = 0$.

Assume, thus, that $\|f\|_\infty > 0$ and let ε be an arbitrary number from $(0, \frac{1}{2}\|f\|_\infty)$. The obvious inequality
$$\|f\|_p \leq \|f\|_\infty (\mu(\Omega))^{\frac{1}{p}}$$
implies
$$\limsup_{p\to\infty} \|f\|_p \leq \|f\|_\infty.$$
Then there exists a set $B \subset \Omega$ of positive measure such that for every $x \in B$, $|f(x)| \geq \|f\|_\infty - \varepsilon$. Moreover,
$$\|f\|_p \geq \left(\int_B |f(x)|^p \, dx\right)^{\frac{1}{p}} \geq (\mu(B))^{\frac{1}{p}} (\|f\|_\infty - \varepsilon).$$
Hence,
$$\liminf_{p\to\infty} \|f\|_p \geq \|f\|_\infty - \varepsilon,$$
and thus
$$\liminf_{p\to\infty} \|f\|_p \geq \|f\|_\infty \geq \limsup_{p\to\infty} \|f\|_p. \quad \square$$

Theorem 3.10.8. *Let $\mu(\Omega) < \infty$. Let $1 \leq p_1 \leq p_2 \leq \ldots$ and suppose*
$$\lim_{k\to\infty} p_k = \infty.$$
Let
$$f \in \bigcap_{k=1}^\infty L^{p_k}(\Omega)$$
and
$$a = \sup_{k\in\mathbb{N}} \|f\|_{p_k} < \infty. \tag{3.10.3}$$
Then $f \in L^\infty(\Omega)$.

Proof. Set
$$M = \{x \in \Omega; |f(x)| \geq a + 1\}.$$
If $\mu(M) = 0$, then the assertion is obvious. Suppose $\mu(M) > 0$. Then
$$\|f\|_{p_k} \geq \left(\int_M |f(x)|^{p_k} \, dx\right)^{\frac{1}{p_k}} \geq (a+1)(\mu(M))^{\frac{1}{p_k}}.$$
Hence, we obtain
$$a = \sup_{k\in\mathbb{N}} \|f\|_{p_k} \geq \lim_{k\to\infty} (a+1)(\mu(M))^{\frac{1}{p_k}} = a+1,$$
which is a contradiction. $\square$

Example 3.10.9. Obviously, $\log x \in L^p(0,1)$ for every $p \in [1,\infty)$ but

$$\log x \notin L^\infty(0,1).$$

This example shows that assumption (3.10.3) is essential for the assertion of Theorem 3.10.8.

The following result follows immediately from the properties of the uniform convergence of functions.

Theorem 3.10.10. $L^\infty(\Omega)$ *is a Banach space.*

The most important feature of the space $L^\infty(\Omega)$ consists in the fact that its elements represent all bounded linear functionals on $L^1(\Omega)$. We shall now study this topic in detail.

Theorem 3.10.11. *Let Ω be a nonempty bounded open subset of $\mathbb{R}^N$.*

(i) *Let $g \in L^\infty(\Omega)$ and define the functional Φ_g on $L^1(\Omega)$ by*

$$\Phi_g(f) = \int_\Omega f(x) g(x) \, dx$$

for $f \in L^1(\Omega)$.

Then,

$$\Phi_g \in [L^1(\Omega)]^* \quad \text{and} \quad \|\Phi_g\| = \|g\|_\infty.$$

(ii) *Let Φ be a bounded linear functional on $L^1(\Omega)$. Then there exists one and only one $g \in L^\infty(\Omega)$ such that*

$$\Phi(f) = \int_\Omega g(x) f(x) \, dx$$

for $f \in L^1(\Omega)$.

Proof. The assertion can be verified in an analogous way as those of Theorems 3.8.1 and 3.8.3. The proof of (ii) uses the following result, analogous to Lemma 3.8.2. □

Lemma 3.10.12. *Let g be a measurable function on Ω. Suppose that there exists an $M > 0$ such that, for arbitrary $f \in L^1(\Omega)$,*

$$fg \in L^1(\Omega); \tag{3.10.4}$$

$$\left| \int_\Omega f(x) g(x) \, dx \right| \leq M \|f\|_1. \tag{3.10.5}$$

Then $g \in L^\infty(\Omega)$ and $\|g\|_\infty \leq M$.

Proof. In the same way as in Lemma 3.8.2 we verify that

$$|g|^n \operatorname{sign} g \in L^1(\Omega)$$

and

$$\left(\int_\Omega |g(x)|^n\right)^{\frac{1}{n}} \leq M(\mu(\Omega))^{\frac{1}{n}}$$

for arbitrary $n \in \mathbb{N}$. Thus,

$$g \in \bigcap_{n=1}^\infty L^n(\Omega), \qquad \sup_{n \in \mathbb{N}} \|g\|_n < \infty$$

and Theorems 3.10.7 and 3.10.8 yield the assertion. □

Theorem 3.10.13. *Let Ω be a nonempty bounded open subset of $\mathbb{R}^N$. Then $L^1(\Omega)$ is not reflexive.*

Proof. Let us suppose that $L^1(\Omega)$ is reflexive. Since it is separable (see Theorem 3.6.1), $[L^1(\Omega)]^*$ is also separable. On the other hand, $[L^1(\Omega)]^*$ is isomorphic with the nonseparable Banach space $L^\infty(\Omega)$, which is a contradiction. □

Theorem 3.10.14. *Let Ω be a nonempty bounded open subset of $\mathbb{R}^N$. Then $L^\infty(\Omega)$ is not reflexive.*

Proof. The assertion follows from the fact that $L^\infty(\Omega)$ contains the nonreflexive Banach space $C(\overline{\Omega})$ as a subspace (see Theorem 2.6.3). □

3.11 Hardy inequalities

In this section we shall collect some very important inequalities involving integrals and suprema under a common title of *Hardy inequalities*. Such inequalities are indispensable for many applications. There is a vast amount of literature available for this topic, thus let us name just a few, for instance [128, 127, 148, 149, 171]. Here we follow the explanation of [149, Section 1.3].

We start with recalling the Minkowski integral inequality.

Theorem 3.11.1 (Minkowski integral inequality). *Let $p \in [1, \infty]$ and let $F(x, y)$ be a measurable function of two variables defined on $(0, \infty) \times (0, \infty)$. Assume that $F(\cdot, y) \in L^p(0, \infty)$ for a.e. fixed $y \in (0, \infty)$ and that*

$$\int_0^\infty \|F(\cdot, y)\|_p \, dy < \infty.$$

Then $\int_0^\infty F(x, y)\,dy$ converges for a.e. $x \in (0, \infty)$ and

$$\left\|\int_0^\infty F(x, y)\,dy\right\|_p \le \int_0^\infty \|F(\cdot, y)\|_p\,dy. \tag{3.11.1}$$

Proof. When $p = \infty$, the assertion is obvious. Let $p < \infty$ and set

$$f(x) := \int_0^\infty F(x, y)\,dy, \quad x \in (0, \infty).$$

Let $g \in L^{p'}(0, \infty)$ be such that $\|g\|_{p'} \le 1$. Then, by the Fubini theorem (1.21.11) and the Hölder inequality (Theorem 3.1.6), we get

$$\left|\int_0^\infty f(x)g(x)\,dx\right| \le \int_0^\infty \int_0^\infty |F(x, y)||g(y)|\,dx\,dy$$

$$\le \int_0^\infty \|F(\cdot, y)\|_p \|g\|_{p'}\,dy$$

$$\le \int_0^\infty \|F(\cdot, y)\|_p\,dy.$$

Thus, by Lemma 3.8.2, we get $f \in L^p(0, \infty)$ and the estimate (3.11.1). The proof is complete. □

Lemma 3.11.2. *Let f be a positive measurable function on $(0, \infty)$ and $p \in (1, \infty)$. Then*

$$\left(\int_0^\infty \left(\int_0^t f(s)\,ds\right)^p \frac{dt}{t^p}\right)^{\frac{1}{p}} \le p'\left(\int_0^\infty f(x)^p\,dx\right)^{\frac{1}{p}}. \tag{3.11.2}$$

Proof. Using the Hölder inequality, we get

$$\int_0^\infty \left(\int_0^t f(s)\,ds\right)^p \frac{dt}{t^p} = \int_0^\infty \left(\int_0^t f(s)s^{\frac{1}{pp'}}s^{-\frac{1}{pp'}}\,ds\right)^p \frac{dt}{t^p}$$

$$\le \int_0^\infty \int_0^t f(s)^p s^{\frac{1}{p'}}\,ds \left(\int_0^t s^{-\frac{1}{p}}\,ds\right)^{p-1}\,dt$$

$$= (p')^{p-1}\int_0^\infty f(s)^p s^{\frac{1}{p'}} \int_s^\infty t^{p(\frac{1}{(p')^2}-1)}\,dt\,ds$$

$$= (p')^p \int_0^\infty f(s)^p\,ds,$$

and the assertion follows on taking pth roots. □

Section 3.11 Hardy inequalities

Remark 3.11.3. The preceding lemma can be trivially extended to the case when $p = \infty$. More precisely, for a positive measurable function f on $(0, \infty)$, one has

$$\operatorname*{ess\,sup}_{t \in (0,\infty)} \frac{1}{t} \int_0^t f(s)\,ds \leq \operatorname*{ess\,sup}_{t \in (0,\infty)} f(t). \tag{3.11.3}$$

Indeed, this follows immediately from setting $K := \operatorname*{ess\,sup}_{t \in (0,\infty)} f(t)$ as then

$$\operatorname*{ess\,sup}_{t \in (0,\infty)} \frac{1}{t} \int_0^t f(s)\,ds \leq \operatorname*{ess\,sup}_{t \in (0,\infty)} \frac{1}{t} \int_0^t K\,ds = K.$$

On the other hand, the assertion of Lemma 3.11.2 does not hold for $p = 1$. This can be observed for example by taking $f = \chi_{(0,1)}$, since then

$$\int_0^\infty f(x)\,dx = 1$$

but

$$\int_0^\infty \frac{1}{t} \int_0^t f(s)\,ds\,dt \geq \int_1^\infty \frac{1}{t} \int_0^1 f(s)\,ds\,dt = \int_1^\infty \frac{dt}{t} = \infty.$$

For $p = 1$, (3.11.2) is not true even when $(0, \infty)$ is replaced with a finite interval. For instance, there is no positive constant C that would render the inequality

$$\int_0^1 \frac{1}{t} \int_0^t f(s)\,ds\,dt \leq C \int_0^1 f(x)\,dx$$

true for all positive functions on $(0, 1)$. To see this, take for example

$$f(t) := \frac{1}{t\left(\log \frac{2}{t}\right)^2}, \quad t \in (0, 1).$$

Then, again, $\int_0^1 f(x)\,dx < \infty$ but, with appropriate c,

$$\int_0^1 \frac{1}{t} \int_0^t f(s)\,ds\,dt = c \int_0^1 \frac{1}{t\left(\log \frac{2}{t}\right)}\,dt = \infty.$$

Definition 3.11.4. Let w be a nonnegative Lebesgue measurable function defined on the interval $(0, \infty)$. We then say that w is a *weight* on $(0, \infty)$.

Lemma 3.11.5. *Assume that f and g are nonnegative functions on $[0, \infty)$ and let $p \in [1, \infty)$. Then*

$$\left(\int_0^\infty g(x) \left(\int_0^x f(t)\,dt\right)^p dx\right)^{\frac{1}{p}} \leq \int_0^\infty f(t) \left(\int_t^\infty g(x)\,dx\right)^{\frac{1}{p}} dt. \tag{3.11.4}$$

Proof. We have

$$\left(\int_0^\infty g(x)\left(\int_0^x f(t)\,dt\right)^p dx\right)^{\frac{1}{p}} = \left(\int_0^\infty \left(\int_0^\infty f(t)g(x)^{\frac{1}{p}}\chi_{[t,\infty)}(x)\,dt\right)^p dx\right)^{\frac{1}{p}}.$$

By the Minkowski integral inequality (Theorem 3.11.1), we obtain

$$\left(\int_0^\infty \left(\int_0^\infty f(t)g(x)^{\frac{1}{p}}\chi_{[t,\infty)}(x)\,dt\right)^p dx\right)^{\frac{1}{p}}$$

$$\leq \int_0^\infty \left(\int_0^\infty [f(t)g(x)^{\frac{1}{p}}\chi_{[t,\infty)}(x)]^p\,dx\right)^{\frac{1}{p}} dt$$

$$= \int_0^\infty f(t)\left(\int_t^\infty g(x)\,dx\right)^{\frac{1}{p}} dt,$$

as desired. The proof is complete. □

Lemma 3.11.6. *Assume that $1 \leq q < p < \infty$, let $b \in (0, \infty]$ and let w be a positive measurable function on $(0, b)$. Then the following two statements are equivalent.*

(i) *There exists a positive constant C such that*

$$\left(\int_0^b w(t)\left(\int_0^t g(s)\,ds\right)^q dt\right)^{\frac{1}{q}} \leq C \left(\int_0^b g(t)^p\,dt\right)^{\frac{1}{p}} \quad (3.11.5)$$

for every positive measurable function g.

(ii) *The estimate*

$$A := \left(\int_0^b \left(\int_t^b w(s)\,ds\right)^{\frac{p}{p-q}} t^{\frac{(q-1)p}{p-q}}\,dt\right)^{\frac{p-q}{pq}} < \infty \quad (3.11.6)$$

holds.

Proof. (ii) ⇒ (i) Assume that $q > 1$. Integrating by parts on the left-hand side of (3.11.5) and using the Hölder inequality (3.1.10) for the parameters $\frac{p}{p-q}, \frac{p}{q-1}$, p (observe that $\frac{p-q}{p} + \frac{q-1}{p} + \frac{1}{p} = 1$), we get

$$\left(\int_0^b w(t)\left(\int_0^t g(s)\,ds\right)^q dt\right)^{\frac{1}{q}}$$

$$= q^{\frac{1}{q}}\left(\int_0^b \int_t^b w(s)\,ds\, g(t)\left(\int_0^t g(y)\,dy\right)^{q-1} dt\right)^{\frac{1}{q}}$$

Section 3.11 Hardy inequalities

$$\leq q^{\frac{1}{q}} \left[\left(\int_0^b g(y)^p \, dy \right)^{\frac{1}{p}} \left(\int_0^b \left(\int_0^t g(s) \, ds \right)^p t^{-p} \, dt \right)^{\frac{q-1}{p}} \right.$$

$$\left. \times \left(\int_0^b t^{\frac{(q-1)p}{p-q}} \left(\int_t^b w(s) \, ds \right)^{\frac{p}{p-q}} dt \right)^{\frac{p-q}{q}} \right]^{\frac{1}{q}}.$$

By (ii) and the power-weight inequality (3.11.2), we see that the last expression is majorized by

$$A(p')^{\frac{1}{q'}} q^{\frac{1}{q}} \left(\int_0^b g(t)^p \, dt \right)^{\frac{1}{p}},$$

as desired.

(i) $\Rightarrow$ (ii) We shall restrict ourselves to the case when $b = \infty$, the proof is similar when $b < \infty$. We first note that when (3.11.5) holds for a weight w, then it also holds for the weight $w_N := w \chi_{(0,N)}$, where $N \in \mathbb{N}$, with unchanged constant. Let

$$g_N(x) := \left(\int_x^\infty w_N(t) \, dt \right)^{\frac{1}{p-q}} x^{\frac{q-1}{p-q}}, \quad x \in (0, \infty),$$

and

$$A_N := \left(\int_0^\infty \left(\int_t^\infty w_N(s) \, ds \right)^{\frac{p}{p-q}} t^{\frac{q-1}{p-q}} \, dt \right)^{\frac{p-q}{p}}.$$

By (i), we get

$$C A_N^{\frac{q}{p-q}} = C \left(\int_0^\infty g_N(x)^p \, dx \right)^{\frac{1}{p}}$$

$$\geq \left(\int_0^\infty w_N(t) \left(\int_0^t g_N(s) \, ds \right)^q dt \right)^{\frac{1}{q}}.$$

By integrating by parts we verify that the last expression is equal to

$$\left(q \int_0^\infty g_N(t) \int_t^\infty w_N(s) \, ds \left(\int_0^t g_N(s) \, ds \right)^{q-1} dt \right)^{\frac{1}{q}}. \tag{3.11.7}$$

Calculation shows that

$$\left(\int_0^t g_N(s) \, ds \right)^{q-1} = \left(\int_0^t x^{\frac{q-1}{p-q}} \left(\int_x^\infty w_N(s) \, ds \right)^{\frac{1}{p-q}} dx \right)^{q-1}$$

$$\geq \left(\frac{p-1}{p-q} \right)^{1-q} \left(\int_t^\infty w_N(s) \, ds \right)^{\frac{(p-1)(q-1)}{p-q}} t^{\frac{q-1}{p-q}}.$$

Hence, the expression in (3.11.7) is no less than

$$q^{\frac{1}{q}} \left(\frac{p-1}{p-q}\right)^{\frac{1-q}{q}} \left(\int_0^\infty \left(\int_t^\infty w_N(s)\,ds\right)^{\frac{p}{p-q}} t^{\frac{p(q-1)}{p-q}}\,dt\right)^{\frac{1}{q}}$$

$$= q^{\frac{1}{q}} \left(\frac{p-1}{p-q}\right)^{\frac{1-q}{q}} A_N^{\frac{p}{p-q}}.$$

We thus get

$$A_N \le q^{-\frac{1}{q}} \left(\frac{p-q}{p-1}\right)^{\frac{1-q}{q}} C,$$

and letting $N \to \infty$ we get the same estimate with A_N replaced by A.

In the case when $q = 1$ the argument is similar and the expressions to work with are even simpler. We omit the details. □

Remark 3.11.7. We note that the assertion of Lemma 3.11.6 can be extended in a suitable way to the case $p = \infty$. The argument of the proof is then similar, using (3.11.3) instead of (3.11.2). We omit the details.

Theorem 3.11.8 (weighted Hardy inequalities). *Let $p, q \in [1, \infty)$ and let w, v be two a.e. positive weights on $(0, \infty)$. Then there exists a positive constant C such that the inequality*

$$\left(\int_0^\infty \left(\int_0^t f(s)\,ds\right)^q w(t)\,dt\right)^{\frac{1}{q}} \le C \left(\int_0^\infty f(t)^p v(t)\,dt\right)^{\frac{1}{p}} \qquad (3.11.8)$$

holds for every nonnegative measurable function f on $(0, \infty)$ if and only if either $1 < p \le q < \infty$ and

$$B := \sup_{t>0} \left(\int_t^\infty w(s)\,ds\right)^{\frac{1}{q}} \left(\int_0^t v(s)^{1-p'}\,ds\right)^{\frac{1}{p'}} < \infty \qquad (3.11.9)$$

or $1 = p \le q < \infty$ and

$$B_1 := \sup_{t>0} \left(\int_t^\infty w(s)\,ds\right)^{\frac{1}{q}} \operatorname*{ess\,sup}_{s \in (0,t)} \frac{1}{v(s)} < \infty \qquad (3.11.10)$$

or $1 \le q < p < \infty$ and

$$M := \left(\int_0^\infty \left(\int_x^\infty w(t)\,dt\right)^{\frac{r}{q}} \left(\int_0^x v(t)^{1-p'}\,dt\right)^{\frac{r}{q'}} v(x)^{1-p'}\,dx\right)^{\frac{1}{r}} < \infty,$$

$$(3.11.11)$$

where

$$\frac{1}{r} = \frac{1}{p} - \frac{1}{q}. \qquad (3.11.12)$$

Section 3.11 Hardy inequalities

Proof. We start with the case $1 < p \leq q < \infty$.

Necessity. Let $a \in (0, \infty)$ and assume that supp $f \subset [0, a]$. Then (3.11.8) implies that

$$\left(\int_a^\infty w(x)\,dx\right)^{\frac{1}{q}} \int_0^a f(t)\,dt \leq C \left(\int_0^a f^p(x)v(x)\,dx\right)^{\frac{1}{p}}.$$

Suppose first that $\int_0^a v^{1-p'}(x)\,dx < \infty$ and set

$$f(x) := \begin{cases} v^{1-p'}(x) & \text{if } x \in [0, a), \\ 0 & \text{if } x \in [a, \infty). \end{cases}$$

Then we get

$$\left(\int_a^\infty w(x)\,dx\right)^{\frac{1}{q}} \left(\int_0^a v^{1-p'}(x)\,dx\right)^{\frac{1}{p'}} \leq C. \tag{3.11.13}$$

Now suppose that $\int_0^a v^{1-p'}(x)\,dx = \infty$. Then we replace the function $v(x)$ in (3.11.8) by $v(x) + \varepsilon$ for $\varepsilon > 0$, run the same argument and then let $\varepsilon \to 0_+$; we thereby obtain (3.11.13) again. In each case, this proves (3.11.9), hence the necessity.

Sufficiency. We denote

$$g(t) := \left(\int_0^t v^{1-p'}(s)\,ds\right)^{\frac{1}{qp'}}, \quad t \in (0, \infty).$$

Then, by the Hölder inequality, applied to the inner integral, we get

$$\left(\int_0^\infty \left(\int_0^t f(s)\,ds\right)^q w(t)\,dt\right)^{\frac{1}{q}}$$

$$\leq \left(\int_0^\infty \left(\int_0^t f(s)^p g(s)^p v(s)\,ds\right)^{\frac{q}{p}} \left(\int_0^t g(s)^{-p'} v(s)^{1-p'}\,ds\right)^{\frac{q}{p'}} w(t)\,dt\right)^{\frac{1}{q}} \tag{3.11.14}$$

By Lemma 3.11.5, the last expression is no bigger than

$$\int_0^\infty f(t)^p g(t)^p v(t) \left(\int_t^\infty \left(\int_0^x g(s)^{-p'} v(s)^{1-p'}\,ds\right)^{\frac{q}{p'}} w(x)\,dx\right)^{\frac{p}{q}} dt.$$

Inserting the formula that defines the function g now yields that the inner integral can be rewritten as

$$\int_t^\infty \left(\int_0^x \left(\int_0^s v^{1-p'}(y)\,dy\right)^{-\frac{1}{q}} v(s)^{1-p'}\,ds\right)^{\frac{q}{p'}} w(x)\,dx. \tag{3.11.15}$$

Calculation now gives

$$\int_0^x \left(\int_0^s v^{1-p'}(y)\,dy\right)^{-\frac{1}{q}} v(s)^{1-p'}\,ds = q' \left(\int_0^x v^{1-p'}(s)\,ds\right)^{\frac{1}{q'}},$$

whence (3.11.15) becomes

$$(q')^{\frac{q}{p'}} \int_t^\infty \left(\int_0^x v(s)^{1-p'}\,ds\right)^{\frac{q}{p'q'}} w(x)\,dx.$$

By (3.11.9), the last expression is no bigger than

$$B^{\frac{q}{q'}}(q')^{\frac{q}{p'}} \int_t^\infty \left(\int_x^\infty w(s)\,ds\right)^{-\frac{1}{q'}} w(x)\,dx = B^{q-1}(q')^{\frac{q}{p'}}q \left(\int_t^\infty w(s)\,ds\right)^{\frac{1}{q}}$$

$$\leq B^q (q')^{\frac{q}{p'}} q \left(\int_0^t v(s)^{1-p'}\,ds\right)^{\frac{-1}{p'}}$$

$$= B^q (q')^{\frac{q}{p'}} q g(t)^{-q}.$$

Altogether, the right-hand side of (3.11.14) is majorized by the expression

$$\int_0^\infty f(t)^p g(t)^p v(t) B^q (q')^{\frac{q}{p'}} q g(t)^{-q}\,dt = B^q (q')^{p-1} q^{\frac{p}{q}} \int_0^\infty f(t)^p v(t)\,dt,$$

establishing the desired Hardy inequality (3.11.8) for the case $1 < p \leq q < \infty$.
When $p = 1$, then we get from Lemma 3.11.5 that

$$\left(\int_0^\infty \left(\int_0^t f(s)\,ds\right)^q w(t)\,dt\right)^{\frac{1}{q}} \leq \int_0^\infty f(t) \left(\int_t^\infty w(x)\,dx\right)^{\frac{1}{q}} \frac{1}{v(t)} v(t)\,dt$$

$$\leq B_1 \int_0^\infty f(t)v(t)\,dt.$$

The necessity follows by setting $f = \frac{1}{v}$. The proof is complete. □

Remark 3.11.9. The assertion of Theorem 3.11.8 can be, again, extended to the case when either $p = \infty$ or $q = \infty$, or both. We will omit the details and just note how the proof goes in these situations.

In the case when $p = q = \infty$, then the appropriate modification of (3.11.8) follows from the estimate

$$\operatorname*{ess\,sup}_{x\in(0,\infty)} w(x) \int_0^x f(t)\,dt \leq \operatorname*{ess\,sup}_{x\in(0,\infty)} w(x) \int_0^x \left[\operatorname*{ess\,sup}_{t\in(0,x)} v(t)f(t)\right] \frac{dt}{v(t)}.$$

Section 3.11 Hardy inequalities

If $p = 1$ and $q = \infty$, we have

$$\operatorname*{ess\,sup}_{x\in(0,\infty)} w(x) \int_0^x f(t)\,dt \le \operatorname*{ess\,sup}_{x\in(0,\infty)} \left(w(x) \operatorname*{ess\,sup}_{t\in(0,x)} \frac{1}{w(t)} \int_0^x f(t)v(t)\,dt \right)$$

$$\le B_1 \int_0^\infty f(t)v(t)\,dt.$$

Finally, when $p > 1$ and $q = \infty$, one has

$$\operatorname*{ess\,sup}_{x\in(0,\infty)} w(x) \int_0^x f(t)\,dt$$

$$\le \operatorname*{ess\,sup}_{x\in(0,\infty)} w(x) \left(\int_0^x v(t)^{1-p'}\,dt \right)^{\frac{1}{p'}} \left(\int_0^x f(t)^p v(t)\,dt \right)^{\frac{1}{p}}$$

$$\le B \int_0^\infty f(t)v(t)\,dt.$$

In the case when $1 \le q < p \le \infty$, we set

$$\tau(x) := \int_0^x v(y)^{1-p'}$$

and define $\tilde{w}$, $\tilde{v}$ by

$$w(x) = \tilde{w}(\tau(x)), \quad v(x) := \tilde{v}(\tau(x)), \quad x \in (0, \infty).$$

Then (3.11.8) becomes

$$\left(\int_0^b \left(\int_0^t g(s)\,ds \right)^q \tilde{w}(t)^q \tilde{v}(t)^{p'}\,dt \right)^{\frac{1}{q}} \le C \left(\int_0^b g(t)^p\,dt \right)^{\frac{1}{p}},$$

where

$$g(\tau(x)) = \int_0^x f(y)\,dy \quad \text{and} \quad b = \int_0^\infty v(y)^{1-p'}\,dy.$$

Thus, for $p < \infty$, the sufficiency follows from Lemma 3.11.6.

Finally, when $p = \infty$, one has

$$M = \left(\int_0^\infty \left(\int_0^x \frac{dy}{v(y)} \right)^{q-1} \int_x^\infty w(y)\,dy \frac{dx}{v(x)} \right)^{\frac{1}{q}}$$

$$= q^{-\frac{1}{q}} \left(\int_0^\infty w(x) \left(\int_0^x \frac{dy}{v(y)} \right)^q dx \right)^{\frac{1}{q}}.$$

Therefore,

$$\left(\int_0^\infty \left(\int_0^t f(s)\,ds \right)^q w(t)\,dt \right)^{\frac{1}{q}} \le M q^{\frac{1}{q}} \operatorname*{ess\,sup}_{x\in(0,\infty)} f(x)v(x),$$

as desired.

3.12 Sequence spaces

Let $\mathbb{N}$ be the set of all positive integers endowed with the arithmetic measure. As we already know from Example 1.20.20, this is a typical completely atomic measure space. We can then define Lebesgue spaces of functions acting on this measure space in the same way as we have built the spaces $L^p(\Omega)$ for $p \in (0, \infty]$ on a nonatomic measure space. The "functions" on such a measure space however turn out to be infinite *sequences* of real numbers. The resulting spaces are traditionally denoted as $\ell^p, 0 < p \leq \infty$.

Definition 3.12.1. Let $0 < p < \infty$. We then define the sequence space ℓ^p by

$$\ell^p := \left\{ \{a_n\}_{n=1}^\infty, a_n \in \mathbb{R} \text{ for every } n \in \mathbb{N}, \sum_{n=1}^\infty |a_n|^p < \infty \right\}.$$

We equip the set ℓ^p with the functional $\|\cdot\|_{\ell^p}$, defined for every sequence $\{a_n\}_{n=1}^\infty$ of real numbers as

$$\|\{a_n\}\|_{\ell^p} := \left(\sum_{n=1}^\infty |a_n|^p \right)^{\frac{1}{p}}.$$

Definition 3.12.2. The sequence space ℓ^∞ is defined by

$$\ell^\infty := \left\{ \{a_n\}_{n=1}^\infty, a_n \in \mathbb{R} \text{ for every } n \in \mathbb{N}, \sup_{n \in \mathbb{N}} |a_n| < \infty \right\}.$$

We equip the set ℓ^∞ with the functional $\|\cdot\|_{\ell^\infty}$, defined for every sequence $\{a_n\}_{n=1}^\infty$ of real numbers as

$$\|\{a_n\}\|_{\ell^\infty} := \sup_{n \in \mathbb{N}} |a_n|.$$

Definition 3.12.3. The sequence spaces c and c_0 are defined by

$$c := \left\{ \{a_n\}_{n=1}^\infty, a_n \in \mathbb{R} \text{ for every } n \in \mathbb{N}, \lim_{n \in \mathbb{N}} a_n < \infty \text{ exists} \right\}$$

and

$$c_0 := \left\{ \{a_n\}_{n=1}^\infty, a_n \in \mathbb{R} \text{ for every } n \in \mathbb{N}, \lim_{n \in \mathbb{N}} a_n = 0 \right\}.$$

We equip both the sets c and c_0 with the functional $\|\cdot\|_{\ell^\infty}$.

Remark 3.12.4. The space $(\ell^p, \|\cdot\|_{\ell^p})$ is a Banach space for every $p \in [1, \infty]$ and is a quasi-normed space when $p \in (0, 1)$. Moreover, for every $1 \leq p \leq q \leq \infty$, one has the relations

$$\ell^1 \hookrightarrow \ell^p \hookrightarrow \ell^q \hookrightarrow c_0 \hookrightarrow c \hookrightarrow \ell^\infty.$$

3.13 Relations between various types of convergence

We have seen relations between various types of convergence of a sequence of measurable functions $\{f_n\}_{n=1}^{\infty}$ on a general measure space in Section 1.22.

Definition 3.13.1. Let $1 \leq p < \infty$. We say that a sequence of measurable function $\{f_n\}_{n=1}^{\infty}$ on Ω *converges in $L^p(\Omega)$* (or *in the L^p-mean*) to a function $f \in L^p\Omega$ if
$$\lim_{n \to \infty} \|f_n - f\|_{L^p(\Omega)} = 0.$$

Theorem 3.13.2. *Let $1 \leq p < \infty$.*

(i) *Let a sequence of functions $\{f_n\}$ in $L^p(\Omega)$ converge in $L^p(\Omega)$ to a function $f \in L^p(\Omega)$. Then*
$$\|f_n\|_p \to \|f\|_p.$$

(ii) *The convergence in the $L^p(\Omega)$-mean does not in general imply convergence almost everywhere.*

(iii) *The convergence in the $L^p(\Omega)$-mean always implies convergence in measure.*

(iv) *In the special case $\mu(\Omega) < \infty$ the uniform convergence implies convergence in the L^p-mean but none of uniform convergence up to small sets, convergence in measure or convergence a.e. in general need not imply.*

(v) *If $\mu(\Omega) = \infty$, then even the uniform convergence in general need not imply the convergence in the L^p-mean.*

The diagram in Figure 3.2 describes the mutual relations between the types of convergence for the case when Ω is of finite measure.

Remarks 3.13.3. (i) The implications $(\alpha), (\beta), (\gamma), (\delta)$ are immediate consequences of the definitions, see also Remark 1.22.2, Propositions 1.22.5 and 1.22.6 and Theorem 3.13.2.

(ii) The implication (ε) holds in general, but only under the assumption that $f_n \in L^p(\Omega)$ for every $n \in \mathbb{N}$. It can be proved by means of the Lebesgue dominated convergence theorem (Theorem 1.21.5).

(iii) The implication (ψ) follows from the Lebesgue dominated convergence theorem (Theorem 1.21.5).

(iv) The implication (ω) is the Egorov theorem (Theorem 1.22.3).

(v) The implication (ν) is the Riesz theorem (Theorem 1.22.8).

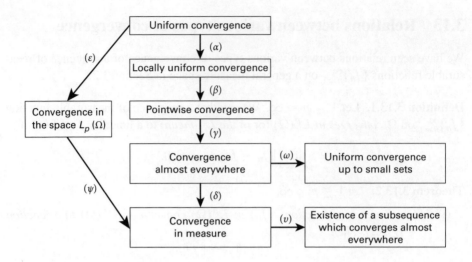

Figure 3.2. Modes of convergence.

3.14 Relatively compact subsets of $L^p(\Omega)$

In this section we shall use the convention stated at the beginning of Section 3.3.

Definition 3.14.1. Let $p \in [1, \infty)$. We say that a set $K \subset L^p(\Omega)$ is *p-mean equicontinuous* if for every $\varepsilon > 0$ there exists a $\delta > 0$ such that

$$\int_\Omega |f(x+h) - f(x)|^p \, dx < \varepsilon^p$$

for each $f \in K$ and $h \in \mathbb{R}^N$ with $|h| < \delta$.

Theorem 3.14.2 (Riesz compactness theorem). *Let $p \in [1, \infty)$ and let $K \subset L^p(\Omega)$. Then K is relatively compact in $L^p(\Omega)$ if and only if the following two conditions are satisfied:*

  (i) *the set K is bounded, i.e. there exists a $c > 0$ such that $\|f\|_p \leq c$ for every $f \in K$;*

  (ii) *the set K is p-mean equicontinuous.*

Proof. Let K be a relatively compact subset of $L^p(\Omega)$. Thus K is bounded, whence (i) follows. Let $\varepsilon > 0$ and let $f_1, \ldots, f_n$ be an $\frac{\varepsilon}{3}$-net of the set K. According to Theorem 3.3.3, $f_1, \ldots, f_n$ are p-mean continuous. Thus there exists a $\delta > 0$ such that

$$\int_\Omega |f_i(x+h) - f_i(x)|^p \, dx < \left(\frac{\varepsilon}{3}\right)^p$$

if $i = 1, \ldots, n$ and $h \in \mathbb{R}^N$, $|h| < \delta$.

Let $f \in K$, $h \in \mathbb{R}^N$, $|h| < \delta$, $\|f_i - f\|_p < \frac{\varepsilon}{3}$. The triangle inequality now implies (ii).

Conversely, let us assume that (i) and (ii) are satisfied. Let $\eta > 0$. It follows from (3.4.2), (3.4.5) and the Arzelà–Ascoli theorem (Theorem 2.5.3) that for a sufficiently small $\varepsilon > 0$, the set $R_\varepsilon(K) = \{R_\varepsilon u; u \in K\}$ is relatively compact in the space $C(\overline{\Omega})$. Thus, there exists a finite η'-net ($\eta' = \frac{1}{2}\eta \, (\mu(\Omega))^{\frac{1}{p}}$) of $R_\varepsilon(K)$ in $C(\overline{\Omega})$, which is an η-net of K in $L^p(\Omega)$. By virtue of Remark 1.13.3 (iii), the proof is complete. □

Remarks 3.14.3. (i) The theorem just proved is due to Riesz [187] in the case of $\Omega = (0, 1)$. Essentially the same proof as that given here was given by Kolmogorov [119].

(ii) Further criteria for relative compactness of subsets of the spaces $L^p(\Omega)$ can be found, e.g., in the books by Natanson [164] and Krasnosel'skiĭ et al. [124], where the following theorem is also proved.

Theorem 3.14.4. *Let M be a relatively compact subset of the space $L^2(\Omega)$ and let $p > 2$. Suppose that for every $\varepsilon > 0$ there exists a $\delta > 0$ such that*

$$\int_F |u(x)|^p \, dx < \varepsilon$$

for every $u \in M$ and $F \subset \Omega$, $\mu(F) < \delta$. Then M is a subset of $L^p(\Omega)$ and is relatively compact in $L^p(\Omega)$.

Exercise 3.14.5. Let $1 \leq q < p$. Then the embedding

$$L^p(\Omega) \hookrightarrow L^q(\Omega)$$

is not compact. (Hint: It suffices to show that the embedding $L^\infty(\Omega) \hookrightarrow L^1(\Omega)$ is not compact. If $\Omega = (0, \pi)$ then the sequence $\{\sin nx\}_{n=1}^\infty$ does not contain a subsequence which converges in $L^1(0, \pi)$. For a general nonempty bounded open set $\Omega \subset \mathbb{R}^N$, an analogous example may be constructed.)

3.15 Weak convergence

Theorem 3.15.1. *Let $p \in (1, \infty)$ and let $\{f_n\}_{n=1}^\infty \subset L^p(0, 1)$ be a sequence of functions. Then $\{f_n\}_{n=1}^\infty$ converges weakly to a function $f \in L^p(0, 1)$ if and only if the following two conditions are satisfied:*

(i) *the sequence $\{\|f_n\|_p\}_{n=1}^\infty$ is bounded;*

(ii) *for every $\tau \in (0, 1)$, we have*

$$\lim_{n \to \infty} \int_0^\tau f_n(t) \, dt = \int_0^\tau f(t) \, dt.$$

Proof. The proof is based on the Banach–Steinhaus theorem (Theorem 1.17.5). Set

$$\alpha_\tau(t) = \begin{cases} 1 & \text{for } 0 \le t \le \tau, \\ 0 & \text{for } \tau < t \le 1. \end{cases}$$

It follows from Exercise 3.5.2 (iv) that the linear hull of all such functions forms a dense subset of $L^{p'}(0, 1)$. Thus, according to the second condition in Theorem 1.17.5 and Section 3.8 it is necessary and sufficient to show that

$$\lim_{n \to \infty} \int_0^1 f_n(t) \alpha_\tau(t) \, dt = \int_0^1 f(t) \alpha_\tau(t) \, dt$$

for arbitrary α_τ, which is nothing else than (ii). $\square$

In the same way it is possible to prove the following assertion dealing with the space $L^p(\Omega)$.

Theorem 3.15.2. *Let Ω be a nonempty bounded open subset of $\mathbb{R}^N$. Let $p \in (1, \infty)$ and let $f_n \in L^p(\Omega)$, $n = 0, 1, \ldots$. Let $\sum_0$ be the collection of measurable subsets of Ω such that the linear hull of their characteristic functions forms a dense subset in $L^{p'}(\Omega)$. Then the sequence $\{f_n\}_{n=1}^\infty$ converges weakly to f if and only if the following two conditions are satisfied:*

(i) *the sequence $\{\|f_n\|_p\}_{n=1}^\infty$ is bounded;*

(ii) $\lim_{n \to \infty} \int_E f_n(x) \, dx = \int_E f(x) \, dx$ *for every $E \in \sum_0$.*

3.16 Isomorphism of $L^p(\Omega)$ and $L^p(0, \mu(\Omega))$

Theorem 3.16.1. *Let Ω be a nonempty bounded open subset of $\mathbb{R}^N$. Then there exists a one-to-one bounded linear mapping from $L^p(\Omega)$ onto $L^p(0, \mu(\Omega))$, i.e. the spaces $L^p(\Omega)$ and $L^p(0, \mu(\Omega))$ are isomorphic. Moreover, it is possible to choose this mapping to be an isometry.*

Sketch of the proof. We explain the idea of the proof in the case $N = 2$ only.

Since Ω is a bounded set, there exists a square $B_0 \subset \mathbb{R}^2$ such that $\Omega \subset B_0$. Consider the sequence of equidistant divisions $\{T_n\}_{n=1}^\infty$ of the set B_0 by the lines parallel to the coordinate axes such that every T_n contains 4^n equal squares B_n^k, $k = 1, 2, \ldots, 4^n$. Passing from n to $n+1$ each square of T_n is divided into 4 equal squares of T_{n+1}.

A mapping of the set of all squares obtained onto the set of subsegments of the interval $I_0 = [0, \mu(\Omega)]$ in which the length of the segment I_n^k corresponding to the square B_n^k equals $\mu(\Omega \cap B_n^k)$ is constructed as follows: Let $[\alpha, \beta] \subset I_0$ correspond by

Section 3.17 Schauder bases

the mapping in question to a square B_n^k of T_n. Denote the four squares obtained from B_n^k when passing to the division T_{n+1} by B^{I}, B^{II}, B^{III}, B^{IV} in the same manner as the quadrants in the plane are numbered. Now divide the segment $[\alpha, \beta]$ by the points γ_1, γ_2, γ_3 such that $\alpha \leq \gamma_1 \leq \gamma_2 \leq \gamma_3 \leq \beta$, and

$$\gamma_1 - \alpha = \mu(\Omega \cap B^{\text{I}}), \qquad \gamma_3 - \gamma_2 = \mu(\Omega \cap B^{\text{III}}),$$
$$\gamma_2 - \gamma_1 = \mu(\Omega \cap B^{\text{II}}), \qquad \beta - \gamma_3 = \mu(\Omega \cap B^{\text{IV}}).$$

Define the segments $[\alpha, \gamma_1]$, $[\gamma_1, \gamma_2]$, $[\gamma_2, \gamma_3]$, $[\gamma_3, \beta]$ as the values of our mapping corresponding successively to B^{I}, B^{II}, B^{III}, B^{IV}.

For every $P \subset \Omega$, let $\sigma(P) \subset I_0$ be the intersection of the intervals corresponding to all squares from the cover of P in the mapping defined above. We have

$$\mu(P) = \mu(\sigma(P)).$$

Setting now $\tilde{u}(t) = u(\sigma^{-1}(t))$ for every $u \in L^p \Omega$ we obtain the desired isomorphism. (Be careful about the sets of measure zero.) □

Remark 3.16.2. For unbounded domains the space $L^p(\Omega)$ is isomorphic with $L^p(0, 1)$, too (see Triebel [227]).

3.17 Schauder bases in $L^p(\Omega)$

First we shall consider the space $L^p(0, 1)$, $p \geq 1$. Since we proved in Theorem 3.6.1 that these spaces are separable, the investigation of the existence of Schauder bases makes sense.

Definition 3.17.1. On the interval $[0, 1]$ we define the following family of functions:

$$f_0^0(t) = 1 \quad \text{for } t \in [0, 1],$$

$$f_n^k(t) = \begin{cases} 2^{\frac{n}{2}} & \text{if } t \in \left[\frac{2k-2}{2^{n+1}}, \frac{2k-1}{2^{n+1}}\right], \\ -2^{\frac{n}{2}} & \text{if } t \in \left(\frac{2k-1}{2^{n+1}}, \frac{2k}{2^{n+1}}\right], \\ 0 & \text{otherwise,} \end{cases}$$

$n = 0, 1, \ldots$; $k = 1, 2, \ldots, 2^n$. We order this family into groups so that every group contains the functions with the same index n; ordering in the groups is done according to increasing k, the groups are ordered with increasing n. The sequence $\{h_i\}_{i=1}^{\infty}$ obtained in this way is called the *Haar system*.

Remark 3.17.2. Using elementary computation one can verify that the Haar system is orthogonal, i.e.

$$\int_0^1 h_i(t) h_j(t) \, dt = 0$$

for $i \neq j$.

Theorem 3.17.3. Let $p \in [0, 1)$, $f \in L^p(0, 1)$ and set

$$c_i = \int_0^1 f(x) h_i(x)\, dx.$$

Then

$$\lim_{n \to \infty} \int_0^1 \left| f(x) - \sum_{i=1}^n c_i h_i(x) \right|^p dx = 0.$$

Proof. Denote

$$s_n^k(f(x)) = f_0^0(x) \int_0^1 f_0^0(t) f(t)\, dt + \cdots + f_n^k(x) \int_0^1 f_n^k(t) f(t)\, dt$$

and

$$K_n^k(s, t) = f_0^0(s) f_0^0(t) + \cdots + f_n^k(s) f_n^k(t)$$

for $n = 0, 1, \ldots$, and $k = 1, 2, \ldots, 2^n$. Clearly,

$$s_n^k(f(x)) = \int_0^1 K_n^k(x, t) f(t)\, dt. \tag{3.17.1}$$

The function $K_n^k(s, t)$ is equal almost everywhere on $[0, 1] \times [0, 1]$ to the function $K(s, t)$ defined by

$$K(s, t) = \begin{cases} 2^{n+1} & \text{for } (t, s) \in \left[\frac{m}{2^{n+1}}, \frac{m+1}{2^{n+1}}\right] \times \left[\frac{m}{2^{n+1}}, \frac{m+1}{2^{n+1}}\right], \\ & m = 0, 1, \ldots, 2k - 1; \\ 2^n & \text{for } (t, s) \in \left[\frac{m}{2^n}, \frac{m+1}{2^n}\right] \times \left[\frac{m}{2^n}, \frac{m+1}{2^n}\right], \\ & m = 2k, \ldots, 2^n - 1; \\ 0 & \text{otherwise.} \end{cases}$$

Let $t \in [0, 1]$, $n \in \mathbb{N}$ and $k \in \{1, 2, \ldots, 2^n\}$ be fixed. Then there exists an integer m, $0 \leq m \leq 2^n - 1$, such that $t \in [\frac{m}{2^n}, \frac{m+1}{2^n}]$. The following possibilities may occur:

(a) $m \leq k$,

(b) $m > k$.

Assume (a): Then either (α):

$$t \in \left[\frac{2m}{2^{n+1}}, \frac{2m+1}{2^{n+1}}\right]$$

or (β):

$$t \in \left[\frac{2m+1}{2^{n+1}}, \frac{2m+2}{2^{n+1}}\right].$$

Section 3.17 Schauder bases

In the case (α)

$$s_n^k(f(t)) = 2^{n+1} \int_{\frac{2m}{2^{n+1}}}^{\frac{2m+1}{2^{n+1}}} f(\tau)\,d\tau,$$

and for the case (β) we have

$$s_n^k(f(t)) = 2^{n+1} \int_{\frac{2m+1}{2^{n+1}}}^{\frac{2m+2}{2^{n+1}}} f(\tau)\,d\tau.$$

Assume (b):

$$s_n^k(f(t)) = 2^n \int_{\frac{m}{2^n}}^{\frac{m+1}{2^n}} 2^n f(\tau)\,d\tau.$$

In all cases the function $s_n^k(f(t))$ is independent of t, thus there exist intervals $[a_0, a_1]$, ..., $[a_{r-1}, a_r]$ covering the interval $[0, 1]$ and on the $(i+1)$st interval $s_n^k(f(t))$ is constant and equal to

$$(a_{i+1} - a_i)^{-1} \int_{a_i}^{a_{i+1}} f(\tau)\,d\tau. \tag{3.17.2}$$

By virtue of the Hölder inequality (Theorem 3.1.6) it follows that

$$\int_0^1 \left|s_n^k(f(t))\right|^p dt \leq \int_0^1 |f(t)|^p\,dt. \tag{3.17.3}$$

Further,

$$\lim_{n\to\infty} s_n^k(f(t)) = f(t) \tag{3.17.4}$$

for almost all $t \in [0, 1]$.

Suppose first that f is bounded, i.e. there exists an $M > 0$ such that $|f(t)| \leq M$ for all $t \in [0, 1]$. From (3.17.3) we obtain

$$\sup_{x\in[0,1]} \left|s_n^k(f(x))\right| \leq \sup_{x\in[0,1]} |f(x)| \leq M \tag{3.17.5}$$

and thus

$$\sup_{x\in[0,1]} \left|s_n^k(f(x)) - f(x)\right| \leq (2M)^p.$$

So

$$\lim_{n\to\infty} \left|s_n^k(f(x)) - f(x)\right|^p = 0$$

and the Lebesgue dominated convergence theorem (Theorem 1.21.5) yields

$$\lim_{n\to\infty} \int_0^1 \left|s_n^k(f(x)) - f(x)\right|^p dx = 0.$$

In this way the assertion is proved under the additional assumption that the function f is bounded.

To prove the claim in a general case, let $\varepsilon > 0$. According to Theorem 3.5.1, there exists a bounded function f_1 and a function $f_2 \in L^p(0,1)$ such that $f = f_1 - f_2$ and
$$\int_0^1 |f_2(x)|^p \, dx < \varepsilon.$$

From (3.17.3) we have
$$\int_0^1 \left|s_n^k(f_2(x))\right|^p \, dx < \varepsilon$$

and, further,
$$\left(\int_0^1 \left|f(x) - s_n^k(f(x))\right|^p \, dx\right)^{\frac{1}{p}}$$
$$\leq \left(\int_0^1 \left|f_1(x) - s_n^k(f_1(x))\right|^p \, dx\right)^{\frac{1}{p}} + \left(\int_0^1 |f_2(x)|^p \, dx\right)^{\frac{1}{p}}$$
$$+ \left(\int_0^1 \left|s_n^k(f_2(x))\right|^p \, dx\right)^{\frac{1}{p}}$$
$$\leq \left(\int_0^1 \left|f_1(x) - s_n^k(f_1(x))\right|^p \, dx\right)^{\frac{1}{p}} + 2\varepsilon^{\frac{1}{p}}.$$

From the first part of the proof we have
$$\limsup_{n \to \infty} \left(\int_0^1 \left|f(x) - s_n^k(f(x))\right|^p \, dx\right)^{\frac{1}{p}} \leq 2\varepsilon^{\frac{1}{p}}$$

which immediately implies the assertion in the general case, since $\varepsilon > 0$ was arbitrary. □

Theorem 3.17.4. *Let $p \in [1, \infty)$. Then the Haar orthogonal system forms a Schauder basis in the space $L^p(0,1)$.*

Proof. Theorem 3.17.3 shows that every $f \in L^p(0,1)$ may be written as
$$f = \sum_{i=1}^{\infty} c_i h_i$$

(the convergence being considered in the space $L^p(0,1)$). Now it suffices to show the uniqueness of c_i. Let $\{c_i\}_{i=1}^{\infty}$ be a sequence of real numbers such that
$$\lim_{n \to \infty} \sum_{i=1}^{n} c_i h_i = 0 \quad \text{in } L^p(0,1).$$

Thus
$$\lim_{n\to\infty} \int_0^1 h_j(t) \left(\sum_{i=1}^n c_i h_i(t) \right) dt = 0$$
for arbitrary $j = 1, 2, \ldots$, which implies
$$0 = \lim_{n\to\infty} \int_0^1 h_j(t) \left(\sum_{i=1}^n c_i h_i(t) \right) dt = c_j.$$

So $c_j = 0$ for $j = 1, 2, \ldots$, and the uniqueness is proved. □

Remarks 3.17.5. (i) The Haar system is an unconditional basis in $L^p(0,1)$ for $1 < p < \infty$ (see [174] and [85]), but not in $L^1(0,1)$.

(ii) The trigonometric system $1, \cos x, \sin x, \cos(2x), \sin(2x), \ldots$ forms a Schauder basis in $L^p(0,1)$; $1 < p < \infty$ (see [247]).

(iii) The trigonometric system from (ii) is not an unconditional basis in $L^p(0,1)$, $p \neq 2$ (see, e.g., [133]).

Theorem 3.16.1 together with Theorem 3.17.4 and Remark 1.15.4 enables us to state a general theorem on the existence of Schauder bases in $L^p(\Omega)$.

Theorem 3.17.6. *Let $p \in [1, \infty)$ and let Ω be a nonempty open subset of $\mathbb{R}^N$. Then there exists a Schauder basis in $L^p(\Omega)$.*

3.18 Weak Lebesgue spaces

Notation 3.18.1. Let f be a (Lebesgue) measurable function on a nonempty bounded open set $\Omega \subset \mathbb{R}^N$. For $\sigma > 0$, define
$$S(f, \sigma) = \{x \in \Omega; \ |f(x)| > \sigma\}.$$
For every $\sigma > 0$ the set $S(f, \sigma)$ is measurable.

Lemma 3.18.2. *Let $f \in L^p(\Omega)$, $1 \leq p < \infty$. Then*
$$\int_\Omega |f(x)|^p \, dx = p \int_0^\infty \sigma^{p-1} \mu(S(f, \sigma)) \, d\sigma. \tag{3.18.1}$$

Proof. Define $F : (0, \infty) \times \Omega \to \mathbb{R}$ by
$$F(\sigma, x) = \begin{cases} 1 & \text{if } \sigma < |f(x)|, \\ 0 & \text{if } \sigma \geq |f(x)|. \end{cases}$$

In other words, the function $F(\sigma, \cdot)$ is the characteristic function of the set $S(f, \sigma)$. The assertion follows from the chain

$$\int_\Omega |f(x)|^p \, dx = \int_\Omega \left(\int_0^{|f(x)|} p\sigma^{p-1} \, d\sigma \right) dx$$

$$= \int_\Omega \left(\int_0^\infty p\sigma^{p-1} F(\sigma, x) \, d\sigma \right) dx = \int_0^\infty \left(\int_\Omega p\sigma^{p-1} F(\sigma, x) \, dx \right) d\sigma$$

$$= p \int_0^\infty \sigma^{p-1} \left(\int_\Omega F(\sigma, x) \, dx \right) d\sigma = p \int_0^\infty \sigma^{p-1} \mu(S(f, \sigma)) \, d\sigma,$$

by virtue of the Fubini theorem (Theorem 1.21.11). □

Definition 3.18.3. Let $p \in [1, \infty)$. We shall then say that the collection of all measurable functions f on Ω such that there exists a $c = c(f) > 0$ satisfying

$$\mu(S(f, t)) \leq \frac{c}{t^p}$$

for every $t > 0$, is the *weak Lebesgue space* or the *Marcinkiewicz space* denoted by $L^{p,\infty}(\Omega)$.

Lemma 3.18.4. *Let $1 \leq p < \infty$. Then*

$$L^p(\Omega) \subset L^{p,\infty}(\Omega).$$

Proof. Let $t > 0$. Then $f \in L^p(\Omega)$ satisfies

$$t^p \mu(S(f, t)) = \mu(S(f, t)) p \int_0^t \sigma^{p-1} \, d\sigma \leq p \int_0^t \sigma^{p-1} \mu(S(f, \sigma)) \, d\sigma$$

$$\leq p \int_0^\infty \sigma^{p-1} \mu(S(f, \sigma)) \, d\sigma = \|f\|_p^p,$$

according to Lemma 3.18.2. □

Example 3.18.5. The inclusion in Lemma 3.18.4 is sharp: Let $p = 1$, $\Omega = (0, 1)$ and $f(x) = \frac{1}{x}$. Then $f \in L^{p,\infty}(\Omega)$ but $f \notin L^p(\Omega)$.

Definition 3.18.6. Let $p \in [1, \infty)$. For $u \in L^{p,\infty}(\Omega)$, set

$$\|u\|_{p,\infty} := \sup_{t > 0} t (\mu(S(u, t)))^{\frac{1}{p}}. \tag{3.18.2}$$

Remarks 3.18.7. Let $p \in [1, \infty)$.

(i) For $u, v \in L^{p,\infty}(\Omega)$ and $\lambda \in \mathbb{R}$, the following relations hold:

(α) $\|u\|_{p,\infty} = 0$ if and only if $u = 0$ almost everywhere on Ω;

(β) $\|u + v\|_{p,\infty} \leq 2(\|u\|_{p,\infty} + \|v\|_{p,\infty})$;

(γ) $\|\lambda u\|_{p,\infty} = |\lambda| \|u\|_{p,\infty}$.

It turns out that $\|\cdot\|_{p,\infty}$ is not a norm. For $p \in (1, \infty)$, there is a partial remedy for this, because it is at least equivalent to a norm. However, when $p = 1$, then a norm on $L^{1,\infty}(\Omega)$ that would be equivalent to $\|\cdot\|_{1,\infty}$ does not exist. We shall return to these questions in detail in Chapter 8, see in particular Theorem 8.2.2 and Corollary 8.2.4. In particular, the proof of (β) will be given in Corollary 8.6.4.

(ii) $L^{p,\infty}(\Omega)$ *is a vector space.* We consider equality in $L^{p,\infty}(\Omega)$ to be in the sense of almost everywhere. For $p = 1$, expression (3.18.2) does not define a norm on $L^{1,\infty}(\Omega)$ since the triangle inequality is not satisfied (for example, let $\Omega = (0, 1)$, $u(t) = t$, $v(t) = 1 - t$; $\|u\|_{1,\infty} = \frac{1}{4}$, $\|v\|_{1,\infty} = \frac{1}{4}$ and $\|u + v\|_{1,\infty} = 1$).

Theorem 3.18.8. *Let $p > 1$ and let $0 < \varepsilon \leq p - 1$. Then*

$$L^p(\Omega) \subset L^{p,\infty}(\Omega) \subset L^{p-\varepsilon}(\Omega). \tag{3.18.3}$$

Proof. The first inclusion follows from Lemma 3.18.4. In order to verify the second one, let $f \in L^{p,\infty}(\Omega)$ and let $\varepsilon > 0$. Then

$$t^{p-1-\varepsilon} \mu(S(f,t)) \leq \begin{cases} \mu(\Omega) & \text{if } t \in (0, 1), \\ \frac{\|u\|_{p,\infty}^p}{t^{1+\varepsilon}} & \text{if } t \in [1, \infty). \end{cases}$$

From this and Lemma 3.18.2, we have

$$\int_\Omega |f(x)|^{p-\varepsilon} \, dx = (p - \varepsilon) \int_0^\infty t^{p-\varepsilon-1} \mu(S(f,t)) \, dt$$

$$\leq (p - \varepsilon)\mu(\Omega) + (p - \varepsilon) \|u\|_{p,\infty}^p \int_1^\infty \frac{1}{t^{1+\varepsilon}} \, dt < \infty,$$

establishing $f \in L^{p-\varepsilon}(\Omega)$, as desired. The proof is complete. □

We shall prove in Chapter 8, where we will have the operation of a nonincreasing rearrangement of a function at our disposal, that although $L^{1,\infty}$ is not equivalently normable, the functional $\|\cdot\|_{1,\infty}$ satisfies certain weaker conditions.

Remark 3.18.9. Weak Lebesgue spaces constitute a particular instant in the scale of two-parameter Lorentz spaces which will be studied in Chapter 8, where also more details about these spaces will be provided, framed in a more general context. They are also a special case of the endpoint Marcinkiewicz spaces, studied in Section 7.10. Weak Lebesgue spaces will also appear in Chapter 5.

3.19 Miscellaneous remarks and related results

Remark 3.19.1 (*The space $L^2(\Omega)$*). The most important space in applications is $L^2(\Omega)$ because it is a Hilbert space with respect to the inner product

$$\langle f, g \rangle = \int_\Omega f(x)g(x)\,dx; \qquad f, g \in L^2(\Omega).$$

For $p \neq 2$, the space $L^p(\Omega)$ is not Hilbert.

Remark 3.19.2 (*The dual space of $L^\infty(\Omega)$*). Let $\mathfrak{M}$ be the family of all measurable subsets of Ω. The space $[L^\infty(\Omega)]^*$ is isometrically isomorphic to the space of all finitely additive measures on $\mathfrak{M}$ which are absolutely continuous with respect to the Lebesgue measure μ. The norm in this space is the total variation of a measure.

Remark 3.19.3 (*Weak convergence in $L^\infty(\Omega)$*). (i) The sequence $\{f_n\}_{n=1}^\infty$ is *weakly convergent* to f_0 in $L^\infty(\Omega)$ if and only if

$$\int_\Omega f_n \, d\lambda \to \int_\Omega f_0 \, d\lambda$$

for every $\lambda \in L^\infty(\Omega)$ (see [66]).

(ii) If the sequence $\{f_n\}_{n=1}^\infty$ weakly converges to f_0 in $L^\infty(\Omega)$ then $f_n \to f_0$ in every $L^p(\Omega)$, $1 \leq p < \infty$. For the proof see [246].

Remark 3.19.4 (*Weak completeness of $L^p(\Omega)$*). (i) A sequence $\{x_n\}_{n=1}^\infty$ of elements of a Banach space X is said to be *weakly Cauchy* if for every $x^* \in X^*$ the sequence $\{x^*(x_n)\}_{n=1}^\infty$ is a Cauchy sequence of scalars. We say that a Banach space is *weakly complete* if every weakly Cauchy sequence is weakly convergent.

(ii) Every reflexive Banach space is weakly complete (see, e.g., [66]).

(iii) Thus the spaces $L^p(\Omega)$, $1 < p < \infty$, are weakly complete. Moreover, the space $L^1(\Omega)$, which is not reflexive, still is weakly complete (see again [66]).

Remark 3.19.5 (*The case* $\mu(\Omega) = \infty$). If the set Ω has infinite measure, some of the properties of $L^p(\Omega)$ with $\mu(\Omega) < \infty$ are preserved. For instance, the assertions concerning separability or reflexivity of these spaces still hold. However, for example, the inclusions

$$L^\infty(\Omega) \subset L^{p_1}(\Omega) \subset L^{p_2}(\Omega) \subset L^1(\Omega)$$

for $\infty > p_1 > p_2 \geq 1$ are no longer true when $\mu(\Omega) = \infty$ (notice that any constant nonzero function serves as a counterexample for the first inclusion). In fact, once $\mu(\Omega) = \infty$, then there is no inclusion between Lebesgue spaces $L^p(\Omega)$ and $L^r(\Omega)$ unless $p = r$.

Sufficient conditions for $f \in L^p(-\infty, \infty)$ to be an element of $L^q(-\infty, \infty)$, $q > p$, are given in [84, 224]. In the latter, the following result is proved:
Let $1 \leq p < q$. Let $f \in L^p(-\infty, \infty)$ be an even function. Suppose

$$\sum_{n=1}^{\infty} n^{\frac{q}{p-2}} \omega_p^q \left(\frac{1}{n}; f\right) < \infty,$$

where

$$\omega_p(\delta; f) = \sup_{0 < h \leq \delta} \left(\int |f(t+h) - f(t)|^p \, dt\right)^{\frac{1}{p}}.$$

Then $f \in L^q(-\infty, \infty)$.

Analogous conditions for $f \in L^p(0, 1)$ to be an element of $f \in L^q(0, 1)$, $q > p$, are given in [231]. A complete characterization of embeddings between Lebesgue spaces with respect to different measures was given by Kabaila [112].

Remark 3.19.6. *The spaces* $L^p(\Omega)$ *for* $0 < p < 1$. Let $p \in (0, 1)$. Then we can construct the space $L^p(\Omega)$ analogously as in Section 3.2 starting with $\mathcal{L}^p(\Omega)$ (see Remark 3.1.10). The space obtained is a linear topological space which is not locally convex (see [60, 121]). Moreover, the only continuous linear functional on the topological vector space $L^p(\Omega)$, $0 < p < 1$ is the zero functional.

Remark 3.19.7. *Uniformly convex Banach spaces.* We say that a Banach space is *uniformly convex* if for every $\varepsilon > 0$ there exists a $\delta = \delta(\varepsilon) > 0$ such that

$$\inf_{\substack{\|x\|=\|y\|=1 \\ \|x-y\| \geq \varepsilon}} \left(1 - \left\|\frac{1}{2}(x+y)\right\|\right) \geq \delta(\varepsilon).$$

It follows immediately from Clarkson inequalities (see Remark 3.1.10 (ii)) that the spaces $L^p(\Omega)$ ($1 < p < \infty$) are uniformly convex. We recall that every uniformly convex Banach space is reflexive (see [4, p. 286], [152, 177]; for a short proof using the so-called James characterization of reflexivity see, for example, [75, Theorem 9.12]).

In every uniformly convex Banach space X, the following assertion holds: Let $\{x_n\}_{n=1}^\infty$, $\{y_n\}_{n=1}^\infty$ be sequences in X such that $\|x_n\| = \|y_n\| = 1$, $\lim_{n\to\infty} \|x_n + y_n\| = 2$. Then
$$\lim_{n\to\infty} \|x_n - y_n\| = 0.$$

Spaces in which the above assertion holds are called *locally uniformly convex Banach spaces*. For such spaces, the implication

$$x_n \rightharpoonup x, \quad \|x_n\| \to \|x\| \Rightarrow x_n \to x \qquad (3.19.1)$$

holds (see [76]). Thus in $L^p(\Omega)$, $1 < p < \infty$, the implication (3.19.1) is true.

Remark 3.19.8. *Weighted Lebesgue spaces.* In approximation theory and also in the theory of partial differential equations, the spaces with weights are of interest.

Let ϱ be a *weight*, that is, a measurable almost everywhere positive function on Ω. Let $p \geq 1$. Then $L^p(\Omega, \varrho)$ denotes the set of all measurable functions f defined almost everywhere on Ω and such that

$$\mathcal{N}_{p,\varrho}(f) = \left(\int_\Omega \varrho(x) |f(x)|^p \, dx \right)^{\frac{1}{p}} < \infty.$$

Defining equality in $L^p(\Omega, \varrho)$ as equality almost everywhere we obtain analogously as in Section 3.2 that $L^p(\Omega, \varrho)$ is a vector space. $\mathcal{N}_{p,\varrho}(f)$ is then a norm on $L^p(\Omega, \varrho)$. The space $L^p(\Omega, \varrho)$ is complete for $p \in [1, \infty]$ and separable for $p \in [1, \infty)$.

The main tool for working with weighted spaces is the Hardy inequality (see, e.g., Theorem 3.11.8).

Remark 3.19.9. *Spaces with a mixed norm.* Consider a set of functions defined on $\Omega = \Omega_1 \times \cdots \times \Omega_n$, where the Ω_i are nonempty open subsets of $\mathbb{R}^{N_i}$ ($i = 1, \ldots, n$), which are elements of $L^{p_i}(\Omega_i)$ "with respect to the variable x_i". More precisely, let $\mathfrak{p} = (p_1, \ldots, p_n)$, $1 \leq p_i \leq \infty$, $i = 1, \ldots, n$. We denote by $L^{\mathfrak{p}}(\Omega)$ the set (with equality in the sense almost everywhere) of all measurable functions f on $\Omega \subset \mathbb{R}^{N_1 + \cdots + N_n}$ such that

$$\|f\|_{\mathfrak{p}} = \mathcal{N}_{p_n}(\mathcal{N}_{p_{n-1}}(\ldots(\mathcal{N}_{p_1}(f(x_1, \ldots, x_n)))\ldots))) < \infty. \qquad (3.19.2)$$

Formula (3.19.2) should be understood as follows: For fixed

$$(x_2, \ldots, x_n) \in \Omega_2 \times \cdots \times \Omega_n,$$

set

$$\mathcal{N}_{p_1}(f(\cdot, x_2, \ldots, x_n)) = \phi_1(x_2, \ldots, x_n).$$

Thus we obtain a function ϕ_1 on $\Omega_2 \times \cdots \times \Omega_n$ and set

$$\mathcal{N}_{p_2}(\phi_1(\cdot, x_3, \ldots, x_n)) = \phi_2(x_3, \ldots, x_n),$$

etc. In particular, if $p_i \in [1, \infty)$, $i = 1, \ldots, n$, then

$$\|f\|_{\mathbf{p}} = \left(\int_{\Omega_n} \cdots \left(\int_{\Omega_1} |f(x_1, \ldots, x_n)|^{p_1} \, dx_1 \right)^{\frac{p_2}{p_1}} \cdots dx_n \right)^{\frac{1}{p_n}}.$$

Given the vector $\mathbf{p} = (p_1, \ldots, p_n)$ let us define the conjugate vector $\mathbf{p}' = (p_1', \ldots, p_n')$. We immediately obtain an analogue of the Hölder inequality:

$$\left| \int_\Omega f(x) g(x) \, dx \right| \leq \|f\|_{\mathbf{p}} \cdot \|g\|_{\mathbf{p}'}$$

for all $f \in L^{\mathbf{p}}(\Omega)$, $g \in L^{\mathbf{p}'}(\Omega)$.

It is easy to see that $L^{\mathbf{p}}(\Omega)$ is a vector space and $\|f\|_{\mathbf{p}}$ forms the so-called mixed norm on $L^{\mathbf{p}}(\Omega)$. Moreover, this norm makes $L^{\mathbf{p}}(\Omega)$ a Banach space which is separable if the coordinates of the vector $\mathbf{p}$ are finite.

The Banach space $L^{\mathbf{p}}(\Omega)$ is reflexive if and only if $p_i \in (1, \infty)$ for every $i = 1, \ldots, n$.

The above assertions and, in fact, a discussion of the whole theory of the spaces $L^{\mathbf{p}}(\Omega)$ can be found in [11].

Chapter 4

Orlicz spaces

4.1 Introduction

Let us return for a moment to the Lebesgue spaces $L^p(\Omega)$, $p \geq 1$, which were studied in Chapter 3. Again, Ω is supposed throughout this chapter to be an open set in $\mathbb{R}^N$. A function u defined on an open set $\Omega \subset \mathbb{R}^N$ belongs to $L^p(\Omega)$ if

$$\int_\Omega |u(x)|^p \, dx < \infty \tag{4.1.1}$$

and the corresponding Lebesgue space $L^p(\Omega)$ is normed by

$$\|u\|_p := \left(\int_\Omega |u(x)|^p \, dx \right)^{\frac{1}{p}}. \tag{4.1.2}$$

If we define the function Φ for $t \in [0, \infty)$ by $\Phi(t) := t^p$, then we can rewrite (4.1.1) and (4.1.2) in the form

$$\int_\Omega \Phi(|u(x)|) \, dx < \infty, \tag{4.1.3}$$

and

$$\|u\|_p = \Phi^{-1} \left(\int_\Omega \Phi(|u(x)|) \, dx \right), \tag{4.1.4}$$

respectively, where Φ^{-1} is the inverse function of Φ, that is, $\Phi^{-1}(t) = t^{\frac{1}{p}}$.

Now, we can ask whether we can replace the power function $\Phi(t) = t^p$ by a more general function.

Definition 4.1.1. Let Φ be a nonnegative measurable function defined on $[0, \infty)$. We shall denote by

$$\mathcal{L}^\Phi(\Omega) \tag{4.1.5}$$

the set of all Lebesgue measurable functions u defined almost everywhere on Ω such that

$$\int_\Omega \Phi(|u(x)|) \, dx < \infty. \tag{4.1.6}$$

The set $\mathcal{L}^\Phi(\Omega)$ will be called an *Orlicz class* and we shall use the notation

$$\varrho(u, \Phi) := \int_\Omega \Phi(|u(x)|) \, dx. \tag{4.1.7}$$

Convention 4.1.2. In a way similar to that in the case of $L^p(\Omega)$, the equality relation in $\mathcal{L}^\Phi(\Omega)$ is introduced to be *equality almost everywhere* (see Notation 3.1.1) and the elements of $\mathcal{L}^\Phi(\Omega)$ are called "functions" in this text.

Examples 4.1.3. (i) The Lebesgue spaces $L^p(\Omega)$ with $p \geq 1$ are special cases of Orlicz classes $\mathcal{L}^\Phi(\Omega)$. Indeed, setting $\Phi(t) = ct^p$ with an arbitrary positive constant c, we obtain $L^p(\Omega) = \mathcal{L}^\Phi(\Omega)$.

(ii) A case that occurs very frequently is that of the function

$$\Phi(t) = \begin{cases} t \log^+ t, & \text{for } t \in (0, \infty), \\ 0 & \text{for } t = 0, \end{cases}$$

where $\log^+ t = \max\{0, \log t\}$. The corresponding Orlicz class $\mathcal{L}^\Phi(\Omega)$ is in some sense "close" to the Lebesgue space $L^1(\Omega)$ which, as we know from Chapter 3, is a rather exceptional case among Lebesgue spaces. In many tasks of analysis in which for some reason $L^1(\Omega)$ "does not work," the class $\mathcal{L}^\Phi(\Omega)$ with this special Φ serves as its certain appropriate "replacement".

(iii) If we set

$$\Phi(t) = |\sin t|, \qquad t \in [0, \infty),$$

then obviously *every* measurable function on Ω belongs to the class $\mathcal{L}^\Phi(\Omega)$ if $\mu(\Omega) < \infty$.

(iv) Let us consider $N = 1$, $\Omega = (0, 1)$ and $\Phi(t) = e^t$. Then the function $u(x) = -\frac{1}{2} \log x$ belongs to $\mathcal{L}^\Phi(\Omega)$ while the function $v(x) = -\log x = 2u(x)$ does not.

Remark 4.1.4. Example 4.1.3 (iv) shows that $\mathcal{L}^\Phi(\Omega)$ *need not be a linear set*. And, even in the cases when it happens to be a linear set, it is not clear whether and how it is possible to introduce a norm on it. The analogy with Lebesgue spaces might suggest a possibility of using formula (4.1.4), but our assumptions on Φ in Definition 4.1.1 do not imply the existence of the inverse function Φ^{-1}. Therefore it is necessary first of all to restrict the set of functions Φ in Definition 4.1.1.

4.2 Young function, Jensen inequality

Definition 4.2.1. We shall say that Φ is a *Young function* if there exists a function $\varphi : [0, \infty) \to [0, \infty)$ such that

$$\Phi(t) := \int_0^t \varphi(s)\, ds, \quad t \geq 0, \qquad (4.2.1)$$

and φ has the following properties:

(i) $\varphi(0) = 0$;

(ii) $\varphi(s) > 0$ for $s > 0$;

(iii) φ is right-continuous at any point $s \geq 0$;

(iv) φ is nondecreasing on $[0, \infty)$;

(v) $\lim_{s \to \infty} \varphi(s) = \infty$.

The properties of Young functions are given in the following lemma.

Lemma 4.2.2. *A Young function Φ is continuous, nonnegative, strictly increasing and convex on $[0, \infty)$. Moreover,*

$$\Phi(0) = 0, \quad \Phi(\infty) := \lim_{t \to \infty} \Phi(t) = \infty, \tag{4.2.2}$$

$$\lim_{t \to 0+} \frac{\Phi(t)}{t} = 0, \tag{4.2.3}$$

$$\lim_{t \to \infty} \frac{\Phi(t)}{t} = \infty, \tag{4.2.4}$$

$$\Phi(\alpha t) \leq \alpha \Phi(t) \text{ for every } \alpha \in [0, 1] \text{ and } t \geq 0, \tag{4.2.5}$$

$$\Phi(\beta t) \geq \beta \Phi(t) \text{ for every } \beta > 1 \text{ and } t \geq 0. \tag{4.2.6}$$

Proof. The continuity, nonnegativity, monotonicity of Φ and the properties in (4.2.2) are obvious. Regarding convexity, let us consider $\lambda \in (0, 1)$ and $0 \leq s \leq t$. We shall prove that

$$\Phi(\lambda s + (1 - \lambda)t) \leq \lambda \Phi(s) + (1 - \lambda)\Phi(t). \tag{4.2.7}$$

We have

$$\Phi(\lambda s + (1 - \lambda)t) = \int_0^{\lambda s + (1-\lambda)t} \varphi(r)\, dr$$

$$= \int_0^s \varphi(r)\, dr + \int_s^{\lambda s + (1-\lambda)t} \varphi(r)\, dr$$

$$= \lambda \int_0^s \varphi(r)\, dr + (1 - \lambda) \int_0^s \varphi(r)\, dr + \int_s^{\lambda s + (1-\lambda)t} \varphi(r)\, dr. \tag{4.2.8}$$

Since φ is nondecreasing and right-continuous, we have

$$\int_s^{\lambda s + (1-\lambda)t} \varphi(r)\, dr \leq (1 - \lambda)(t - s)\varphi(\lambda s + (1 - \lambda)t)$$

and

$$\int_{\lambda s + (1-\lambda)t}^t \varphi(r)\, dr \geq \lambda(t - s)\varphi(\lambda s + (1 - \lambda)t).$$

Section 4.2 Young function, Jensen inequality

The comparison of the foregoing two inequalities yields

$$\lambda \int_s^{\lambda s+(1-\lambda)t} \varphi(r)\,dr \le (1-\lambda)\int_{\lambda s+(1-\lambda)t}^t \varphi(r)\,dr,$$

that is

$$\int_s^{\lambda s+(1-\lambda)t} \varphi(r)\,dr = \lambda \int_s^{\lambda s+(1-\lambda)t} \varphi(r)\,dr + (1-\lambda)\int_s^{\lambda s+(1-\lambda)t} \varphi(r)\,dr$$

$$\le (1-\lambda)\int_s^t \varphi(r)\,dr.$$

Substituting into (4.2.8) we obtain

$$\Phi(\lambda s + (1-\lambda)t) \le \lambda \int_0^s \varphi(r)\,dr + (1-\lambda)\int_s^t \varphi(r)\,dr + (1-\lambda)\int_0^s \varphi(r)\,dr$$

$$= \lambda \int_0^s \varphi(r)\,dr + (1-\lambda)\int_0^t \varphi(r)\,dr$$

which is inequality (4.2.7).
Further,

$$\lim_{t\to 0+} \frac{\Phi(t)}{t} = \lim_{t\to 0+} \frac{1}{t}\int_0^t \varphi(s)\,ds = \varphi(0) = 0$$

by Definition 4.2.1 (i) and (iii). Thus we have shown (4.2.3).
From 4.2.1 (iv) we have for $t > 0$

$$\frac{\Phi(t)}{t} = \frac{1}{t}\int_0^t \varphi(s)\,ds \ge \frac{1}{t}\int_{\frac{t}{2}}^t \varphi(s)\,ds \ge \frac{1}{t}\frac{t}{2}\varphi\left(\frac{t}{2}\right) = \frac{1}{2}\varphi\left(\frac{t}{2}\right)$$

and hence property (4.2.4) follows by Definition 4.2.1 (iv).
Rewriting (4.2.7) in the form

$$\Phi((1-\alpha)s + \alpha t) \le (1-\alpha)\Phi(s) + \alpha\Phi(t), \quad 0 \le \alpha \le 1, \quad 0 \le s \le t,$$

and setting $s = 0$ we obtain property (4.2.5) from (4.2.2).
Let $\beta > 1$. Inequality (4.2.5) for $\alpha = \frac{1}{\beta}$ and $t = \beta s$ implies

$$\Phi\left(\frac{1}{\beta}\beta s\right) \le \frac{1}{\beta}\Phi(\beta s),$$

which is nothing else than (4.2.6). □

From Lemma 4.2.2 we can derive some properties of the Orlicz class $\mathcal{L}^\Phi(\Omega)$ provided Φ is a Young function.

Theorem 4.2.3. *Let Φ be a Young function. Then $\mathcal{L}^\Phi(\Omega)$ is a convex set and*

$$\mathcal{L}^\Phi(\Omega) \subset L^1(\Omega) \tag{4.2.9}$$

for $\mu(\Omega) < \infty$.

Proof. If we set $s = |u(x)|$ and $t = |v(x)|$ in (4.2.7) and integrate over Ω, we obtain the convexity of $\mathcal{L}^\Phi(\Omega)$ from the monotonicity and convexity of Φ.

To prove (4.2.9) let $u \in \mathcal{L}^\Phi(\Omega)$. According to (4.2.4), there exists a $k > 0$ such that for $|u(x)| > k$

$$\frac{\Phi(|u(x)|)}{|u(x)|} > 1, \quad \text{i.e.} \quad |u(x)| < \Phi(|u(x)|).$$

Denoting

$$\Omega_k = \{x \in \Omega; \ |u(x)| > k\},$$

we conclude that

$$\int_\Omega |u(x)| \, dx = \int_{\Omega_k} |u(x)| \, dx + \int_{\Omega \setminus \Omega_k} |u(x)| \, dx$$
$$\leq \int_{\Omega_k} \Phi(|u(x)|) \, dx + k\mu(\Omega \setminus \Omega_k)$$
$$\leq \varrho(u, \Phi) + k\mu(\Omega) < \infty,$$

hence $u \in L^1(\Omega)$. □

Remark 4.2.4. The inclusion (4.2.9) is strict because there exists a function u defined on Ω such that

$$u \in L^1(\Omega) \quad \text{but} \quad u \notin \mathcal{L}^\Phi(\Omega). \tag{4.2.10}$$

Such a function u can be constructed as follows: From (4.2.4) we have $t_n > 1, n \in \mathbb{N}$, such that

$$t_n^{-1} \Phi(t_n) \geq 2^n, \quad n \in \mathbb{N}.$$

We choose a sequence of disjoint subsets $\Omega_n \subset \Omega, n \in \mathbb{N}$, such that

$$\mu(\Omega_n) = \frac{1}{t_n 2^n} \mu(\Omega), \quad n \in \mathbb{N}.$$

Then we have that

$$\sum_{n=1}^\infty \mu(\Omega_n) = \mu(\Omega) \sum_{n=1}^\infty \frac{1}{t_n 2^n} < \mu(\Omega).$$

Now let us define

$$u(x) = \begin{cases} t_n & \text{for } x \in \Omega_n, \ n \in \mathbb{N}, \\ 0 & \text{otherwise in } \Omega. \end{cases}$$

Section 4.2 Young function, Jensen inequality

Then

$$\int_\Omega |u(x)|\, dx = \sum_{n=1}^\infty t_n \mu(\Omega_n) = \sum_{n=1}^\infty t_n \frac{1}{t_n 2^n} \mu(\Omega) = \mu(\Omega) < \infty,$$

but

$$\int_\Omega \Phi(|u(x)|)\, dx = \sum_{n=1}^\infty \Phi(t_n) \mu(\Omega_n) \geq \sum_{n=1}^\infty t_n 2^n \frac{1}{t_n 2^n} \mu(\Omega) = \infty,$$

so that u satisfies (4.2.10).

The Lebesgue space $L^1(\Omega)$ can be considered as the union of all Orlicz classes $\mathcal{L}^\Phi(\Omega)$ where Φ varies over all Young functions.

Theorem 4.2.5. *Let $\mu(\Omega) < \infty$ and let $u \in L^1(\Omega)$. Then there exists a Young function Φ such that $u \in \mathcal{L}^\Phi(\Omega)$.*

Proof. We denote

$$\Omega_n = \{x \in \Omega;\ n-1 \leq |u(x)| < n\}, \quad n \in \mathbb{N}.$$

Then

$$\int_\Omega |u(x)|\, dx = \sum_{n=1}^\infty \int_{\Omega_n} |u(x)|\, dx$$

$$\geq \sum_{n=1}^\infty (n-1)\mu(\Omega_n)$$

$$= \sum_{n=1}^\infty n\mu(\Omega_n) - \mu(\Omega).$$

The series $\sum_{n=1}^\infty n\mu(\Omega_n)$ converges, because $u \in L^1(\Omega)$ and $\mu(\Omega) < \infty$. Moreover, there exists a nondecreasing sequence of numbers α_n such that

$$\alpha_n > 1, \quad \lim_{n\to\infty} \alpha_n = \infty \quad \text{and} \quad \sum_{n=1}^\infty \alpha_n n \mu(\Omega_n) < \infty.$$

If we define

$$\varphi(t) = \begin{cases} t & \text{for } t \in [0,1), \\ \alpha_n & \text{for } t \in [n, n+1),\ n \in \mathbb{N}, \end{cases}$$

then

$$\Phi(t) = \int_0^t \varphi(s)\, ds$$

is a Young function such that $\Phi(n) \leq n\alpha_n$ for $n \in \mathbb{N}$. Hence, we have

$$\int_\Omega \Phi(|u(x)|) \, dx = \sum_{n=1}^\infty \int_{\Omega_n} \Phi(|u(x)|) \, dx \leq \sum_{n=1}^\infty \Phi(n) \mu(\Omega_n)$$

$$\leq \sum_{n=1}^\infty n\alpha_n \mu(\Omega_n) < \infty,$$

i.e. $u \in \mathcal{L}^\Phi(\Omega)$. □

Examples 4.2.6. (i) For $\varphi(t) = t^{p-1}$ with $p > 1$, we obtain the Young function $\Phi(t) = p^{-1} t^p$.

(ii) For $\varphi(t) = e^t - 1$, we obtain the Young function $\Phi(t) = e^t - t - 1$.

(iii) For $\alpha > 0$ and $\varphi(t) = \alpha t^{\alpha-1} e^{t^\alpha}$, we obtain the Young function $\Phi(t) = e^{t^\alpha} - 1$.

(iv) The function $\Phi(t) = t$, corresponding to the Lebesgue space $L^1(\Omega)$, is not a Young function as it satisfies neither (4.2.3) nor (4.2.4).

(v) The functions Φ in Examples 4.1.3 (ii) and (iii) are not Young functions.

Remark 4.2.7. It is shown in [123] that if Φ is continuous, increasing and convex in $[0, \infty)$ and if (4.2.3) and (4.2.4) hold, then Φ is a Young function.

Remark 4.2.8. In some literature, the definition of a Young function is relaxed in the sense that assumptions (4.2.3) and (4.2.4) are not required. Sometimes the notion of an *N-function* for such functions is used.

Remark 4.2.9. We can relax the definition of Young function by replacing conditions (ii), (iii) in Definition 4.2.1 by the following ones:

(ii)* $\varphi(s) \geq 0$ for $s > 0$;

(iii)* $\varphi(s)$ is left-continuous at any point $s > 0$.

See, for example [141, 241]. In general, after this modification all results concerning Orlicz classes remain valid.

Remark 4.2.10. Let us mention that the function in Example 4.1.3 (ii) is not a Young function, but it is a relaxed Young function in the sense of the preceding remark.

Theorem 4.2.11 (Jensen inequality). *Let Φ be a convex function on $\mathbb{R}$.*

(i) *Let $t_1, \ldots, t_n \in \mathbb{R}$ and let $\alpha_1, \ldots, \alpha_n$ be positive numbers. Then*

$$\Phi\left(\frac{\alpha_1 t_1 + \cdots + \alpha_n t_n}{\alpha_1 + \cdots + \alpha_n}\right) \leq \frac{\alpha_1 \Phi(t_1) + \cdots + \alpha_n \Phi(t_n)}{\alpha_1 + \alpha_2 + \cdots + \alpha_n}. \quad (4.2.11)$$

(ii) *Let $\alpha = \alpha(x)$ be defined and positive almost everywhere on Ω.*

Then,
$$\Phi\left(\frac{\int_\Omega u(x)\alpha(x)\,dx}{\int_\Omega \alpha(x)\,dx}\right) \leq \frac{\int_\Omega \Phi(u(x))\alpha(x)\,dx}{\int_\Omega \alpha(x)\,dx} \qquad (4.2.12)$$

for every nonnegative function u provided all the integrals in (4.2.12) are meaningful.

Proof. We shall give only a sketch.

(i) For $n = 2$ inequality (4.2.11) is just the definition of the convexity of Φ (see also (4.2.7)). For $n > 2$ we obtain inequality (4.2.11) by induction.

(ii) Let $\gamma > 0$ be fixed. From the convexity of Φ it follows that there exists a $k \in \mathbb{R}$ such that
$$\Phi(t) - \Phi(\gamma) \geq k(t - \gamma) \qquad \text{for all} \quad t \geq 0.$$

Setting $t = u(x)$ and multiplying the resultant inequality by $\alpha(x)$ we obtain after integration over Ω that

$$\int_\Omega \Phi(u(x))\alpha(x)\,dx - \Phi(\gamma)\int_\Omega \alpha(x)\,dx$$
$$\geq k\left\{\int_\Omega u(x)\alpha(x)\,dx - \gamma \int_\Omega \alpha(x)\,dx\right\}.$$

Inequality (4.2.12) now follows by setting

$$\gamma = \frac{\int_\Omega u(x)\alpha(x)\,dx}{\int_\Omega \alpha(x)\,dx}. \qquad (4.2.13)$$
□

4.3 Complementary functions

Definition 4.3.1. Let Φ be a Young function generated by a function φ, that is,

$$\Phi(t) = \int_0^t \varphi(s)\,ds, \quad t \in [0, \infty). \qquad (4.3.1)$$

We set
$$\psi(t) = \sup_{\varphi(s) \leq t} s \qquad (4.3.2)$$

and
$$\Psi(t) = \int_0^t \psi(s)\,ds, \quad t \in [0, \infty). \qquad (4.3.3)$$

The function Ψ is called the *complementary function* to Φ.

Exercises 4.3.2. (i) Prove that Ψ is also a Young function (i.e. prove that ψ satisfies the corresponding conditions of Definition 4.2.1).

(ii) Prove that if ψ is given by (4.3.2), then

$$\varphi(t) = \sup_{\psi(s) \leq t} s, \quad t \geq 0. \tag{4.3.4}$$

Remark 4.3.3. If φ is continuous and strictly increasing in $[0, \infty)$ then ψ is the inverse function φ^{-1} and vice versa. In the general case, φ and ψ are mutually "inverse" in a generalized sense. Consequently, if Ψ is complementary to Φ then Φ is complementary to Ψ. We can also call Φ, Ψ a *pair of complementary Young functions*.

We shall now introduce the Young inequality. Note that a special form of it appeared in Theorem 3.1.4.

Theorem 4.3.4 (Young inequality). *Let Φ, Ψ be a pair of complementary Young functions. Then for all $a, b \in [0, \infty)$ we have*

$$ab \leq \Phi(a) + \Psi(b). \tag{4.3.5}$$

Equality holds in (4.3.5) if and only if

$$b = \varphi(a) \quad \text{or } a = \psi(b). \tag{4.3.6}$$

For the idea how to prove (4.3.5) see Theorem 3.1.4. A detailed proof of a more general assertion is given in [241].

Examples 4.3.5. (i) Let $p \in (1, \infty)$ and let

$$\Phi(t) := \frac{t^p}{p}, \quad t \in [0, \infty).$$

Then the complementary function Ψ of Φ satisfies

$$\Psi(t) = \frac{t^{p'}}{p'}, \quad t \in [0, \infty),$$

where p' is the Lebesgue conjugate index (see Definition 3.1.3).

(ii) Let $\alpha > 1$. Then the function

$$\Phi(t) := e^{t^\alpha} - 1, \quad t \in [0, \infty),$$

is a Young function generated by

$$\varphi(s) = \alpha s^{\alpha-1} e^{s^\alpha}, \quad s \in [0, \infty).$$

We are unable to give an analytical expression for either its complementary function Ψ or the function ψ (see (4.3.2)). It can be however shown that $\Psi(t)$ is asymptotically (for $t \to \infty$) equivalent to

$$\widetilde{\Psi}(t) = t\sqrt{\log t}$$

(for details see Section 4.5).

(iii) Let

$$\Phi(t) := e^t - t - 1, \quad t \in [0, \infty).$$

Then Φ is a Young function and its complementary function Ψ satisfies

$$\Psi(t) = (1+t)\log(1+t) - t, \quad t \in [0, \infty)$$

(see Example 4.2.6 (ii)).

Example 4.3.6. Let

$$\Phi(t) := \begin{cases} t \log^+ t & \text{for } t \in (0, \infty), \\ 0 & \text{for } t = 0. \end{cases} \quad (4.3.7)$$

Then Φ is a relaxed Young function and φ (see (4.3.1)) is defined by

$$\varphi(t) = \begin{cases} 0 & \text{if } 0 < t < 1, \\ \log t + 1 & \text{if } t \geq 1. \end{cases}$$

Set

$$\psi(t) := \begin{cases} 0 & \text{if } t = 0, \\ 1 & \text{if } 0 < t < 1, \\ e^{t-1} & \text{if } t \geq 1, \end{cases}$$

and

$$\Psi(t) := \int_0^t \psi(s)\,ds = \begin{cases} t & \text{if } 0 < t < 1, \\ e^{t-1} & \text{if } t \geq 1. \end{cases}$$

Then Φ, Ψ form a pair of complementary relaxed Young functions (in the sense of Remark 4.2.8). Let us note that Ψ does not satisfy condition (4.2.3).

Exercise 4.3.7. Let Φ, Ψ be a pair of complementary Young functions, let a, b be positive constants. Show that for

$$\Phi_1(t) = a\Phi(bt), \quad t \in [0, \infty),$$

the complementary function is given by

$$\Psi_1(t) = a\Psi\left(\frac{t}{ab}\right), \quad t \in [0, \infty).$$

Let us recall that the Hölder inequality for Lebesgue spaces (Theorem 3.1.6) states that if $u \in L^p(\Omega)$ and $v \in L^{p'}(\Omega)$, then $uv \in L^1(\Omega)$. The following theorem represents an analogous result for general Orlicz classes.

Theorem 4.3.8. *Let Φ, Ψ be a pair of complementary Young functions, $u \in \mathcal{L}^\Phi(\Omega)$, $v \in \mathcal{L}^\Psi(\Omega)$. Then*

$$\int_\Omega |u(x)v(x)| \, dx \le \varrho(u, \Phi) + \varrho(v, \Psi) \tag{4.3.8}$$

and consequently

$$uv \in L^1(\Omega).$$

Proof. The assertion follows by integrating inequality (4.3.5), applied to $a = |u(x)|$ and $b = |v(x)|$, over Ω. □

Remark 4.3.9. It follows from (4.3.6) that equality occurs in (4.3.8) if

$$|v(x)| = \varphi(|u(x)|) \quad \text{or} \quad |u(x)| = \psi(|v(x)|)$$

for almost all $x \in \Omega$.

Exercise 4.3.10. Let Φ, Ψ and Φ_1, Ψ_1 be two pairs of complementary Young functions. Assume that there exists a $T \in [0, \infty)$ such that

$$\Phi(t) \le \Phi_1(t) \quad \text{for} \quad t \ge T. \tag{4.3.9}$$

Show that then there exists a $T_0 \ge 0$ such that

$$\Psi(t) \ge \Psi_1(t) \quad \text{for} \quad t \ge T_0. \tag{4.3.10}$$

(Hint: Use the fact that (4.3.5) and (4.3.6) imply

$$\psi_1(t)t = \Phi_1(\psi_1(t)) + \Psi_1(t)$$

and use the Young inequality for Φ, Ψ with $a = \psi_1(t), b = t$.)

Exercise 4.3.11. Show that for two complementary Young functions Φ, Ψ

$$t \le \Phi^{-1}(t)\Psi^{-1}(t) \le 2t \quad \text{for} \quad t \ge 0.$$

(Hint: The latter inequality follows from the Young inequality; for the former, use the fact that $\Psi\left(\frac{\Phi(t)}{t}\right) \le \Phi(t)$ (sketch a figure).)

4.4 The Δ_2-condition

In this section we shall investigate a special class of Young functions which will be important later.

Definition 4.4.1. A Young function Φ is said to satisfy the Δ_2-*condition* (we shall write $\Phi \in \Delta_2$) if there exist constants $k > 0$ and $T \geq 0$ such that

$$\Phi(2t) \leq k\Phi(t) \quad \text{for all} \quad t \geq T. \tag{4.4.1}$$

Example 4.4.2. The function $\Phi(t) = ct^p$ with $c > 0$, $p > 1$ satisfies the Δ_2-condition: we can set $T = 0$ and $k = 2^p$.

Exercise 4.4.3. Show that we cannot have $k \leq 2$ for $\Phi \in \Delta_2$.
 (Hint: Use (4.2.6) with $\beta = 2$.)

The following assertion is a useful criterion for a function Φ to satisfy the Δ_2-condition.

Theorem 4.4.4. *Let Φ be a Young function and let $\Phi(t) = \int_0^t \varphi(s)\,ds$, $t \in (0, \infty)$. Then Φ satisfies the Δ_2-condition if and only if*

$$\limsup_{t \to \infty} \frac{t\varphi(t)}{\Phi(t)} < \infty. \tag{4.4.2}$$

Proof. Suppose first that (4.4.2) holds. Then there exist $T_0 > 0$ and $c > 0$ such that, for all $t \geq T_0$,

$$\frac{t\varphi(t)}{\Phi(t)} \leq c, \tag{4.4.3}$$

i.e.

$$\frac{c}{t} \geq \frac{\varphi(t)}{\Phi(t)} = \frac{\Phi'(t)}{\Phi(t)} = (\log \Phi(t))',$$

and, consequently,

$$\log \frac{\Phi(2t)}{\Phi(t)} = \int_t^{2t} \frac{\varphi(s)}{\Phi(s)}\,ds \leq c \int_t^{2t} \frac{ds}{s} = c \log 2, \tag{4.4.4}$$

i.e. $\Phi(2t) \leq 2^c \Phi(t)$ for $t \geq T_0$. Thus $\Phi \in \Delta_2$ with $k = 2^c$.

Conversely, let $\Phi \in \Delta_2$ and let k, T_0 be the constants from Definition 4.4.1. Then we have

$$k\Phi(t) \geq \Phi(2t) = \int_0^{2t} \varphi(s)\,ds > \int_t^{2t} \varphi(s)\,ds \geq t\varphi(t)$$

for $t \geq T$ which implies (4.4.3) (with $c = k$) and also (4.4.2). □

Example 4.4.5. It follows from Theorem 4.4.4 that $\Phi \in \Delta_2$ for

$$\Phi(t) = (1+t)\log(1+t) - t$$

(see Example 4.3.5 (iii)) and $\Phi \notin \Delta_2$ for

$$\Phi(t) = e^t - t - 1 \quad \text{and} \quad \Phi(t) = e^{t^\alpha} - 1$$

(see Example 4.3.5 (ii) and (iii)).

Remark 4.4.6. Let $\Phi \in \Delta_2$. Theorem 4.4.4 implies (see (4.4.3)) that there exists a $T \geq 0$ such that for every $t \geq T$ we have

$$\frac{t\Phi'(t)}{\Phi(t)} \leq c, \quad \text{i.e.} \quad (\log \Phi(t))' \leq (\log t^c)'. \tag{4.4.5}$$

Hence, there exist $c_0 > 0$ and $T_0 > 0$ such that

$$\Phi(t) \leq c_0 t^c \quad \text{for } t \geq T_0.$$

Moreover, by Exercise 4.4.3, we necessarily have $c > 1$ in (4.4.5).

Remark 4.4.7. Remark 4.4.6 shows that in some sense the growth of the function $\Phi \in \Delta_2$ is dominated, for large t, by that of a power function $c_0 t^p$ with $p > 1$. Note that this is consistent with the fact that the Young functions $e^t - t - 1$ and $e^{t^\beta} - 1$, $\beta > 1$, do not belong to Δ_2.

We shall now present a necessary and sufficient condition for a Young function Φ to satisfy the Δ_2-condition expressed in terms of properties of the complementary function Ψ of Φ.

Theorem 4.4.8. *A Young function Φ satisfies the Δ_2-condition if and only if there exist $k_0 > 1$ and $T_0 > 0$ such that*

$$\Psi(t) \leq \frac{1}{2k_0}\Psi(k_0 t) \quad \text{for } t \geq T_0, \tag{4.4.6}$$

where Ψ is the complementary Young function to Φ.

Proof. Let (4.4.6) hold and denote

$$\Psi_1(t) := \frac{1}{2k_0}\Psi(k_0 t), \quad t \in [0, \infty).$$

Then Ψ_1 is a Young function whose complementary function, Φ_1, satisfies, in view of Exercise 4.3.7,

$$\Phi_1(t) = \frac{1}{2k_0}\Phi(2t) \quad t \in [0, \infty).$$

Section 4.4 The Δ_2-condition

We rewrite (4.4.6) in the form

$$\Psi(t) \leq \Psi_1(t), \quad t \in [T_0, \infty),$$

and use Exercise 4.3.10 to show that, for some $T_1 > 0$,

$$\Phi(t) \geq \Phi_1(t) \quad t \geq T_1.$$

This, however, is (4.4.1) with $k = 2k_0$ and $T_0 = T_1$.
The reverse implication can be proved analogously. □

Examples 4.4.9. (i) We can use Theorem 4.4.8 to obtain an alternative proof of the fact that the complementary function Ψ to the Young function Φ from (4.3.7) does not satisfy the Δ_2-condition without even knowing the explicit formula for Ψ. Indeed, (4.4.6) would require to find a $k_0 > 1$ such that

$$2 \log t \leq \log(k_0 t) \quad \text{for large values of } t,$$

which is impossible.

(ii) By means of Theorem 4.4.8 we can also show that the function Ψ complementary to the Young function $\Phi(t) = e^{t^\beta} - 1$, where $\beta > 1$, satisfies the Δ_2-condition, although we do not know the explicit form of Ψ (see Example 4.3.5 (iii)). Indeed, it suffices to find a $k_0 > 1$ and a $T_0 \in [0, \infty)$ so that

$$e^{t^\beta} - 1 \leq \frac{1}{2k_0} \left(e^{(k_0 t)^\beta} - 1 \right), \quad t \geq T_0.$$

In fact, an appropriate T_0 can be found for *any* $k_0 > 1$.

All examples of Young functions we have seen thus far had the property that at least one of the functions from the pair (Φ, Ψ), where Ψ is complementary to Φ, satisfies the Δ_2-condition. This is however not true in general, as the following example shows.

Example 4.4.10. Set

$$\varphi(t) = \begin{cases} t & \text{for } t \in [0, 1), \\ n! & \text{for } t \in [(n-1)!, n!), \ n = 2, 3, \ldots, \end{cases}$$

and

$$\Phi(t) = \int_0^t \varphi(s) \, ds, \quad t \in [0, \infty).$$

Then

$$\psi(s) = \begin{cases} s & \text{for } s \in [0, 1), \\ (n-1)! & \text{for } s \in [(n-1)!, n!), \ n = 2, 3, \ldots \end{cases}$$

(see the notation introduced in Definition 4.3.1). For the corresponding pair Φ, Ψ of complementary functions we have

$$\Phi \notin \Delta_2, \qquad \Psi \notin \Delta_2,$$

since

$$\Phi(2n!) > n\Phi(n!) \quad \text{and} \quad \Psi(2n!) > n\Psi(n!), \qquad n \in \mathbb{N}.$$

Remark 4.4.11. It was noted in Remark 4.4.7 that every Young function which satisfies the condition Δ_2 is in certain sense dominated by some power function. Example 4.4.10 shows that a converse implication is not valid, since the function Φ from Example 4.4.10 does not satisfy $\Phi \in \Delta_2$ but it is easy to verify that

$$\Phi(t) \leq t^2, \quad t \in [0, \infty).$$

4.5 Comparison of Orlicz classes

Remark 4.5.1. Let Φ_1, Φ_2 be two Young functions and let us suppose that

$$\Phi_1(t) \leq \Phi_2(t) \qquad \text{for all} \quad t \geq 0. \tag{4.5.1}$$

Then obviously

$$\mathcal{L}^{\Phi_2}(\Omega) \subset \mathcal{L}^{\Phi_1}(\Omega). \tag{4.5.2}$$

If we suppose in addition that $\mu(\Omega) < \infty$, then in order to guarantee (4.5.2) it suffices that (4.5.1) holds only for $t \geq T$ with some $T > 0$. Indeed, for $u \in \mathcal{L}^{\Phi_2}(\Omega)$ we can set

$$\Omega_1 = \{x \in \Omega; \ |u(x)| \leq T\},$$

and so

$$\int_\Omega \Phi_1(|u(x)|) \, dx = \int_{\Omega_1} \Phi_1(|u(x)|) \, dx + \int_{\Omega \setminus \Omega_1} \Phi_1(|u(x)|) \, dx$$

$$\leq \Phi_1(T)\mu(\Omega_1) + \int_{\Omega \setminus \Omega_1} \Phi_2(|u(x)|) \, dx$$

$$\leq \Phi_1(T)\mu(\Omega_1) + \varrho(u, \Phi_2) < \infty. \tag{4.5.3}$$

We shall now show that it is also possible to derive from (4.5.2) a certain inequality of the type (4.5.1).

Theorem 4.5.2. *Let $\mu(\Omega) < \infty$ and let Φ_1, Φ_2 be two Young functions. Then the inclusion (4.5.2) is true if and only if there exist $c > 0$ and $T > 0$ such that*

$$\Phi_1(t) \leq c\Phi_2(t) \qquad \text{for all } t \geq T. \tag{4.5.4}$$

Proof. The sufficiency of (4.5.4) follows by the same argument as above, only instead of (4.5.3) we obtain the estimate

$$\varrho(u, \Phi_1) \leq \Phi_1(T)\mu(\Omega_1) + c\varrho(u, \Phi_2). \qquad (4.5.5)$$

To prove that (4.5.4) is also a necessary condition for (4.5.2), let us suppose that (4.5.4) does not hold. Then there exists a strictly increasing sequence $\{t_n\}_{n=1}^\infty$ of positive real numbers such that $\lim_{n\to\infty} t_n = \infty$ and

$$\Phi_1(t_n) > 2^n \Phi_2(t_n), \quad n \in \mathbb{N}.$$

We can find in Ω a sequence of pairwise disjoint subsets $\{\Omega_n\}_{n=1}^\infty$ such that

$$\mu(\Omega_n) = \frac{\Phi_2(t_1)\mu(\Omega)}{2^n \Phi_2(t_n)}, \quad n \in \mathbb{N},$$

and set

$$u(x) = \begin{cases} t_n & \text{for } x \in \Omega_n, \\ 0 & \text{otherwise in } \Omega. \end{cases}$$

Then

$$\int_\Omega \Phi_2(|u(x)|)\,dx = \sum_{n=1}^\infty \int_{\Omega_n} \Phi_2(|u(x)|)\,dx = \sum_{n=1}^\infty \Phi_2(t_n)\mu(\Omega_n)$$

$$= \sum_{n=1}^\infty \frac{1}{2^n} \Phi_2(t_1)\mu(\Omega) < \infty,$$

so that $u \in \mathcal{L}^{\Phi_2}(\Omega)$. On the other hand, $u \notin \mathcal{L}^{\Phi_1}(\Omega)$, since

$$\int_\Omega \Phi_1(|u(x)|)\,dx = \sum_{n=1}^\infty \int_{\Omega_n} \Phi_1(|u(x)|)\,dx = \sum_{n=1}^\infty \Phi_1(t_n)\mu(\Omega_n)$$

$$\geq \sum_{n=1}^\infty \Phi_2(t_1)\mu(\Omega) = \infty.$$

Consequently, (4.5.2) does not hold. □

By virtue of Theorem 4.5.2, we can show that if $\Phi \in \Delta_2$ then the Orlicz class $\mathcal{L}^\Phi(\Omega)$ is a linear set.

Theorem 4.5.3. (i) *Let $\mu(\Omega) < \infty$. Then $\mathcal{L}^\Phi(\Omega)$ is a linear set if and only if $\Phi \in \Delta_2$.*

(ii) *Let $\mu(\Omega) = \infty$ and let $\Phi \in \Delta_2$ with $T = 0$. Then $\mathcal{L}^\Phi(\Omega)$ is a linear set.*

Proof. (i) Let $\Phi \in \Delta_2$ and let $\gamma \geq 0$. Then there exists an $n \in \mathbb{N}$ such that $\gamma \leq 2^n$. The monotonicity of Φ together with the repeated use of the Δ_2-condition imply

$$\Phi(\gamma t) \leq \Phi(2^n t) \leq k^n \Phi(t) \qquad \text{for } t \geq T.$$

It follows from these inequalities that if $u \in \mathcal{L}^\Phi(\Omega)$ then also $cu \in \mathcal{L}^\Phi(\Omega)$ for arbitrary $c \in \mathbb{R}$: For $\mu(\Omega) < \infty$ we use Theorem 4.5.2 with $\Phi_1(t) = \Phi(\gamma t)$, $\Phi_2(t) = \Phi(t)$ and $c = k^n$, for $\mu(\Omega) = \infty$ we use the same argument and the fact that $T = 0$ so that $\Phi_1(T) = \Phi(\gamma T) = \Phi(0) = 0$.

Let $u, v \in \mathcal{L}^\Phi(\Omega)$. The same reasoning as before yields that $2u$ and $2v$ also belong to $\mathcal{L}^\Phi(\Omega)$. Using the convexity of Φ (namely formula (4.2.7) with $\lambda = \frac{1}{2}$, $s = 2|u(x)|$ and $t = 2|v(x)|$) and the monotonicity of Φ we obtain after integrating over Ω

$$\int_\Omega \Phi(|u(x) + v(x)|) \, dx \leq \frac{1}{2} \int_\Omega \Phi(|2u(x)|) \, dx + \frac{1}{2} \int_\Omega \Phi(|2v(x)|) \, dx < \infty.$$

Consequently, $u + v \in \mathcal{L}^\Phi(\Omega)$ and hence $\mathcal{L}^\Phi(\Omega)$ is a linear set.

(ii) Let $\mathcal{L}^\Phi(\Omega)$ be a linear set and let $\mu(\Omega) < \infty$. Then for every $u \in \mathcal{L}^\Phi(\Omega)$ we also have $2u \in \mathcal{L}^\Phi(\Omega)$ which means that

$$\mathcal{L}^\Phi(\Omega) \subset \mathcal{L}^{\Phi_1}(\Omega) \quad \text{with} \quad \Phi_1(t) = \Phi(2t).$$

From Theorem 4.5.2 it follows that there exist constants $T > 0$ and $c > 0$ such that

$$\Phi(2t) = \Phi_1(t) \leq c\Phi(t) \quad \text{for } t \geq T,$$

i.e. $\Phi \in \Delta_2$. $\square$

Remark 4.5.4. Let us suppose that $\mu(\Omega) < \infty$ and that the Δ_2-condition is *not* satisfied for Φ. It follows from Theorem 4.5.3 that there exists a $u \in \mathcal{L}^\Phi(\Omega)$ such that a part of the ray

$$\{\gamma u; \gamma \in [0, \infty)\}$$

lies *outside* of $\mathcal{L}^\Phi(\Omega)$. Let us denote

$$\gamma_u = \sup\{\gamma; \gamma > 0, \gamma u \in \mathcal{L}^\Phi(\Omega)\}.$$

We have $\gamma_u \geq 1$ because $\mathcal{L}^\Phi(\Omega)$ is convex (see Theorem 4.2.3); consequently, all functions γu with $0 \leq \gamma \leq 1$ belong to $\mathcal{L}^\Phi(\Omega)$. We have that

$$\gamma u \in \mathcal{L}^\Phi(\Omega) \quad \text{for} \quad \gamma < \gamma_u, \qquad \gamma u \notin \mathcal{L}^\Phi(\Omega) \quad \text{for } \gamma > \gamma_u.$$

Exercise 4.5.5. Let $\alpha > 1$. Show that $\Phi \in \Delta_2$ if and only if there exist $k(\alpha) > 0$ and $T(\alpha) > 0$ such that
$$\Phi(\alpha t) \leq k(\alpha)\Phi(t) \quad \text{for } t \geq T(\alpha). \tag{4.5.6}$$
(Hint: Proceed in a way analogous to that in Theorem 4.5.2.)

We shall now turn our attention to the study of comparison relations between Young functions.

Definition 4.5.6. Let Φ_1, Φ_2 be two Young functions. If there exist two positive constants c and T such that
$$\Phi_1(t) \leq \Phi_2(ct) \quad \text{for } t \geq T, \tag{4.5.7}$$
we say that Φ_2 *dominates* Φ_1 *near* ∞ and write
$$\Phi_1 \prec \Phi_2. \tag{4.5.8}$$
If
$$\Phi_1 \prec \Phi_2 \quad \text{and} \quad \Phi_2 \prec \Phi_1$$
hold simultaneously, we say that Φ_1 and Φ_2 are *equivalent* and write $\Phi_1 \approx \Phi_2$.

Examples 4.5.7. (i) Let $1 < p \leq q < \infty$ and let $\Phi_1(t) := t^p$ and $\Phi_2(t) := t^q$. Then $\Phi_1 \prec \Phi_2$. The evaluation of c and T is left to the reader.

(ii) Let $1 < p < \infty$ and $\varepsilon > 0$. Let $\Phi_1(t) := t^p(|\log t| + 1)$ and $\Phi_2(t) := t^{p+\varepsilon}$. Then $\Phi_1 \prec \Phi_2$.

(iii) For every Young function Φ, the functions $\Phi(t)$ and $\Phi_1(t) = \Phi(kt)$ with $k > 0$ are equivalent.

Exercise 4.5.8. Show that if $\Phi_1 \prec \Phi_2$, then their complementary functions Ψ_1, Ψ_2 satisfy
$$\Psi_2 \prec \Psi_1.$$
(Hint: Use Exercise 4.3.7 and Exercise 4.3.10.)

Remarks 4.5.9. (i) A comparison of relations (4.5.7) and (4.5.4) shows that, in general, the ordering $\Phi_1 \prec \Phi_2$ does not provide any information about the inclusion between $\mathcal{L}^{\Phi_1}(\Omega)$ and $\mathcal{L}^{\Phi_2}(\Omega)$. Nonetheless, as will be seen in Section 4.17, the ordering can also be useful when comparison of Orlicz spaces is discussed.

(ii) Theorem 4.5.2 suggests that if Ω has finite measure, then the character of a particular Young function Φ is "determined" by its behavior for large values of the argument t.

For the purpose of a comparison of Young functions, we need also an essentially stronger relation than the "dominance" relation $\prec$.

Definition 4.5.10. Let Φ_1, Φ_2 be two Young functions. If

$$\lim_{t \to \infty} \frac{\Phi_1(t)}{\Phi_2(\lambda t)} = 0$$

for all $\lambda > 0$, we say that Φ_2 *grows essentially faster near* ∞ than Φ_1 and write

$$\Phi_1 \lll \Phi_2.$$

Examples 4.5.11. (i) Let $1 < p \le q < \infty$ and let $\Phi_1(t) := t^p$ and $\Phi_2(t) := t^q$. Then $\Phi_1 \lll \Phi_2$.

(ii) Let $1 < p < \infty$ and $\varepsilon > 0$. Then

$$t^p \lll t^p(|\log t| + 1) \lll t^{p+\varepsilon}.$$

(iii) Let Φ_1, Φ be two Young functions and set $\Phi_2(t) := \Phi_1(\Phi(t))$. Then

$$\Phi_1 \lll \Phi_2.$$

Remark 4.5.12. If $\Phi_1 \lll \Phi_2$ then obviously $\Phi_1 \prec \Phi_2$. The converse does not hold: If we take $\Phi_1(t) = \Phi(\lambda t)$ with $\lambda > 0$ then $\Phi_1 \prec \Phi$ but it is not true that $\Phi_1 \lll \Phi$.

Exercise 4.5.13. Show that if $\Phi_1 \lll \Phi_2$ then $\Psi_2 \lll \Psi_1$ for the corresponding complementary functions Ψ_1, Ψ_2.
(Hint: Use Exercises 4.3.10 and 4.3.7.)

4.6 Orlicz spaces

Definition 4.6.1. Let Φ be a Young function and Ψ its complementary Young function. Let u be a measurable function defined almost everywhere on $\Omega \subset \mathbb{R}^N$. The number

$$\|u\|_\Phi := \sup_v \int_\Omega |u(x)v(x)| \, dx, \tag{4.6.1}$$

where the supremum is taken over all functions $v \in \mathcal{L}^\Psi(\Omega)$ such that

$$\varrho(v, \Psi) \le 1,$$

is called the *Orlicz norm* of u. The set $L^\Phi(\Omega)$ of all measurable functions u with $\|u\|_\Phi < \infty$ is called the *Orlicz space*.

… Section 4.6 Orlicz spaces

Remark 4.6.2. The functional $\|u\|_\Phi$ is indeed a norm on $L^\Phi(\Omega)$ for every Young function Φ, as we shall see in Theorem 4.6.4.

We shall now establish a simple relation between an Orlicz class and the corresponding Orlicz space.

Proposition 4.6.3. *Let Φ be a Young function. Then*

$$\mathcal{L}^\Phi(\Omega) \subset L^\Phi(\Omega), \tag{4.6.2}$$

and

$$\|u\|_\Phi \leq \varrho(u, \Phi) + 1. \tag{4.6.3}$$

Proof. It follows from formula (4.3.8) that whenever $v \in \mathcal{L}^\Psi(\Omega)$ and

$$\varrho(v, \Psi) \leq 1,$$

then

$$\int_\Omega |u(x)v(x)| \, dx \leq \varrho(u, \Phi) + \varrho(v, \Psi) \leq \varrho(u, \Phi) + 1.$$

Thus, (4.6.3) follows from the definition of $\|u\|_\Phi$. The inclusion in (4.6.2) is its simple consequence. □

Theorem 4.6.4. *The set $L^\Phi(\Omega)$ is a vector space and the formula (4.6.1) defines a norm on $L^\Phi(\Omega)$.*

Proof. Let $u, w \in L^\Phi(\Omega)$ (i.e. $\|u\|_\Phi < \infty$, $\|v\|_\Phi < \infty$) and let $\lambda \in \mathbb{R}$. Then

$$\|\lambda u\|_\Phi = \sup_v \int_\Omega |\lambda u(x)v(x)| \, dx = |\lambda| \sup_v \int_\Omega |u(x)v(x)| \, dx = |\lambda| \|u\|_\Phi < \infty,$$

i.e. $\lambda u \in L^\Phi(\Omega)$. Further,

$$\|u + w\|_\Phi = \sup_v \int_\Omega |u(x) + w(x)| |v(x)| \, dx$$
$$\leq \sup_v \int_\Omega |u(x)| |v(x)| \, dx + \sup_v \int_\Omega |w(x)| |v(x)| \, dx$$
$$= \|u\|_\Phi + \|w\|_\Phi < \infty,$$

showing the triangle inequality and $u + w \in L^\Phi(\Omega)$.

Thus $L^\Phi(\Omega)$ is a vector space. To prove that (4.6.1) is a norm we have still to show that $\|u\|_\Phi = 0$ holds if and only if u is zero almost everywhere in Ω. Obviously $u = 0$ almost everywhere implies $\|u\|_\Phi = 0$.

Let $\|u\|_\Phi = 0$ and let Ω_1 be a subset of Ω such that $0 < \mu(\Omega_1) < \infty$. We have $\lim_{t \to 0} \Psi(t) = 0$ since Ψ is a Young function, and consequently, there exists a $k > 0$ such that $\Psi(k) < \frac{1}{\mu(\Omega_1)}$. If we define the function v_0 by

$$v_0(x) := \begin{cases} k & \text{for } x \in \Omega_1, \\ 0 & \text{for } x \in \Omega \setminus \Omega_1, \end{cases}$$

then

$$\varrho(v_0, \Psi) = \int_\Omega \Psi(|v_0(x)|) \, dx = \int_{\Omega_1} \Psi(k) \, dx < 1.$$

From Definition 4.6.1 we have

$$\|u\|_\Phi \geq \int_\Omega |u(x)| \, v_0(x) \, dx = k \int_{\Omega_1} |u(x)| \, dx,$$

whence $\|u\|_\Phi = 0$ implies $u(x) = 0$ for almost all $x \in \Omega_1$. Since $\Omega_1 \subset \Omega$ was arbitrary, we obtain that $u(x) = 0$ for almost all $x \in \Omega$. □

Convention 4.6.5. In the preceding proof, the notation

$$\sup_v \quad \text{instead of} \quad \sup_{v, \varrho(v, \Psi) \leq 1}$$

was used. We shall use this brief notation also in this text if there is no danger of misunderstanding.

Exercise 4.6.6. It is not difficult to verify that, for any Young function Φ and any finite measurable function u on Ω, we have

$$\|u\|_\Phi = \sup_v \left| \int_\Omega u(x) v(x) \, dx \right|. \tag{4.6.4}$$

Example 4.6.7. Let $p \in (1, \infty)$ and

$$\Phi(t) = \frac{t^p}{p}, \quad t \in [0, \infty).$$

Then

$$\Psi(t) = \frac{t^{p'}}{p'}, \quad t \in [0, \infty),$$

and, consequently,

$$\|u\|_\Phi = \sup_v \int_\Omega |u(x) v(x)| \, dx, \tag{4.6.5}$$

where the $v \in \mathscr{L}^\Psi(\Omega)$ are such that

$$\varrho(v, \Psi) = \frac{1}{p'} \int_\Omega |v(x)|^{p'} \, dx \leq 1.$$

Section 4.6 Orlicz spaces

Denoting by $\|\cdot\|_p$ the ordinary norm in the Lebesgue space $L^p(\Omega)$ (see, e.g., Lemma 3.2.3), we obtain

$$\|u\|_\Phi = (p')^{\frac{1}{p'}} \|u\|_p. \tag{4.6.6}$$

Indeed, if $u_1 \in L^\Phi(\Omega) = L^p(\Omega)$ is such that $\|u_1\|_p = 1$, then the Hölder inequality (Theorem 3.1.6) yields

$$\int_\Omega |u_1(x)v(x)|\,dx \leq \left(\int_\Omega |v(x)|^{p'}\,dx\right)^{\frac{1}{p'}}.$$

If $\varrho(v, \Psi) \leq 1$, then

$$\int_\Omega |u_1(x)v(x)|\,dx \leq (p')^{\frac{1}{p'}}$$

and, consequently,

$$\|u_1\|_\Phi \leq (p')^{\frac{1}{p'}}. \tag{4.6.7}$$

Choosing $v_1(x) = (p')^{\frac{1}{p'}} |u_1(x)|^{p-1} \operatorname{sign} u_1(x)$, we obtain

$$\varrho(v_1, \Psi) = \frac{1}{p'} \int_\Omega |v_1(x)|^{p'}\,dx = \frac{1}{p'} \int_\Omega p' |u_1(x)|^p\,dx = \|u_1\|_p^p = 1$$

and

$$\int_\Omega u_1(x)v_1(x)\,dx = \int_\Omega (p')^{\frac{1}{p'}} |u_1(x)|^{p-1} |u_1(x)|\,dx = (p')^{\frac{1}{p'}} \|u_1\|_p^p = (p')^{\frac{1}{p'}}.$$

From (4.6.1) we have that $\|u_1\|_\Phi \geq (p')^{\frac{1}{p'}}$ which together with (4.6.7) implies

$$\|u_1\|_\Phi = (p')^{\frac{1}{p'}}. \tag{4.6.8}$$

If u is an arbitrary function from $L^\Phi(\Omega)$ such that $\|u\|_p \neq 0$, we set

$$u_1(x) = \frac{u(x)}{\|u\|_p}$$

and (4.6.6) follows from (4.6.8).

Now, (4.6.6) shows that if we consider the Lebesgue space $L^p(\Omega)$ as an Orlicz space $L^\Phi(\Omega)$, we obtain $L^p(\Omega) = L^\Phi(\Omega)$ with the equality of sets and with different but equivalent norms. The difference between $\|u\|_\Phi$ and $\|u\|_p$ also follows from a comparison of (4.6.5) to

$$\|u\|_p = \sup_{\tilde{v}} \int_\Omega |u(x)\tilde{v}(x)|\,dx, \tag{4.6.9}$$

where the $\tilde{v}$ are such that

$$\int_\Omega |\tilde{v}(x)|^{p'}\,dx \leq 1$$

(for (4.6.9) see Exercise 3.9.2).

Remark 4.6.8. Let Φ be a Young function. Then the Orlicz norm $\|\cdot\|_\Phi$ is *monotone* in the following sense: If $u, w \in L^\Phi(\Omega)$ and if

$$|u(x)| \leq |w(x)| \quad \text{for almost all } x \in \Omega,$$

then

$$\|u\|_\Phi \leq \|w\|_\Phi.$$

This means, in other words, that the space $L^\Phi(\Omega)$ is a *lattice*.

To calculate precisely the Orlicz norm of a given function might be very difficult, or even impossible. However, it is possible to calculate the norm of the characteristic function of a given measurable set. It turns out that this information is very important and useful.

Proposition 4.6.9. *Let Φ be a Young function. Assume that $E \subset \Omega$ is such that $0 < \mu(E) < \infty$. Then*

$$\|\chi_E\|_\Phi = \mu(E)\Psi^{-1}\left(\frac{1}{\mu(E)}\right), \tag{4.6.10}$$

where Ψ is the complementary Young function to Φ and Ψ^{-1} is the inverse function of Ψ.

Proof. We have, by the definition of the Orlicz norm,

$$\|\chi_E\|_\Phi = \sup_{\varrho(v,\Psi)\leq 1} \int_E v(x)\,dx.$$

Next, by the Jensen integral inequality (4.2.12), applied to Ψ (which, as we know, is convex) and to $\alpha(x) = \chi_E(x)$, we obtain, for every $v \in \mathcal{L}^\Psi(\Omega)$ satisfying $\varrho(v, \Psi) \leq 1$,

$$\Psi\left(\frac{1}{\mu(E)}\int_E |v(x)|\,dx\right) \leq \frac{1}{\mu(E)}\int_E \Psi(|v(x)|)\,dx, \tag{4.6.11}$$

hence

$$\int_E |v(x)|\,dx \leq \mu(E)\Psi^{-1}\left(\frac{1}{\mu(E)}\right). \tag{4.6.12}$$

Altogether, this shows

$$\|\chi_E\|_\Phi \leq \mu(E)\Psi^{-1}\left(\frac{1}{\mu(E)}\right).$$

Conversely, let

$$v_0(x) := \chi_E(x)\Psi^{-1}\left(\frac{1}{\mu(E)}\right), \quad x \in \Omega.$$

Then obviously $\varrho(v_0, \Psi) = 1$, hence

$$\|\chi_E\|_\Phi \geq \int_E v_0(x)\,dx = \mu(E)\Psi^{-1}\left(\frac{1}{\mu(E)}\right).$$

Formula (4.6.10) now follows by combining the estimates obtained. □

4.7 Hölder inequality in Orlicz spaces

Remark 4.7.1. It follows from (4.6.2) that if $u \in \mathcal{L}^\Phi(\Omega)$, then $u \in L^\Phi(\Omega)$. This inclusion can be generalized in the following sense: if there exists a $\lambda > 0$ such that $\lambda u \in \mathcal{L}^\Phi(\Omega)$, then $u \in L^\Phi(\Omega)$.

Indeed, if $\lambda u \in \mathcal{L}^\Phi(\Omega)$ then $\lambda u \in L^\Phi(\Omega)$ and since $L^\Phi(\Omega)$ is a vector space, we have $\lambda^{-1}(\lambda u) = u \in L^\Phi(\Omega)$.

Moreover, a statement in some sense converse is true as well: if $u \in L^\Phi(\Omega)$, then there exists a $\lambda > 0$ such that $\lambda u \in \mathcal{L}^\Phi(\Omega)$. The explicit form of the constant λ will be given in the following lemma.

Lemma 4.7.2. *Let Φ be a Young function and let $u \in L^\Phi(\Omega)$ be such that $\|u\|_\Phi \neq 0$. Then*

$$\int_\Omega \Phi\left(\frac{1}{\|u\|_\Phi}|u(x)|\right)dx \leq 1. \quad (4.7.1)$$

Proof. For $u \in L^\Phi(\Omega)$,

$$\int_\Omega |u(x)v(x)|\,dx \leq \begin{cases} \|u\|_\Phi & \text{for } \varrho(v, \Psi) \leq 1, \\ \|u\|_\Phi \varrho(v, \Psi) & \text{for } \varrho(v, \Psi) > 1. \end{cases} \quad (4.7.2)$$

The first part of inequality (4.7.2) follows immediately from the definition of the Orlicz norm. For the second part we will use (4.2.5). We have $\Psi(\alpha t) \leq \alpha \Psi(t)$ for $t \geq 0$ and $\alpha \in [0, 1]$. Setting $t = |v(x)|$, $\alpha = \frac{1}{\varrho(v,\Psi)}$ and integrating over Ω we arrive at

$$\int_\Omega \Psi\left(\frac{1}{\varrho(v, \Psi)}|v(x)|\right)dx \leq \frac{1}{\varrho(v, \Psi)}\varrho(v, \Psi) = 1.$$

By the definition of the Orlicz norm,

$$\int_\Omega |u(x)|\frac{|v(x)|}{\varrho(v, \Psi)}\,dx \leq \|u\|_\Phi$$

which proves the second part of inequality (4.7.2).

Let us first suppose that $u \in L^\Phi(\Omega)$ is bounded and that $u(x) = 0$ for $x \in \Omega \setminus \Omega_0$ for some $\Omega_0 \subset \Omega$ with $\mu(\Omega_0) < \infty$. If we take

$$v(x) = \varphi\left(\frac{1}{\|u\|_\Phi}|u(x)|\right),$$

where φ is the derivative of Φ, then both the functions $\Phi\left(\frac{1}{\|u\|_\Phi}|u(x)|\right)$ and $\Psi(|v(x)|)$ are bounded and, consequently, integrable over Ω_0; furthermore, they belong to $L^1(\Omega)$ because they are zero outside of Ω_0.

Inequality (4.3.8) together with Remark 4.3.9 implies

$$\int_\Omega \frac{1}{\|u\|_\Phi} |u(x)v(x)|\, dx = \varrho\left(\frac{|u|}{\|u\|_\Phi}, \Phi\right) + \varrho(v, \Psi).$$

Then we have from the above inequality and (4.7.2) (written for $\frac{u}{\|u\|_\Phi}, v$) that

$$\max\{\varrho(v, \Psi), 1\} \geq \int_\Omega \Phi\left(\frac{1}{\|u\|_\Phi}|u(x)|\right) dx + \varrho(v, \Psi).$$

If $\varrho(v, \Psi) > 1$ then necessarily

$$\int_\Omega \Phi\left(\frac{1}{\|u\|_\Phi}|u(x)|\right) dx = 0.$$

If $\varrho(v, \Psi) \leq 1$ then

$$\int_\Omega \Phi\left(\frac{1}{\|u\|_\Phi}|u(x)|\right) dx + \varrho(v, \Psi) \leq 1$$

and (4.7.1) holds.

Now let $u \in L^\Phi(\Omega)$ be arbitrary. We define a sequence of subsets $\{\Omega_n\}_{n=1}^\infty$ of Ω such that $\Omega_n \subset \Omega_{n+1}$, $\mu(\Omega_n) < \infty$ and $\Omega = \bigcup_{n=1}^\infty \Omega_n$, and the functions $\{u_n\}_{n=1}^\infty$ by

$$u_n(x) = \begin{cases} u(x) & \text{for } x \in \Omega_n \text{ and } |u(x)| \leq n, \\ n & \text{for } x \in \Omega_n \text{ and } |u(x)| > n, \\ 0 & \text{for } x \in \Omega \setminus \Omega_n. \end{cases}$$

From the first part of the proof it follows that (4.7.1) holds for every function u_n, i.e.

$$\int_\Omega \Phi\left(\frac{1}{\|u_n\|_\Phi}|u_n(x)|\right) dx \leq 1.$$

Further, $|u_n(x)| \leq |u(x)|$ for almost all $x \in \Omega$, and according to Remark 3.6.8 we have $\|u_n\|_\Phi \leq \|u\|_\Phi$ for all $n \in \mathbb{N}$. Thus

$$\frac{|u_n(x)|}{\|u\|_\Phi} \leq \frac{|u_n(x)|}{\|u_n\|_\Phi} \quad \text{and} \quad \Phi\left(\frac{|u_n(x)|}{\|u\|_\Phi}\right) \leq \Phi\left(\frac{|u_n(x)|}{\|u_n\|_\Phi}\right)$$

(Φ is increasing) and, consequently,

$$\int_\Omega \Phi\left(\frac{1}{\|u\|_\Phi}|u_n(x)|\right) dx \leq 1.$$

Section 4.7 Hölder inequality in Orlicz spaces

The sequences $\{|u_n(x)|\}_{n=1}^{\infty}$ and $\{\Phi(|u_n(x)|\,\|u\|_\Phi^{-1})\}_{n=1}^{\infty}$ are nondecreasing in $n \in \mathbb{N}$ for every fixed $x \in \Omega$ and, therefore, the monotone convergence theorem (Theorem 1.21.4) implies

$$\int_\Omega \Phi\left(\frac{1}{\|u\|_\Phi}|u(x)|\right) dx = \lim_{n\to\infty} \int_\Omega \Phi\left(\frac{1}{\|u\|_\Phi}|u_n(x)|\right) dx \leq 1. \quad \square$$

By virtue of Lemma 4.7.2 we can now characterize when an Orlicz class coincides with the corresponding Orlicz space.

Theorem 4.7.3. *Let Φ satisfy the Δ_2-condition (with $T = 0$ if $\mu(\Omega) = \infty$). Then*

$$L^\Phi(\Omega) = \mathcal{L}^\Phi(\Omega). \tag{4.7.3}$$

Proof. We know that $\mathcal{L}^\Phi(\Omega) \subset L^\Phi(\Omega)$. Let $u \in L^\Phi(\Omega)$, $\|u\|_\Phi \neq 0$. Lemma 4.7.2 implies

$$w = \frac{1}{\|u\|_\Phi} u \in \mathcal{L}^\Phi(\Omega).$$

Since $\mathcal{L}^\Phi(\Omega)$ is a linear set (see Theorem 4.5.3) we have

$$w\,\|u\|_\Phi = u \in \mathcal{L}^\Phi(\Omega),$$

and hence $L^\Phi(\Omega) \subset \mathcal{L}^\Phi(\Omega)$. $\square$

Remark 4.7.4. It follows from Lemma 4.7.2 that the Orlicz space $L^\Phi(\Omega)$ is the *linear hull* of the Orlicz class $\mathcal{L}^\Phi(\Omega)$. Indeed, if $u \in L^\Phi(\Omega)$, $\|u\|_\Phi \neq 0$, then $u = cv$ with $v \in \mathcal{L}^\Phi(\Omega)$, $v = \frac{u}{\|u\|_\Phi}$, and $c = \|u\|_\Phi$.

In the environment of Lebesgue spaces, the classical Hölder inequality (see Theorem 3.1.6) has the form

$$\int_\Omega |u(x)v(x)|\,dx \leq \|u\|_p \|v\|_{p'},$$

where $p \in [1, \infty]$. The following theorem provides an analogous inequality for Orlicz spaces.

Theorem 4.7.5. *Let Φ, Ψ be a pair of complementary Young functions. If $u \in L^\Phi(\Omega)$ and $v \in L^\Psi(\Omega)$ then $uv \in L^1(\Omega)$ and*

$$\int_\Omega |u(x)v(x)|\,dx \leq \|u\|_\Phi \|v\|_\Psi. \tag{4.7.4}$$

Proof. For $\|v\|_\Psi = 0$, the inequality (4.7.4) is obvious. If $\|v\|_\Psi \neq 0$, we apply Lemma 4.7.2 to the Young function Ψ, obtaining thereby $\varrho(\frac{v}{\|v\|_\Psi}, \Psi) \leq 1$. Inequality (4.7.4) now follows from the definition of the Orlicz norm of u, since

$$\int_\Omega |u(x)v(x)|\,dx = \|v\|_\Psi \int_\Omega \left|u(x)\frac{v(x)}{\|v\|_\Psi}\right|\,dx \leq \|u\|_\Psi \cdot \|v\|_\Phi. \quad \square$$

Remark 4.7.6. Inequality (4.7.4) can be viewed as an extension of the Hölder inequality, but it should be noted that the usual Hölder inequality is not a special case of (4.7.4). Indeed, if we regard the Lebesgue spaces $L^p(\Omega)$ and $L^{p'}(\Omega)$, where $p \in (1, \infty)$, as Orlicz spaces $L^\Phi(\Omega)$ and $L^\Psi(\Omega)$ with $\Phi(t) = \frac{t^p}{p}$, $\Psi(t) = \frac{t^{p'}}{p'}$, then the inequality (4.7.4) has the form

$$\int_\Omega |u(x)v(x)|\,dx \leq p^{\frac{1}{p}}(p')^{\frac{1}{p'}} \|u\|_p \cdot \|v\|_{p'}$$

(see Example 4.6.7, formula (4.6.6)), which differs from the usual Hölder inequality.

4.8 The Luxemburg norm

The evaluation of the Orlicz norm $\|u\|_\Phi$ requires knowledge of the expression for the complementary Young function Ψ which is not, in general, merely a routine matter. Another equivalent norm in $L^\Phi(\Omega)$ expressed only in terms of Φ was introduced by Luxemburg in [141] and independently by Weiss in [233].

Definition 4.8.1. Let Φ be a Young function and let u be a measurable function defined on Ω. The number

$$\|\|u\|\|_\Phi = \inf\left\{k > 0; \int_\Omega \Phi\left(\frac{|u(x)|}{k}\right)\,dx \leq 1\right\}$$

is called the *Luxemburg norm* of u.

Remark 4.8.2. From Lemma 4.7.2 it follows that if $u \in L^\Phi(\Omega)$ then

$$\|\|u\|\|_\Phi \leq \|u\|_\Phi. \tag{4.8.1}$$

If we let k tend to $\|\|u\|\|_\Phi$ in

$$\varrho\left(\frac{u}{k}, \Phi\right) = \int_\Omega \Phi\left(\frac{|u(x)|}{k}\right)\,dx \leq 1,$$

we obtain

$$\varrho\left(\frac{u}{\|\|u\|\|_\Phi}, \Phi\right) = \int_\Omega \Phi\left(\frac{|u(x)|}{\|\|u\|\|_\Phi}\right)\,dx \leq 1 \tag{4.8.2}$$

by the Fatou lemma (Theorem 1.21.6).

Section 4.8 The Luxemburg norm

Exercise 4.8.3. Show that $\|\cdot\|_\Phi$ is a norm on $L^\Phi(\Omega)$. (Use (4.2.7) to verify the triangle inequality.)

Lemma 4.8.4. *Let Φ be a Young function and let $u \in L^\Phi(\Omega)$. Then*

(i) $\varrho(u, \Phi) \leq \|u\|_\Phi$ *if* $\|u\|_\Phi \leq 1$;

(ii) $\varrho(u, \Phi) \geq \|u\|_\Phi$ *if* $\|u\|_\Phi > 1$.

Proof. The assertion follows from Proposition 1.5.8, since $\varrho(u, \Phi)$ is a modular in the sense of Definition 1.5.1 □

Theorem 4.8.5. *Let Φ be a Young function. Then, for each $u \in L^\Phi(\Omega)$,*

$$\|u\|_\Phi \leq \|u\|_\Phi \leq 2\|u\|_\Phi, \tag{4.8.3}$$

i.e. the norms $\|\cdot\|_\Phi$ and $\|\cdot\|_\Phi$ are equivalent.

Proof. For the first inequality see (4.8.1). The inequality (4.6.3) applied to $w = \frac{u}{\|u\|_\Phi}$ implies

$$\|u\|_\Phi \leq \varrho(w, \Phi) + 1 \leq 2,$$

since we have $\varrho(w, \Phi) = \varrho(\frac{u}{\|u\|_\Phi}, \Phi) \leq 1$ (see (4.8.2)), from which the other inequality follows. □

Remark 4.8.6. Lemma 4.8.4 (see also Proposition 1.5.7) implies that $\varrho(u, \Psi) \leq 1$ if and only if

$$\|u\|_\Psi \leq 1.$$

Using this fact, we can express the Orlicz norm in the following form:

$$\|u\|_\Phi = \sup_{\|v\|_\Psi \leq 1} \int_\Omega |u(x)v(x)| \, dx. \tag{4.8.4}$$

By the same argument as in Theorem 4.7.5 and by virtue of (4.8.4) we can obtain the following modified versions of the Hölder inequality.

Theorem 4.8.7. *Let Φ, Ψ be a pair of complementary Young functions. Then*

$$\int_\Omega |u(x)v(x)| \, dx \leq \|u\|_\Phi \|v\|_\Psi \tag{4.8.5}$$

and

$$\int_\Omega |u(x)v(x)| \, dx \leq \|u\|_\Phi \|v\|_\Psi \tag{4.8.6}$$

for all $u \in L^\Phi(\Omega)$ and $v \in L^\Psi(\Omega)$.

Example 4.8.8. Let $p \in (1, \infty)$ and let $\Phi(t) = \frac{t^p}{p}$. Then

$$\|u\|_\Phi = \left(\frac{1}{p}\right)^{\frac{1}{p}} \|u\|_p, \tag{4.8.7}$$

since

$$\varrho\left(\frac{u}{k}, \Phi\right) \leq 1$$

means in this case that

$$\frac{1}{p}\int_\Omega |u(x)|^p \, dx \leq k^p.$$

Inequality (4.8.5) then reads as

$$\int_\Omega |u(x)v(x)| \, dx \leq (p')^{\frac{1}{p'}} \left(\frac{1}{p'}\right)^{\frac{1}{p'}} \|u\|_p \|v\|_{p'} = \|u\|_p \|v\|_{p'},$$

which is nothing more than the classical Hölder inequality (3.1.6).

Remarks 4.8.9. Let Φ be a Young function.

(i) There are other equivalent norms in Orlicz spaces and also there exist alternative expressions for the Orlicz norm such as, for example,

$$\|u\|_\Phi = \inf_{k>0} \frac{1}{k}\left[1 + \int_\Omega \Phi(k\,|u(x)|)\,dx\right].$$

For details, see [123].

(ii) It is shown in [123] that

$$\|u\|_\Phi = \int_\Omega \varphi(k_0\,|u(x)|)\,|u(x)|\,dx,$$

where φ is the function which generates the Young function Φ (see (4.2.1)) and k_0 is a positive constant such that

$$\int_\Omega \Psi(\varphi(k_0\,|u(x)|))\,dx = 1$$

(Ψ is complementary to Φ).

For equivalent norms in $L^\Phi(\Omega)$ also see [247].

4.9 Completeness of Orlicz spaces

Theorem 4.9.1. *Let Φ be a Young function. Then the Orlicz space $L^\Phi(\Omega)$ is a Banach space.*

Proof. Let $\{u_n\}_{n=1}^\infty$ be a Cauchy sequence in $L^\Phi(\Omega)$, i.e. for every $\varepsilon > 0$ there exists an $n_0 = n_0(\varepsilon) \in \mathbb{N}$ such that for arbitrary $v \in \mathcal{L}^\Psi(\Omega)$, $\varrho(v, \Psi) \leq 1$ and for every $n > n_0, m > n_0$,

$$\int_\Omega |u_m(x) - u_n(x)| \, |v(x)| \, dx < \varepsilon. \tag{4.9.1}$$

Let us decompose Ω into a sequence of its subsets $\{\Omega_n\}_{n=1}^\infty$, $n \in \mathbb{N}$, with $0 < \mu(\Omega_n) < \infty$ such that $\Omega = \bigcup_{n=1}^\infty \Omega_n$, $\Omega_i \cap \Omega_j = \emptyset$ for $i \neq j$. Let us choose $k > 0$ such that

$$\Psi(k) \leq \frac{1}{\mu(\Omega_1)},$$

and define a function v by

$$v(x) = \begin{cases} k & \text{for } x \in \Omega_1, \\ 0 & \text{for } x \in \Omega \setminus \Omega_1. \end{cases}$$

Then

$$\varrho(v, \Psi) = \int_\Omega \Psi(|v(x)|) \, dx = \int_{\Omega_1} \Psi(k) \, dx = \Psi(k)\mu(\Omega_1) \leq 1,$$

and using this function v in (4.9.1) we obtain

$$\int_{\Omega_1} |u_m(x) - u_n(x)| \, dx < \frac{\varepsilon}{k} \quad \text{for } n > n_0, \quad m > n_0.$$

This means that $\{u_n\}_{n=1}^\infty$ is a Cauchy sequence in $L^1(\Omega_1)$. This space is complete by Theorem 3.7.1 and it follows from the proof of this theorem that there exists a subsequence $\{u_{n,1}\}$ of $\{u_n\}_{n=1}^\infty$ which converges to a function $u \in L^1(\Omega_1)$ almost everywhere on Ω_1.

Now we use the same procedure for Ω_2 and $\{u_{n,1}\}$ to obtain a subsequence $\{u_{n,2}\}$ of $\{u_{n,1}\}$ which converges almost everywhere to a function (let us denote it by u again) in $L^1(\Omega_2)$. Repeating this procedure we obtain a sequence of subsequences

$$\{u_n\} \supset \{u_{n,1}\} \supset \{u_{n,2}\} \supset \ldots$$

and each subsequence $\{u_{n,k}\}$ converges to u on Ω_k almost everywhere.

Replacing u_m by $u_{m,m}$ in (4.9.1) we obtain from the convergence of $u_{m,m}(x)$ to $u(x)$ for almost all $x \in \Omega$ and from the Fatou lemma (Theorem 1.21.6) that

$$\int_\Omega |u(x) - u_n(x)| \, |v(x)| \, dx \leq \varepsilon$$

provided $v \in \mathcal{L}^\Psi(\Omega)$, $\varrho(v, \Psi) \leq 1$ and $n \geq n_0$. This means that

(i) $u_n - u \in L^\Phi(\Omega)$ and, consequently, also
$$u = u_n - (u_n - u) \in L^\Phi(\Omega);$$
moreover,

(ii) $\lim_{n\to\infty} \|u_n - u\|_\Phi = 0$. □

4.10 Convergence in Orlicz spaces

The usual (norm) convergence in the Orlicz space $L^\Phi(\Omega)$ can be introduced in terms of the Orlicz norm $\|\cdot\|_\Phi$ as follows:

$$u_n \to u \text{ in } L^\Phi(\Omega) \quad \text{means} \quad \lim_{n\to\infty} \|u_n - u\|_\Phi = 0.$$

We shall now introduce another important type of convergence in Orlicz spaces.

Definition 4.10.1. Let Φ be a Young function. A sequence $\{u_n\}_{n=1}^\infty$ in $L^\Phi(\Omega)$ is said to *converge in Φ-mean* to $u \in L^\Phi(\Omega)$ if

$$\lim_{n\to\infty} \varrho(u_n - u, \Phi) = \lim_{n\to\infty} \int_\Omega \Phi(|u_n(x) - u(x)|)\,dx = 0. \tag{4.10.1}$$

Remark 4.10.2. Let Φ be a Young function. The Φ-mean convergence is also sometimes called the *modular convergence*, since it is the convergence with respect to the (convex) modular $\varrho(\cdot, \Phi)$. By Lemma 4.8.4 (i) and by Theorem 4.8.5, one has, for $w \in L^\Phi(\Omega)$,

$$\varrho(w, \Phi) \le \|w\|_\Phi \tag{4.10.2}$$

once $\|w\|_\Phi \le 1$. Hence, for a sequence $\{u_n\}_{n=1}^\infty$ and a function $u \in L^\Phi(\Omega)$, we always have

$$\|u_n - u\|_\Phi \to 0 \quad \Rightarrow \quad \varrho(u_n - u, \Phi) \to 0.$$

In other words, in an Orlicz space, the norm topology is stronger than the modular one and, therefore, also the norm convergence is stronger than the modular one. This is a very important fact which is worth formulating as a proposition.

Proposition 4.10.3. *Let Φ be a Young function. Then convergence in $L^\Phi(\Omega)$ always implies Φ-mean convergence.*

The assertion of Proposition 4.10.3 cannot be in general reversed as can be demonstrated with the following counterexample.

Example 4.10.4. Assume that Φ is a Young function which does not satisfy the condition Δ_2. Then there exists a strictly increasing sequence of positive numbers $\{t_n\}_{n=1}^\infty$ such that

$$\Phi(2t_n) > 2^n \Phi(t_n), \quad n \in \mathbb{N}.$$

Section 4.10 Convergence in Orlicz spaces

In Ω one can find a sequence of disjoint sets $\{E_n\}_{n=1}^{\infty}$ such that their union

$$E := \bigcup_{n=1}^{\infty} E_n$$

satisfies $\mu(E) < \infty$. Next, for each fixed $n \in \mathbb{N}$ and for all $k \in \mathbb{N}$, $1 \leq k \leq n$, there is a finite sequence $\{E_{n,k}\}_{k=1}^{n}$ of disjoint subsets of E_n such that

$$\mu(E_{n,k}) = \frac{\mu(E)}{n 2^k \Phi(t_k)}, \qquad n \in \mathbb{N}, \ k = 1, \ldots, n.$$

We then set

$$u_n(x) := \begin{cases} t_k & \text{if } x \in E_{n,k}, \ k = 1, \ldots, n \\ 0 & \text{otherwise.} \end{cases}$$

Then we have

$$\int_E \Phi(u_n(x))\, dx = \sum_{k=1}^{n} \int_{E_{n,k}} \Phi(u_n(x))\, dx = \sum_{k=1}^{n} \Phi(t_k) \mu(E_{n,k}) \leq \frac{\mu(E)}{n}.$$

In other words, we get

$$\lim_{n \to \infty} \int_E \Phi(u_n(x))\, dx = 0,$$

so the sequence $\{u_n\}$ converges to zero in Φ-mean. However, it does not converge to zero in norm, since, if it was so, then, by (4.10.2), we would have

$$\lim_{n \to \infty} \int_E \Phi(2 u_n(x))\, dx \leq \lim_{n \to \infty} \|2 u_n\|_\Phi = 0,$$

which is impossible as

$$\int_E \Phi(2 u_n(x))\, dx = \sum_{k=1}^{n} \int_{E_{n,k}} \Phi(2 u_n(x))\, dx = \sum_{k=1}^{n} \Phi(2 t_k) \mu(E) > \mu(E),$$

which is a contradiction.

The fact that the Young function Φ in Example 4.10.4 does not satisfy the Δ_2-condition is of crucial importance. In fact, a certain converse of (4.10.2) is available when $\Phi \in \Delta_2$.

Proposition 4.10.5. *Let Φ be a Young function which satisfies the Δ_2-condition (with $T = 0$ if $\mu(\Omega) = \infty$). Let $w \in L^\Phi(\Omega)$. If there exists an $m \in \mathbb{N}$ such that*

$$\varrho(w, \Phi) \leq \frac{1}{k^m} \qquad (4.10.3)$$

where k is the constant in (4.4.1), then

$$\|w\|_\Phi \le \frac{c}{2^m}, \qquad (4.10.4)$$

where $c = 2$ if $\mu(\Omega) = \infty$ and $c = \Phi(T)\mu(\Omega) + 2$ if $\mu(\Omega) < \infty$.

Proof. Let $m \in \mathbb{N}$ be fixed. If $\mu(\Omega) < \infty$, we denote

$$\Omega_1 = \{x \in \Omega; \ 2^m |w(x)| \le T\}.$$

Then,

$$\Phi(2^m |w(x)|) \le \Phi(T) \quad \text{if } x \in \Omega_1, \qquad (4.10.5)$$
$$\Phi(2^m |w(x)|) \le k^m \Phi(|w(x)|) \quad \text{if } x \in \Omega \setminus \Omega_1, \qquad (4.10.6)$$

(to obtain the latter inequality the Δ_2-condition was used m-times). So we have

$$\int_\Omega \Phi(2^m |w(x)|) \, dx = \int_{\Omega_1} \Phi(2^m |w(x)|) \, dx + \int_{\Omega \setminus \Omega_1} \Phi(2^m |w(x)|) \, dx$$
$$\le \int_{\Omega_1} \Phi(T) \, dx + k^m \int_{\Omega - \Omega_1} \Phi(|w(x)|) \, dx$$
$$\le \Phi(T)\mu(\Omega_1) + k^m \int_\Omega \Phi(|w(x)|) \, dx,$$

i.e.

$$\varrho(2^m w, \Phi) \le \Phi(T)\mu(\Omega) + k^m \varrho(w, \Phi). \qquad (4.10.7)$$

If $\mu(\Omega) = \infty$, we set $\Omega_1 = \emptyset$ and by virtue of inequality (4.10.6) we obtain immediately that

$$\varrho(2^m w, \Phi) \le k^m \varrho(w, \Phi).$$

If now (4.10.3) holds then we have in both cases ($\mu(\Omega) = \infty$ as well as $\mu(\Omega) < \infty$) that

$$\varrho(2^m w, \Phi) \le c - 1,$$

where $c = 2$ for $\mu(\Omega) = \infty$ and $c = \Phi(T)\mu(\Omega) + 2$ for $\mu(\Omega) < \infty$. Thus (4.3.8) yields

$$\int_\Omega |2^m w(x)| |v(x)| \, dx \le \varrho(2^m w, \Phi) + \varrho(v, \Psi) \le c$$

where $v \in \mathcal{L}^\Psi(\Omega)$ is such that $\varrho(v, \Psi) \le 1$ and, consequently, $\|2^m w\|_\Phi \le c$, which is (4.10.4). □

Now we are able to prove the equivalence of convergence in $L^\Phi(\Omega)$ and Φ-mean convergence.

Theorem 4.10.6. *Let Φ be a Young function which satisfies the Δ_2-condition (with $T = 0$ if $\mu(\Omega) = \infty$). Then u_n converges to u in $L^\Phi(\Omega)$ if and only if u_n converges to u in Φ-mean.*

Proof. In view of Proposition 4.10.3 it suffices to show that Φ-mean convergence implies convergence in $L^\Phi(\Omega)$. So, let $\{u_n\}_{n=1}^\infty \subset L^\Phi(\Omega)$, $u \in L^\Phi(\Omega)$ and let
$$\lim_{n\to\infty} \varrho(u_n - u, \Phi) = 0.$$
Given $\varepsilon > 0$, we choose an $m \in \mathbb{N}$ such that $\varepsilon > c2^{-m}$, where $c = 2$ if $\mu(\Omega) = \infty$ and $c = 2 + \Phi(T)\mu(\Omega)$ if $\mu(\Omega) < \infty$. Then we choose an $n_0 = n_0(\varepsilon)$ so that $\varrho(u_n - u, \Phi) \leq \frac{1}{k^m}$ for $n \geq n_0$. By (4.10.4) (for $w = u_n - u$), we have
$$\|u_n - u\|_\Phi \leq c2^{-m} < \varepsilon \quad \text{for } n \geq n_0,$$
and the proof is complete. $\square$

We shall now present an analogue of the Lebesgue dominated convergence theorem (Theorem 1.21.5) for Orlicz spaces. The proof is left to the reader as an exercise.

Theorem 4.10.7. *Let Φ be a Young function. Let $\{u_n\}_{n=1}^\infty$ be a sequence in $L^\Phi(\Omega)$. Suppose that there exists $c > 0$ such that $|u_n(x)| \leq c$ for all $n \in \mathbb{N}$ and almost all $x \in \Omega$. Let $\lim_{n\to\infty} u_n(x) = u(x)$ almost everywhere in Ω. Then $u_n \to u$ in Φ-mean.*

With regard to Theorem 4.10.7, let us mention that Φ-mean convergence does not imply pointwise convergence. This is shown by the following example.

Example 4.10.8. Let Φ be an arbitrary Young function. Let $N = 1$, $\Omega = (0, 1)$, $\Omega_1 = (0, \frac{1}{2})$, $\Omega_2 = (\frac{1}{2}, 1)$, $\Omega_3 = (0, \frac{1}{3})$, $\Omega_4 = (\frac{1}{3}, \frac{2}{3})$, $\Omega_5 = (\frac{2}{3}, 1)$, $\Omega_6 = (0, \frac{1}{4}), \ldots$ and define, for $n \in \mathbb{N}$,
$$u_n(x) = \begin{cases} 1 & \text{for } x \in \Omega_n, \\ 0 & \text{elsewhere in } \Omega. \end{cases}$$
Then
$$\lim_{n\to\infty} \int_0^1 \Phi(|u_n(x)|) \, dx = 0,$$
but u_n does not converge pointwise to zero.

Remark 4.10.9. Let $u \in \mathcal{L}^\Phi(\Omega)$. Then $u(x)$ is a finite number for almost all $x \in \Omega$. If we denote
$$\Omega_n = \{x \in \Omega; \, |u(x)| > n\}$$

then $\Omega_n \supset \Omega_{n+1}$ for $n \in \mathbb{N}$ and $\lim_{n\to\infty} \mu(\Omega_n) = 0$. If we introduce further the sequence $\{u_n\}_{n=1}^{\infty}$ in $\mathcal{L}^\Phi(\Omega)$ by

$$u_n(x) = \begin{cases} 0 & \text{for } x \in \Omega_n, \\ u(x) & \text{elsewhere in } \Omega, \end{cases} \tag{4.10.8}$$

then

$$\lim_{n\to\infty} \varrho(u_n - u, \Phi) = \lim_{n\to\infty} \int_{\Omega_n} \Phi(|u(x)|) \, dx = 0. \tag{4.10.9}$$

We thus get the following assertion.

Corollary 4.10.10. *Let Φ be a Young function. Then*

(i) *every function $u \in \mathcal{L}^\Phi(\Omega)$ can be approximated in the sense of the Φ-mean convergence by a sequence of bounded measurable functions;*

(ii) *if moreover $\mu(\Omega) < \infty$, then the set of all bounded measurable functions on Ω is a dense subset of $\mathcal{L}^\Phi(\Omega)$ in the topology of Φ-mean convergence.*

Proof. The assertion (i) follows from the observations made in Remark 4.10.9. If $\mu(\Omega) < \infty$, then every bounded measurable function belongs to $\mathcal{L}^\Phi(\Omega)$, and (ii) follows. □

Exercise 4.10.11. With the help of Proposition 4.10.3, formulate assertions analogous to those in Corollary 4.10.10 (i) and (ii) for convergence in $L^\Phi(\Omega)$.

We shall now introduce the notion of a Φ-mean boundedness.

Definition 4.10.12. *Let Φ be a Young function. A subset K of $L^\Phi(\Omega)$ is said to be Φ-mean bounded if there exists a $c > 0$ such that*

$$\int_\Omega \Phi(|u(x)|) \, dx \le c \qquad \text{for all} \quad u \in K.$$

Remark 4.10.13. Let Φ be a Young function. Then it follows from (4.6.3) that each Φ-mean bounded set K in $L^\Phi(\Omega)$ is automatically bounded (we shall call it *norm bounded*), as we have

$$\|u\|_\Phi \le c + 1 \qquad \text{for all } u \in K.$$

Evidently, a norm bounded set in $L^\Phi(\Omega)$ need not be Φ-mean bounded.

As expected in view of Theorem 4.10.6, we have the following assertion.

Proposition 4.10.14. *If Φ satisfies the Δ_2-condition (with $T = 0$ if $\mu(\Omega) = \infty$) then norm boundedness and Φ-mean boundedness are equivalent.*

Proof. Let $K \subset L^\Phi(\Omega)$ be a norm bounded set, that is, there exists a $c_1 > 0$ such that $\|u\|_\Phi \leq c_1$ for all $u \in K$. We choose an $m \in \mathbb{N}$ such that $c_1 \leq 2^m$, take $w = \frac{u}{2^m}$ in (4.10.7) and use the fact that

$$\varrho\left(\frac{u}{2^m}, \Phi\right) \leq \varrho\left(\frac{u}{\|u\|_\Phi}, \Phi\right) \leq 1;$$

see Lemma 4.7.2. Thus,

$$\varrho(u, \Phi) \leq \begin{cases} \Phi(T)\mu(\Omega) + k^m & \text{if } \mu(\Omega) < \infty, \\ k^m & \text{if } \mu(\Omega) = \infty, \end{cases}$$

for all $u \in K$, i.e. K is Φ-mean bounded. □

4.11 Separability

In this section we shall characterize when an Orlicz space is separable.

Theorem 4.11.1. *Let Φ be a Young function which satisfies the Δ_2-condition (with $T = 0$ if $\mu(\Omega) = \infty$). Then the Orlicz space $L^\Phi(\Omega)$ is separable.*

Proof. Let S be the set of all open cubes in $\mathbb{R}^N$ with vertices at points with rational coordinates and with edges parallel to the coordinate axes.

The set S is evidently countable. Further, let $\mathfrak{F}$ denote the set of all functions on $\mathbb{R}^N$ of the type

$$f(x) = \sum_{i=1}^{\sigma} r_i \chi_{C_i}(x), \quad x \in \mathbb{R}^N,$$

where $\sigma = \sigma(f) \in \mathbb{N}$, r_i are rational numbers, $C_1, C_2, \ldots, C_\sigma$ are pairwise disjoint cubes from S and χ_{C_i} is the characteristic function of C_i. Evidently $\mathfrak{F}$ is also countable. We shall prove that a subset of $\mathfrak{F}$ forms a dense subset in the Orlicz space $L^\Phi(\Omega)$.

Step 1. Given $u \in L^\Phi(\Omega)$ and $\varepsilon > 0$, let $c = 2$ if $\mu(\Omega) = \infty$ and $c = 2 + \Phi(T)\mu(\Omega)$ if $\mu(\Omega) < \infty$. We choose $m \in \mathbb{N}$ such that $\frac{c}{2^m} < \frac{\varepsilon}{3}$ and $\eta > 0$ such that $\eta \leq \frac{1}{k^m}$ (k, T are the constants in the Δ_2-condition).

Step 2. By Theorem 4.7.3, $L^\Phi(\Omega) = \mathcal{L}^\Phi(\Omega)$, and consequently, $u \in \mathcal{L}^\Phi(\Omega)$. Therefore there exists a function u_0 on Ω such that

(i) $u_0(x) = u(x)$ for $x \in \Omega_0$ with some $\Omega_0 \subset \Omega$, $\mu(\Omega_0) < \infty$;

(ii) $u_0(x) = 0$ for $x \in \Omega \setminus \Omega_0$;

(iii) $\varrho(u_0 - u, \Phi) < \eta$.

The term Ω_0 can be taken to be intersection of Ω with a sufficiently large ball in $\mathbb{R}^N$. Then $\varrho(u_0 - u, \Phi) < \frac{1}{k^m}$, and from Proposition 4.10.3 it follows that

$$\|u_0 - u\|_\Phi \leq \frac{c}{2^m} < \frac{1}{3}\varepsilon. \tag{4.11.1}$$

Step 3. Now we can write u_0 in the form

$$u_0 = u_1 - u_2,$$

where u_1, u_2 are nonnegative functions in $\mathcal{L}^\Phi(\Omega)$ which vanish outside of Ω_0. Let us consider, e.g., the function u_1. Then there exists a nondecreasing sequence of simple functions $\{f_n\}_{n=1}^\infty$, such that every f_n is nonnegative, vanishes outside of Ω_0, the range of f_n contains only rational points, and

$$u_1(x) = \lim_{n \to \infty} f_n(x)$$

for almost all $x \in \Omega_0$.

Since

$$\Phi(|u_1(x) - f_n(x)|) \leq \Phi(|u_1(x)|)$$

and

$$\lim_{n \to \infty} \Phi(|u_1(x) - f_n(x)|) = 0 \quad \text{almost everywhere in } \Omega,$$

we have by Theorem 4.10.7 that

$$\lim_{n \to \infty} \int_\Omega \Phi(|u_1(x) - f_n(x)|) \, dx = 0.$$

Using Proposition 4.10.3 again with $w = u_1 - f_n$, we obtain that for sufficiently large n

$$\|u_1 - f_n\|_\Phi < \frac{1}{12}\varepsilon.$$

Proceeding analogously with u_2, we can eventually construct a simple function v_0 with rational values, which vanishes outside of Ω_0 and is such that

$$\|v_0 - u_0\|_\Phi < \frac{1}{3}\varepsilon. \tag{4.11.2}$$

So the function v_0 has the form

$$v_0(x) = \sum_{i=1}^m r_i \chi_i(x),$$

where the χ_i are the characteristic functions of the subsets Ω_i of Ω_0 ($i = 1, 2, \ldots, m$) such that $\Omega_i \cap \Omega_j = \emptyset$ for $i \neq j$ and $\bigcup_{i=1}^m \Omega_i = \Omega_0$; $r_i \in \mathbb{Q}$, $r_i \neq 0$.

Step 4. For the characteristic function χ_M of the set $M \subset \Omega$ we have

$$\varrho(\chi_M, \Phi) = \int_M \Phi(1) \, dx = \Phi(1) \mu(M).$$

When $\mu(M) \to 0$ then also $\varrho(\chi_M, \Phi) \to 0$, and consequently, by Proposition 4.10.5, $\|\chi_M\|_\Phi \to 0$. This implies that for each set Ω_i in Step 3 there exists a set S_i which is the union of a finite number of cubes from S and

$$\|\chi_i - \chi_{S_i}\| \leq \frac{\varepsilon}{3m |r_i|}, \quad i = 1, 2, \ldots, m.$$

If we set

$$w_0(x) = \sum_{i=1}^{m} r_i \chi_{S_i}(x)$$

then $w_0 \in \mathfrak{F}$ and

$$\|w_0 - v_0\|_\Phi = \sum_{i=1}^{m} |r_i| \, \|\chi_{S_i} - \chi_i\|_\Phi \leq \frac{1}{3}\varepsilon. \tag{4.11.3}$$

Step 5. From (4.11.1), (4.11.2), (4.11.3) we conclude that

$$\|u - w_0\|_\Phi \leq \|u - u_0\|_\Phi + \|u_0 - v_0\|_\Phi + \|v_0 - w_0\|_\Phi < \varepsilon,$$

and since $\mathfrak{F}$ is countable, the proof is complete. □

Remark 4.11.2. The fact that Φ satisfies the Δ_2-condition is not only sufficient but also necessary for $L^\Phi(\Omega)$ to be separable. This will be shown in Section 6.5.

4.12 The space $E^\Phi(\Omega)$

In this section we shall introduce an important new function space related to a given Young function Φ.

Definition 4.12.1. Let Φ be a Young function. The space $E^\Phi(\Omega)$ is defined by

$$E^\Phi(\Omega) := \left\{ u \in L^\Phi(\Omega); \; ku \in \mathcal{L}^\Phi(\Omega) \text{ for every } k > 0 \right\}.$$

Remark 4.12.2. Let Φ be a Young function. It is obvious that the space $E^\Phi(\Omega)$ is a vector space and that the inclusions

$$E^\Phi(\Omega) \subset \mathcal{L}^\Phi(\Omega) \subset L^\Phi(\Omega) \tag{4.12.1}$$

hold.

Converse inclusions to (4.12.1) are not valid in general, but we have the following result.

Proposition 4.12.3. *If Φ is a Young function which satisfies the Δ_2-condition (with $T = 0$ if $\mu(\Omega) = \infty$), then*

$$E^\Phi(\Omega) = \mathcal{L}^\Phi(\Omega) = L^\Phi(\Omega). \tag{4.12.2}$$

Proof. Let $u \in L^\Phi(\Omega)$. Then (see Remark 4.7.1 and Lemma 4.7.2) there exists some $\lambda > 0$ such that $\lambda u \in \mathcal{L}^\Phi(\Omega)$. Let $k > 0$. Then it follows from $\Phi \in \Delta_2$ that there exists a $C > 0$ such that

$$\Phi(kt) \leq C\Phi(\lambda t) \quad \text{for every } t \in [T, \infty).$$

Thus, in particular,

$$\int_\Omega \Phi(k|u(x)|)\,dx \leq C \int_\Omega \Phi(\lambda|u(x)|)\,dx < \infty,$$

hence $u \in E^\Phi(\Omega)$. This shows that $L^\Phi(\Omega) \subset E^\Phi(\Omega)$, and (4.12.2) follows from (4.12.1). □

Remark 4.12.4. If Φ is a Young function such that $\Phi \notin \Delta_2$ then

$$E^\Phi(\Omega) \subsetneqq \mathcal{L}^\Phi(\Omega) \subsetneqq L^\Phi(\Omega),$$

since $E^\Phi(\Omega)$ and $L^\Phi(\Omega)$ are vector spaces and $\mathcal{L}^\Phi(\Omega)$ is not a linear set according to Theorem 4.5.3.

Exercise 4.12.5. Show that $E^\Phi(\Omega)$ is the maximal subspace (i.e. maximal closed linear subset) of $L^\Phi(\Omega)$ which is contained in $\mathcal{L}^\Phi(\Omega)$.

It is of interest to study the distance of a given function from the set $E^\Phi(\Omega)$.

Remark 4.12.6. For $u \in L^\Phi(\Omega)$ let us denote

$$d(u; E^\Phi(\Omega)) := \inf_{w \in E^\Phi(\Omega)} \|u - w\|_\Phi. \tag{4.12.3}$$

For $r > 0$, let

$$B_{r,\Phi} := \{u \in L^\Phi(\Omega) : d(u; E^\Phi(\Omega)) < r\}. \tag{4.12.4}$$

If $\Phi \notin \Delta_2$ then

$$B_{1,\Phi} \subsetneqq \mathcal{L}^\Phi(\Omega) \subsetneqq \overline{B_{1,\Phi}}, \tag{4.12.5}$$

where the closure $\overline{B_{1,\Phi}}$ is taken with respect to the Orlicz norm.

Formula (4.12.5) has various interesting consequences. For example, since equality is excluded in (4.12.5), the Orlicz class is neither open nor closed subset of the corresponding Orlicz space.

For the proof of (4.12.5), see [123].

We now aim for yet another important characterization of the set $E^\Phi(\Omega)$.

Section 4.12 The space $E^\Phi(\Omega)$

Notation 4.12.7. Let us denote by $S(\Omega)$ the set of all simple functions defined on Ω. We recall that simple functions were introduced in Definition 1.20.23.

Theorem 4.12.8. *Let Φ be a Young function and let $\mu(\Omega) < \infty$. Then $E^\Phi(\Omega)$ coincides with the closure $\overline{S(\Omega)}$ of $S(\Omega)$ in the norm $\|\cdot\|_\Phi$.*

Proof. Let $u \in E^\Phi(\Omega)$. With no loss of generality, assume that $u \geq 0$. Consequently, there exists a sequence of simple functions $\{u_n\}_{n=1}^\infty$ such that

$$0 \leq u_n \uparrow u \quad \mu - \text{a.e. on } \Omega.$$

Then, for every $\lambda > 0$ and $n \in \mathbb{N}$, one has

$$\lambda u_n \in E^\Phi(\Omega)$$

and

$$\Phi(\lambda u) \geq \Phi(\lambda(u - u_n)).$$

Therefore, by the Lebesgue dominated convergence theorem (Theorem 1.21.5),

$$\int_\Omega \Phi(\lambda(u - u_n))\,dx \to 0, \quad n \to \infty.$$

Fix $\varepsilon > 0$ and take $\lambda > \frac{2}{\varepsilon}$. Then there is a $n_0 \in \mathbb{N}$ such that for every $n \in \mathbb{N}, n \geq n_0$, one has

$$\int_\Omega \Phi(\lambda(u - u_n))\,dx \leq 1.$$

Let Ψ be the complementary Young function of Φ. Then, by (4.6.3),

$$\|u - u_n\|_\Phi = \tfrac{1}{\lambda}\|\lambda(u - u_n)\|_\Phi$$
$$\leq \tfrac{1}{\lambda}\left(\int_\Omega \Phi(\lambda(u - u_n))\,dx + 1\right)$$
$$\leq \tfrac{2}{\lambda} < \varepsilon.$$

In other words, $u \in \overline{S(\Omega)}$.

Conversely, assume that $u \in \overline{S(\Omega)}$ and fix $\lambda > 0$. By (4.8.3), the Orlicz and the Luxemburg norms are equivalent on $L^\Phi(\Omega)$, hence, there exists a simple function u_0 such that

$$\|u - u_0\|_\Phi \leq \tfrac{1}{2\lambda}.$$

By Lemma 4.8.4 (i), we obtain

$$\varrho(2\lambda(u - u_0), \Phi) \leq 1.$$

Thus, in particular, we have

$$2\lambda(u - u_0) \in \mathcal{L}^\Phi(\Omega).$$

The set of simple functions $S(\Omega)$ is obviously contained in the Orlicz class $\mathcal{L}^\Phi(\Omega)$, hence also
$$2\lambda u_0 \in \mathcal{L}^\Phi(\Omega).$$
Finally, by the convexity of the Orlicz class $\mathcal{L}^\Phi(\Omega)$, we get
$$\lambda u = \frac{2\lambda(u - u_0) + 2\lambda u_0}{2} \in \mathcal{L}^\Phi(\Omega).$$

Because λ was an arbitrary positive number, this implies that $u \in E^\Phi(\Omega)$, as desired. The proof is complete. □

Based on the characterization of the space $E^\Phi(\Omega)$ given in Theorem 4.12.8, we can now establish certain approximative property of this space.

Lemma 4.12.9. *Let Φ be a Young function and let $\mu(\Omega) < \infty$. Let*
$$u \in L^\Phi(\Omega).$$
Then
$$d(u; E^\Phi(\Omega)) = \lim_{n\to\infty} \|u - u_n\|_\Phi, \tag{4.12.6}$$
where
$$u_n(x) = \begin{cases} u(x) & \text{if } |u(x)| \leq n, \\ 0 & \text{if } |u(x)| > n. \end{cases}$$

Proof. The sequence $\{|u(x) - u_n(x)|\}_{n=1}^\infty$ is nonincreasing for almost every $x \in \Omega$. Consequently, the sequence
$$\{\|u - u_n\|_\Phi\}_{n=1}^\infty$$
is nonincreasing as well (see Remark 4.6.8), therefore $\lim_{n\to\infty} \|u - u_n\|_\Phi$ exists and we have
$$\lim_{n\to\infty} \|u - u_n\|_\Phi \geq d(u; E^\Phi(\Omega)) \tag{4.12.7}$$
for u_n bounded.

Let $\varepsilon \in (0, 2)$ and
$$\frac{1}{d(u; E^\Phi(\Omega)) + 2\varepsilon} < \alpha < \frac{1}{d(u; E^\Phi(\Omega)) + \varepsilon}.$$
Then
$$d(\alpha u; E^\Phi(\Omega)) = \inf_{w \in E^\Phi(\Omega)} \|\alpha u - w\|_\Phi = \alpha \inf_{w \in E^\Phi(\Omega)} \|u - w\|_\Phi$$
$$< \frac{d(u; E^\Phi(\Omega))}{d(u; E^\Phi(\Omega)) + \varepsilon} < 1$$

and (4.12.5) implies

$$\alpha u \in \mathcal{L}^\Phi(\Omega), \quad \text{i.e.} \quad \int_\Omega \Phi(\alpha|u(x)|)\,dx < \infty.$$

Choosing $n \in \mathbb{N}$ sufficiently large we obtain

$$\int_\Omega \Phi(\alpha|u(x) - u_n(x)|)\,dx < \alpha\varepsilon$$

(see, e.g., (4.10.9)) and from formula (4.6.3) it follows that

$$\|\alpha u - \alpha u_n\|_\Phi < 1 + \alpha\varepsilon, \quad \text{i.e.} \quad \|u - u_n\|_\Phi < \alpha^{-1} + \varepsilon < d(u; E^\Phi(\Omega)) + 3\varepsilon.$$

Since $\varepsilon > 0$ is arbitrary we have

$$\lim_{n\to\infty} \|u - u_n\|_\Phi \leq d(u; E^\Phi(\Omega)). \tag{4.12.8}$$

Inequalities (4.12.7) and (4.12.8) imply (4.12.6). □

The following proposition, whose proof is left to the reader as an exercise, can be found useful in evaluating the Orlicz norm of a given function.

Proposition 4.12.10. *Let Φ be a Young function and let $u \in L^\Phi(\Omega)$. Then we have*

$$\|u\|_\Phi = \sup\left\{\int_\Omega |u(x)v(x)|\,dx,\ v \in E^\Psi(\Omega),\ \varrho(v, \Psi) \leq 1\right\}.$$

Essentially in the same way as in the proof of Theorem 4.11.1, we can prove the following theorem. We emphasize that it does not require the Δ_2-condition.

Theorem 4.12.11. *Let Φ be a Young function. Then the space $E^\Phi(\Omega)$ is separable.*

Definition 4.12.12. Let Φ be a Young function, let $\mu(\Omega) < \infty$ and let $u \in L^\Phi(\Omega)$. We say that u has an *absolutely continuous norm* in $L^\Phi(\Omega)$ if for every $\varepsilon > 0$ there exists a $\delta > 0$ such that

$$\|u\chi_E\|_\Phi < \varepsilon$$

whenever $\mu(E) < \delta$.

We shall denote by $L_a^\Phi(\Omega)$ the subspace of $L^\Phi(\Omega)$ containing all functions with absolutely continuous norms in $L^\Phi(\Omega)$.

Theorem 4.12.13. *Let Φ be a Young function, let $\mu(\Omega) < \infty$. Then*

$$L_a^\Phi(\Omega) = E^\Phi(\Omega).$$

Proof. Assume first that $u \in E^\Phi(\Omega)$ and let $\varepsilon > 0$. According to Theorem 4.12.8, there is a bounded function v on Ω such that

$$\|u - v\|_\Phi < \frac{\varepsilon}{2}.$$

Let us denote

$$M := \sup_{x \in \Omega} |v(x)|.$$

Then there exists a uniquely determined $\delta > 0$ such that

$$\delta \Psi^{-1}\left(\frac{1}{\delta}\right) = \frac{\varepsilon}{2M},$$

since the function

$$t \to t \Psi^{-1}\left(\frac{1}{t}\right)$$

is nondecreasing. If now $E \subset \Omega$ is such that $\mu(E) < \delta$, then, by (4.6.10),

$$\|u \chi_E\|_\Phi \leq \|u - v \chi_E\|_\Phi + M \|\chi_E\|_\Phi \leq \frac{\varepsilon}{2} + M \mu(E) \Psi^{-1}\left(\frac{1}{\mu(E)}\right)$$

$$\leq \frac{\varepsilon}{2} + M \delta \Psi^{-1}\left(\frac{1}{\delta}\right) = \varepsilon,$$

hence $u \in L_a^\Phi(\Omega)$.

Conversely, let $u \in L_a^\Phi(\Omega)$. Denote

$$E_n := \{x \in \Omega;\ |u(x)| \leq n\}, \quad n \in \mathbb{N}.$$

Since $u \in L^1(\Omega)$ and $\mu(\Omega) < \infty$, we have

$$\lim_{n \to \infty} \mu(\Omega \setminus E_n) = 0.$$

Thus, by the absolute continuity of the norm,

$$\lim_{n \to \infty} \|u - u \chi_{E_n}\|_\Phi = 0.$$

In other words, u is a limit in $L^\Phi(\Omega)$ of a sequence of bounded functions, whence, by Theorem 4.12.8, $u \in E^\Phi(\Omega)$. The proof is complete. □

Definition 4.12.14. Let X be a Banach space containing functions defined on Ω and let $K \subset X$. We say that functions in K have *uniformly absolutely continuous* norms in X if for every $\varepsilon > 0$ there exists a $\delta > 0$ such that for every measurable subset $E \subset \Omega$ with $\mu(E) < \delta$ we have

$$\|u \chi_E\|_X < \varepsilon \quad \text{for all } u \in K. \tag{4.12.9}$$

Theorem 4.12.15 (de la Vallée-Poussin)**.** *Let Φ be a Young function and let K be a Φ-mean bounded subset of $L^\Phi(\Omega)$. Then the functions $u \in K$ have uniformly absolutely continuous norms in $L^1(\Omega)$.*

Proof. Assume that K is a Φ-mean bounded set in $L^\Phi(\Omega)$. Then there exists a positive constant C such that $\varrho(u, \Phi) \leq C$ for every $u \in K$. By (4.6.3), this implies that
$$\|u\|_\Phi \leq C + 1, \quad u \in K.$$

Let $E \subset \Omega$. Then, by the complementary version of (4.6.10), we have
$$\|\chi_E\|_\Psi = \mu(E)\Phi^{-1}\left(\frac{1}{\mu(E)}\right).$$

However, in view of (4.2.3), one has
$$\lim_{t \to 0+} t\Phi(\tfrac{1}{t}) = 0.$$

Thus, given $\varepsilon > 0$, there exists a $\delta > 0$ such that
$$\|\chi_E\|_\Psi < \varepsilon \quad \text{for every } E \subset \Omega \text{ satisfying } \mu(E) < \delta.$$

Altogether, then, for every such $E \subset \Omega$ and every $u \in K$, we have by (4.7.4)
$$\|u\chi_E\|_{L^1} = \int_E |u(x)|\,dx \leq \|u\|_\Phi \|\chi_E\|_\Psi \leq (C+1)\varepsilon,$$

and the assertion follows. The proof is complete. □

4.13 Continuous linear functionals

In Section 3.8 we characterized bounded linear functionals on the spaces $L^p(\Omega)$. If $L^p(\Omega)$ is replaced by $L^\Phi(\Omega)$, then only a partial analogue of the Riesz representation theorem (see Theorems 3.8.3 and 3.8.1) holds.

Theorem 4.13.1. *Let Φ, Ψ be a pair of complementary Young functions, let v be a fixed function in $L^\Psi(\Omega)$. Then the formula*
$$F(u) = \int_\Omega u(x)v(x)\,dx, \quad u \in L^\Phi(\Omega), \tag{4.13.1}$$

defines a continuous linear functional F on $L^\Phi(\Omega)$ and
$$\frac{1}{2}\|v\|_\Psi \leq \|F\| \leq \|v\|_\Psi, \tag{4.13.2}$$

where $\|F\|$ is the norm of F in $\left[L^\Psi(\Omega)\right]^*$.

Proof. From formula (4.7.4) we have that

$$|F(u)| \le \|u\|_\Phi \|v\|_\Psi. \tag{4.13.3}$$

Thus the second inequality in (4.13.2) clearly holds. The first inequality in (4.13.2) follows from (4.6.3) and from the definition of the Orlicz norm $\|v\|_\Psi$. Indeed, for $\varrho(u, \Phi) \le 1$ we have $\|u\|_\Phi \le \varrho(u, \Phi) + 1 \le 2$ and

$$\|v\|_\Psi = \sup_{\varrho(u,\Phi) \le 1} \left| \int_\Omega u(x)v(x)\, dx \right| \le \sup_{\|u\|_\Phi \le 2} |F(u)| = \sup_{\|u\|_\Phi \le 1} |F(2u)| \le 2\|F\|.$$

We have also used Exercise 4.6.6. □

Remark 4.13.2. The analogue of Theorem 3.8.3 does not hold. Firstly, formula (4.13.1) need not be the general form of a bounded linear functional (see Theorem 4.13.5) and, secondly, we do not generally have

$$\|F\| = \|v\|_\Psi$$

as would be reasonable to require. Setting

$$k(v) = \frac{\|v\|_\Psi}{\|F\|}, \quad v \in L^\Psi(\Omega), \quad \|v\|_\Psi \ne 0 \tag{4.13.4}$$

we have from (4.13.2) only the estimate

$$1 \le k(v) \le 2.$$

Example 4.13.3. Let $p \in (1, \infty)$ and set $\Phi(t) := \frac{t^p}{p}$. Then we have

$$k(v) = (p)^{\frac{1}{p}} (p')^{\frac{1}{p'}}$$

for every $v \in L^\Psi(\Omega)$, where k is the function from (4.13.4). In case of the Lebesgue space we had

$$\tilde{k}(v) = \frac{1}{p} + \frac{1}{p'} = 1.$$

Remark 4.13.4. The function k in (4.13.4) defines a mapping of the set $L^\Psi(\Omega) \setminus \{0\}$ into $\mathbb{R}$, more precisely into the interval $[1, 2]$. In the foregoing example the function $k(v)$ was constant. Salekhov [193] investigated the functions $k(v)$ in detail and has shown that $k(v)$ is constant only if $\Phi(t) = ct^p$ with $c > 0$, $p > 1$. One can show (see, e.g., [123]) that the range of the function k contains all numbers of the form

$$\frac{1}{\alpha} \Phi^{-1}(\alpha) \Psi^{-1}(\alpha), \quad \alpha > \frac{1}{\mu(\Omega)}.$$

According to Theorem 4.13.1 we can consider the space $L^\Psi(\Omega)$ to be a subset of $[L^\Phi(\Omega)]^*$. But, in general

$$L^\Psi(\Omega) \neq [L^\Phi(\Omega)]^*.$$

Theorem 4.13.5. *Assume that Φ does not satisfy the Δ_2-condition. Then there exists a continuous linear functional on $L^\Phi(\Omega)$ which cannot be expressed in the form* (4.13.1).

Proof. Under the assumptions of the theorem, $E^\Phi(\Omega)$ is a proper subspace $L^\Phi(\Omega)$ (see Remark 4.12.4), and thus there exists a function $u_0 \in L^\Phi(\Omega)$ such that $u_0 \notin E^\Phi(\Omega)$. Let F be a bounded linear functional on $L^\Phi(\Omega)$ with

$$F(u_0) = 1, \quad F(u) = 0 \quad \text{for } u \in E^\Phi(\Omega). \tag{4.13.5}$$

The existence of at least one functional F with property (4.13.5) follows immediately from the Hahn–Banach theorem (Theorem 1.16.4).

Let us suppose that F can be expressed in the form (4.13.1) with a function $v \in L^\Psi(\Omega)$ and let us define the sequence of functions $\{v_n\}_{n=1}^\infty$ by

$$v_n(x) = \begin{cases} v(x) & \text{for } |v(x)| \leq n, \\ 0 & \text{for } |v(x)| > n. \end{cases}$$

Obviously $v_n \in E^\Phi(\Omega)$, and hence

$$F(v_n) = \int_\Omega v_n(x) v(x)\, dx = 0 \quad \text{for } n \in \mathbb{N}.$$

This means that $v(x) = 0$ almost everywhere on Ω and thus also $F(u_0) = 0$, which contradicts the first condition in (4.13.5). $\square$

The following theorem is an "almost-complete" analogue of Theorem 3.8.3.

Theorem 4.13.6. *Let F be a bounded linear functional on $E^\Phi(\Omega)$. Then there exists a uniquely determined $v \in L^\Psi(\Omega)$ such that*

$$F(u) = \int_\Omega u(x) v(x)\, dx, \quad u \in E^\Phi(\Omega). \tag{4.13.6}$$

Proof. Since the uniqueness of v is obvious, it is enough to show its existence. For a measurable subset M of Ω, set

$$\nu(M) = F(\chi_M).$$

Using Example 4.6.9 we have

$$|\nu(M)| = |F(\chi_M)| \leq \|F\| \, \|\chi_M\|_\Phi = \|F\| \mu(M) \Psi^{-1}\left(\frac{1}{\mu(M)}\right),$$

and hence
$$\lim_{\mu(M)\to 0} \nu(M) = 0,$$

($\lim_{t\to\infty} \frac{\Psi^{-1}(t)}{t} = 0$ because $\lim_{t\to\infty} \frac{\Psi(t)}{t} = \infty$). Hence ν is a measure which is absolutely continuous with respect to the Lebesgue measure μ. The Radon–Nikodým Theorem (Theorem 1.21.15) implies the existence of a function $v \in L^1(\Omega)$ such that

$$\nu(M) = \int_M v(x)\,dx. \qquad (4.13.7)$$

If u is a simple function, i.e. $u(x) = \sum_{i=1}^m \alpha_i \chi_{M_i}(x)$ with

$$M_i \subset \Omega, \quad M_i \cap M_j = \emptyset \quad \text{for } i \neq j,$$

we have, according to (4.13.7),

$$\begin{aligned}
F(u) &= \sum_{i=1}^m \alpha_i F(\chi_{M_i}) = \sum_{i=1}^m \alpha_i \nu(\chi_{M_i}) = \sum_{i=1}^m \alpha_i \int_{M_i} v(x)\,dx \\
&= \sum_{i=1}^m \alpha_i \int_\Omega v(x)\chi_{M_i}(x)\,dx = \int_\Omega u(x)v(x)\,dx.
\end{aligned} \qquad (4.13.8)$$

Let $u \in E^\Phi(\Omega)$. Then there exists a sequence of simple functions $\{u_n\}_{n=1}^\infty$ such that

$$u(x) = \lim_{n\to\infty} u_n(x), \quad |u_n(x)| \leq |u(x)|$$

for almost all $x \in \Omega$ (see the proof of Theorem 4.11.1). Hence (see Remark 4.6.8) $\|u_n\|_\Phi \leq \|u\|_\Phi$ and, moreover,

$$\lim_{n\to\infty} |u_n(x)v(x)| = |u(x)v(x)|$$

almost everywhere on Ω. The Fatou lemma 1.21.6 implies that

$$\begin{aligned}
\left|\int_\Omega u(x)v(x)\,dx\right| &\leq \sup_{n\in\mathbb{N}} \int_\Omega |u_n(x)v(x)|\,dx \\
&= \sup_{n\in\mathbb{N}} F(|u_n|\operatorname{sign} v) \leq \|F\| \sup_{n\in\mathbb{N}} \|u_n\|_\Phi \leq \|F\|\,\|u\|_\Phi.
\end{aligned}$$

Thus $v \in L^\Psi(\Omega)$.

Now let F_1 be the functional which corresponds to this function v by formula (4.13.1), that is,

$$F_1(u) := \int_\Omega u(x)v(x)\,dx, \quad u \in E^\Phi(\Omega).$$

In view of (4.13.8) we obtain $F_1(u) = F(u)$ for simple functions u. However, the set of simple functions is dense in $E^\Phi(\Omega)$ due to Theorem 4.12.8, and consequently $F_1(u) = F(u)$ for all $u \in E^\Phi(\Omega)$. The proof is complete. $\square$

Remark 4.13.7. Let F be a bounded linear functional on $L^\Phi(\Omega)$ with norm

$$\|F\| = \sup\left\{|F(u)|\,;\, u \in L^\Phi(\Omega),\, \|u\|_\Phi \le 1\right\}.$$

Since the restriction of F to $E^\Phi(\Omega)$ (denote it again by F) is a bounded linear functional on $E^\Phi(\Omega)$ with norm

$$\|F\|_E = \sup\left\{|F(u)|\,;\, u \in E^\Phi(\Omega),\, \|u\|_\Phi \le 1\right\},$$

there exists (according to Theorem 4.13.6) a function $v \in L^\Psi(\Omega)$ such that

$$F(u) = \int_\Omega u(x)v(x)\,dx, \quad u \in E^\Phi(\Omega). \tag{4.13.9}$$

Obviously,
$$\|F\|_E \le \|F\|. \tag{4.13.10}$$

Moreover, if we suppose that (4.13.9) holds not only for $u \in E^\Phi(\Omega)$ but also for all $u \in L^\Phi(\Omega)$ then $\|F\| = \|F\|_E$. (The proof proceeds by contradiction.)

Remark 4.13.8. Using the isomorphism given by (4.13.6), it is possible to rewrite the assertions of Theorems 4.13.6 and 4.13.1 as

$$L^\Psi(\Omega) = \left[E^\Phi(\Omega)\right]^*. \tag{4.13.11}$$

If $\Phi \in \Delta_2$, then, according to Proposition 4.12.3,

$$L^\Psi(\Omega) = \left[L^\Phi(\Omega)\right]^*$$

and if $\Psi \in \Delta_2$, then

$$L^\Phi(\Omega) = \left[L^\Psi(\Omega)\right]^*.$$

The last two equalities are of the same type as in the case of Lebesgue spaces and we can prove the following theorem in the same way as Theorem 3.9.1.

Theorem 4.13.9. *Let Φ, Ψ be a pair of complementary Young functions. Then the Orlicz space $L^\Phi(\Omega)$ is reflexive if and only if both Φ and Ψ satisfy the Δ_2-condition.*

4.14 Compact subsets of Orlicz spaces

In this section we shall give necessary and sufficient conditions for K to be a relatively compact subset of $L^\Phi(\Omega)$. First we introduce some notions which are analogous to those introduced in Definition 3.14.1.

Definition 4.14.1. A subset K of $L^\Phi(\Omega)$ is said to be Φ-*equicontinuous* for every $\varepsilon > 0$ there exists a $\delta = \delta(\varepsilon) > 0$ such that for $h \in \mathbb{R}^N$ with $|h| < \delta$ and for every $u \in K$ we have
$$\|u - u_h\|_\Phi < \varepsilon, \tag{4.14.1}$$
where u_h is the function defined by
$$u_h(x) := \begin{cases} u(x+h) & \text{for } x \in \Omega \text{ with } x+h \in \Omega, \\ 0 & \text{otherwise in } \mathbb{R}^N. \end{cases} \tag{4.14.2}$$

Notation 4.14.2. For $r > 0$ and $x \in \mathbb{R}^N$ let $B(x, r)$ be the ball
$$\{y \in \mathbb{R}^N; |x - y| < r\}$$
and m_r the measure of $B(x, r)$. Further, for $u \in L^\Phi(\Omega)$ let us set $u(y) = 0$ if $y \notin \Omega$. The so-called *Steklov function* $S_r(u)$ corresponding to u is defined as follows:
$$S_r(u)(x) := \frac{1}{m_r} \int_{B(x,r)} u(y)\,dy = \frac{1}{m_r} \int_{|y|<r} u(x+y)\,dy. \tag{4.14.3}$$

Remark 4.14.3. The term "Steklov function" is used, e.g., in [123]. The meaning of the function $S_r(u)$ is analogous to that of a *mollifier* introduced in Section 3.4. One can prove in the same way as in Theorems 3.4.3 and 3.4.4 that $S_r(u)$ is continuous on $\mathbb{R}^N$ and has compact support.

Remark 4.14.4 (further properties of Steklov functions).   (i) Let us investigate the Orlicz norm of $S_r(u)$. We have
$$\|S_r(u)\|_\Phi \leq \frac{1}{m_r} \sup_{\varrho(v,\Psi)\leq 1} \int_\Omega \int_{B(x,r)} |u(y)v(x)|\,dy\,dx$$
$$= \frac{1}{m_r} \sup_{\varrho(v,\Psi)\leq 1} \int_\Omega \int_{|y|<r} |u(x+y)v(x)|\,dy\,dx.$$

From the Fubini theorem (Theorem 1.21.11), we have
$$\|S_r(u)\|_\Phi \leq \frac{1}{m_r} \int_{|y|<r} \left(\sup_{\varrho(v,\Psi)\leq 1} \int_\Omega |u(x+y)v(x)|\,dx \right) dy = \|u_y\|_\Phi,$$
where u_y is the function defined in (4.14.2). Using the fact that $u(x) = 0$ for $x \notin \Omega$ we obtain
$$\|u_y\|_\Phi = \sup_{\varrho(v,\Psi)\leq 1} \int_\Omega |u(x+y)v(x)|\,dx$$
$$\leq \sup_{\varrho(v,\Psi)\leq 1} \int_\Omega |u(x)v(x-y)|\,dx \leq \|u\|_\Phi,$$

Section 4.14 Compact subsets of Orlicz spaces 157

and, finally, we have
$$\|S_r(u)\|_\Phi \le \|u\|_\Phi. \tag{4.14.4}$$

(ii) Let K be a bounded subset of $L^\Phi(\Omega)$, i.e. let there exist a constant $C > 0$ such that
$$\|u\|_\Phi \le C \quad \text{for all } u \in K. \tag{4.14.5}$$

For $r \in (0, \infty)$ fixed we denote by K_r the set of all Steklov functions $S_r(u)$ corresponding to $u \in K$. As was mentioned in Remark 4.14.3 we have $K_r \subset C_0^\infty(\mathbb{R}^N)$.

Proposition 4.14.5. *Let Φ be a Young function, let K be a bounded subset of $L^\Phi(\Omega)$ and let $r \in (0, \infty)$. Then K_r is relatively compact in $C(\overline{\Omega})$.*

Proof. We give only a sketch. Using (4.14.4), the monotonicity of Φ and Lemma 4.7.2, we have
$$\int_\Omega \Phi\left(\frac{1}{C}|u(x)|\right) dx \le \int_\Omega \Phi\left(\frac{1}{\|u\|_\Phi}|u(x)|\right) dx \le 1 \tag{4.14.6}$$
for all $u \in K$, and (4.14.6) together with the Young inequality implies
$$|S_r(u)(x)| \le \frac{1}{m_r} \int_{B(x,r)} |u(y)|\,dy = \frac{C}{m_r} \int_{B(x,r)} \frac{|u(y)|}{C}\,dy$$
$$\le \frac{C}{m_r} \int_\Omega \frac{|u(y)|}{C}\,dy \le \frac{C}{m_r}\left(\varrho\left(\frac{u}{C},\Phi\right) + \varrho(1,\Psi)\right)$$
$$\le \frac{C}{m_r}(1 + \Psi(1)\mu(\Omega)),$$

i.e. the functions $S_r(u) \in K_r$ are uniformly bounded, or, in other words, K_r is a bounded subset of $C(\overline{\Omega})$. Now, according to (4.14.6), K_r is Φ-mean bounded. Thus, by the de la Vallée–Poussin theorem (Theorem 4.12.15), for every $\varepsilon > 0$, there exists a $\delta > 0$ such that, for every $E \subset \Omega$ such that $\mu(E) < \delta$, we have
$$\int_E |u(x)|\,dx < \varepsilon \quad \text{for all } u \in K_r.$$

Let us denote
$$B_{x,y} = (B(x,r) \cup B(y,r)) \setminus (B(x,r) \cap B(y,r))$$
and let us choose an $\eta > 0$ such that $\mu(B_{x,y}) < \delta$ for $|x - y| < \eta$. With respect to (4.12.9) we have for such x, y
$$|S_r(u)(x) - S_r(u)(y)| \le \frac{1}{m_r} \int_{B_{x,y}} |u(z)|\,dz \le \frac{\varepsilon}{m_r}$$
provided $S_r(u) \in K_r$, i.e. the set K_r is equicontinuous (see Definition 1.5.1). The assertion now follows from the Arzelà–Ascoli theorem (Theorem 2.5.3). □

Now we are able to prove a criterion for relative compactness in $E^\Phi(\Omega)$.

Theorem 4.14.6. *A subset K of $E^\Phi(\Omega)$ is relatively compact if and only if the following conditions are satisfied:*

(i) *the set K is bounded in $E^\Phi(\Omega)$;*

(ii) *for every $\varepsilon > 0$ there exists a $\delta = \delta(\varepsilon) > 0$ such that*

$$\|u - S_r(u)\|_\Phi < \varepsilon \tag{4.14.7}$$

for $0 < r < \delta$ and every $u \in K$.

Proof. Conditions (i), (ii) are sufficient for K to be relatively compact as follows from Proposition 4.14.5. Indeed, the corresponding ε-net in $E^\Phi(\Omega)$ can be taken to be the ε-net in $C(\overline{\Omega})$, the existence of which is guaranteed by Proposition 4.14.5, and then we use (4.14.7).

Conversely, let K be relatively compact in $E^\Phi(\Omega)$. Then there exists a finite $\frac{1}{3}\varepsilon$-net $\{w^1, w^2, \ldots, w^s\}$. Because $K \subset E^\Phi(\Omega)$, we can take the functions $\{w^i\}$, $i = 1, \ldots, s$, to be continuous on $\overline{\Omega}$. Let $c = \|\chi_\Omega\|_\Phi$ and let us choose a number $r > 0$ such that

$$\left| w^i(x) - w^i(y) \right| < \frac{\varepsilon}{3c} \quad (i = 1, \ldots, s) \tag{4.14.8}$$

for $x, y \in \Omega$ with $|x - y| < r$. Then in view of (4.14.8) the corresponding Steklov functions $S_r(w^i)$ satisfy

$$\left| S_r(w^i)(x) - w^i(x) \right| \leq \left| \frac{1}{m_r} \int_{B_{x,y}} w(y)\, dy - \frac{1}{m_r} \int_{B_{x,y}} w^i(x)\, dy \right|$$

$$= \frac{1}{m_r} \left| \int_{B_{x,y}} (w^i(y) - w^i(x))\, dy \right| < \frac{\varepsilon}{3c},$$

and the monotonicity of the Orlicz norm (see Remark 4.6.8) yields

$$\left\| S_r(w^i) - w^i \right\|_\Phi < \left\| \frac{\varepsilon}{3c} \chi_\Omega \right\|_\Phi = \frac{\varepsilon}{3}. \tag{4.14.9}$$

Let u be an arbitrary function in K. Then there exists an index σ, $1 \leq \sigma \leq s$ such that

$$\|u - w^\sigma\|_\Phi < \frac{\varepsilon}{3}. \tag{4.14.10}$$

On the other hand, it follows from (4.14.4) that also

$$\|S_r(u - w^\sigma)\|_\Phi = \|S_r(u) - S_r(w^\sigma)\|_\Phi < \frac{\varepsilon}{3}. \tag{4.14.11}$$

Using (4.14.9), (4.14.10) and (4.14.11), we have

$$\|u - S_r u\|_\Phi \leq \|u - w^\sigma\|_\Phi + \|w^\sigma - S_r(w^\sigma)\|_\Phi + \|S_r(w^\sigma) - S_r(u)\|_\Phi < \varepsilon,$$

so that condition (ii) is necessary for K to be relatively compact. As the necessity of (i) is obvious, the theorem is proved. □

Remark 4.14.7. If Φ satisfies the Δ_2-condition then $E^\Phi(\Omega) = L^\Phi(\Omega)$ so that Theorem 4.14.6 gives necessary and sufficient conditions for K to be relatively compact in $L^\Phi(\Omega)$. For $\Phi \in \Delta_2$ norm convergence and norm boundedness are equivalent to Φ-mean convergence and Φ-mean boundedness, respectively (see Theorem 4.10.6, Remark 4.10.13) so that we can reformulate Theorem 4.14.6 as follows.

Theorem 4.14.8. *Let Φ satisfy the Δ_2-condition. Then a subset K of $L^\Phi(\Omega)$ is relatively compact in $L^\Phi(\Omega)$ if and only if the following conditions are satisfied:*

(i) *there exists a $C > 0$ such that*

$$\int_\Omega \Phi(|u(x)|)\,dx \leq C \qquad \text{for all} \quad u \in K;$$

(ii) *for every $\varepsilon > 0$ there exists a $\delta = \delta(\varepsilon) > 0$ such that*

$$\int_\Omega \Phi(|u(x) - S_r(u)(x)|)\,dx < \varepsilon$$

for $0 < r < \delta$ and for all $u \in K$.

Another criterion for relative compactness, more similar to that from the Riesz theorem (Theorem 3.14.2) and using the concept of Φ-equicontinuity introduced in Definition 4.14.1, is given by the following theorem.

Theorem 4.14.9. *A subset K of $E^\Phi(\Omega)$ is relatively compact if and only if the following conditions are satisfied:*

(i) *the set K is bounded in $E^\Phi(\Omega)$;*

(ii) *the set K is Φ-equicontinuous.*

Proof. Let $u \in K$ and let $S_r(u)$ be the corresponding Steklov function. We have

$$|u(x) - S_r(u)(x)| = \left|\frac{1}{m_r}\int_{|y|<r}(u(x) - u(x+y))\,dy\right|$$

$$\leq \frac{1}{m_r}\int_{|y|<r}|u(x) - u(x+y)|\,dy.$$

Multiplying this inequality by $|v(x)|$ where $v \in \mathcal{L}^\Psi(\Omega)$ with $\varrho(v, \Psi) \leq 1$ and integrating over Ω, we obtain in view of the Fubini theorem (Theorem 1.21.11) that

$$\int_\Omega |u(x) - S_r(u)(x)| \, |v(x)| \, dx$$

$$\leq \frac{1}{m_r} \int_{|y|<r} \left[\int_\Omega |u(x+y) - u(x)| \, |v(x)| \, dx \right] dy$$

$$\leq \frac{1}{m_r} \int_{|y|<r} \|u_y - u\|_\Phi \, dy,$$

and consequently also

$$\|u - S_r(u)\|_\Phi \leq \frac{1}{m_r} \int_{|y|<r} \|u_y - u\|_\Phi \, dy. \tag{4.14.12}$$

If condition (ii) is satisfied then (4.14.12) implies that condition (ii) in Theorem 4.14.6 is also satisfied. Since the conditions (i) of both theorems coincide, it follows from Theorem 4.14.6 that conditions (i), (ii) are sufficient for relative compactness.

Conversely, the necessity of (i), (ii) follows by an argument analogous to that in the proof of Theorem 4.14.6: Given $\varepsilon > 0$, denote by $w^1, \ldots, w^s$ the finite ε-net of K with $w^i \in C(\overline{\Omega})$ and choose a $\delta = \delta(\varepsilon) > 0$ such that

$$|w^i(x+h) - w^i(x)| < \tfrac{\varepsilon}{3c} \text{ for } h \in \mathbb{R}^N, |h| < \delta, i = 1, \ldots, s$$

with $c = \|\chi_\Omega\|_\Phi$. Then $\|(w^i)_h - w^i\|_\Phi < \tfrac{1}{3}\varepsilon$ and the obvious inequality

$$\|u_h\|_\Phi \leq \|u\|_\Phi,$$

which holds for all $u \in L^\Phi(\Omega)$ finally yields

$$\|u - u_h\|_\Phi \leq \|u - w^\sigma\|_\Phi + \|w^\sigma - (w^\sigma)_h\|_\Phi + \|(w^\sigma - u)_h\|_\Phi < \varepsilon$$

which is in fact (ii). Hence the theorem is proved. □

In concluding this section we give a sufficient condition for compactness using the concept of absolutely continuous L^Φ-norms (see Definition 4.12.12).

Theorem 4.14.10. *Let K be a subset of $E^\Phi(\Omega)$ which is relatively compact in the sense of convergence in measure. Let $u \in K$ have uniformly absolutely continuous norms in $L^\Phi(\Omega)$. Then K is relatively compact in $L^\Phi(\Omega)$.*

Proof. Let $\{u_n\}_{n=1}^\infty$ be a sequence in K and let $\varepsilon > 0$. The uniform absolute continuity of the L^Φ-norms of the set K implies that there exists a $\delta = \delta(\varepsilon) > 0$ such that

$$\|u_n \chi_M\|_\Phi < \frac{1}{3}\varepsilon \tag{4.14.13}$$

for every $M \subset \Omega$ with $\mu(M) < \delta$ and for all $n \in \mathbb{N}$. Since K is relatively compact in the sense of convergence in measure there exist a subsequence $\{\tilde{u}_n\}_{n=1}^\infty$ of $\{u_n\}_{n=1}^\infty$, a function $u \in L^\Phi(\Omega)$ and $n_0 = n_0(\delta) \in \mathbb{N}$ such that

$$\mu(\Omega_n) < \delta \quad \text{for } n > n_0,$$

where

$$\Omega_n = \left\{ x \in \Omega; \ |\tilde{u}_n(x) - u(x)| > \frac{1}{3}\varepsilon \left[\mu(\Omega) \Psi^{-1}\left(\frac{1}{\mu(\Omega)}\right) \right]^{-1} \right\}.$$

Consequently, we have

$$\|\tilde{u}_n - u\|_\Phi \leq \|(\tilde{u}_n - u)\chi_{\Omega_n}\|_\Phi + \|(\tilde{u}_n - u)(1 - \chi_{\Omega_n})\|_\Phi$$

$$\leq \|\tilde{u}_n \chi_{\Omega_n}\|_\Phi + \|u \chi_{\Omega_n}\|_\Phi + \frac{1}{3}\varepsilon \left[\mu(\Omega) \Psi^{-1}\left(\frac{1}{\mu(\Omega)}\right) \right]^{-1} \|\chi_{\Omega_n}\|_\Phi$$

and using formulas (4.6.10) and (4.14.13) we obtain

$$\|\tilde{u}_n - u\|_\Phi < \varepsilon \quad \text{for } n > n_0,$$

so that $\tilde{u}_n$ converges to u in $L^\Phi(\Omega)$. $\square$

4.15 Further properties of Orlicz spaces

Definition 4.15.1. Let Φ be a Young function. A function $u \in L^\Phi(\Omega)$ is said to be Φ-*mean continuous* if for every $\varepsilon > 0$ there exists a $\delta = \delta(\varepsilon) > 0$ such that

$$\|u_h - u\|_\Phi < \varepsilon \tag{4.15.1}$$

for $h \in \mathbb{R}^N$ with $|h| < \delta$, where

$$u_h(x) = \begin{cases} u(x+h) & \text{if } x \in \Omega \text{ and } x+h \in \Omega, \\ 0 & \text{otherwise in } \mathbb{R}^N. \end{cases}$$

Remark 4.15.2. Let $p \in (1, \infty)$. For $\Phi(t) = \frac{t^p}{p}$, inequality (4.15.1) means the same (possible up to a multiplicative constant) as

$$\int_\Omega \Phi(|u(x+h) - u(x)|) \, dx < \varepsilon,$$

i.e.

$$\varrho(u_h - u, \Phi) < \varepsilon.$$

Moreover, every function $u \in L^p(\Omega) = L^\Phi(\Omega)$ is Φ-mean continuous (i.e. in the terminology introduced in Definition 3.3.2, $u \in L^p(\Omega)$ is *p-mean continuous* – see Theorem 3.3.3).

These assertions do not hold in general for Orlicz spaces. Nonetheless, the following analogous assertion, whose proof is left to the reader, holds for $E^\Phi(\Omega)$.

Theorem 4.15.3. *Let $u \in E^\Phi(\Omega)$. Then for every $\varepsilon > 0$ there exists a $\delta = \delta(\varepsilon) > 0$ such that for $h \in \mathbb{R}^N$ with $|h| < \delta$ we have*

$$\varrho(u_h - u, \Phi) < \varepsilon.$$

If Φ satisfies the Δ_2-condition then $L^\Phi(\Omega) = E^\Phi(\Omega)$. Using Theorem 4.10.6, we obtain immediately from Theorem 4.15.3 the following result.

Theorem 4.15.4. *If Φ satisfies the Δ_2-condition, then every function $u \in L^\Phi(\Omega)$ is Φ-mean continuous.*

For $u \in E^\Phi(\Omega)$, the same assertion as above holds without any restriction on Φ:

Theorem 4.15.5. *If $u \in E^\Phi(\Omega)$, then u is Φ-mean continuous.*

Proof. The assertion follows from Theorem 4.15.3. Let $u \in E^\Phi(\Omega)$ and let $\varepsilon > 0$. Let $k > 0$ be arbitrary but fixed, and let us denote $v = \frac{u}{k}$. Then $v_h = \frac{u_h}{k}$, $v \in E^\Phi(\Omega)$ and Theorem 4.15.3 used for v yields that there exists a $\delta = \delta(\varepsilon) > 0$ such that for $h \in \mathbb{R}^N$ with $|h| < \delta$ we have

$$\varrho(v_h - v, \Phi) = \varrho(\frac{u_h - u}{k}, \Phi) \le 1.$$

If we choose $k = \frac{1}{2}\varepsilon$ we conclude from the definition of the Luxemburg norm that

$$\|u_h - u\|_\Phi < \frac{1}{2}\varepsilon,$$

and from (4.8.3) we immediately obtain

$$\|u_h - u\|_\Phi \le 2\|u_h - u\|_\Phi < \varepsilon. \quad \square$$

As the general representation of functionals on general Orlicz spaces is not known, we shall now introduce the notion of a modified weak convergence:

Definition 4.15.6. A sequence $\{u_n\}_{n=1}^\infty$ in $L^\Phi(\Omega)$ is said to *converge E^Ψ-weakly* to $u \in L^\Phi(\Omega)$ if

$$\lim_{n \to \infty} \int_\Omega (u_n(x) - u(x))v(x)\,dx = 0$$

for all $v \in E^\Psi(\Omega)$.

Remark 4.15.7. If both complementary Young functions Φ, Ψ satisfy the Δ_2-condition then the concept of E^Ψ-weak convergence coincide with that of the usual weak convergence.

Standard theorems concerning weak convergence enable us to derive various assertions, see, e.g., [123], where also further properties connected with the concept of E^Φ-weak convergence can be found. We shall give only the following example.

Example 4.15.8. For $u \in L^\Phi(\Omega)$ we define a sequence $\{u_n\}_{n=1}^\infty$ of bounded functions as follows:

$$u_n(x) = \begin{cases} u(x) & \text{for } x \in \Omega_n = \{x \in \Omega; \ |u(x)| \le n\}, \\ 0 & \text{for } x \in \Omega \setminus \Omega_n. \end{cases}$$

Then

$$\lim_{n \to \infty} \int_\Omega (u(x) - u_n(x)) v(x) \, dx = \lim_{n \to \infty} \int_{\Omega \setminus \Omega_n} u(x) v(x) \, dx = 0$$

for every $v \in L_\Psi(\Omega)$ and thus also for every $v \in E^\Psi(\Omega)$ since

$$\lim_{n \to \infty} \mu(\Omega \setminus \Omega_n) = 0.$$

This means that $\{u_n\}_{n=1}^\infty$ converges E^Ψ-weakly to u.

Note that the sequence $\{u_n\}_{n=1}^\infty$ constructed above does *not* in general converge to u in $L^\Phi(\Omega)$.

Remark 4.15.9. In Example 4.15.8 we have constructed for every $u \in L^\Phi(\Omega)$ a sequence of bounded functions u_n, i.e. $u_n \in E^\Phi(\Omega)$, such that $\{u_n\}_{n=1}^\infty$ converges E^Ψ-weakly to u. This shows that *$L^\Phi(\Omega)$ is the closure of $E^\Phi(\Omega)$ with respect to E^Ψ-weak convergence*.

4.16 Isomorphism properties, Schauder bases

Let Ω be bounded and let us denote by J the open interval $(0, t_0)$ where $t_0 = \mu(\Omega)$. The connection between the space $L^\Phi(\Omega)$ of functions of N variables and the space $L^\Phi(J)$ of functions of *one variable* is described by the following theorem.

Theorem 4.16.1. *The spaces $L^\Phi(\Omega)$ and $L^\Phi(J)$ are isometrically isomorphic.*

The proof follows literally the ideas given in the proof of Theorem 3.16.1 and is omitted.

Remark 4.16.2. The mapping which realizes the isometric isomorphism between $L^\Phi(\Omega)$ and $L^\Phi(J)$ simultaneously maps the space $E^\Phi(\Omega)$ onto $E^\Phi(J)$.

Theorem 4.16.1 enables us to formulate and prove various results for the simpler cases of the space $L^\Phi(J)$. These results can then be transferred to the general space $L^\Phi(\Omega)$.

Remark 4.16.3. We know from Remark 4.11.2 that the space $L^\Phi(\Omega)$ need not be separable, so that in general it makes no sense to speak of Schauder bases in Orlicz spaces. However, the space $E^\Phi(\Omega)$ is separable (see Theorem 4.12.11) so that the problem of the existence of a Schauder basis in $E^\Phi(\Omega)$ is meaningful. First we shall consider the case $N = 1$, $\Omega = (0, 1)$, denoting the space $E^\Phi(\Omega)$ by $E^\Phi(0, 1)$ instead of by $E^\Phi((0, 1))$.

Recall that the *Haar system* $\{h_i\}_{i=1}^\infty$ was introduced in Definition 3.17.1.

Theorem 4.16.4. *The Haar system $\{h_i\}_{i=1}^\infty$ forms a Schauder basis in the space $E^\Phi(0, 1)$, i.e.*

$$\lim_{n\to\infty} \left\| u - \sum_{i=1}^n c_i h_i \right\|_\Phi = 0 \tag{4.16.1}$$

for $u \in E^\Phi(0, 1)$, where the numbers c_i, $i = 1, \ldots, n$, are uniquely determined by u:

$$c_i = \int_0^1 u(x) h_i(x)\, dx, \quad i \in \mathbb{N}. \tag{4.16.2}$$

Proof. Let the c_i, $i \in \mathbb{N}$, be defined by the relations (4.16.2). Denote

$$s_n(u, x) = \sum_{i=1}^n c_i h_i(x)$$

for $x \in (0, 1)$ and $n \in \mathbb{N}$. Let the numbers $a_1, a_2, \ldots, a_r$ ($r = r(n)$) be defined as follows:

$$0 = a_1 < a_2 < \cdots < a_{r-1} < a_r = 1$$

and $a_1, a_2, \ldots, a_{r-1}$ are precisely all the points at which the Haar functions $h_1, \ldots, h_n$ have jumps. Then, for $x \in (a_\sigma, a_{\sigma+1})$, one has

$$s_n(u, x) = \frac{1}{a_{\sigma+1} - a_\sigma} \int_{a_\sigma}^{a_{\sigma+1}} u(y)\, dy. \tag{4.16.3}$$

(We note that the proof is the same as that of Theorem 3.17.3.)

Now using the Jensen integral inequality (4.2.12) with $|u|$ instead of u and α the characteristic function of the interval $[a_\sigma, a_{\sigma+1}]$, we obtain

$$\Phi(|s_n(u, x)|) \leq \Phi\left(\frac{1}{a_{\sigma+1} - a_\sigma} \int_{a_\sigma}^{a_{\sigma+1}} |u(y)|\, dy\right)$$

$$\leq \frac{1}{a_{\sigma+1} - a_\sigma} \int_{a_\sigma}^{a_{\sigma+1}} \Phi(|u(y)|)\, dy$$

for every $x \in (a_\sigma, a_{\sigma+1})$ and hence

$$\int_{a_\sigma}^{a_{\sigma+1}} \Phi(|s_n(u, x)|)\, dx \leq \int_{a_\sigma}^{a_{\sigma+1}} \Phi(|u(x)|)\, dx$$

Section 4.16 Isomorphism properties, Schauder bases

and
$$\int_0^1 \Phi(|s_n(u,x)|)\,dx \le \int_0^1 \Phi(|u(x)|)\,dx.$$
If $\|u\|_\Phi \le 1$ then also
$$\int_0^1 \Phi(|u(x)|)\,dx \le 1$$
(see (4.10.2)) and from inequality (4.6.3) for such a function u we conclude that
$$\|s_n(u,x)\|_\Phi \le \varrho(s_n(u,x), \Phi) + 1 \le \varrho(u, \Phi) + 1 \le 2.$$
This means that the operator norm (in $E^\Phi(0,1)$) of s_n satisfies
$$\|s_n\| = \sup_{\|u\|_\Phi \le 1} \|s_n(u,x)\|_\Phi \le 2 \qquad (4.16.4)$$
for every $n \in \mathbb{N}$. Denote by H the set of all jump points of all Haar functions h_i. Then H is obviously a countable set, and because (4.16.3) implies
$$\lim_{n \to \infty} s_n(u,x) = u(x) \quad \text{uniformly on } [0,1] \setminus H$$
for $u \in C([0,1])$, we have $\lim_{n \to \infty} s_n(u,x) = u(x)$ in $L^\Phi(0,1)$ for every $u \in C([0,1])$, i.e. the operators s_n converge to the identity operator I on the set $C([0,1])$ which is dense in $E^\Phi(0,1)$. Since, in view of (4.16.4), all operators are uniformly bounded, we have by the Banach–Steinhaus theorem (Theorem 1.17.5) that
$$s_n(u,x) \to u(x) \quad \text{in } L^\Phi(0,1)$$
for all $u \in E^\Phi(0,1)$, which is the same as (4.16.1).

The uniqueness of the decomposition of u into a series
$$\sum_{i=1}^\infty c_i h_i$$
follows in the same way as in Theorem 3.17.4, which completes the proof of our theorem. □

In the general case we have the following theorem.

Theorem 4.16.5. *Let $\mu(\Omega) < \infty$. Then there exists a Schauder basis in the space $E^\Phi(\Omega)$.*

Proof. The assertion follows from Theorems 4.16.4 and 4.16.1. Indeed, the functions
$$g_i(t) = h_i\left(\frac{t}{t_0}\right)$$
with $t_0 = \mu(\Omega)$ form a Schauder basis in $E^\Phi(0,t_0)$ and the mapping from Theorem 4.16.1 maps $\{g_i\}_{i=1}^\infty$ onto a basis in $E^\Phi(\Omega)$. If $\Phi \in \Delta_2$ then it suffices to use Lemma 4.12.3. □

Corollary 4.16.6. *Let $\mu(\Omega) < \infty$ and let Φ be a Young function satisfying the Δ_2-condition. Then a Schauder basis exists in the space $L^\Phi(\Omega)$.*

Remark 4.16.7. Let us suppose $\Phi \notin \Delta_2$. Formula (4.16.2) which defines the coefficients c_i is meaningful for all $u \in L^\Phi(0, 1)$, not only for $u \in E^\Phi(0, 1)$. However if we suppose for $u \in L^\Phi(0, 1)$ that

$$u = \sum_{i=1}^{\infty} c_i h_i, \tag{4.16.5}$$

where the series converges in $L^\Phi(0, 1)$, then necessarily $u \in E^\Phi(0, 1)$. Consequently, for $u \in L^\Phi(0, 1) \setminus E^\Phi(0, 1)$ the series $\sum_{i=1}^{\infty} c_i h_i$ cannot converge in $L^\Phi(0, 1)$.

4.17 Comparison of Orlicz spaces

In Exercise 3.2.4 (i) we have shown that if $\mu(\Omega) < \infty$ and $1 < p < q$ then

$$L^q(\Omega) \hookrightarrow L^p(\Omega).$$

We can obtain a similar result for the Orlicz spaces $L^{\Phi_1}(\Omega)$ and $L^{\Phi_2}(\Omega)$ making use of the ordering $\prec$ introduced in Definition 4.5.6.

Theorem 4.17.1. *Let Φ_1, Φ_2 be two Young functions. Then the inclusion*

$$L^{\Phi_1}(\Omega) \subset L^{\Phi_2}(\Omega) \tag{4.17.1}$$

holds if and only if

$$\Phi_2 \prec \Phi_1. \tag{4.17.2}$$

Proof. Let (4.17.2) hold, i.e. let $c > 0$ and $T \geq 0$ ($T = 0$ if $\mu(\Omega) = \infty$) be such that

$$\Phi_2(t) \leq \Phi_1(ct) \quad \text{for} \quad t \geq T. \tag{4.17.3}$$

Let $u \in L^{\Phi_1}(\Omega)$. Then there exists a $\gamma > 0$ such that

$$\gamma u \in \mathcal{L}^{\Phi_1}(\Omega),$$

i.e. $\varrho(\gamma u, \Phi_1) < \infty$ (see Lemma 4.7.2). Let us denote

$$\Omega_1 = \left\{ x \in \Omega; |u(x)| < c\frac{T}{\gamma} \right\}.$$

It follows from (4.17.3) that for $x \in \Omega \setminus \Omega_1$ we have

$$\Phi_2\left(\frac{\gamma}{c}|u(x)|\right) \leq \Phi_1(\gamma|u(x)|)$$

Section 4.17 Comparison of Orlicz spaces

and consequently,

$$\int_\Omega \Phi_2\left(\frac{\gamma}{c}|u(x)|\right) dx = \int_{\Omega_1} \Phi_2\left(\frac{\gamma}{c}|u(x)|\right) dx + \int_{\Omega\setminus\Omega_1} \Phi_2\left(\frac{\gamma}{c}|u(x)|\right) dx$$

$$\leq \Phi_2(T)\mu(\Omega) + \int_{\Omega\setminus\Omega_1} \Phi_1(\gamma|u(x)|) dx \leq \Phi_2(T)\mu(\Omega) + \varrho(\gamma u, \Phi_1) < \infty,$$

i.e.
$$\frac{\gamma}{c}u \in \mathcal{L}^{\Phi_2}(\Omega),$$

which means that $u \in L^{\Phi_2}(\Omega)$ (see, e.g., the generalization mentioned in Remark 4.7.1). So we have proved inclusion (4.17.1).

Conversely, let us suppose that condition (4.17.2) is not satisfied. Then there exists a sequence $\{t_n\}_{n=1}^\infty$ such that

$$0 < t_1 < t_2 < \ldots, \quad \lim_{n\to\infty} t_n = \infty,$$

and
$$\Phi_2(t_n) > \Phi_1(2^n n t_n), \quad n \in \mathbb{N}. \tag{4.17.4}$$

From the formula (4.2.5) used for the Young function Φ_1 with $t = 2^n n t_n, \alpha = 2^{-n}$ we obtain

$$\Phi_1(n t_n) \leq 2^{-n} \Phi_1(2^n n t_n)$$

which together with (4.17.4) yields

$$\Phi_2(t_n) > 2^n \Phi_1(n t_n). \tag{4.17.5}$$

Let us choose a sequence of disjoint subsets Ω_n in $\Omega, n \in \mathbb{N}$ such that

$$\mu(\Omega_n) = \frac{\Phi_1(t_1)\mu(\Omega)}{2^n \Phi_1(t_n)}.$$

This is possible since

$$\sum_{n=1}^\infty \mu(\Omega_n) < \sum_{n=1}^\infty 2^{-n} \mu(\Omega) = \mu(\Omega).$$

The function u, defined by

$$u(x) = \begin{cases} n t_n & \text{for } x \in \Omega_n, \ n \in \mathbb{N}, \\ 0 & \text{otherwise in } \Omega, \end{cases}$$

is an element of the space $L^{\Phi_1}(\Omega)$, since

$$\int_\Omega \Phi_1(|u(x)|)\,dx = \sum_{n=1}^\infty \int_{\Omega_n} \Phi_1(nt_n)\,dx = \sum_{n=1}^\infty \Phi_1(nt_n)\mu(\Omega_n)$$

$$= \sum_{n=1}^\infty \frac{1}{2^n}\Phi_1(t_1)\mu(\Omega) < \infty.$$

On the other hand, u is not an element of the space $L^{\Phi_2}(\Omega)$. Indeed, for an arbitrary γ with $0 < \gamma \leq 1$ we have that

$$\gamma u \notin \mathcal{L}^{\Phi_2}(\Omega)$$

because it suffices to choose an $m \in \mathbb{N}$ such that $\gamma > \frac{1}{m}$ and we have, in virtue of (4.17.5), that

$$\int_\Omega \Phi_2(\gamma|u(x)|)\,dx = \sum_{n=1}^\infty \Phi_2(\gamma n t_n)\mu(\Omega_n)$$

$$\geq \sum_{n=m}^\infty \Phi_2(t_n)\mu(\Omega_n) > \sum_{n=m}^\infty 2^n \Phi_1(nt_n)\mu(\Omega_n)$$

$$= \sum_{n=m}^\infty \Phi_1(t_1)\mu(\Omega) = \infty.$$

Consequently, condition (4.17.2) is necessary for the inclusion (4.17.1) and the theorem is proved. □

The following theorem follows directly from Theorem 4.17.1.

Theorem 4.17.2. *Let Φ_1, Φ_2 be two Young functions. Then*

$$L^{\Phi_1}(\Omega) = L^{\Phi_2}(\Omega) \tag{4.17.6}$$

if and only if Φ_1 and Φ_2 are equivalent (in the sense of Definition 4.5.6).

Remark 4.17.3. Equality (4.17.6) is a set identity. Nevertheless, we can also show that the norms of the two spaces considered are equivalent. This fact will follow from the next theorem which states that the inclusion $L^{\Phi_1}(\Omega) \subset L^{\Phi_2}(\Omega)$ implies the embedding

$$L^{\Phi_1}(\Omega) \hookrightarrow L^{\Phi_2}(\Omega).$$

Theorem 4.17.4. *Let $L^{\Phi_1}(\Omega) \subset L^{\Phi_2}(\Omega)$. Then there exists a $k > 0$ such that*

$$\|u\|_{\Phi_2} \leq k\|u\|_{\Phi_1} \tag{4.17.7}$$

for all $u \in L^{\Phi_1}(\Omega)$.

Proof. Suppose that (4.17.7) fails. Then there exists a sequence $\{u_n\}_{n=1}^\infty$ of functions $u_n \geq 0$ in $L^{\Phi_1}(\Omega)$ such that

$$\|u_n\|_{\Phi_1} \leq 1 \quad \text{but} \quad \|u_n\|_{\Phi_2} > n^3 \text{ for all } n \in \mathbb{N}.$$

We know that Orlicz spaces are complete, hence they have the Riesz–Fischer property. Thus, $\sum \frac{u_n}{n^2}$ converges in $L^{\Phi_1}(\Omega)$ to some function $u \in L^{\Phi_1}(\Omega)$. By the hypothesis $L^{\Phi_1}(\Omega) \subset L^{\Phi_2}(\Omega)$ we get also $u \in L^{\Phi_2}(\Omega)$. However, since $0 \leq n^{-2} u_n \leq u$ and so $\|u\|_{\Phi_2} \geq n^{-2} \|u_n\|_{\Phi_2} > n$, for all n, we get a contradiction. □

Example 4.17.5. If Φ_1, Φ_2 are equivalent Young functions then the norms $\|\cdot\|_{\Phi_1}$ and $\|\cdot\|_{\Phi_2}$ are equivalent. In example 4.5.7 (iii) it was shown that the functions $\Phi(t)$, $\Phi_k(t) = \Phi(kt)$ with $k > 0$ are equivalent provided Φ is a Young function. In this case we even have that

$$\|u\|_{\Phi_k} = k \|u\|_{\Phi},$$

which is a consequence of the fact that the corresponding complementary functions Ψ, Ψ_k satisfy $\Psi_k(t) = \Psi(\frac{t}{k})$ (see Exercise 4.3.7).

If we use the ordering $\Phi_2 \ll \Phi_1$ introduced in Definition 4.5.10 we obtain sometimes more than the inclusion $L^{\Phi_1}(\Omega) \subset L^{\Phi_2}(\Omega)$:

Theorem 4.17.6. *If $\Phi_2 \ll \Phi_1$ then*

$$L^{\Phi_1}(\Omega) \hookrightarrow L^{\Phi_2}(\Omega). \tag{4.17.8}$$

Proof. Let $u \in L^{\Phi_1}(\Omega)$. Then there exists a $\gamma > 0$ such that $\gamma u \in \mathcal{L}^{\Phi_1}(\Omega)$. Since $\Phi_2 \ll \Phi_1$ we have that

$$\lim_{t \to \infty} \frac{\Phi_2(\kappa t)}{\Phi_1(\gamma t)} = 0 \quad \text{for every} \quad \kappa > 0. \tag{4.17.9}$$

Let us fix κ; then (4.17.9) means that there exists a $T_\kappa \geq 0$ such that

$$\Phi_2(\kappa t) < \Phi_1(\gamma t) \quad \text{for every} \quad t \geq T_\kappa.$$

However, Theorem 4.5.2 implies that $\kappa u \in \mathcal{L}^{\Phi_2}(\Omega)$. This result is true for all $\kappa > 0$ so that we have

$$\kappa u \in \mathcal{L}^{\Phi_2}(\Omega) \quad \text{for all} \quad \kappa > 0,$$

which means that $u \in E^{\Phi_2}(\Omega)$. So we have

$$L^{\Phi_1}(\Omega) \subset E^{\Phi_2}(\Omega).$$

In view of $E^{\Phi_2}(\Omega) \subset L^{\Phi_2}(\Omega)$, the embedding in (4.17.8) follows from Theorem 4.17.4. □

Theorem 4.17.7. *Let $\Phi_2 \prec\prec \Phi_1$ and let K be a bounded subset of $L^{\Phi_1}(\Omega)$. Then the functions $u \in K$ have uniformly absolutely continuous norms in the space $L^{\Phi_2}(\Omega)$.*

Proof. From $\Phi_2 \prec\prec \Phi_1$ we conclude by Exercise 4.5.13 that the complementary functions Ψ_1, Ψ_2 satisfy
$$\Psi_1 \prec\prec \Psi_2. \tag{4.17.10}$$
Let us denote by V the set of all $v \in \mathcal{L}^{\Psi_2}(\Omega)$ such that $\varrho(v, \Psi_2) \leq 1$. Moreover, let W be the set of all functions w of the form
$$w(x) = \Psi_1(\lambda |v(x)|), \qquad v \in V, \quad \lambda > 0 \quad \text{fixed}.$$
Then it follows from (4.17.10) and the de la Vallée-Poussin theorem (Theorem 4.12.15) that all $w \in W$ have uniformly absolutely continuous L^1-norms, i.e. for every $\varepsilon > 0$ there exists a $\delta = \delta(\varepsilon) > 0$ such that for $M \subset \Omega$ with $\mu(M) < \delta$ we have
$$\int_M |w(x)|\,dx = \int_M \Psi_1(\lambda |v(x)|)\,dx < \frac{\varepsilon}{2} \tag{4.17.11}$$
for all $v \in V$. Let K be a bounded subset of $L^{\Phi_1}(\Omega)$, i.e. let $c > 0$ be such that $\|u\|_{\Phi_1} \leq c$ for all $u \in K$. Let $0 < \varepsilon < 1$ and let us take $\lambda = \frac{2c}{\varepsilon}$ in (4.17.11). For $u \in K$ and $v \in V$ we have by the Young inequality and (4.10.2) that
$$\int_M |u(x)v(x)|\,dx \leq \int_M \Phi_1\left(\frac{|u(x)|}{\lambda}\right)\,dx + \int_M \Psi_1(\lambda|v(x)|)\,dx$$
$$\leq \left\|\frac{u}{\lambda}\right\|_{\Phi_1} + \int_M \Psi_1(\lambda|v(x)|)\,dx.$$
We have $\left\|\frac{u}{\lambda}\right\|_{\Phi_1} = \frac{\varepsilon}{2c}\|u\|_{\Phi_1} \leq \frac{\varepsilon}{2}$; if we choose $M \subset \Omega$ so that $\mu(M) < \delta$ we can use (4.17.11) and thus we obtain for every $u \in K$
$$\|u\chi_M\|_{\Phi_2} = \sup_{v \in V} \int_M |u(x)v(x)|\,dx < \frac{\varepsilon}{2} + \frac{\varepsilon}{2} = \varepsilon. \quad \square$$

Remark 4.17.8. In view of Theorem 4.14.10 the preceding theorem shows that for $\Phi_2 \prec\prec \Phi_1$ the embedding $L^{\Phi_1}(\Omega) \hookrightarrow E^{\Phi_2}(\Omega)$ is "almost-compact". Nonetheless, this embedding still fails to be compact (see Exercise 3.14.5 and Exercise 4.5.11 (i)).

An "almost-compact embedding" turns out to be an important relation between function spaces. In the literature it is also often called an *absolutely continuous embedding*. We shall study this type of relation in a far more general context in Section 7.11.

It was shown in Remark 4.5.12 that the relation $\Phi_2 \prec \Phi_1$ does not generally imply the relation $\Phi_2 \prec\prec \Phi_1$. Some additional conditions guaranteeing this implication are given in the following theorem which is in a sense the converse of Theorem 4.17.7.

Theorem 4.17.9. *Let $\Phi_2 \prec \Phi_1$ and let for every bounded subset K of $L^{\Phi_1}(\Omega)$ all $u \in K$ have uniformly absolutely continuous L^{Φ_2}-norms. Then*

$$\Phi_2 \lll \Phi_1. \tag{4.17.12}$$

Proof. Assume that (4.17.12) does not hold. Then Exercise 4.5.13 implies that $\Psi_1 \lll \Psi_2$ is impossible, i.e. there exists a constant $\lambda > 0$ and a sequence $\{t_n\}_{n=1}^\infty$ with $\lim_{n\to\infty} t_n = \infty$ so that

$$\Psi_1(t_n) > \Phi_2(\lambda t_n) \quad \text{for all } n \in \mathbb{N}. \tag{4.17.13}$$

If we denote $s_n = \Psi_1(t_n)$ then (4.17.13) implies that the inverse functions to Ψ_1 and Ψ_2 satisfy

$$\Psi_2^{-1}(s_n) > \lambda t_n = \lambda \Psi_1^{-1}(s_n) \quad \text{for all } n \in \mathbb{N}. \tag{4.17.14}$$

Let $\Omega_n (n \in \mathbb{N})$ be subsets of Ω with $\mu(\Omega_n) = \frac{1}{s_n}$ and let us define a sequence of functions u_n:

$$u_n(x) = \begin{cases} \dfrac{s_n}{\Psi_1^{-1}(s_n)} & \text{for } x \in \Omega_n, \\ 0 & \text{otherwise in } \Omega. \end{cases}$$

From Example 4.6.9 we have

$$\|u_n\|_{\Phi_1} = \frac{s_n}{\Psi_1^{-1}(s_n)} \|\chi_{\Omega_n}\|_{\Phi_1} = \frac{s_n}{\Psi_1^{-1}(s_n)} \frac{1}{s_n} \Psi_1^{-1}(s_n) = 1,$$

i.e. the sequence $\{u_n\}_{n=1}^\infty$ is bounded in $L^{\Phi_1}(\Omega)$. By the assumptions of our theorem, all u_n have uniformly absolutely continuous L^{Φ_2}-norms, i.e.

$$\lim_{\mu(M)\to 0} \|u_n \chi_M\|_{\Phi_2} = 0$$

uniformly with respect to $n \in \mathbb{N}$. We can set $M = \Omega_n$; as $\lim_{n\to\infty} \mu(\Omega_n) = 0$ we obtain

$$\lim_{n\to\infty} \|u_n\|_{\Phi_2} = 0.$$

However, this leads to a contradiction, because (4.17.14) yields

$$\|u_n\|_{\Phi_2} = \frac{s_n}{\Psi_1^{-1}(s_n)} \|\chi_{\Omega_n}\|_{\Phi_2} = \frac{s_n}{\Psi_1^{-1}(s_n)} \cdot \frac{1}{s_n} \Psi_2^{-1}(s_n) > \lambda > 0$$

for all $n \in \mathbb{N}$. Because of this contradiction, (4.17.12) must hold. $\square$

In conclusion we shall prove that, for $\Phi_2 \lll \Phi_1$, Φ_1-mean convergence implies norm convergence in $L^{\Phi_2}(\Omega)$:

Theorem 4.17.10. *Let $\Phi_2 \prec\prec \Phi_1$ and let $\{u_n\}_{n=1}^\infty$ be a sequence in $L^{\Phi_1}(\Omega)$ such that*

$$\lim_{n\to\infty} \int_\Omega \Phi_1(|u_n(x)|)\,dx = 0. \tag{4.17.15}$$

Then

$$\lim_{n\to\infty} \|u_n\|_{\Phi_2} = 0. \tag{4.17.16}$$

Proof. By virtue of $\|u_n\|_{\Phi_1} \leq \varrho(u_n, \Phi_1) + 1$ it follows from (4.17.15) that the sequence $\{u_n\}_{n=1}^\infty$ is bounded in $L^{\Phi_1}(\Omega)$. According to Theorem 4.17.7 all the u_n have uniformly absolutely continuous L^{Φ_2}-norms.

Simultaneously, it follows from (4.17.15) that u_n converges in measure to zero: If we denote

$$\Omega_n = \{x \in \Omega; |u_n(x)| \geq \eta\}$$

and if we had $\mu(\Omega_n) \geq \lambda > 0$ for some fixed $\eta > 0$ then we should have

$$\int_\Omega \Phi_1(|u_n(x)|)\,dx \geq \int_{\Omega_n} \Phi_1(|u_n(x)|)\,dx \geq \Phi_1(\eta)\mu(\Omega_n) \geq \Phi_1(\eta)\lambda > 0,$$

which contradicts (4.17.15).

Now relation (4.17.16) follows from Theorem 4.14.10. □

Chapter 5

Morrey and Campanato spaces

5.1 Introduction

In this chapter we shall study systematically the subspaces of real Lebesgue spaces characterized by the property that the "mean oscillation" of their elements are bounded in some sense. These spaces were studied in close connection with the theory of partial differential equations and helped to obtain many interesting results. Some special cases of these spaces were introduced by Morrey in 1938 in [157]. The early 1960s may be considered the beginning of an intense development of the general theory; we mention here the papers by John and Nirenberg [110], Meyers [151], Campanato [25], [26], Stampacchia [211], and a survey paper by Peetre [176]. In our exposition of the theory we follow the lecture notes of Campanato [28]. Our notation and terminology differ slightly from that commonly used (we denote Morrey spaces by $L_M^{p,\lambda}$ instead of $L^{p,\lambda}$-spaces, Campanato spaces by $L_C^{p,\lambda}$ instead of $\mathcal{L}^{p,\lambda}$-spaces, etc.).

5.2 Marcinkiewicz spaces and their connection with the spaces $L^{p,\infty}(\Omega)$

Definition 5.2.1. We suppose in this section that $\Omega \subset \mathbb{R}^N$ is a domain with $\mu(\Omega) < \infty$. Let $\beta \in (0, 1)$. Denote by $M_\beta(\Omega)$ the set of all measurable functions on Ω for which

$$\|u\|_{M_\beta} = \sup \frac{1}{(\mu(E))^\beta} \int_E |u(x)| \, dx < \infty \tag{5.2.1}$$

where the supremum is taken over all measurable sets $E \subset \Omega$ with positive measure. The set $M_\beta(\Omega)$ is called the *Marcinkiewicz space* (see also Section 3.18).

Theorem 5.2.2. *The quantity* $\|\cdot\|_{M_\beta}$ *defined by the relation* (5.2.1) *is a norm on the vector space* $M_\beta(\Omega)$. $M_\beta(\Omega)$ *with this norm is a Banach space.*

Proof. The first part of the theorem is obvious. We shall prove the completeness. Let $\{u_n\}_{n=1}^\infty$ be a Cauchy sequence in the space $M_\beta(\Omega)$. From the relation

$$\|f\|_1 = \int_\Omega |f(x)| \, dx \le (\mu(\Omega))^\beta \|f\|_{M_\beta},$$

valid for every $f \in M_\beta$, we conclude that $\{u_n\}_{n=1}^\infty$ is a Cauchy sequence in the space $L^1(\Omega)$. The completeness of the space $L^1(\Omega)$ (see Theorem 3.7.1) guarantees the

existence of a function $u \in L^1(\Omega)$ such that
$$\lim_{n\to\infty} \|u_n - u\|_1 = 0.$$
It suffices now to prove that $u \in M_\beta(\Omega)$ and
$$\lim_{n\to\infty} \|u_n - u\|_{M_\beta} = 0.$$
For each measurable set $E \subset \Omega$, $\mu(E) > 0$ we have
$$\frac{1}{(\mu(E))^\beta} \int_E |u(x)| \, dx$$
$$\leq \frac{1}{(\mu(E))^\beta} \int_E |u_n(x) - u(x)| \, dx + \frac{1}{(\mu(E))^\beta} \int_E |u_n(x)| \, dx$$
$$\leq \frac{1}{(\mu(E))^\beta} \int_E |u_n(x) - u(x)| \, dx + \|u_n\|_{M_\beta}. \tag{5.2.2}$$

The sequence $\{u_n\}_{n=1}^\infty$ in $M_\beta(\Omega)$ is Cauchy and hence bounded, i.e. there exists a constant $c > 0$ such that $\|u_n\|_{M_\beta} \leq c$ for all $n \in \mathbb{N}$. This together with the convergence $u_n \to u$ in $L^1(\Omega)$ and relation (5.2.2) implies
$$\frac{1}{(\mu(E))^\beta} \int_E |u(x)| \, dx \leq c,$$
hence $\|u\|_{M_\beta} \leq c$. Consequently, $u \in M_\beta(\Omega)$.

Further, we obtain for E as above that
$$\frac{1}{(\mu(E))^\beta} \int_E |u(x) - u_n(x)| \, dx$$
$$\leq \frac{1}{(\mu(E))^\beta} \int_E |u_n(x) - u_m(x)| \, dx + \frac{1}{(\mu(E))^\beta} \int_E |u(x) - u_m(x)| \, dx$$
$$\leq \|u_n - u_m\|_{M_\beta} + \frac{1}{(\mu(E))^\beta} \int_E |u(x) - u_m(x)| \, dx. \tag{5.2.3}$$

For every $\varepsilon > 0$ there exists an $n_0 \in \mathbb{N}$ such that
$$\|u_n - u_m\|_{M_\beta} < \varepsilon$$
for all $n, m \geq n_0$. From this and inequality (5.2.3) it follows that
$$\frac{1}{(\mu(E))^\beta} \int_E |u(x) - u_n(x)| \, dx \leq \varepsilon + \frac{1}{(\mu(E))^\beta} \int_E |u(x) - u_m(x)| \, dx$$
for all $n, m \geq n_0$. Passing to the limit as $m \to \infty$ and using relation (5.2.1) we obtain
$$\|u - u_n\|_{M_\beta} \leq \varepsilon$$
for all $n \geq n_0$. Thus the sequence $\{u_n\}_{n=1}^\infty$ converges to u in the space $M_\beta(\Omega)$. □

Section 5.2 Marcinkiewicz spaces

In Section 3.18 we introduced the spaces $L^{p,\infty}(\Omega)$. Let us recall here the notation. For a measurable function f on Ω and for any positive number σ we define

$$S(f,\sigma) = \{x \in \Omega;\ |f(x)| > \sigma\}.$$

For $1 \le p < \infty$ we denote

$$\|f\|_{p,\infty} = \sup_{t>0}(t[\mu(S(f,t))]^{\frac{1}{p}}).$$

Then the weak Lebesgue space $L^{p,\infty}(\Omega)$ is the set of all functions u which are measurable on Ω and for which $\|u\|_{p,\infty} < \infty$.

We shall now establish the following theorem.

Theorem 5.2.3. *Let $p \in (1,\infty)$ and let $\beta = 1 - \frac{1}{p}$. Then the sets $M_\beta(\Omega)$ and $L^{p,\infty}(\Omega)$ coincide.*

Proof. (I) Suppose $u \in L^{p,\infty}(\Omega)$. Then for measurable E with $\mu(E) > 0$ and for $\alpha \in \mathbb{R}$ we have (see Lemma 2.18.2)

$$\int_E |u(x)|\,dx = \int_0^\infty \mu(\{x \in E;\ |u(x)| > t\})\,dt \tag{5.2.4}$$

$$\le \int_0^{(\mu(E))^\alpha} \mu(\{x \in E;\ |u(x)| > t\})\,dt + \int_{(\mu(E))^\alpha}^\infty \mu(S(u,t))\,dt$$

$$\le (\mu(E))^{\alpha+1} + (\|u\|_{p,\infty})^p \int_{(\mu(E))^\alpha}^\infty \left(\frac{1}{t}\right)^p dt$$

$$= (\mu(E))^{\alpha+1} + (\mu(E))^{\alpha(1-p)}(p-1)^{-1}(\|u\|_{p,\infty})^p.$$

Setting $\alpha = -\frac{1}{p}$ in (5.2.4) we arrive at

$$\int_E |u(x)|\,dx \le \left(1 + \frac{(\|u\|_{p,\infty})^p}{p-1}\right)(\mu(E))^\beta$$

which implies

$$\|u\|_{M_\beta} \le 1 + (\|u\|_{p,\infty})^p(p-1)^{-1}.$$

Consequently, $u \in M_\beta(\Omega)$.

(II) Suppose now that $u \in M_\beta(\Omega)$, $t > 0$. Then

$$t\mu(S(u,t)) = t\int_{S(u,t)} dx \le \int_{S(u,t)} |u(x)|\,dx \le \|u\|_{M_\beta}(\mu(S(u,t)))^\beta,$$

so that

$$\|u\|_{p,\infty} \le \|u\|_{M_\beta}. \qquad \square$$

Remark 5.2.4. In the subsequent sections we shall investigate certain generalizations of the spaces $M_\beta(\Omega)$. The method of generalization consists in taking the supremum in relation (5.2.1) not over the class of all measurable sets $E \subset \Omega$ with positive measure but over some narrower class of more special subsets of Ω. Moreover, the integrand $|u(x)|$ will be replaced by another expression depending on the function $u(x)$.

5.3 Morrey and Campanato spaces: Definitions and basic properties

Definition 5.3.1. For a bounded domain $\Omega \subset \mathbb{R}^N$ denote

$$\delta = \mathrm{diam}\,\Omega = \sup\{|x-y|;\ x,y \in \Omega\}.$$

The bounded domain $\Omega \subset \mathbb{R}^N$ is said to be *of type* $\mathcal{A}$ if there exists a constant $A > 0$ such that for every $x \in \overline{\Omega}$ and all $\varrho \in (0, \delta)$

$$\mu(\Omega(x,\varrho)) \geq A\varrho^N,$$

where

$$\Omega(x,\varrho) = \{y \in \Omega;\ |x-y| < \varrho\}.$$

Example 5.3.2. A square in the plane is a set of type $\mathcal{A}$ with $A = \frac{1}{2}$. On the other hand, the domain

$$\Omega = \{(x,y) \in \mathbb{R}^2;\ 0 < x < 1,\ 0 < y < x^2\}$$

is not of type $\mathcal{A}$ for any $A > 0$. (It should be noted that the origin is a cuspidal point of the boundary of Ω.)

In the rest of this chapter we shall consider only domains of type $\mathcal{A}$.

Definition 5.3.3. Denote by Ω_δ the Cartesian product $\Omega \times (0, \delta)$ (with δ from Definition 5.3.1). For $\lambda \geq 0$ and $p \in (1, \infty)$, set

$$L_M^{p,\lambda}(\Omega) := \left\{ u \in L^p(\Omega);\ \sup \frac{1}{r^\lambda} \int_{\Omega(x,r)} |u(y)|^p\,dy < \infty \right\}$$

where the supremum is taken over all ordered pairs $(x,r) \in \Omega_\delta$. Define

$$^M\|u\|_{p,\lambda} := \left\{ \sup_{(x,r)\in\Omega_\delta} \frac{1}{r^\lambda} \int_{\Omega(x,r)} |u(y)|^p\,dy \right\}^{\frac{1}{p}}.$$

The set $L_M^{p,\lambda}(\Omega)$ equipped with the above norm is called the *Morrey space*.

Section 5.3 Morrey and Campanato spaces

Remarks 5.3.4. (i) The reader will easily verify that $^M\|\cdot\|_{p,\lambda}$ is a norm on the linear space $L_M^{p,\lambda}(\Omega)$. Using the assumption that Ω is of type $\mathcal{A}$, we obtain that the expression

$$\left\{ \sup_{(x,r)\in\Omega_\delta} \mu(\Omega(x,r))^{-\frac{\lambda}{N}} \int_{\Omega(x,r)} |u(y)|^p \, dy \right\}^{\frac{1}{p}}$$

defines another norm in the space $L_M^{p,\lambda}(\Omega)$ which is equivalent to $^M\|\cdot\|_{p,\lambda}$.

(ii) Let $u \in L^1(\Omega)$. Denote by $\mathcal{G}$ the set of all points $x \in \Omega$ for which

$$u(x) = \lim_{r\to 0+} \frac{1}{\mu(B(x,r))} \int_{B(x,r)} u(y) \, dy. \tag{5.3.1}$$

(The set $B(x,r)$ is a ball with its center at the point x and radius r.) The set $\mathcal{G}$ is called the *Lebesgue set* of u and its elements are called the *Lebesgue points* of u. Recall that

$$\mu(\Omega \setminus \mathcal{G}) = 0. \tag{5.3.2}$$

It is proved in [212] that the equality (5.3.2) remains valid if we replace the limit in formula (5.3.1) by the expression

$$\lim_{\substack{E\in\mathcal{F}\\ \mu(E)\to 0}} \frac{1}{\mu(E)} \int_E u(x-y) \, dy,$$

where $\mathcal{F}$ is the so-called regular family of subsets in $\mathbb{R}^N$ (as a simple example of $\mathcal{F}$ we can take the family of all N-dimensional cubes with their centers at the origin).

We shall now introduce yet another class of function spaces.

Definition 5.3.5. Let $\lambda \geq 0$ and $p \in (1, \infty)$. Define

$$L_C^{p,\lambda}(\Omega) := \left\{ u \in L^p(\Omega); \sup_{(x,r)\in\Omega_\delta} \frac{1}{r^\lambda} \int_{\Omega(x,r)} |u(y) - u_{x,r}|^p \, dy < \infty \right\},$$

where

$$u_{x,r} = \frac{1}{\mu(\Omega(x,r))} \int_{\Omega(x,r)} u(y) \, dy.$$

Denote

$$[u]_{p,\lambda} := \left\{ \sup_{(x,r)\in\Omega_\delta} \frac{1}{r^\lambda} \int_{\Omega(x,r)} |u(y) - u_{x,r}|^p \, dy \right\}^{\frac{1}{p}}.$$

The expression $[u]_{p,\lambda}$ is a seminorm in the vector space $L_C^{p,\lambda}(\Omega)$ and $[u]_{p,\lambda} = 0$ if and only if u is constant almost everywhere on Ω. Define the norm in the space $L_C^{p,\lambda}(\Omega)$ by

$$^C\|u\|_{p,\lambda} := \|u\|_p + [u]_{p,\lambda}.$$

The set $L_C^{p,\lambda}(\Omega)$ with the above norm will be called the *Campanato space*.

5.4 Completeness

Theorem 5.4.1. *The spaces $L_M^{p,\lambda}(\Omega)$ and $L_C^{p,\lambda}(\Omega)$ are Banach spaces.*

The proof is essentially the same as that of Theorem 5.2.2.

5.5 Relations to Lebesgue spaces

We shall now study in detail connections between Morrey, Campanato and Lebesgue spaces as well as their embedding properties.

Theorem 5.5.1. (i) *Let $p \in (1, \infty)$. Then,*

$$L_M^{p,0}(\Omega) \rightleftarrows L^p(\Omega).$$

(ii) *Let $p \in (1, \infty)$. Then,*

$$L_M^{p,N}(\Omega) \rightleftarrows L^\infty(\Omega).$$

(iii) *Let $1 \leq p \leq q < \infty$, let λ and ν be nonnegative numbers. If*

$$\frac{\lambda - N}{p} \leq \frac{\nu - N}{q},$$

then

$$L_M^{q,\nu}(\Omega) \hookrightarrow L_M^{p,\lambda}(\Omega).$$

Proof. Assertion (i) follows immediately from the equality

$$^M\|u\|_{p,0} = \|u\|_p.$$

(ii) The space $L_M^{p,N}(\Omega)$ is complete (see Theorem 5.4.1), and so is the space $L^\infty(\Omega)$. We shall prove that the identity operator maps $L^\infty(\Omega)$ continuously onto $L_M^{p,N}(\Omega)$. Then we complete the proof by making use of Theorem 1.14.8. Thus, let $u \in L^\infty(\Omega)$. For every ordered pair $(x, r) \in \Omega_\delta$ the inequality

$$\frac{1}{r^N} \int_{\Omega(x,r)} |u(y)|^p \, dy \leq \frac{\mu(\Omega(x,r))}{r^N} \|u\|_\infty^p \leq c_N \|u\|_\infty^p$$

Section 5.5 Relations to Lebesgue spaces

holds, where c_N is the volume of the unit ball in the space $\mathbb{R}^N$ and where $\|\cdot\|_\infty$ is a norm in $L^\infty(\Omega)$. Thus,

$$^M\|u\|_{p,N} \leq c_N^{\frac{1}{p}}\|u\|_\infty$$

and the continuity of the identity operator $\mathrm{Id} : L^\infty(\Omega) \to L_M^{p,N}(\Omega)$ follows from the last inequality.

Suppose now that the operator Id does not act onto $L_M^{p,N}(\Omega)$. Let

$$u \in L_M^{p,N}(\Omega) \setminus L^\infty(\Omega).$$

Inasmuch as $u \in L^p(\Omega)$, almost all points of Ω are Lebesgue points of $|u(x)|^p$ (see Remark 5.3.4 (ii)). Denote the set of all Lebesgue points of the function $|u(x)|^p$ by $\mathcal{G}$. It follows from the fact that $u \notin L^\infty(\Omega)$, i.e. $\|u\|_\infty = \infty$, that for every $K > 0$ the set $S(u, K) = \{x \in \Omega; |u(x)| > K\}$ has positive measure. So $\mathcal{G} \cap S(u, K) \neq \emptyset$.

If $x \in \mathcal{G} \cap S(u, K)$, we have

$$\lim_{r \to 0+} \frac{1}{\mu(\Omega(x,r))} \int_{\Omega(x,r)} |u(y)|^p \, dy = |u(x)|^p > K^p.$$

From this fact it follows easily that for every $C > 0$ there exists $(x, r) \in \Omega_\delta$ such that

$$\frac{1}{\mu(\Omega(x,r))} \int_{\Omega(x,r)} |u(x)|^p \, dx > C,$$

and so $^M\|u\|_{p,N} = \infty$ (see Remark 5.3.4 (i)). This contradicts the assumption that $u \in L_M^{p,N}(\Omega)$.

(iii) By the Hölder inequality for Lebesgue norms (3.1.6) we obtain for $p \leq q$ and $(x, r) \in \Omega_\delta$

$$\int_{\Omega(x,r)} |u(y)|^p \, dy \leq (\mu(\Omega(x,r)))^{1-\frac{p}{q}} \left(\int_{\Omega(x,r)} |u(y)|^q \, dy \right)^{\frac{p}{q}}$$

$$\leq c_N^{1-\frac{p}{q}} r^{N(1-\frac{p}{q})+\nu\frac{p}{q}} \left(\frac{1}{r^\nu} \int_{\Omega(x,r)} |u(y)|^q \, dy \right)^{\frac{p}{q}}. \quad (5.5.1)$$

According to the inequality

$$\lambda \leq N\left(1 - \frac{p}{q}\right) + \nu \frac{p}{q}$$

we have

$$\left(\frac{r}{\delta}\right)^{N(1-\frac{p}{q})+\nu\frac{p}{q}} \leq \left(\frac{r}{\delta}\right)^\lambda,$$

so that
$$r^{N(1-\frac{p}{q})+v\frac{p}{q}} \leq r^\lambda \delta^{N(1-\frac{p}{q})+v\frac{p}{q}-\lambda}.$$

Using the last relation we transform the formula (5.5.1) into

$$\left(\frac{1}{r^\lambda}\int_{\Omega(x,r)}|u(y)|^p\,dy\right)^{\frac{1}{p}} \leq c\left(\frac{1}{r^v}\int_{\Omega(x,r)}|u(y)|^q\,dy\right)^{\frac{1}{q}}$$

and taking the supremum first on the right-hand side and then on the left-hand side of the previous inequality we obtain

$$^M\|u\|_{p,\lambda} \leq c\,^M\|u\|_{q,v},$$

which is the assertion occurring in (iii). □

Theorem 5.5.2. (i) *Let* $p \in (1,\infty)$. *Then*

$$L_C^{p,0}(\Omega) \rightleftarrows L^p(\Omega).$$

(ii) *Let* $1 \leq p \leq q < \infty$ *and let* λ *and* v *be nonnegative numbers. If*

$$\frac{\lambda - N}{p} \leq \frac{v - N}{q}$$

then

$$L_C^{q,v}(\Omega) \hookrightarrow L_C^{p,\lambda}(\Omega).$$

Proof. Both of these assertions can be proved in a way similar to that of the corresponding assertions (i) and (iii) of Theorem 5.5.1. □

Remarks 5.5.3. (i) It is easy to see that $L_M^{p,\lambda}(\Omega) = \{0\}$ for $\lambda > N$. Further, it follows from Theorem 5.5.1 that the collection $\{L_M^{p,\lambda}(\Omega)\}_{\lambda \in [0,N]}$ for fixed $p \in [1,\infty)$ generates a certain "scale of spaces" between $L^p(\Omega)$ and $L^\infty(\Omega)$.

(ii) As will be proved later (Lemma 5.8.1) the spaces $L_C^{p,N}(\Omega)$ and $L^\infty(\Omega)$ do not coincide, so that the analogue to assertion (ii) of Theorem 5.5.1 for Campanato spaces is not valid.

5.6 Some lemmas

Now we prove several lemmas to be used in the next section. Unless otherwise stated we shall suppose that Ω is a set of type $\mathcal{A}$ and, moreover, $1 \leq p < \infty$ and $\lambda \geq 0$.

Lemma 5.6.1. *A function u belongs to the space $L_C^{p,\lambda}(\Omega)$ if and only if $u \in L^p(\Omega)$ and*
$$\|u\|_{p,\lambda}^p = \sup_{(x,r) \in \Omega_\delta} \frac{1}{r^\lambda} \left(\inf_{c \in \mathbb{R}} \int_{\Omega(x,r)} |u(y) - c|^p \, dy \right) < \infty. \tag{5.6.1}$$

Proof. Evidently
$$\|u\|_{p,\lambda}^p \leq [u]_{p,\lambda}^p.$$

Let $u \in L^p(\Omega)$ and assume (5.6.1). Then the inequalities

$$\int_{\Omega(x,r)} |u(y) - u_{x,r}|^p \, dy$$
$$\leq 2^{p-1} \left(\int_{\Omega(x,r)} |u(y) - c|^p \, dy \right.$$
$$\left. + \int_{\Omega(x,r)} (\mu(\Omega(x,r)))^{-p} \left| \int_{\Omega(x,r)} (c - u(y)) \, dy \right|^p dz \right)$$
$$\leq 2^p \int_{\Omega(x,r)} |u(y) - c|^p \, dy$$

hold for every $c \in \mathbb{R}$. Hence we conclude that
$$[u]_{p,\lambda} \leq 2\|u\|_{p,\lambda},$$
which proves the lemma. $\square$

Remark 5.6.2. In the course of the proof of Lemma 5.6.1 we have established the equivalence of the norms $^C\|u\|_{p,\lambda}$ and $\|u\|_{p,\lambda} + \|u\|_p$ on the space $L_C^{p,\lambda}(\Omega)$.

Lemma 5.6.3. *Let A be the constant from Definition 5.3.1. There exists a constant $C = C(p, A)$ such that the implication*
$$0 < \sigma < \varrho < \delta \Rightarrow |u_{x,\varrho} - u_{x,\sigma}| \leq C(p, A) \left(\frac{\varrho^\lambda + \sigma^\lambda}{\sigma^N} \right)^{\frac{1}{p}} [u]_{p,\lambda} \tag{5.6.2}$$

holds for all $u \in L_C^{p,\lambda}(\Omega)$ and all $x \in \Omega$.

Proof. For almost all $y \in \Omega(x,\sigma) \subset \Omega(x,\varrho)$ the inequality

$$|u_{x,\varrho} - u_{x,\sigma}|^p \leq 2^{p-1}(|u_{x,\varrho} - u(y)|^p + |u_{x,\sigma} - u(y)|^p) \tag{5.6.3}$$

holds. Integrating (5.6.3) with respect to the variable y over the set $\Omega(x,\sigma)$ we obtain

$$\int_{\Omega(x,\sigma)} |u_{x,\varrho} - u_{x,\sigma}|^p \, dy$$

$$\leq 2^{p-1} \left(\int_{\Omega(x,\sigma)} |u_{x,\sigma} - u(y)|^p \, dy + \int_{\Omega(x,\varrho)} |u_{x,\varrho} - u(y)|^p \, dy \right)$$

$$\leq 2^{p-1}(\varrho^\lambda + \sigma^\lambda)[u]_{p,\lambda}^p.$$

Since Ω is of type $\mathcal{A}$ (see Definition 5.3.1) we have

$$\int_{\Omega(x,\sigma)} |u_{x,\varrho} - u_{x,\sigma}|^p \, dy = |u_{x,\varrho} - u_{x,\sigma}|^p \mu(\Omega(x,\sigma)) \geq |u_{x,\varrho} - u_{x,\sigma}|^p A \sigma^N.$$

Thus,

$$|u_{x,\varrho} - u_{x,\sigma}|^p \leq \frac{2^{p-1}}{A} \frac{\varrho^\lambda + \sigma^\lambda}{\sigma^N}[u]_{p,\lambda}^p,$$

and the inequality in (5.6.2) follows. □

Lemma 5.6.4. *There exists a constant $C = C(p,\lambda,A)$ such that*

$$|u_{x,\varrho} - u_{x,\frac{\varrho}{2^n}}| \leq C(p,\lambda,A)[u]_{p,\lambda} \varrho^{\frac{\lambda-N}{p}} \sum_{m=0}^{n-1} 2^{m\frac{N-\lambda}{p}} \tag{5.6.4}$$

whenever $u \in L_C^{p,\lambda}(\Omega)$, $(x,\varrho) \in \Omega_\delta$ and $n \in \mathbb{N}$.

Proof. Fix $u \in L_C^{p,\lambda}(\Omega)$ and $(x,\varrho) \in \Omega_\delta$. Lemma 5.6.3 implies for $m = 0, 1, 2, \ldots$ that

$$|u_{x,\frac{\varrho}{2^{m+1}}} - u_{x,\frac{\varrho}{2^m}}| \leq C[u]_{p,\lambda} \frac{((\frac{\varrho}{2^m})^\lambda + (\frac{\varrho}{2^{m+1}})^\lambda)^{\frac{1}{p}}}{(\frac{\varrho}{2^{m+1}})^{\frac{N}{p}}}$$

$$= C'[u]_{p,\lambda} \varrho^{\frac{\lambda-N}{p}} \cdot 2^{m\frac{N-\lambda}{p}}, \tag{5.6.5}$$

where C' is a constant which is independent of m. Hence we obtain (5.6.4) by summing (5.6.5) over $m = 0, 1, \ldots, n-1$. □

Lemma 5.6.5. *Let $\lambda > N$. Then for every $u \in L_C^{p,\lambda}(\Omega)$ there exists a function $\tilde{u}$ defined on $\overline{\Omega}$ such that u equals $\tilde{u}$ almost everywhere on Ω and*

$$\lim_{\varrho \to 0+} u_{x,\varrho} = \tilde{u}(x)$$

for all $x \in \overline{\Omega}$, the convergence being uniform on $\overline{\Omega}$.

Section 5.6 Some lemmas

Proof. According to Remark 5.3.4 (ii) we have

$$\lim_{\varrho \to 0+} u_{x,\varrho} = u(x)$$

almost everywhere in Ω. It remains prove that the convergence of $u_{x,\varrho}$ is uniform with respect to x.

Let $u \in L_C^{p,\lambda}(\Omega)$. Fix $\varrho \in (0, \delta)$. By Lemma 5.6.4 we have

$$|u_{x,\frac{\varrho}{2^n}} - u_{x,\frac{\varrho}{2^{n+q}}}| \leq C''[u]_{p,\lambda} \left(\frac{\varrho}{2^n}\right)^{\frac{\lambda-N}{p}},$$

where the constant C'' is independent of x and q. We see that the sequence $\{u_{x,\frac{\varrho}{2^n}}\}_{n=1}^{\infty}$ is Cauchy uniformly with respect to x. Let

$$\tilde{u}(x) = \lim_{n \to \infty} u_{x,\frac{\varrho}{2^n}}, \quad x \in \overline{\Omega}.$$

We shall prove that $\tilde{u}$ does not depend on the choice of ϱ. Let $0 < \sigma < \delta$. Since

$$|u_{x,\frac{\sigma}{2^n}} - \tilde{u}(x)| \leq |u_{x,\frac{\varrho}{2^n}} - \tilde{u}(x)| + |u_{x,\frac{\varrho}{2^n}} - u_{x,\frac{\sigma}{2^n}}|$$

$$\leq C''[u]_{p,\lambda} \left(\frac{\varrho}{2^n}\right)^{\frac{\lambda-N}{p}} + |u_{x,\frac{\varrho}{2^n}} - u_{x,\frac{\sigma}{2^n}}|$$

$$\leq C''[u]_{p,\lambda} \left(\frac{\varrho}{2^n}\right)^{\frac{\lambda-N}{p}} + C[u]_{p,\lambda} \left(\frac{\sigma^\lambda + \varrho^\lambda}{[\min(\sigma,\varrho)]^N}\right)^{\frac{1}{p}} \cdot 2^{n\frac{N-\lambda}{p}}$$

$$\leq C(u,\varrho,\sigma) \cdot 2^{n\frac{N-\lambda}{p}} \to 0 \quad (\text{as } n \to \infty),$$

we conclude that $\tilde{u}$ is the uniform limit of any sequence of the type $\{u_{x,\frac{\sigma}{2^n}}\}_{n=1}^{\infty}$, where σ is an arbitrary real number in the interval $(0, \delta)$. Using Lemma 5.6.4 once more we obtain

$$|u_{x,\sigma} - u_{x,\frac{\sigma}{2^n}}| \leq C'[u]_{p,\lambda} \sigma^{\frac{\lambda-N}{p}} \sum_{m=0}^{n-1} 2^{\frac{m(n-\lambda)}{p}}.$$

Letting $n \to \infty$, we obtain

$$|u_{x,\sigma} - \tilde{u}(x)| \leq C[u]_{p,\lambda} \sigma^{\frac{\lambda-N}{p}}, \tag{5.6.6}$$

and consequently

$$\lim_{\sigma \to 0+} u_{x,\sigma} = \tilde{u}(x)$$

uniformly on $\overline{\Omega}$. □

Lemma 5.6.6. *Let $0 \le \lambda < N$. Then there exists a constant $C(A,p,\lambda,N) > 0$ such that for any $u \in L_C^{p,\lambda}(\Omega)$ and all $(x,\varrho) \in \Omega_\delta$ the inequality*

$$|u_{x,\varrho}| \le |u_\Omega| + C(A,p,\lambda,N)[u]_{p,\lambda} \varrho^{\frac{\lambda-N}{p}}$$

holds, where

$$u_\Omega = \frac{1}{\mu(\Omega)} \int_\Omega u(x)\,dx.$$

Proof. Fix $u \in L_C^{p,\lambda}(\Omega)$ and $\varrho \in (0,\delta)$. Then

$$|u_{x,\varrho}| \le |u_\Omega| + |u_\Omega - u_{x,\frac{\delta}{2^n}}| + |u_{x,\frac{\delta}{2^n}} - u_{x,\varrho}|, \tag{5.6.7}$$

Choose $n \in \mathbb{N}$ with

$$\frac{\delta}{2^{n+1}} \le \varrho < \frac{\delta}{2^n}.$$

Applying Lemma 5.6.3 to $|u_{x,\frac{\delta}{2^n}} - u_{x,\varrho}|$, we obtain

$$|u_{x,\frac{\delta}{2^n}} - u_{x,\varrho}| \le C[u]_{p,\lambda} \left(\frac{(\frac{\delta}{2^n})^\lambda + \varrho^\lambda}{(\frac{\delta}{2^n})^N}\right)^{\frac{1}{p}} \le C'[u]_{p,\lambda} \varrho^{\frac{\lambda-N}{p}}. \tag{5.6.8}$$

Since $u_\Omega = u_{x,\delta}$ for $x \in \Omega$, Lemma 5.6.4 applied to $|u_{x,\frac{\delta}{2^n}} - u_\Omega|$ yields

$$|u_{x,\frac{\delta}{2^n}} - u_\Omega| \le C[u]_{p,\lambda} \delta^{\frac{\lambda-N}{p}} \sum_{m=0}^{n-1} 2^{m\frac{N-\lambda}{p}}$$

$$= C[u]_{p,\lambda} \delta^{\frac{\lambda-N}{p}} \frac{1 - 2^{n\frac{N-\lambda}{p}}}{1 - 2^{\frac{N-\lambda}{p}}}$$

$$\le C[u]_{p,\lambda} \varrho^{\frac{\lambda-N}{p}} \frac{1}{1 - 2^{\frac{N-\lambda}{p}}} \left(\frac{1}{2^{(N-\lambda)\frac{n}{p}}} - 1\right). \tag{5.6.9}$$

Substituting (5.6.8) and (5.6.9) into (5.6.7) we obtain the desired inequality. □

Remark 5.6.7. Lemma 5.6.5 implies the following important assertion: For $\lambda > N$, the function u can be replaced in our considerations by the uniform limit $\tilde{u}$ of the mean values $u_{x,\varrho}$.

The inequality in Lemma 5.6.6 describes the situation in the case $0 \le \lambda < N$. It yields a local estimate

$$u_{x,\varrho} = O(\varrho^{\frac{\lambda-N}{p}}) \quad (\varrho \to 0+)$$

and a global estimate for $u_{x,\varrho}$ on the interval $0 < \varrho < \delta$.

Section 5.7 Embeddings

Remark 5.6.8. The case $\lambda = N$ is not dealt with in this section, because the methods of proofs of Lemmas 5.6.5 and 5.6.6 cannot be used. This case will be studied in Section 5.8.

Now we prove another lemma.

Lemma 5.6.9. *There exists a constant $\overline{C} = \overline{C}(A, N, \lambda)$ such that the implication*
$$\varrho = 2|x - y| \Rightarrow |u_{x,\varrho} - u_{y,\varrho}| \le \overline{C}(A, N, \lambda)[u]_{1,\lambda}|x - y|^{\lambda - N}$$
holds for every $u \in L_C^{1,\lambda}(\Omega)$ and all $x, y \in \overline{\Omega}$.

Proof. Fix $x, y \in \overline{\Omega}$, $u \in L_C^{1,\lambda}(\Omega)$ and set
$$J_\varrho = \Omega(x, \varrho) \cap \Omega(y, \varrho).$$

We have
$$|u_{x,\varrho} - u_{y,\varrho}| \le |u_{x,\varrho} - u(t)| + |u(t) - u_{y,\varrho}|$$
for almost all $t \in J_\varrho$. Since
$$\Omega(x, \tfrac{1}{2}\varrho) \subset J_\varrho \text{ and } \mu(J_\varrho) \ge A(\tfrac{1}{2}\varrho)^N,$$
integration with respect to the variable t over J_ϱ yields
$$|u_{x,\varrho} - u_{y,\varrho}| \le C\varrho^{\lambda - N}[u]_{1,\lambda} = \overline{C}[u]_{1,\lambda}|x - y|^{\lambda - N}.$$
The proof is complete. □

5.7 Relations between the spaces $L_C^{p,\lambda}(\Omega)$, $L_M^{p,\lambda}(\Omega)$ and $C^{0,\alpha}(\overline{\Omega})$

Theorem 5.7.1. *Let $1 \le p < \infty$. We have that*

(i) $L_C^{p,\lambda}(\Omega) \rightleftarrows L_M^{p,\lambda}(\Omega)$ *provided $\lambda \in [0, N)$,*

(ii) $L_C^{p,\lambda}(\Omega) \rightleftarrows C^{0,\alpha}(\overline{\Omega})$ *with $\alpha = \frac{\lambda - N}{p}$ provided $\lambda \in (N, N + p]$.*

Proof. Let $\lambda \in [0, N)$ and $u \in L_M^{p,\lambda}(\Omega)$. According to Remark 5.6.2, there exists a constant $Q > 0$ such that
$$^C\|u\|_{p,\lambda}^p \le Q(\|u\|_p^p + \|u\|_p^p)$$
$$= Q\left(\|u\|_p^p + \sup_{(x,r) \in \Omega_\delta} \frac{1}{r^\lambda}\left(\inf_{c \in \mathbb{R}} \int_{\Omega(x,r)} |u(y) - c|^p \, dy\right)\right)$$
$$\le Q(\|u\|_p^p + {}^M\|u\|_{p,\lambda}^p) \le Q_1{}^M\|u\|_{p,\lambda}^p.$$

Hence $u \in L_C^{p,\lambda}(\Omega)$ and the identity operator

$$\text{Id} : L_M^{p,\lambda}(\Omega) \to L_C^{p,\lambda}(\Omega)$$

is continuous.

Now let $u \in L_C^{p,\lambda}(\Omega)$. We have (using Lemma 5.6.6) that

$$\int_{\Omega(x,r)} |u(y)|^p \, dy \leq 2^{p-1} \left(\int_{\Omega(x,r)} |u(y) - u_{x,r}|^p \, dy + \int_{\Omega(x,r)} |u_{x,r}|^p \, dy \right)$$
$$\leq 2^{p-1} (r^\lambda [u]_{p,\lambda}^p + C r^N (r^{\lambda-n}[u]_{p,\lambda}^p + |u_\Omega|^p))$$
$$\leq C_1 (r^\lambda [u]_{p,\lambda}^p + r^N \|u\|_p^p).$$

Now the inequality

$$^M\|u\|_{p,\lambda}^p \leq C_2 {}^C\|u\|_{p,\lambda}^p$$

easily follows. Thus $u \in L_M^{p,\lambda}(\Omega)$ and the identity operator

$$\text{Id} : L_C^{p,\lambda}(\Omega) \to L_M^{p,\lambda}(\Omega)$$

is continuous, which proves assertion (i).

Let $\lambda > N$ and $\alpha = \frac{\lambda - N}{p}$, $u \in C^{0,\alpha}(\overline{\Omega})$. We have

$$\int_{\Omega(x,r)} |u(y) - u_{x,r}|^p \, dy$$
$$= \int_{\Omega(x,r)} (\mu(\Omega(x,r)))^{-p} \left(\left| \int_{\Omega(x,r)} (u(y) - u(t)) \, dt \right|^p \right) dy$$
$$\leq \int_{\Omega(x,r)} \frac{1}{\mu(\Omega(x,r))} \left(\int_{\Omega(x,r)} |u(y) - u(t)|^p \, dt \right) dy. \quad (5.7.1)$$

Recall the notation introduced in Notation 2.2.4 and Lemma 2.3.1:

$$H_{0,\alpha}(u) = \sup_{\substack{x,y \in \Omega \\ x \neq y}} \frac{|u(x) - u(y)|}{|x-y|^\alpha} \quad \text{and} \quad \|u\|_{C^{0,\alpha}(\overline{\Omega})} = \|u\|_{C(\overline{\Omega})} + H_{0,\alpha}(u).$$
$$(5.7.2)$$

From formula (5.7.1) we now conclude that

$$\int_{\Omega(x,r)} |u(y) - u_{x,r}|^p \, dy$$
$$\leq \int_{\Omega(x,r)} \frac{1}{\mu(\Omega(x,r))} \left(\int_{\Omega(x,r)} \frac{|u(y) - u(t)|^p}{|y-t|^{\alpha p}} |y-t|^{\alpha p} \, dt \right) dy$$
$$\leq \int_{\Omega(x,r)} \frac{1}{\mu(\Omega(x,r))} \left(\int_{\Omega(x,r)} H_{0,\alpha}^p(u) r^{\alpha p} \, dt \right) dy$$
$$= \mu(\Omega(x,r)) H_{0,\alpha}^p(u) r^{p\alpha} \leq C_N H_{0,\alpha}^p(u) r^\lambda$$

and consequently

$$[u]_{p,\lambda} + \|u\|_p \leq C(H_{0,\alpha}(u) + \|u\|_{C(\overline{\Omega})}).$$

Thus $u \in L_C^{p,\lambda}(\Omega)$ and the identity operator

$$\text{Id} : C^{0,\alpha}(\overline{\Omega}) \to L_C^{p,\lambda}(\Omega) \tag{5.7.3}$$

is continuous.

Let $u \in L_C^{p,\lambda}(\Omega)$. Since, by Theorem 5.5.2 (ii), we have

$$L_C^{p,\lambda}(\Omega) \hookrightarrow L_C^{1,\alpha+N}(\Omega),$$

it suffices to prove that $u \in C^{0,\alpha}(\overline{\Omega})$ for $\alpha \in (0,1)$ and $u \in L_C^{1,\alpha+N}(\Omega)$. By virtue of (5.7.3) the continuity of the mapping

$$\text{Id} : L_C^{p,\lambda}(\Omega) \to C^{0,\alpha}(\overline{\Omega})$$

follows from the Banach theorem (Theorem 1.14.8).

Let $u \in L_C^{1,\alpha+N}(\Omega)$. We prove that

$$\lim_{\varrho \to 0+} u_{x,\varrho} = \tilde{u} \in C^{0,\alpha}(\overline{\Omega})$$

and use Remark 5.6.7. Let $x, y \in \Omega$, $\varrho = 2|x-y|$. From (5.6.6) and Lemma 5.6.9 we have

$$|\tilde{u}(x) - \tilde{u}(y)| \leq |\tilde{u}(x) - u_{x,\varrho}| + |u_{x,\varrho} - u_{y,\varrho}| + |u_{y,\varrho} - \tilde{u}(y)|$$
$$\leq C[u]_{1,N+\alpha}|x-y|^\alpha.$$

Thus

$$H_{0,\alpha}(u) \leq C[u]_{1,N+\alpha}$$

and the proof of the theorem is complete. $\square$

Remark 5.7.2. For $\lambda > N + p$ the spaces $L_C^{p,\lambda}(\Omega)$ and $C^{0,\frac{\lambda-N}{p}}(\overline{\Omega})$ contain only constant functions on Ω.

5.8 $L_C^{p,N}(\Omega)$ and the John–Nirenberg space

In Section 5.3 we postponed the proof of the assertion

$$L_C^{p,N}(\Omega) \neq L^\infty(\Omega)$$

(see Remark 5.5.3 (ii)). We shall finish this proof in this section.

Lemma 5.8.1. *Let $p \in [1, \infty)$. Then*
$$L^\infty(\Omega) \subsetneq L_C^{p,N}(\Omega).$$

Proof. Let $u \in L^\infty(\Omega)$ and $(x, \varrho) \in \Omega_\delta$. Then
$$\int_{\Omega(x,\varrho)} |u(y) - u_{x,\varrho}|^p \, dy \leq 2^{p-1} \left(\int_{\Omega(x,\varrho)} |u(y)|^p \, dy + \mu(\Omega(x,\varrho))|u_{x,\varrho}|^p \right)$$
$$\leq 2^p \mu(\Omega(x,\varrho)) \| |u|^p \|_\infty \leq c \varrho^N \| |u|^p \|_\infty,$$

whence
$$[u]_{p,N} \leq c \|u\|_\infty.$$

In order to prove that
$$L_C^{p,N}(\Omega) \setminus L^\infty(\Omega) \neq \emptyset,$$
note that, for $N = 1$ and $\Omega = (0, 1)$, one has $\log x \in L_C^{p,1}(\Omega)$ for any $p \geq 1$, but $\log x \notin L^\infty(\Omega)$. It is easy to construct a similar example for $N > 1$ and general $\Omega \subset \mathbb{R}^N$. □

Remark 5.8.2. Lemma 5.8.1 and Theorem 5.7.1 suggest that the spaces $L_C^{p,N}(\Omega)$ play an important part in the family of Campanato spaces. If Ω is an N-dimensional cube then the space $L_C^{p,N}(\Omega)$ will be shown to coincide with a certain vector space $JN(\Omega)$, which was introduced by John and Nirenberg [110]. The space $JN(\Omega)$ is more familiar under the notation BMO(Ω) (the space of functions with bounded mean oscillation). There is a vast literature available on this space, see, e.g., [14, 79].

Definition 5.8.3. Denote by Q the N-dimensional cube whose edges are parallel with the coordinate axes. We shall denote by Q' the cubes contained in Q and homothetic with Q. It is easy to see that the norm
$$^C\|u\|_{p,N} = [u]_{p,N} + \|u\|_p$$
which was used until now is equivalent for $\Omega = Q$ to the norm
$$\|u\|_p + \langle u \rangle_{p,N}, \tag{5.8.1}$$
where
$$\langle u \rangle_{p,N} = \left(\sup_{Q' \subset Q} \frac{1}{\mu(\Omega')} \int_{\Omega'} |u(y) - u_{Q'}|^p \, dy \right)^{\frac{1}{p}},$$
$$u_{Q'} = \frac{1}{\mu(Q')} \int_{Q'} u(x) \, dx. \tag{5.8.2}$$

From now on we shall denote by $^C\|u\|_{p,N}$ the norm defined by relation (5.8.1), since it is more suitable for our purpose.

Section 5.8 The John–Nirenberg space

Definition 5.8.4. For any measurable function u on the cube Q and for $b > 0$ denote

$$S(u, b, Q') = \{x \in Q'; \ |u(x) - u_{Q'}| > b\}. \tag{5.8.3}$$

Let $u \in L^1(Q)$ be a function for which there exist two positive constants $\beta > 0$ and $H > 0$ such that for every $b > 0$ and every cube $Q' \subset Q$ we have

$$\mu(S(u, b, Q')) \leq H e^{-\beta b} u(Q').$$

We shall denote the set of all functions with this property by $JN(Q)$.

Lemma 5.8.5. $JN(Q) \subset L_C^{p,N}(Q)$ for every $p \geq 1$. Moreover, the inequality

$$\langle u \rangle_{p,N} \leq \frac{1}{\beta}(H \cdot \Gamma(p+1))^{\frac{1}{p}}$$

holds for all $u \in JN(Q)$ where Γ is the Gamma function.

Proof. Let $u \in JN(Q)$. Using Lemma 3.18.2, we obtain

$$\int_{Q'} |u(x) - u_{Q'}|^p \, dx = p \int_0^\infty t^{p-1} \mu(S(u, t, Q')) \, dt$$

$$\leq p H \mu(Q') \int_0^\infty t^{p-1} e^{-\beta t} \, dt = H \beta^{-p} \mu(Q') \Gamma(p+1)$$

and this implies the assertion. □

Our next aim will be to prove that $L_C^{p,N}(Q) = JN(Q)$ for all $p \geq 1$. We start with some auxiliary material.

Lemma 5.8.6. Let $u \in L^1(Q)$ and $K \in (0, \infty)$. Suppose

$$\frac{1}{\mu(Q)} \int_Q |u(x)| \, dx \leq K.$$

Then there exists a countable collection $\{Q'_i\}$ of pairwise disjoint open cubes $Q'_i \subset Q$ such that

(i) $|u(x)| \leq K$ almost everywhere on $Q \setminus \bigcup Q'_i$;

(ii) $|u|_{Q'_i} \leq 2^N K$ for each cube Q'_i in the collection $\{Q'_i\}$;

(iii) $\sum \mu(Q'_i) \leq \frac{1}{K} \int_Q |u(x)| \, dx.$

Proof. Halving the edges of Q we decompose it into 2^N equal cubes Q'. Because the measures of all these cubes are the same the following holds:

$$\frac{1}{\mu(Q)} \int_Q |u(x)|\,dx = \frac{1}{2^N} \sum \frac{1}{\mu(Q')} \int_{Q'} |u(x)|\,dx \qquad (5.8.4)$$

(we sum over all cubes obtained by the decomposition). Denote by $Q'_{1,1}, Q'_{1,2}, \ldots$ those of the cubes Q' for which

$$\frac{1}{\mu(Q')} \int_{Q'} |u(x)|\,dx \geq K.$$

This inequality together with (5.8.4) implies

$$K\mu(Q'_{1,i}) \leq \int_{Q'_{1,i}} |u(x)|\,dx \leq K 2^N \mu(Q'_{1,i}), \quad i \in \mathbb{N}.$$

Now we decompose each of the remaining cubes Q' for which the mean value $|u|_{Q'}$ is less than K into 2^N equal cubes and then repeat the process described above. Proceeding this way we obtain a countable collection of cubes. Let us arrange this collection in a sequence $\{Q'_j\}$. The interiors of the members of $\{Q'_j\}$ form a collection satisfying (i)–(iii). Indeed, these sets Int Q'_j, $j = 1, 2, \ldots$ are pairwise disjoint open cubes and

$$K\mu(Q'_j) \leq \int_{Q'_j} |u(x)|\,dx \leq 2^N K\mu(Q'_j), \quad j = 1, 2, \ldots.$$

Hence assertion (ii) is valid. Summing the inequalities

$$K\mu(Q'_j) \leq \int_{Q'_j} |u(x)|\,dx$$

we obtain inequality (iii).

It remains to prove (i). Inasmuch as $u \in L^1(Q)$ we have that almost all the points of Q are Lebesgue points of the function u (see Remark 5.3.4 (ii)). Let x be a Lebesgue point of u belonging to $Q \setminus \bigcup Q'_j$. It follows from the construction of $\{Q'_j\}$ that for every $\varepsilon > 0$ there exists a cube Q_ε with its edge shorter than ε containing the point $x \in Q_\varepsilon$ and such that

$$\frac{1}{\mu(Q_\varepsilon)} \int_{Q_\varepsilon} |u(x)|\,dx < K.$$

Thus

$$|u(x)| = \lim_{\varepsilon \to 0+} \frac{1}{\mu(Q_\varepsilon)} \int_{Q_\varepsilon} |u(y)|\,dy \leq K. \quad \square$$

Section 5.8 The John–Nirenberg space

In the following lemma, we shall use the symbols $\langle u \rangle_{1,N}$ from (5.8.2) and $S(u, b, Q')$ from (5.8.3).

Lemma 5.8.7. *Let $u \in L_C^{1,N}(Q)$. Then $u \in JN(Q)$ and there exist positive constants A, α such that*

$$\mu(S(u, b, Q')) \leq A \exp\left(-\frac{\alpha b}{\langle u \rangle_{1,N}}\right) \mu(Q') \tag{5.8.5}$$

for any $Q' \subset Q$.

Proof. Let u be a nonconstant function in $L_C^{1,N}(Q)$. Suppose that

$$u_Q = 0 \quad \text{and} \quad \langle u \rangle_{1,N} = 1. \tag{5.8.6}$$

Define the function $F : (0, \infty) \to \mathbb{R}$ by

$$F(b) = \sup \frac{\mu(S(u, b, Q))}{\int_Q |u(x)| \, dx}$$

(here the supremum is taken over the set of all functions satisfying (5.8.6)).

Notice that the definition of F does not depend on the length of the edge of Q. In fact, if Q' is a cube, $Q' \subset Q$ and if we define

$$\widetilde{F}(b) = \sup \frac{\mu(S(u, b, Q'))}{\int_{Q'} |u(x)| \, dx}$$

where the supremum is taken over the set of all functions in $L_C^{1,N}(Q')$ for which $u_{Q'} = 0$ and $\langle u \rangle_{1,N} = 1$ (calculated on Q'), then $\widetilde{F}(b) = F(b)$.

The function F has the following property:
For every $b \geq 2^N e$ and for every $K \in [1, 2^{-N}b]$,

$$F(b) \leq K^{-1} F(b - 2^N K). \tag{5.8.7}$$

This inequality will be proved later. We show first of all how assertion of Lemma 5.8.7 follows from it.

Fix a positive number ε. Let A be a positive number such that for all $b \in [\varepsilon, 2^N e + \varepsilon]$,

$$F(b) \leq A e^{-\alpha b}, \quad \alpha = \frac{1}{2^N e}. \tag{5.8.8}$$

Such an A exists. (Its existence follows from the definition of F: We have

$$F(\varepsilon) = \sup \frac{\mu(S(u, \varepsilon, Q))}{\int_Q |u(x)| \, dx} = \sup \frac{\mu(S(u, \varepsilon, Q))}{\int_0^\infty \mu(S(u, t, Q)) \, dt}$$

$$\leq \sup \frac{\mu(S(u, \varepsilon, Q))}{\int_0^\varepsilon \mu(S(u, t, Q)) \, dt} \leq \frac{1}{\varepsilon}$$

and the function F is nonincreasing.) Then (5.8.8) is valid for all $b \in (\varepsilon, \infty)$. Indeed, set $K = \mathrm{e}$ in (5.8.7). Hence

$$F(b) \leq \mathrm{e}^{-1} F(b - 2^N \mathrm{e})$$

provided $b \in [2^N \mathrm{e}, \infty)$. For $b \in [2^N \mathrm{e} + \varepsilon, 2^{N+1} \mathrm{e} + \varepsilon)$, it follows from here and from (5.8.8) that

$$F(b) \leq A \exp\left(-1 - \frac{1}{2^N \mathrm{e}}(b - 2^N \mathrm{e})\right) = A \mathrm{e}^{-\alpha b}.$$

We have thus proved that (5.8.8) is satisfied in the interval $[\varepsilon, 2^{N+1} \mathrm{e} + \varepsilon]$. Using the induction on k we can easily prove the validity of (5.8.8) in each interval $[\varepsilon, 2^{N+k} \mathrm{e} + \varepsilon]$. So the formula (5.8.8) is valid in $[\varepsilon, \infty)$.

It follows from the estimate (5.8.8) and the definition of F that

$$\mu(S(u,b,Q)) \leq A\mu(Q) \mathrm{e}^{-\alpha b} \tag{5.8.9}$$

for all $b \in [\varepsilon, \infty)$. In the general case, for a function $u \in L_C^{1,N}(Q)$ we deduce (substituting

$$U(x) = \frac{u(x) - u_Q}{\langle u \rangle_{1,N}}$$

into (5.8.9)) that

$$\mu(S(u,b,Q)) \leq A\mu(Q) \exp\left(\frac{-\alpha b}{\langle u \rangle_{1,N}}\right) \tag{5.8.10}$$

for all $b \in [\varepsilon, \infty)$. Regarding $\mu(S(u,b,Q)) \leq \mu(Q)$ it is possible, increasing A if necessary, to achieve estimate (5.8.10) on the entire interval $[0, \infty)$. Finally, by virtue of the fact that F is independent of the choice of Q', it is easy to see that (5.8.10) is valid (with the same values of A and α) for any cube $Q' \subset Q$.

It remains to prove assertion (5.8.7). Let u satisfy (5.8.6) and let $b \geq 2^N \mathrm{e}$, $K \in [1, 2^{-N} b]$. According to Lemma 5.8.6 there exists an at most countable collection $\{Q'_j\}$ of cubes from Q with the properties:

(i) $|u(x)| \leq K$ almost everywhere on $Q - \bigcup_j Q'_j$;

(ii) $|u|_{Q'_j} \leq 2^N K$, $j = 1, 2, \ldots$;

(iii) $\sum_j \mu(Q'_j) \leq \frac{1}{K} \int_Q |u(x)| \, \mathrm{d}x$.

We can suppose $S(u,b,Q) \neq \emptyset$. (Otherwise everything is obvious.) Let $x \in S(u,b,Q)$. Thus $|u(x)| > b > K$. According to (i) either there exists a Q'_j such that $x \in Q'_j$ or x belongs to a subset of $Q - \bigcup_j Q'_j$ which has measure zero. Hence it follows that

$$\mu(S(u,b,Q)) = \mu\left(\bigcup_j \{x \in Q'_j; |u(x)| > b\}\right). \tag{5.8.11}$$

Section 5.8 The John–Nirenberg space

The sets in the union on the right-hand side of (5.8.11) are pairwise disjoint and

$$\mu(\{x \in Q'_j; |u(x)| > b\}) \le \mu(\{x \in Q'_j; |u(x) - u_{Q'_j}| > b - |u|_{Q'_j}\})$$
$$\le \mu(\{x \in Q'_j; |u(x) - u_{Q'_j}| > b - 2^N K\}) = \mu(S(u, b - 2^N K, Q'_j)).$$

This estimate together with (5.8.11) yields

$$\mu(S(u, b, Q)) \le \sum_j \mu(S(u, b - 2^N K, Q'_j)). \tag{5.8.12}$$

On Q'_j the functions $u - u_{Q'_j}$ satisfy the conditions

$$(u - u_{Q'_j})_{Q'_j} = 0, \ \langle u - u_{Q'_j} \rangle_{1,N} = 1.$$

From this and the definition of the function F we conclude that

$$\sum_j \mu(S(u, b - 2^N K, Q'_j)) \le \mu(Q) F(b - 2^N K).$$

Substituting this inequality into (5.8.12) and dividing the resulting inequality by

$$\int_Q |u(x)| \, dx$$

we obtain (also using Lemma 5.8.6 (iii))

$$\frac{\mu(S(u, b, Q))}{\int_Q |u(x)| \, dx} \le \frac{1}{K} F(b - 2^N K).$$

Finally, taking the supremum over the set of all functions u satisfying (5.8.6) we obtain (5.8.7). □

Theorem 5.8.8. *Let $p \in [1, \infty)$. Then*

$$L_C^{p,N}(Q) = JN(Q).$$

Proof. It follows from the definition of $L_C^{p,N}(Q)$ that if $1 \le p < q$, then $L_C^{q,N}(Q) \subset L_C^{p,N}(Q)$ and

$$\langle u \rangle_{p,N} \le \langle u \rangle_{q,N}.$$

By virtue of Lemmas 5.8.5 and 5.8.7 we obtain immediately the assertion of Theorem 5.8.8. □

5.9 Another definition of the space $JN(Q)$

Theorem 5.9.1. *Let $\widetilde{JN}(Q)$ be the set of all functions $u \in L^1(Q)$ with the following property: There exist $k = k(u) > 0$ and $M = M(u) > 0$ such that for each $Q' \subset Q$*

$$\int_{Q'} \exp(k|u(x) - u_{Q'}|)\,dx \le M\mu(Q') \tag{5.9.1}$$

holds. Then
$$JN(Q) = \widetilde{JN}(Q).$$

Proof. Let $u \in \widetilde{JN}(Q)$. Then we have that

$$e^{kb}\mu(S(u,b,Q')) \le \int_{\Omega'} \exp(k|u(x) - u_{Q'}|)\,dx \le M\mu(Q').$$

Hence $\mu(S(u,b,Q')) \le e^{-kb}M\mu(Q')$ and so $u \in JN(Q)$.

Now let $u \in JN(Q)$. Then

$$\int_{\Omega'} \exp(k|u(x) - u_{Q'}|)\,dx = k\int_0^\infty e^{kt}\mu(S(u,t,Q'))\,dt. \tag{5.9.2}$$

(This assertion can be proved in the same manner as Lemma 3.18.2.) Inasmuch as $\mu(S(u,t,Q')) \le Ae^{-\alpha t}\mu(Q')$, we conclude from (5.9.2) (choosing $k = \frac{1}{2}\alpha$) that

$$\int_{Q'} \exp\left(\frac{\alpha}{2}|u(x) - u_{Q'}|\right) dx \le \frac{\alpha}{2}\left(\int_0^\infty e^{-\frac{\alpha t}{2}}\,dt\right) A\mu(Q').$$

Consequently, $u \in \widetilde{JN}(Q)$. □

Definition 5.9.2. Let X_1 be the set of all functions u which are defined almost everywhere on Q and satisfy the conditions

(i) $u \in \bigcap_{p \in \mathbb{N}} L^p(Q)$;

(ii) to every u there exists a constant C_u such that
$$\|u\|_p \le C_u p$$
for all $p \in \mathbb{N}$.

Theorem 5.9.3. $JN(Q) \subsetneq X_1$.

Proof. Let $u \in JN(Q)$. According to Lemma 3.18.2,

$$\int_Q |u(x) - u_Q|^p\,dx = p\int_0^\infty t^{p-1}\mu(S(u,t,Q))\,dt$$

$$\le Ap\mu(Q)\int_0^\infty t^{p-1}e^{-\alpha t}\,dt = \frac{A\mu(Q)}{\alpha^p}p\Gamma(p).$$

Section 5.9 Another definition of the space $JN(Q)$

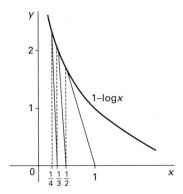

Figure 5.1. A logarithmic function.

Using the well-known formula
$$p\Gamma(p) \sim \frac{p^{p+\frac{1}{2}}}{e^p} \quad (p \to \infty)$$
and the estimate
$$\int_Q |u(x)|^p \, dx \le 2^{p-1} \left(\int_Q |u(x) - u_Q|^p \, dx + \int_Q |u_Q|^p \, dx \right),$$
we obtain that $u \in X_1$.

Now we shall construct a function belonging to $X_1 \setminus JN(Q)$. In the case $N = 1$, $Q = (0, 1)$ set
$$Q_n = \left[\tfrac{1}{n}, \tfrac{1}{n-1}\right), \quad n = 2, 3, \ldots$$
and define
$$g(x) = -n(1 + \log n)[(n-1)x - 1]$$
for $x \in Q_n$, $n = 2, 3, \ldots$ (Figure 5.1).

It is evident that for $p \in \mathbb{N}$
$$0 \le g^p(x) \le (1 - \log x)^p \le 2^{p-1}(|\log x|^p + 1)$$
and so
$$\left(\int_0^1 g^p(x) \, dx \right)^{\frac{1}{p}} \le (2^{p-1}(\Gamma(p+1) + 1))^{\frac{1}{p}} \le 2(\Gamma(p+1))^{\frac{1}{p}}.$$

From the last estimate we obtain, by virtue of
$$\Gamma(p+1) \sim \frac{p^{p+\frac{1}{2}}}{e^p} \quad (p \to \infty),$$
that $g \in X_1$.

In order to prove $g \notin JN((0,1))$ it suffices to show that $g \notin L_C^{1,1}((0,1))$. However,

$$g_{Q_n} = \frac{1}{\frac{1}{n-1} - \frac{1}{n}} \int_{\frac{1}{n}}^{\frac{1}{n-1}} g(x)\,dx = \tfrac{1}{2}(1+\log n)$$

and thus

$$\frac{1}{\mu(Q_n)} \int_{Q_n} |g(x) - g_{Q_n}|\,dx = \tfrac{1}{4}(1+\log n) \to \infty$$

(as $n \to \infty$). □

Theorem 5.9.4. *Let u be a measurable function on Q. Suppose there exists a $C > 0$ such that for all $p \in \mathbb{N}$ and all $Q' \subset Q$*

$$\frac{1}{(\mu(Q'))^2} \int_{Q'} \int_{Q'} |u(x) - u(y)|^p \,dy\,dx \leq C^p p^p. \tag{5.9.3}$$

Then $u \in \widetilde{JN}(Q)$.

Proof. We have

$$\frac{1}{\mu(Q')} \int_{Q'} \exp(k|u(x) - u_{Q'}|)\,dx = \sum_{p=0}^{\infty} \frac{k^p}{p!} \frac{1}{\mu(Q')} \int_{Q'} |u(x) - u_{Q'}|^p\,dx. \tag{5.9.4}$$

If there exists a $C > 0$ such that for all $p \in \mathbb{N}$ and all $Q' \subset Q$

$$\frac{1}{\mu(Q')} \int_{Q'} |u(x) - u_{Q'}|^p\,dx < C^p p^p, \tag{5.9.5}$$

we can choose k such that the series in (5.9.4) converges and so $u \in JN(Q)$. However, from condition (5.9.5) we easily obtain condition (5.9.3) by expressing $u_{Q'}$ explicitly and using the Hölder inequality. □

Exercises 5.9.5. (i) Denote

$$\|u\|_p^* = \left(\frac{1}{\mu(Q)} \int_Q |u(x)|^p\,dx \right)^{\frac{1}{p}}$$

and define for $\alpha \in [0,1]$ the set $X_\alpha(Q)$ as the set of all $u \in L^1(Q)$ for which there exists a $C > 0$ such that $\|u\|_p^* \leq Cp^\alpha$ for every $p \in \mathbb{N}$. Prove that $X_\alpha(Q)$ is a Banach space with the norm

$$\|u\|_{X_\alpha} = \sup_{p \in \mathbb{N}} \frac{\|u\|_p^*}{p^\alpha}.$$

(ii) We have
$$X_0(Q) = L^\infty(Q).$$
For $\alpha < \beta$ prove that
$$X_\alpha(Q) \hookrightarrow X_\beta(Q).$$

(iii) Show that
$$X_\alpha(Q) \setminus JN(Q) \neq \emptyset$$
and
$$JN(Q) \setminus X_\alpha(Q) \neq \emptyset \quad \text{for } \alpha \in (0, 1).$$

(In the case of $\alpha = 0$, $\alpha = 1$ the situation is described in Lemma 5.8.1 and Theorem 5.8.8.)

(Hint: $f(x) = -\log x$ belongs to $JN((0, 1))$ but not to $X_\alpha((0, 1))$. To prove $X_\alpha((0, 1)) \setminus JN((0, 1)) \neq \emptyset$ we construct a function g which is analogous to the function g in the proof of Theorem 5.8.8, but taking the majorant $(-\log x)^\alpha + 1$ instead of the majorant $(-\log x) + 1$.)

5.10 Spaces $N_{p,\lambda}(Q)$ and their relation to the spaces $L_C^{p,\lambda}(Q)$

Let us introduce another type of spaces which are closely related to $L_C^{p,\lambda}(Q)$. As in the previous sections we denote by Q an N-dimensional cube, Q' is an arbitrary cube contained in Q with its edges parallel to the edges of Q. By $\{\Delta\}$ we denote the collection of all finite decompositions Δ of the cube Q into subcubes Q'. The symbol $\sum_\Delta$ denotes the sum over all cubes Q'_i of Δ.

With the exception of the last assertion of this section we omit the proofs.

Theorem 5.10.1. *Let $p \in (1, \infty)$. The function $u \in L^1(Q)$ belongs to $L^p(Q)$ if and only if*
$$\sup_{\Delta \in \{\Delta\}} \sum_\Delta (\mu(Q'_i))^{1-p} \left(\int_{Q'_i} |u(x)| \, dx \right)^p < \infty.$$

Proof. For the proof, see, e.g., [28, 187]. □

Definition 5.10.2. We say that $u \in L^1(Q)$ belongs to the class $N_{p,\lambda}(Q)$ for $p \in (1, \infty)$ and $\lambda \in \mathbb{R}$ if and only if
$$K_{p,\lambda}(u) := \left(\sup_{\Delta \in \{\Delta\}} (\mu(Q'_i))^{1-p-\lambda} \left(\int_{Q'_i} |u(x) - u_{Q'_i}| \, dx \right)^p \right)^{\frac{1}{p}} < \infty. \quad (5.10.1)$$

The following result is stated and proved in [110].

Theorem 5.10.3 (John–Nirenberg). *Let $u \in L^1(Q)$ and suppose that for some $p \in (1, \infty)$ the expression $K_{p,0}(u)$ is finite. Then there exists a constant $A > 0$ such that for all $\sigma > 0$,*

$$\mu(\{x \in Q;\ |u(x) - u_Q| > \sigma\}) \leq A \frac{K_{p,0}^p(u)}{\sigma^p}.$$

Remark 5.10.4. It is easy to prove that for every $p \in (1, \infty)$ the inclusion $u - u_Q \in L^{p,\infty}(Q)$, where $L^{p,\infty}(Q)$ is the weak Lebesgue space, introduced in Definition 3.18.3, implies $u \in L^{p,\infty}(Q)$. Thus Theorem 5.10.3 gives a sufficient condition for u to be in $L^{p,\infty}(Q)$.

Theorem 5.10.5. *Let $u \in L^1(Q)$. Then*

$$\lim_{p \to \infty} K_{p,-p+\frac{\lambda p}{N}}(u) = \sup_{Q' \subset Q} (\mu(Q'))^{-\frac{\lambda}{N}} \int_{Q'} |u(x) - u_{Q'}|\, dx, \qquad (5.10.2)$$

and so

$$u \in L_C^{1;\lambda}(Q) \quad \text{if and only if} \quad \lim_{p \to \infty} K_{p,-p+\frac{\lambda p}{N}}(u) < \infty.$$

Proof. Set

$$M = \sup_{Q' \subset Q} (\mu(Q'))^{-\frac{\lambda}{N}} \int_{Q'} |u(x) - u_{Q'}|\, dx.$$

If $M = 0$, then u is constant and $K_{p,-p+\frac{\lambda p}{N}}(u)$ is also zero. If $M > 0$, choose M' so that $0 < M' < M$. There exists at least one cube $Q' \subset Q$ such that

$$(\mu(Q'))^{-\frac{\lambda}{N}} \int_{Q'} |u(x) - u_{Q'}|\, dx \geq M'$$

and so

$$K_{p,-p+\frac{\lambda p}{N}}(u) \geq \left(\int_{Q'} |u(x) - u_{Q'}|\, dx \right) (\mu(Q'))^{-\frac{\lambda}{N}} (\mu(Q'))^{\frac{1}{p}}$$

$$\geq M'(\mu(Q'))^{\frac{1}{p}}.$$

Hence we obtain

$$\liminf_{p \to \infty} K_{p,-p+\frac{\lambda p}{N}}(u) \geq M. \qquad (5.10.3)$$

On the other hand

$$K_{p,-p+\frac{\lambda p}{N}}(u) \leq M(\mu(Q))^{\frac{1}{p}}$$

and so

$$\limsup_{p \to \infty} K_{p,-p+\frac{\lambda p}{N}}(u) \leq M. \qquad (5.10.4)$$

Relations (5.10.3) and (5.10.4) imply (5.10.2) for $M > 0$. This completes the proof of the theorem. □

5.11 Miscellaneous remarks

Remark 5.11.1 (the spaces $L_C^{p,\lambda,m}(\Omega)$). Let $\Omega \subset \mathbb{R}^N$ be a domain of type $\mathcal{A}$ (see Definition 5.3.1). In Remark 5.6.2 we have seen that $L_C^{p,\lambda}(\Omega)$ can be renormed by the equivalent norm $\|\cdot\|_p + \|\cdot\|_{p,\lambda}$, where the seminorm $\|\cdot\|_{p,\lambda}$ is defined by

$$\|u\|_{p,\lambda} = \left(\sup_{(x,\varrho)\in\Omega_\delta} \frac{1}{\varrho^\lambda} \left(\inf_{c\in\mathbb{R}} \int_{\Omega(x,\varrho)} |u(y) - c|^p \, dy \right) \right)^{\frac{1}{p}}. \tag{5.11.1}$$

So $L_C^{p,\lambda}(\Omega)$ is a subspace of $L^p(\Omega)$ which consists of those functions u for which $\|u\|_{p,\lambda} < \infty$.

This definition can be generalized in the following way: We replace the space of all constant functions over which we take the infimum in formula (5.11.1) by another subspace Y of $L^p(\Omega)$. The case of $Y = \mathcal{P}_m$ (the set of all polynomials in N variables and of degree $\leq m$) is especially important. Therefore, let us define

$$\|u\|_{p,\lambda,m} := \left(\sup_{(x,\rho)\in\Omega_\delta} \frac{1}{\varrho^\lambda} \left(\inf_{P\in\mathcal{P}_m} \int_{\Omega(x,\varrho)} |u(y) - P(y)|^p \, dy \right) \right)^{\frac{1}{p}}, \tag{5.11.2}$$

$$L_C^{p,\lambda,m}(\Omega) := \{u \in L^p(\Omega); \|u\|_{p,\lambda,m} < \infty\} \tag{5.11.3}$$

and let the norm in $L_C^{p,\lambda,m}(\Omega)$ be given by the expression $\|\cdot\|_p + \|\cdot\|_{p,\lambda,m}$. Using this notation we can write $L_C^{p,\lambda,0}(\Omega)$ instead of $L_C^{p,\lambda}(\Omega)$. The spaces $L_C^{p,\lambda,m}(\Omega)$ for $1 \leq p < \infty$, $\lambda \geq 0$, $m \in \mathbb{N}_0$ are Banach spaces.

Remark 5.11.2 (properties of the spaces $L_C^{p,\lambda,m}(\Omega)$). As was proved in Theorem 5.7.1, $L_C^{p,\lambda,0}(\Omega) \rightleftarrows L_M^{p,\lambda}$ for $0 \leq \lambda < N$ and $L_C^{p,\lambda,0}(\Omega) \rightleftarrows C^{0,\alpha}(\overline{\Omega})$ for $N < \lambda \leq N + p$ with $\alpha = \frac{\lambda - N}{p}$ (this result was established independently by Campanato [26] and Meyers [151]). In Theorem 5.8.8 we proved that $L_C^{p,N,0}(Q) = JN(Q)$ provided Q is a cube. $L_C^{p,\lambda,0}(\Omega)$ is the set of constant functions if $\lambda > N + p$.

For $m > 0$, the following result is due to Campanato [27]: Let $m \geq 1$, $p \in (1, \infty)$. Then

- $L_C^{p,\lambda,m}(\Omega) \rightleftarrows L_C^{p,\lambda,m-1}(\Omega)$ for $0 \leq \lambda < N + mp$;

- $L_C^{p,\lambda,m}(\Omega)$ is isomorphic to a certain limit space $\mathcal{E}_m(\Omega)$ for $\lambda = N + mp$;

- $L_C^{p,\lambda,m}(\Omega) \rightleftarrows C^{m,\alpha}(\overline{\Omega})$ with $\alpha = \frac{\lambda - N}{p} - m$ for $N + mp < \lambda \leq N + (m+1)p$;

- for $\lambda > N + (m+1)p$, the space $L_C^{p,\lambda,m}(\Omega)$ consists of the polynomials $P \in \mathcal{P}_m$ only.

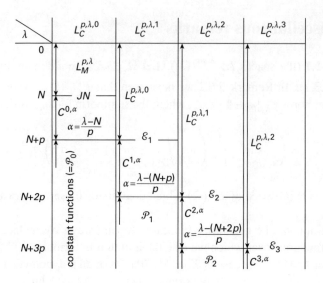

Figure 5.2. Campanato spaces.

Consequently, fixing $m \in \mathbb{N}_0$ we obtain all the spaces $L_M^{p,\lambda}(\Omega)$, $C^{0,\alpha}(\overline{\Omega})$, ..., $C^{m,\alpha}(\overline{\Omega})$ by changing only one parameter, λ. Figure 5.2 is an attempt to describe the situation graphically.

Remark 5.11.3 (the spaces $L_C^{p,\lambda}(\Omega, \sigma)$). We obtain another generalization of the spaces $L_C^{p,\lambda}(\Omega)$ if we replace $\Omega(x, \varrho)$ in (5.11.1) by some other sets; the following choice is of great importance: Let σ be a metric in $\mathbb{R}^N$ with the following properties

(i) $B_\sigma(0, \varrho) := \{y \in \mathbb{R}^N; \sigma(0, y) \leq \varrho\}$ is a convex set for all $\rho > 0$;

(ii) there exist two constants $M_1 > 0$, $M_2 > 0$ and a number $m \geq N$ such that for every $\varrho > 0$
$$M_1 \varrho^m \leq \mu(B_\sigma(0, \varrho)) \leq M_2 \varrho^m.$$

Set again $\delta(\sigma) = \sup\{\sigma(x, y); x, y \in \Omega\}$ and define the seminorm

$$\left(\sup_{x \in \Omega, 0 < \varrho < \delta(\sigma)} \mu(\Omega_\sigma(x, \varrho))^{-\lambda} \left(\inf_{c \in \mathbb{R}} \int_{\Omega_\sigma(x, \varrho)} |u(y) - c|^p \, dy \right) \right)^{\frac{1}{p}}, \quad (5.11.4)$$

where
$$\Omega_\sigma(x, \varrho) := \{y \in \Omega; \sigma(x, y) < \varrho\}$$

Denote by $L_C^{p,\lambda}(\Omega,\sigma)$ the set of all $u \in L^p(\Omega)$ for which expression (5.11.4) is finite. We introduce a norm in $L_C^{p,\lambda}(\Omega,\sigma)$ in a natural way, i.e. by adding $\|u\|_p$ with the seminorm (5.11.4).

If Ω is of type $\mathcal{A}$ with respect to the metric σ, i.e.

$$\mu(\Omega_\sigma(x,r)) \geq A\mu(B_\sigma(x,r))$$

holds with the same constant A for all $x \in \Omega$ and for all $r \leq \delta(\sigma)$, we obtain a result similar to Theorem 5.7.1 (ii) with the only difference that the Hölderian seminorm becomes

$$H_{0,\beta,\sigma}(u) = \sup_{\substack{x,y \in \Omega \\ x \neq y}} \frac{|u(x) - u(y)|}{\sigma^\beta(x,y)}$$

and

$$\|u\|_{C^{0,\beta}(\overline{\Omega},\sigma)} = \|u\|_{C(\overline{\Omega})} + H_{0,\beta,\sigma}(u)$$

instead of (5.7.2).

Theorem 5.11.4. *If $\lambda > 1$ then $L_C^{p,\lambda}(\Omega,\sigma) \rightleftarrows C^{0,\beta}(\overline{\Omega},\sigma)$ with $\beta = \frac{m(\lambda-1)}{p}$.*

In particular, let

$$\sigma(x,y) = \left(\sum_{i=1}^N |x_i - y_i|^{\alpha_i}\right)^{\frac{1}{\alpha}}$$

where the α_i are integers, $\alpha_i \geq 1$ for $i = 1, 2, \ldots, N$ and $\alpha = \max \alpha_i$. It is easy to see that this metric has the properties (i) and (ii). The estimate

$$\frac{|u(x_1, x_2, \ldots, x_k, \ldots, x_N) - u(x_1, x_2, \ldots, y_k, \ldots, x_N)|}{|x_k - y_k|^{\frac{\alpha_k \beta}{\alpha}}} \leq H_{0,\beta,\sigma}(u)$$

for $k = 1, 2, \ldots, N$ means that we have Hölder continuity of u which is nonhomogeneous with respect to different directions. (For details see [30, 58, 88].)

Remark 5.11.5 (the spaces $L_C^\varphi(Q)$). Another important generalization of the spaces $L_C^{p,\lambda}(\Omega)$, $\lambda > N$ was studied by Spanne in [209]. For $p = 1$, $\Omega = Q$ (for the notation see Definition 5.8.3) he introduces the space of all functions u in $L^1(Q)$ for which the seminorm

$$\sup_{Q' \subset Q} \frac{1}{\mu(Q')\varphi((\mu(Q'))^{\frac{1}{N}})} \left(\inf_{c \in \mathbb{R}} \int_{Q'} |u(x) - c|\, dx\right)$$

is finite. Here φ is a positive nondecreasing function. Denoting the space defined above by $L_C^\varphi(Q)$ we see immediately that if $\varphi(t) = t^{\lambda - N}$ for $\lambda > N$ then

$$L_C^\varphi(Q) = L_C^{1,\lambda}(Q).$$

The following results were obtained:

(i) If for some $\delta > 0$ the inequality

$$\int_0^\delta \frac{\varphi(t)}{t}\, dt < \infty$$

holds then every function $u \in L_C^\varphi(Q)$ is continuous and its modulus of continuity satisfies the estimate

$$\omega(u,r) \leq c \int_0^r \frac{\varphi(t)}{t}\, dt$$

(c depends on u, r is sufficiently small).

(ii) If $\frac{\varphi(t)}{t}$ is nonincreasing and the integral

$$\int_0^\delta \frac{\varphi(t)}{t}\, dt$$

is not finite, then there exists a function $u \in L_C^\varphi(Q)$ which is neither continuous nor essentially bounded.

Chapter 6

Banach function spaces

In the previous chapters we have studied in detail several types of function spaces, emphasizing their distinctive properties. In this chapter we shall abandon this approach and focus instead on properties that are common for many classes of function spaces. More precisely, we shall develop an abstract and rather general theory of the so-called *Banach function spaces* of measurable functions defined on a measure space on which only very mild assumptions will be adopted. A Banach function space is an abstract structure that covers many important examples of scales of function spaces including Lebesgue, Orlicz and Morrey spaces and their various modifications. It also includes a very important class of Lorentz spaces and their likes and, more generally, rearrangement-invariant (r.i.) spaces; these classes of function spaces will be studied in subsequent chapters. We will basically follow the exposition in [14, Chapter 1].

6.1 Banach function spaces

Convention 6.1.1. We shall work in this chapter with more general measure spaces than in the preceding chapters. We shall assume once and for all that $(\mathcal{R}, \mu)$ is a σ-finite measure space and that there exists a sequence of sets $\{R_n\}_{n=1}^{\infty}$ such that $\mu(R_n) < \infty$ for every $n \in \mathbb{N}$ and

$$\mathcal{R} = \bigcup_{n=1}^{\infty} R_n.$$

Notation 6.1.2. Let $(\mathcal{R}, \mu)$ be a σ–finite nonatomic measure space. Let $\mathcal{M}(\mathcal{R}, \mu)$ be the set of all μ–measurable real functions on $\mathcal{R}$. Let $\mathcal{M}_0(\mathcal{R}, \mu)$ denote the class of functions in $\mathcal{M}(\mathcal{R}, \mu)$ that are finite μ-a.e. By $\mathcal{M}_+(\mathcal{R}, \mu)$ we denote the subset of $\mathcal{M}_0(\mathcal{R}, \mu)$ consisting of nonnegative functions. When $\mathcal{R}$ is an interval (a, b), $-\infty \leq a < b \leq \infty$, and μ is the one-dimensional Lebesgue measure, then we just write $\mathcal{M}_+(a, b)$.

Remark 6.1.3. As usual, any two functions coinciding μ-a.e. will be identified. The natural vector space operations are well-defined on $\mathcal{M}_0$ (although not on all of $\mathcal{M}$), and when $\mathcal{M}_0$ is given the topology of convergence in measure on sets of finite measure it becomes a metrizable topological vector space. One of the possible corresponding metrics on this space, is given by

$$d(f,g) := \sum_{n=1}^{\infty} 2^{-n} \int_{S_n} \frac{|f(x) - g(x)|}{1 + |f(x) - g(x)|} d\mu(x),$$

where S_n are disjoint sets such that $\mu(S_n) < \infty$ for every $n \in \mathbb{N}$ and $\bigcup_{n \in \mathbb{N}} S_n = \mathcal{R}$. Moreover, the metric space $(\mathcal{M}, d)$ is complete.

Convention 6.1.4. By $A \lesssim B$ and $A \gtrsim B$ we mean that $A \leq CB$ and $B \leq CA$, respectively, where C is a positive constant independent of appropriate quantities involved in A and B. We write $A \approx B$ when both of the estimates $A \lesssim B$ and $B \lesssim A$ are satisfied. We shall use throughout the convention $0 \cdot \infty = 0$, $\frac{0}{0} = 0$ and $\frac{\infty}{\infty} = 0$.

Definition 6.1.5. We say that a function $\varrho : \mathcal{M}_+ \to [0, \infty]$ is a *Banach function norm* if, for all f, g and $\{f_n\}_{n=1}^{\infty}$ in $\mathcal{M}_+$, for every $\lambda \geq 0$ and for all μ-measurable subsets E of $\mathcal{R}$, the following five properties are satisfied:

(P1) $\varrho(f) = 0 \Leftrightarrow f = 0$ μ-a.e.; $\quad \varrho(\lambda f) = \lambda \varrho(f)$;
$\varrho(f + g) \leq \varrho(f) + \varrho(g)$

(P2) $0 \leq g \leq f$ μ-a.e. $\Rightarrow$ $\varrho(g) \leq \varrho(f)$ (the *lattice property*);

(P3) $0 \leq f_n \uparrow f$ μ-a.e. $\Rightarrow$ $\varrho(f_n) \uparrow \varrho(f)$ (the *Fatou property*);

(P4) $\mu(E) < \infty \Rightarrow \varrho(\chi_E) < \infty$;

(P5) $\mu(E) < \infty \Rightarrow \int_E f \, d\mu \leq C_E \varrho(f)$.

for some constant $C_E \in (0, \infty)$ and all f.

Definition 6.1.6. Let ϱ be a Banach function norm. We then say that the set $X = X(\varrho)$ of those functions in $\mathcal{M}$ for which $\varrho(|f|) < \infty$, is a *Banach function space*. For each $f \in X$, we then define

$$\|f\|_X := \varrho(|f|). \tag{6.1.1}$$

Remark 6.1.7. We note that the functional $\|f\|_X$ is defined for every $f \in \mathcal{M}_0$ but it can be infinite. A function f then belongs to $X = X(\varrho)$ if and only if $\|f\|_X < \infty$.

The notion of a simple function was introduced in Definition 1.20.23. We shall now for the sake of completeness recall this definition in the context suitable for this chapter.

Definition 6.1.8. A real-valued function s on the measure space $(\mathcal{R}, \mu)$ is called a μ-*simple function* if it is a finite linear combination of characteristic functions of μ-measurable sets of finite measure, i.e. if there exist an $m \in \mathbb{N}$, a finite sequence of real numbers $\{a_1, \ldots, a_m\}$ and a finite sequence $\{E_1, \ldots, E_m\}$ of disjoint μ-measurable subsets of $\mathcal{R}$ of finite measure such that

$$s(x) = \begin{cases} a_j, & x \in E_j, \ j = 1, \ldots, m, \\ 0, & x \in \mathcal{R} \setminus \bigcup_{j=1}^{m} E_j. \end{cases}$$

The set of all μ-simple functions will be denoted by S. If no confusion can arise, we will say shortly *simple* instead of μ-*simple*.

Theorem 6.1.9. *Let X be a Banach function space generated by a Banach function norm ϱ. Then $(X, \|\cdot\|_X)$ is a normed linear space. Moreover, the inclusions*

$$S \subset X \hookrightarrow \mathcal{M}_0 \tag{6.1.2}$$

hold, where S is the set of simple functions on $\mathcal{R}$.

Proof. Since μ is a σ-finite measure, locally integrable functions are μ-a.e. finite. Hence, $X \subset \mathcal{M}_0$. Thanks to the property (P5) from Definition 6.1.5 of the norm $\|\cdot\|_X$, all functions in X are locally integrable on $\mathcal{R}$. Thus, since $\mathcal{M}_0$ is a vector space, so is X. It follows immediately from (P1) that X is a normed space. By (P4), $\chi_E \in X$ for every set E such that $\mu(E) < \infty$. Consequently, by the linearity of X, we get $S \subset X$.

It remains to show that the embedding $X \hookrightarrow \mathcal{M}_0$ is continuous. Assume that a sequence $\{f_n\}_{n=1}^{\infty}$ satisfies $f_n \to f$ in X. Then, by (6.1.1), $\varrho(|f_n - f|) \to 0$ as $n \to \infty$. Given $\varepsilon > 0$ and a set $E \subset \mathcal{R}$ such that $\mu(E) < \infty$, we get from (P5) that

$$\mu\{x \in E : |f(x) - f_n(x)| > \varepsilon\} \leq \frac{1}{\varepsilon} \int_E |f - f_n| \, d\mu \leq \frac{C_E}{\varepsilon} \varrho(|f - f_n|),$$

which converges to 0 as $n \to \infty$ since C_E is independent of n. Therefore, $f_n \to f$ in measure on every set of finite measure, in other words, $f_n \to f$ in $\mathcal{M}_0$. □

Remark 6.1.10. It follows in particular from Theorem 6.1.9 that if $f_n \to f$ in X, then $f_n \to f$ in measure on sets of finite measure. Hence, it follows for example from [188, p. 92] that there exists a subsequence of $\{f_n\}_{n=1}^{\infty}$ that converges pointwise μ-a.e. to f.

Remark 6.1.11. Let X be a Banach function space and let $f_n \in X$, $n \in \mathbb{N}$. Assume that

$$0 \leq f_n \uparrow f \ \mu\text{-a.e.}$$

for some function $f \in \mathcal{M}$. Then we have the following two possibilities:

- either $f \in X$, hence $\|f\|_X < \infty$, and $\|f_n\|_X \uparrow \|f\|_X$;
- or $\|f_n\|_X \uparrow \infty$ and $f \notin X$.

Indeed, this fact is an immediate consequence of Definition 6.1.6 and (P3).

We shall now show that the Fatou lemma, familiar from the theory of the Lebesgue integral (cf. Theorem 1.21.6), holds for every Banach function space. The key ingredient of the proof is the Fatou property (P3).

Lemma 6.1.12 (Fatou lemma for Banach function spaces). *Let X be a Banach function space and assume that $f_n \in X$, $n \in \mathbb{N}$, and $f_n \to f$ μ-a.e. for some $f \in \mathcal{M}$. Assume further that*
$$\liminf_{n \to \infty} \|f_n\|_X < \infty.$$
Then $f \in X$ and
$$\|f\|_X \le \liminf_{n \to \infty} \|f_n\|_X.$$

Proof. Denote $g_n(x) := \inf_{m \ge n} |f_m(x)|$. Then $0 \le g_n \uparrow |f|$ μ-a.e, whence, by (P2) and (P3),
$$\|f\|_X = \lim_{n \to \infty} \|g_n\|_X \le \lim_{n \to \infty} \inf_{m \ge n} \|f_m\|_X = \liminf_{n \to \infty} \|f_n\|_X < \infty.$$

Hence $f \in X$ and $\|f\|_X \le \liminf_{n \to \infty} \|f_n\|_X$. □

Remark 6.1.13. As a consequence of the Fatou lemma, every Banach function space is complete.

We shall now turn our attention to the Riesz–Fischer property (see Definition 1.9.4).

Theorem 6.1.14. *Every Banach function space has the Riesz–Fischer property.*

Proof. Let X be a Banach function space, let $\{f_n\}_{n=1}^\infty \subset X$ and suppose that
$$\sum_{n=1}^\infty \|f_n\|_X < \infty. \tag{6.1.3}$$

We denote, for every $n \in \mathbb{N}$, $g_n := \sum_{k=1}^n |f_k|$, and $g = \sum_{n=1}^\infty |f_n|$, so that $0 \le g_n \uparrow g$. Since
$$\|g_n\|_X \le \sum_{k=1}^n \|f_k\|_X \le \sum_{n=1}^\infty \|f_n\|_X, \qquad n \in \mathbb{N},$$

it follows from (6.1.3) and Lemma 6.1.12 that g belongs to X. By the embedding $X \hookrightarrow \mathcal{M}_0$ in (6.1.2) and Remark 6.1.3, the series $\sum_{n=1}^\infty |f_n(x)|$ converges pointwise μ-a.e. and hence so does $\sum_{n=1}^\infty f_n(x)$. We set
$$f := \sum_{n=1}^\infty f_n$$

and
$$h_n = \sum_{k=1}^{n} f_k, \quad n \in \mathbb{N},$$
then $h_n \to f$ μ-a.e. Hence, for any $m \in \mathbb{N}$, we have
$$h_n - h_m \to f - h_m \quad \mu\text{-a.e. as } n \to \infty.$$
Furthermore,
$$\liminf_{n \to \infty} \|h_n - h_m\|_X \leq \liminf_{n \to \infty} \sum_{k=m+1}^{n} \|f_k\|_X = \sum_{k=m+1}^{\infty} \|f_k\|_X,$$
which tends to 0 as $m \to \infty$ because of (6.1.3). Thus, by Lemma 6.1.12, we get $f - h_m \in X$, therefore also $f \in X$, and $\|f - h_m\|_X \to 0$ as $m \to \infty$. This implies that, for every $m \in \mathbb{N}$,
$$\|f\|_X \leq \|f - h_m\|_X + \|h_m\|_X \leq \|f - h_m\|_X + \sum_{k=1}^{m} \|f_k\|_X.$$
By letting $m \to \infty$, we get
$$\|f\|_X \leq \sum_{n=1}^{\infty} \|f_n\|_X. \tag{6.1.4}$$
$\square$

Corollary 6.1.15. *Every Banach function space is complete.*

Proof. This is an immediate consequence of Theorems 6.1.14 and 1.9.5. $\square$

Remark 6.1.16. Let us summarize the basic properties of Banach function spaces. Let X be a Banach function space generated by a Banach function norm ϱ. Assume that $\|f\|_X = \varrho(|f|)$ for every $f \in X$. Then $(X, \|\cdot\|_X)$ is a Banach space and the following properties hold for all $f, g, f_n, n \in \mathbb{N}$, in $\mathcal{M}$ and for all measurable subsets $E \subset \mathcal{R}$:

(i) *(The lattice property)* If $|g| \leq |f|$ μ-a.e. and $f \in X$, then $g \in X$ and $\|g\|_X \leq \|f\|_X$.

(ii) In particular, a function $f \in \mathcal{M}$ satisfies $f \in X$ if and only if $|f| \in X$, and $\|f\|_X = \| |f| \|_X$.

(iii) *(The Fatou property)* Suppose that $f_n \in X$, $f_n \geq 0, n \in \mathbb{N}$, and $f_n \uparrow f$ μ-a.e. If $f \in X$, then $\|f_n\|_X \uparrow \|f\|_X$, while if $f \notin X$, then $\|f_n\| \uparrow \infty$.

(iv) *(Fatou lemma)* If $f_n \in X$, $n \in \mathbb{N}$, $f_n \to f$ μ-a.e., and
$$\liminf_{n \to \infty} \|f_n\|_X < \infty,$$
then $f \in X$ and
$$\|f\|_X \leq \liminf_{n \to \infty} \|f_n\|_X.$$

(v) The space X contains the set S of all simple functions.

(vi) To each set E of finite measure there corresponds a positive constant C_E depending only on E such that
$$\int_E |f| \, d\mu \leq C_E \|f\|_X$$
for all $f \in X$.

(vii) If $f_n \to f$ in X, then $f_n \to f$ in measure on every set of finite measure; in particular, some subsequence of $\{f_n\}$ converges to f pointwise μ-a.e.

We conclude this section with a simple, but useful, observation that a set-theoretic inclusion between two Banach function spaces in fact already implies the continuous embedding.

Theorem 6.1.17. *Let X and Y be Banach function spaces over the same measure space and assume that $X \subset Y$. Then $X \hookrightarrow Y$.*

Proof. Suppose $X \hookrightarrow Y$ fails. Then there exists a sequence $\{f_n\}_{n=1}^{\infty}$ of functions $f_n \geq 0$ in X such that
$$\|f_n\|_X \leq 1, \qquad \|f_n\|_Y > n^3, \quad n \in \mathbb{N}.$$

By Theorem 6.1.14, $\sum \frac{f_n}{n^2}$ converges in X to some function $f \in X$. By the hypothesis $X \subset Y$ we get also $f \in Y$. However, since $0 \leq n^{-2} f_n \leq f$ and so $\|f\|_Y \geq n^{-2} \|f_n\|_Y > n$ for all $n \in \mathbb{N}$, we get a contradiction with $f \in Y$. □

Exercise 6.1.18. Let $\Omega \subset \mathbb{R}^N$ be a suitable domain.

(i) Let $1 \leq p \leq \infty$. Then the Lebesgue space $L^p(\Omega)$ is a Banach function space.

(ii) Let $1 \leq p \leq \infty$ and let ϱ be a weight on Ω. Then the weighted Lebesgue space $L^p(\Omega, \varrho)$ (cf. Remark 3.19.8) is a Banach function space.

(iii) Let Φ be a Young function. Then the Orlicz space $L^\Phi(\Omega)$ is a Banach function space.

(iv) Let $\lambda \geq 0$ and $p \in (1, \infty)$. Then the Morrey space $L_M^{p,\lambda}(\Omega)$ is a Banach function space.

(v) Let $\lambda \geq 0$ and $p \in (1, \infty)$. Then the Campanato space $L_C^{p,\lambda}(\Omega)$ is not a Banach function space.

We shall see more examples of Banach function spaces in subsequent chapters.

6.2 Associate space

In this section we shall study the important concept of the associate space to a given Banach function space, modeled upon the example of the duality in Lebesgue spaces between L^p and $L^{p'}$, where $1 \leq p \leq \infty$.

Definition 6.2.1. Let ϱ be a Banach function norm. Then the functional ϱ', defined on $\mathcal{M}^+$ by

$$\varrho'(g) = \sup \left\{ \int_\mathcal{R} fg \, d\mu : f \in \mathcal{M}^+, \varrho(f) \leq 1 \right\}, \qquad g \in \mathcal{M}^+, \qquad (6.2.1)$$

is called the *associate norm* of ϱ.

Theorem 6.2.2. *Let ϱ be a Banach function norm. Then ϱ' is also a Banach function norm.*

Proof. Suppose that $\varrho(f) \leq 1$. Then (6.1.2) implies that $f(x) < \infty$ μ-a.e. If moreover $g = 0$ μ-a.e., then

$$\int_\mathcal{R} fg \, d\mu = 0,$$

hence, by (6.2.1), $\varrho'(g) = 0$.

If $\varrho'(g) = 0$, then $\int_\mathcal{R} fg \, d\mu = 0$ for all $f \in \mathcal{M}^+$ with $\varrho(f) \leq 1$. If $E \subset \mathcal{R}$ is a μ-measurable set with $0 < \mu(E) < \infty$, then $0 < \varrho(\chi_E) < \infty$ by the properties (P1) and (P4) of ϱ. The function $f := \chi_E / \varrho(\chi_E)$ satisfies $\varrho(f) = 1$, therefore

$$\varrho(\chi_E)^{-1} \int_E g \, d\mu = \int_\mathcal{R} fg \, d\mu = 0.$$

Thus, $g = 0$ μ-a.e. on E. Because E was chosen arbitrarily, we get that $g = 0$ μ-a.e. The positive homogeneity and the triangle inequality for ϱ' can be easily verified. This shows (P1). Next, again, (P2) trivially follows from the definition of ϱ'.

We shall show (P3). Let $\{g_n\}_{n=1}^\infty \subset \mathcal{M}$ and assume that $0 \leq g_n \uparrow g$ μ-a.e. for some $g \in \mathcal{M}$. We already know that ϱ' has the lattice property (P2). Thus, for every $m, n \in \mathbb{N}$, $m \leq n$, $\varrho'(g_m) \leq \varrho'(g_n) \leq \varrho'(g)$. We can with no loss of generality assume that $\varrho(g_n) < \infty$ for every $n \in \mathbb{N}$. Let ε be any number satisfying $\varepsilon < \varrho'(g_n)$. By (6.2.1), there is a function f in $\mathcal{M}^+$ with $\varrho(f) \leq 1$ and such that $\int fg \, d\mu > \varepsilon$. Now $0 \leq fg_n \uparrow fg$ μ-a.e. so the monotone convergence theorem (Theorem 1.21.4) implies that $\int fg_n \uparrow \int fg$. Hence there is $n_0 \in \mathbb{N}$ such that $\int fg_n > \varepsilon$ for all

$n \geq n_0$. Thus, by (6.2.1), we obtain $\varrho'(g_n) > \varepsilon$ for all $n \geq n_0$. Consequently, we get $\varrho'(g_n) \uparrow \varrho'(g)$, in other words, ϱ' enjoys the property (P3).

In order to verify (P4) for ϱ', we use (P5) for ϱ, and vice versa. Let $E \subset \mathcal{R}$ satisfy $\mu(E) < \infty$, then, by (P5) for ϱ, there is constant $C_E < \infty$ for which

$$\int_{\mathcal{R}} \chi_E f \, d\mu \leq C_E \varrho(f), \qquad (f \in \mathcal{M}^+).$$

Together with (6.2.1) this gives $\varrho'(\chi_E) \leq C_E < \infty$, proving (P4) for ϱ'.

Finally, let $E \subset \mathcal{R}$ be such that $\mu(E) < \infty$. If $\mu(E) = 0$, then $\int_E f \, d\mu = 0$, hence (P5) holds automatically. When $\mu(E) > 0$, we have by (P4) for ϱ that $\varrho(\chi_E) < \infty$ and by (P1) for ϱ that $\varrho(\chi_E) > 0$. We set $C_E' := \varrho(\chi_E)$ and $f := \frac{\chi_E}{\varrho(\chi_E)}$. Then $\varrho(f) = 1$, whence, for any $g \in \mathcal{M}^+$, we obtain from (6.2.1)

$$\int_E g \, d\mu = C_E' \int_{\mathcal{R}} fg \, d\mu \leq C_E' \varrho'(g),$$

proving (P5) for ϱ'. The proof is complete. $\square$

Exercise 6.2.3. Prove that the functional τ, defined on Lebesgue measurable functions g over the interval $(0, 1)$ by

$$\tau(g) := \sup_{0 < \alpha < 1} \int_0^1 g(t) \alpha t^{\alpha - 1} \, dt,$$

is a Banach function norm on $\mathcal{M}((0, 1), dt)$.

(i) For $0 < a < b < 1$, calculate $\tau(\chi_{(a,b)})$.

(ii) Characterize $\tau(g)$ for g nonincreasing on $(0, 1)$.

(iii) Characterize τ'.

(The norm τ was introduced in [132], where also solutions to (i) and (ii) can be found. The problem (iii) is open.)

Definition 6.2.4. Let ϱ be a Banach function norm and let X be the Banach function space determined by ϱ. Let ϱ' be the associate norm of ϱ and let X' be the Banach function space determined by ϱ'. We say that X' is the *associate space* of X.

Remark 6.2.5. Let X be a Banach function space and X' its associate space. Then

$$\|g\|_{X'} = \sup\left\{\int_{\mathcal{R}} |fg| \, d\mu; \; f \in X, \|f\|_X \leq 1\right\} \quad \text{for every } g \in \mathcal{M}. \tag{6.2.2}$$

Indeed, this immediately follows from (6.1.1) and (6.2.1).

Section 6.2 Associate space

We have seen that the Hölder inequality holds for Lebesgue spaces (Theorem 3.1.6) and also for Orlicz spaces (Theorem 4.7.5). We shall now show a general version of this inequality valid for Banach function spaces.

Theorem 6.2.6 (Hölder inequality). *Let X be a Banach function space and let X' be its associate space. If $f \in X$ and $g \in X'$, then $\int_{\mathcal{R}} fg \, d\mu < \infty$ and, moreover,*

$$\int_{\mathcal{R}} |fg| \, d\mu \leq \|f\|_X \|g\|_{X'}. \tag{6.2.3}$$

Proof. Assume first that $\|f\|_X = 0$. Then $f = 0$ μ-a.e. on $\mathcal{R}$, hence both sides of (6.2.3) are zero. When $\|f\|_X > 0$, then

$$\left\| \frac{f}{\|f\|_X} \right\|_X = 1.$$

Thus, by the definition of X', we get by (6.2.2) that

$$\int_{\mathcal{R}} \left| \left(\frac{f}{\|f\|_X} \right) g \right| d\mu \leq \|g\|_{X'}.$$

Multiplying both sides of the last estimate by $\|f\|_X$ we obtain (6.2.3). □

We note that we know from Theorem 6.2.6 that $g \in X'$ implies $fg \in L^1$ for every $f \in X$. We shall now show that, in fact, the converse is also true.

Theorem 6.2.7 (Landau resonance theorem). *Let X be a Banach function space over a measure space $(\mathcal{R}, \mu)$. Then a measurable function g belongs to the associate space X' if and only if fg is integrable for every f in X.*

Proof. The "only if" part follows from Theorem 6.2.6. As for the "if" part, assume that $\|g\|_{X'} = \infty$ but that fg is integrable for every f in X. By (6.2.1), there exist nonnegative functions f_n satisfying

$$\|f_n\|_X \leq 1, \quad \int_{\mathcal{R}} |f_n g| \, d\mu > n^3, \quad n \in \mathbb{N}.$$

It follows from Theorem 6.1.14 that the function $f = \sum_{n=1}^{\infty} n^{-2} f_n$ belongs to X. However,

$$\int_{\mathcal{R}} |fg| \, d\mu > n^{-2} \int_{\mathcal{R}} |f_n g| \, d\mu > n, \quad n \in \mathbb{N},$$

which is a contradiction. □

Notation 6.2.8. Let X be a Banach function space and let X' be its associate space. Then the associate space $(X')'$ of X is called the *second associate space* of X and is denoted by X''.

Theorem 6.2.9 (Lorentz–Luxemburg theorem). *Let X be a Banach function space. Then X coincides with its second associate space X'' in the sense that*

$$f \in X \quad \Leftrightarrow \quad f \in X''$$

and

$$\|f\|_X = \|f\|_{X''}, \quad f \in \mathcal{M}. \tag{6.2.4}$$

Proof. If $f \in X$, then by the Hölder inequality (6.2.3) $fg \in L^1$ for every $g \in X'$. It follows therefore from Theorem 6.2.7 (applied to X' instead of X) that $f \in X''$. We thus get

$$X \subset X''.$$

We also obtain from (6.2.2) and (6.2.3) that

$$\|f\|_{X''} = \sup\left\{\int_{\mathcal{R}} |fg|\, d\mu : \|g\|_{X'} \leq 1\right\} \leq \|f\|_X.$$

Hence, in order to complete the proof we need only show $X'' \subset X$ and

$$\|f\|_X \leq \|f\|_{X''}, \quad f \in X''. \tag{6.2.5}$$

Conversely, let $f \in X''$ and $n \in \mathbb{N}$. Set

$$f_n(x) := \min(|f(x)|, n)\chi_{R_n}(x), \quad n \in \mathbb{N}, \tag{6.2.6}$$

where $\{R_n\}_{n=1}^\infty$ is the sequence of sets from Convention 6.1.1. Now,

$$0 \leq f_n \leq n\chi_{R_n},$$

hence, by Remark 6.1.16 (i) and (v), $f_n \in X$ and likewise $f_n \in X''$. We further know that both X and X'' have the Fatou property and that

$$0 \leq f_n \uparrow |f|.$$

Hence, in order to verify (6.2.5), we just need show that

$$\|f_n\|_X \leq \|f_n\|_{X''}, \quad n \in \mathbb{N}. \tag{6.2.7}$$

From now on we shall suppose that f and n are fixed. With no loss of generality we may assume that $\|f_n\|_X > 0$.

We note that the space

$$M_n := \{f \in \mathcal{M}, \text{ supp } f \subset R_n\},$$

endowed with the norm

$$\|g\|_{M_n} := \int_{R_n} |g(x)|\, d\mu(x),$$

Section 6.2 Associate space

is a Banach space. Let S_X denote the closed unit ball in X. Then the set

$$U := S_X \cap M_n$$

is a convex subset of M_n. If $\{h_k\}_{k=1}^\infty \subset U$ and $h_k \to h$ in M_n, then there exists a subsequence $\{h_{k_j}\}_{j=1}^\infty$ such that $h_{k_j} \to h$ μ-a.e. on $\mathcal{R}$. Since $h_{k_j} \in S_X$ for every $j \in \mathbb{N}$, it follows from the Fatou lemma (Remark 6.1.16 (iv)) that $h \in S_X$. Hence also $h \in U$, which means that U is a closed subset of M_n.

Let now $\lambda > 1$. Then

$$g := \lambda \frac{f_n}{\|f_n\|_X} \in (M_n \setminus U).$$

Therefore, by the Hahn–Banach theorem, there is a closed hyperplane that separates g and U. In other words, there exists a nonzero function $\varphi \in L^\infty$ on $\mathcal{R}$ with $\operatorname{supp} \varphi \subset R_n$ and some $\gamma \in \mathbb{R}$ such that

$$\int_{R_n} \varphi h \, d\mu < \gamma < \int_{R_n} \varphi g \, d\mu \tag{6.2.8}$$

for every h in U. Writing $\varphi = |\varphi| \psi$ in polar form, where $|\psi| = 1$, and since $\psi |h| \in U$ if and only if $h \in U$, we get

$$\sup_{h \in U} \int_{R_n} |\varphi h| \, d\mu \leq \gamma < \int_{R_n} \varphi g \, d\mu \leq \int_{R_n} |\varphi g| \, d\mu. \tag{6.2.9}$$

For every $h \in S$ we have

$$h(x) = \lim_{k \to \infty} h_k(x) := \min(h(x), k) \chi_{R_n}(x), \quad x \in R_n.$$

Obviously, $h_k \in M_n$, so $h_k \in U$ for every $k \in \mathbb{N}$. A function $h \in S$ can thus be approximated by the sequence $\{h_k\} \subset U$. Since $U \subset S$, we get from the monotone convergence theorem (Theorem 1.21.4) that

$$\sup_{h \in U} \int_{R_n} |\varphi h| \, d\mu = \sup_{h \in S} \int_{R_n} |\varphi h| \, d\mu.$$

Thus, by the definition of X' and (6.2.8), one obtains (recall that $\operatorname{supp} \varphi \subset R_n$)

$$\|\varphi\|_{X'} = \sup_{h \in S} \int_{R_n} |\varphi h| \, d\mu \leq \gamma < \frac{\lambda}{\|f_n\|_X} \int_{R_n} |\varphi f_n| \, d\mu.$$

This together with the Hölder inequality (6.2.3), applied to X' and X'', implies

$$\|f_n\|_X < \lambda \int_{\mathcal{R}} \left| f_n \frac{\varphi}{\|\varphi\|_{X'}} \right| d\mu \leq \lambda \|f_n\|_{X''}.$$

On letting $\lambda \to 1$, we obtain (6.2.7). The proof is complete. $\square$

Given a Banach function space X, it is of interest to compare its associate space X' with its dual space X^*. It is clear from the definitions that one always has $X' \subset X^*$. On the other hand, it is equally clear from Theorem 6.2.9 that the converse inclusion is not always true (take $X = L^\infty$ for instance). The following lemma is the first step in our study of relations between X' and X^*.

Lemma 6.2.10. *Let X be a Banach function space and X' its associate space. Then*

$$\|g\|_{X'} = \sup\left\{\left|\int_{\mathcal{R}} fg \, d\mu\right|; \, f \in X, \|f\|_X \leq 1\right\}, \quad g \in \mathcal{M}. \tag{6.2.10}$$

Proof. By Theorem 6.2.6, we have for every $f \in X$, $\|f\|_X \leq 1$, and every $g \in X'$,

$$\left|\int_{\mathcal{R}} fg \, d\mu\right| \leq \int_{\mathcal{R}} |fg| \, d\mu \leq \|f\|_X \|g\|_{X'} \leq \|g\|_{X'},$$

which yields immediately

$$\sup\left\{\left|\int_{\mathcal{R}} fg \, d\mu\right|; \, f \in X, \|f\|_X \leq 1\right\} \leq \|g\|_{X'}.$$

In order to prove the reverse inequality, denote by S_X the unit ball in X. Given $g \in \mathcal{M}$, denote

$$E := \{x \in \mathcal{R}, \, g(x) \neq 0\}$$

and write g in the polar form:

$$g(x) = |g(x)|\psi(x), \quad x \in E,$$

where $|\psi| = 1$. Let further $f \in S_X$. Denote

$$h(x) := |f(x)|\psi(x)\chi_E(x), \quad x \in \mathcal{R}.$$

Then

$$|h(x)| \leq |f(x)| \quad \text{for } x \in \mathcal{R},$$

hence, by the lattice property of X, also $h \in S_X$. Altogether,

$$\int_{\mathcal{R}} |fg| \, d\mu = \int_{E} |fg| \, d\mu = \int_{E} |f|\psi g \, d\mu.$$

Thus,

$$\int_{\mathcal{R}} |fg| \, d\mu = \int_{\mathcal{R}} hg \, d\mu \leq \left|\int_{\mathcal{R}} hg \, d\mu\right| \leq \sup_{f \in S_X} \left|\int_{\mathcal{R}} fg \, d\mu\right|.$$

Passing on the left-hand side to the supremum over all f in S_X and using (6.2.2), we obtain the inequality

$$\|g\|_{X'} \leq \sup\left\{\left|\int_{\mathcal{R}} fg \, d\mu\right|; \, f \in X, \|f\|_X \leq 1\right\}, \quad g \in \mathcal{M},$$

finishing the proof. $\square$

Remark 6.2.11. Let X be a Banach function space and X' its associate space. It might be useful to note that there exists an alternative way of expressing $\|g\|_{X'}$ for $g \in \mathcal{M}$, namely,

$$\|g\|_{X'} = \sup_{f \neq 0} \frac{\left|\int_{\mathcal{R}} fg \, d\mu\right|}{\|f\|_X}. \tag{6.2.11}$$

We shall now recall the notion of a norm-fundamental subspace of the dual space.

Definition 6.2.12. Let $B \subset X^*$ be a closed linear subspace. We say that B is *norm-fundamental* in X, if

$$\|f\|_X = \sup\{|\Lambda(f)|;\ \Lambda \in B,\ \|\Lambda\|_{X^*} \leq 1\}, \quad f \in X.$$

In other words, a space is norm-fundamental if it is rich enough so that it can reproduce the norm of each $f \in X$.

Theorem 6.2.13. *Let X be a Banach function space and let X' be its associate space. Then X' is isometrically isomorphic to a norm-fundamental subspace of X^*.*

Proof. Given $g \in X'$, we define the functional Λ_g on X by

$$\Lambda_g(f) := \int_{\mathcal{R}} fg \, d\mu, \quad f \in X.$$

Then Λ_g is obviously linear and by the Hölder inequality it is bounded on X. Next, $\Lambda_g(f) = 0$ if and only if $f \equiv 0$ on $\mathcal{R}$. Therefore, the mapping

$$g \mapsto \Lambda_g \tag{6.2.12}$$

is an isomorphism between X' and some subspace Y of X^*. Moreover, by the definition of the norm in X^*, by the definition of Λ_g and by Lemma 6.2.10, we subsequently get

$$\|\Lambda_g\|_{X^*} = \sup\{|\Lambda_g(f)|;\ \|f\|_X \leq 1\}$$
$$= \sup\left\{\left|\int_{\mathcal{R}} fg \, d\mu\right|;\ \|f\|_X \leq 1\right\}$$
$$= \|g\|_{X'},$$

hence the mapping in (6.2.12) is an isometry. Since X' is a Banach space, this mapping has closed range in X^*.

It remains to verify the norm-fundamental property of X'. By Theorem 6.2.9 and Lemma 6.2.10, we have

$$\|f\|_X = \|f\|_{X''} = \sup\left\{\left|\int_{\mathcal{R}} fg \, d\mu\right|;\ \|g\|_{X'} \leq 1\right\} = \sup\{|\Lambda_g(f)|;\ \|\Lambda_g\|_{X^*} \leq 1\},$$

in other words, X' is norm-fundamental. The proof is complete. $\square$

Proposition 6.2.14. *Let X and Y be Banach function spaces. Then*
$$X \hookrightarrow Y \quad \Leftrightarrow \quad Y' \hookrightarrow X',$$
and the embedding constants coincide.

Proof. By (6.2.11), we have

$$\|\operatorname{Id}\|_{X \hookrightarrow Y} = \sup_{f \neq 0} \frac{\|f\|_Y}{\|f\|_X} = \sup_{f \neq 0} \sup_{g \neq 0} \frac{|\int_{\mathcal{R}} fg \, d\mu|}{\|f\|_X \|g\|_{Y'}} = \sup_{g \neq 0} \sup_{f \neq 0} \frac{|\int_{\mathcal{R}} fg \, d\mu|}{\|f\|_X \|g\|_{Y'}}$$

$$= \sup_{g \neq 0} \frac{\|g\|_{X'}}{\|f\|_{Y'}} = \|\operatorname{Id}\|_{Y' \hookrightarrow X'},$$

and the assertion follows. □

6.3 Absolute continuity of the norm

Absolute continuity of the norm, which will be now studied in detail, is one of the basic concepts in the theory of Banach function spaces and it is indispensable for example for the investigation of basic properties such as reflexivity or separability of the spaces.

Definition 6.3.1. Let $\{E_n\}_{n=1}^{\infty}$ be a sequence of μ-measurable subsets of $\mathcal{R}$. We say that E_n *tends to the empty set* and write $E_n \to \emptyset$ if $\chi_{E_n} \to 0$ μ-a.e. on $\mathcal{R}$. We write $E_n \downarrow \emptyset$ when $E_n \to \emptyset$ and moreover the sequence $\{E_n\}_{n=1}^{\infty}$ is monotonically decreasing, that is, $E_n \supset E_{n+1}$.

Remarks 6.3.2. (i) Let $\{E_n\}_{n=1}^{\infty}$ be a sequence of μ-measurable subsets of $\mathcal{R}$. Then $E_n \to \emptyset$ μ-a.e. if and only if the set

$$\limsup_{n \to \infty} E_n := \bigcap_{m=1}^{\infty} \bigcup_{n=m}^{\infty} E_n$$

has measure zero.

(ii) It follows from (i) that $E_n \to \emptyset$ μ-a.e. if and only if E_n can be modified on a set of measure zero in order that the resulting sequence converges to $\emptyset$ everywhere.

(iii) It is important to note that the sets E_n can have infinite measure.

Definition 6.3.3. Let X be a Banach function space and let $f \in X$. We say that f has an *absolutely continuous norm* in X if

$$\|f \chi_{E_n}\|_X \to 0$$

for every sequence $\{E_n\}_{n=1}^{\infty}$ satisfying $E_n \to \emptyset$ μ-a.e. We shall denote

$$X_a := \{f \in X;\ f \text{ has an absolute continuous norm in } X\}.$$

If $X_a = X$, we say that the space X has an *absolutely continuous norm*.

We first note that we can restrict ourselves in Definition 6.3.3 to nonincreasing sequences.

Proposition 6.3.4. *Let X be a Banach function space and let $f \in X$. Then f has an absolutely continuous norm in X if and only if*

$$\|f \chi_{E_n}\|_X \downarrow 0 \tag{6.3.1}$$

for every sequence $\{E_n\}_{n=1}^{\infty}$ satisfying $E_n \downarrow \emptyset$ μ-a.e.

Proof. The "only if" part is clear. As for the "if" part, assume that $f \in X$ satisfies (6.3.1) and let $\{F_n\}_{n=1}^{\infty} \subset \mathcal{R}$ be a sequence for which $F_n \to \emptyset$ μ-a.e. For $n \in \mathbb{N}$, denote

$$E_n := \bigcup_{m=n}^{\infty} F_m.$$

Then the sequence $\{E_n\}$ is decreasing and satisfies

$$\limsup_{n \to \infty} F_n = \limsup_{n \to \infty} E_n.$$

However, we know that $\limsup_{n \to \infty} F_n$ is a set of measure zero, hence so is $\limsup_{n \to \infty} E_n$. Moreover, $E_n \supset E_{n+1}$. By (6.3.1), we get

$$\|f \chi_{E_n}\|_X \downarrow 0.$$

Since $F_n \subset E_n$ for all n we also have $\|f \chi_{F_n}\|_X \to 0$. In other words, f has an absolutely continuous norm in X, as desired. $\square$

Examples 6.3.5. (i) Let $X = L^p$, $p \in [1, \infty)$. Then $X = X_a$, that is, every function in L^p has an absolutely continuous norm.

(ii) Let $X = L^{\infty}$ and let $(\mathcal{R}, \mu)$ be a nonatomic measure space. Then

$$(L^{\infty}(\mathcal{R}, \mu))_a = \{0\},$$

that is, in this space, only the function which is identically equal to zero, has an absolutely continuous norm.

(iii) Let $(\mathcal{R}, \mu) = (\mathbb{N}, m)$ where m is the arithmetic measure on $\mathbb{N}$, and let $X = \ell^{\infty}$. Then $X_a = c_0$.

(iv) Let Φ be a Young function and let L^Φ be the corresponding Orlicz space. Then
$$L_a^\Phi = E^\Phi.$$

Our next aim is to search for a reasonable characterizations of the space of functions with absolute continuous norms.

Lemma 6.3.6. *Let X be a Banach function space and let $f \in X$. Then $f \in X_a$ if and only if for every $\varepsilon > 0$ there exists a $\delta > 0$ such that whenever $E \subset \mathcal{R}$ satisfies $\mu(E) < \delta$, then, necessarily, $\|f\chi_E\|_X < \varepsilon$.*

Proof. Assume that there exists an $\varepsilon > 0$ and a sequence $\{E_n\}_{n=1}^\infty$ of measurable sets in $\mathcal{R}$ such that $\mu(E_n) < \frac{1}{2^n}$ and
$$\|f\chi_{E_n}\|_X \geq \varepsilon.$$

Then, for every $m \in \mathbb{N}$,
$$\mu\left(\bigcup_{n=m}^\infty E_n\right) \leq \sum_{n=m}^\infty \mu(E_n) < 2^{1-m},$$

whence $E_n \searrow \emptyset$ μ-a.e. on $\mathcal{R}$. Consequently, $\|f\chi_{E_n}\|_X \geq \varepsilon$, which is a contradiction, since $f \in X_a$. □

Now we are in a position to characterize the set X_a in terms of assertions analogous to the monotone convergence theorem and the dominated convergence theorem which we know from the theory of Lebesgue spaces (cf. Theorems 1.21.4 and 1.21.5).

Theorem 6.3.7. *Let X be a Banach function space and let $f \in X$. Then $f \in X_a$ if and only if, for every sequence $\{f_n\}_{n=1}^\infty$ satisfying $f_n \leq |f_n|$ and $f_n \downarrow 0$ μ-a.e., we have $\|f_n\| \downarrow 0$.*

Proof. $\Leftarrow$ This implication is obvious (it suffices to take $f_n = f\chi_{E_n}$).

$\Rightarrow$ Assume that $f \in X_a$ and let a sequence $\{f_n\}_{n=1}^\infty$ satisfy $|f| \geq f_n$ and $f_n \searrow 0$ μ-a.e. on $\mathcal{R}$. Let $\varepsilon > 0$. Recall that $\{R_k\}_{k=1}^\infty$ is the fixed sequence of sets in $\mathcal{R}$ satisfying $0 < \mu(\mathcal{R}) < \infty$ and $\bigcup_{k=1}^\infty R_k = \mathcal{R}$. Set $Q_k := \mathcal{R} \setminus R_k$, $k \in \mathbb{N}$. Since $f \in X_a$, there exists a $k_0 \in \mathbb{N}$ such that for every $k \in \mathbb{N}$, $k \geq k_0$, one has
$$\|f\chi_{Q_k}\|_X < \frac{\varepsilon}{2}.$$

Now fix $k \geq k_0$ and let
$$\alpha := \frac{\varepsilon}{4}\|\chi_{R_k}\|_X$$
and
$$E_n := \{x \in R_k, \ f_n(x) \geq \alpha\}.$$

Section 6.3 Absolute continuity of the norm

Then
$$\lim_{k \to \infty} \mu(E_k) = 0,$$
since $f_n \searrow 0$ μ-a.e. on $\mathcal{R}$ and R_k is of finite measure. Therefore, by Lemma 6.3.6, there exists an $n_0 \in \mathbb{N}$ such that for every $n \in \mathbb{N}$, $n \geq n_0$, we have
$$\|f\chi_{E_n}\|_X < \frac{\varepsilon}{4}.$$

Finally, for such n, one gets
$$\begin{aligned}
\|f_n\|_X &\leq \|f_n \chi_{Q_k}\|_X + \|f_n \chi_{R_k}\|_X \\
&\leq \|f_n \chi_{Q_k}\|_X + \|f_n \chi_{E_n}\|_X + \|f_n \chi_{R_k \setminus E_n}\|_X \\
&\leq \|f_n \chi_{Q_k}\|_X + \|f \chi_{E_n}\|_X + \alpha \|\chi_{R_k \setminus E_n}\|_X \\
&< \frac{\varepsilon}{2} + \frac{\varepsilon}{4} + \frac{\varepsilon}{4} \\
&= \varepsilon.
\end{aligned}$$

The proof is complete. □

Remark 6.3.8. In the theory of Lebesgue spaces, we have monotone convergence theorem and the dominated convergence theorem. The former result holds also in the theory of Banach function spaces because the Fatou axiom is postulated. As for the latter theorem, we have the following theorem which characterizes the space X_a as the biggest space contained in X on which the dominated convergence theorem holds.

Theorem 6.3.9. *Let X be a Banach function space and let $f \in X$. Then $f \in X_a$ if and only if, for every sequence $\{f_n\}_{n=1}^\infty$ and every $g \in X$, satisfying $f_n \to g$ μ-a.e., we have*
$$\lim_{n \to \infty} \|f_n - g\| = 0.$$

Proof. ⇐ Again, this implication is clear (just take $f_n = f\chi_{E_n}$ and $g \equiv 0$). ⇒ Suppose that $f \in X_a$ and let a sequence $\{f_n\}_{n=1}^\infty$ satisfy $|f| \geq |f_n|$ and $f_n \to g$ μ-a.e. on $\mathcal{R}$. Set
$$h_n(x) := \sup_{m \geq n} |f_m(x) - g(x)|, \quad n \in \mathbb{N}.$$
Then $2|f| \geq h_n \searrow 0$ μ-a.e. By Theorem 6.3.7, we get $\|h_n\|_X \searrow 0$. Thus, since $\|f_n - g\|_X \leq \|h_n\|_X$, also
$$\lim_{n \to \infty} \|f_n - g\|_X = 0,$$
as desired. The proof is complete. □

Definition 6.3.10. Let X be a Banach function space and let $Y \subset X$ be a closed linear subspace of X. We say that Y is an *ideal* in X if for every $f \in Y$ and for every $g \in \mathcal{M}$, $|g| \leq |f|$ μ-a.e. we have $g \in Y$.

Example 6.3.11. Let X be a Banach function space. Then both $Y = X$ and $Y = \{0\}$ are ideals in X.

Theorem 6.3.12. *Let X be a Banach function space. Then the space X_a is an ideal in X.*

Proof. We clearly have $X_a \subset X$. Moreover, it is also obvious that if $f \in X_a$ and $g \in \mathcal{M}$, $|g| \leq |f|$ μ-a.e., then, necessarily, $g \in X_a$. The only part of the assertion which is not entirely clear is the closedness of X_a in X.

Suppose that $\{f_n\}_{n=1}^{\infty}$ is a sequence of functions in X_a and let

$$\lim_{n \to \infty} \|f_n - f\|_X = 0.$$

Let $\varepsilon > 0$. Then there exists $n_0 \in \mathbb{N}$ such that for every $n \in \mathbb{N}$, $n \geq n_0$, one has

$$\|f - f_n\|_X < \frac{\varepsilon}{2}.$$

Let $\{E_k\}_{k=1}^{\infty}$ be a sequence of μ-measurable subsets of $\mathcal{R}$ such that $E_k \searrow \emptyset$ μ-a.e. as $k \to \infty$. Let $n \geq n_0$. Since $f_n \in X_a$, hence there is a $k_0 \in \mathbb{N}$ such that for every $k \in \mathbb{N}$, $k \geq k_0$,

$$\|f_n \chi_{E_k}\|_X < \frac{\varepsilon}{2}.$$

Thus, altogether, we have for every $k \geq k_0$

$$\|f \chi_{E_k}\|_X \leq \|(f - f_n)\chi_{E_k}\|_X + \|f_n \chi_{E_k}\|_X$$
$$\leq \|f - f_n\|_X + \|f_n \chi_{E_k}\|_X$$
$$\leq \frac{\varepsilon}{2} + \frac{\varepsilon}{2} = \varepsilon.$$

It follows that $\lim_{k \to \infty} \|f \chi_{E_k}\|_X = 0$, and therefore $f \in X_a$. In other words, X_a is a closed subspace of X. □

Theorems 6.3.7 and 6.3.9 obviously yield the following result.

Corollary 6.3.13. *Let X be a Banach function space and let $f \in X_a$. Assume that $\{f_n\}_{n=1}^{\infty} \subset X$ is a sequence of functions in X satisfying*

$$0 \leq f_n \uparrow f \quad \mu - a.e.$$

Then

$$\|f - f_n\|_X \to 0.$$

Definition 6.3.14. Let X be a Banach function space and let S denote the space of simple functions. We will denote by X_b the closure in X of S.

Theorem 6.3.15. *Let X be a Banach function space. Then*

$$X_b = \overline{\{f \in X;\ f \text{ bounded},\ \mu(\operatorname{supp} f) < \infty\}},$$

where the closure is understood in the norm topology of the space X.

Proof. Only the inclusion "$\supset$" needs proof. Let f be a bounded function on $\mathcal{R}$, and let $\mu(\operatorname{supp} f) < \infty$. Assume first that $f \geq 0$. Let $\{f_n\}_{n=1}^\infty$ be a sequence of simple functions on $\mathcal{R}$ such that $\operatorname{supp} f_n = \operatorname{supp} f$ and f_n converge uniformly to f. Then

$$\|f_n - f\|_X = \|(f_n - f)\chi_{\operatorname{supp} f}\|_X \leq \|f_n - f\|_{L^\infty}\|\chi_{\operatorname{supp} f}\|_X.$$

It follows from $\mu(\operatorname{supp} f) < \infty$ and the property (P4) of X that $\|\chi_{\operatorname{supp} f}\|_X < \infty$. Thus,

$$\lim_{n \to \infty} \|f_n - f\|_X = 0,$$

as desired. The proof is complete. $\square$

Theorem 6.3.16. *Let X be a Banach function space. Then X_b is an ideal in X and*

$$X_a \subset X_b \subset X.$$

Proof. The closedness of X_b in X follows immediately from the definition. Assume that $f \in X_b$ and let $g \in \mathcal{M}_0(\mathcal{R})$, $|g| \leq |f|$ μ-a.e. on $\mathcal{R}$. Suppose further that $\{f_n\}_{n=1}^\infty$ is a sequence of simple functions on $\mathcal{R}$ such that

$$\lim_{n \to \infty} \|f_n - f\|_X = 0.$$

We then define, for $n \in \mathbb{N}$ and $x \in \mathcal{R}$,

$$g_n(x) := \operatorname{sign}(g(x)) \min\{|f_n(x)|, |g(x)|\}.$$

Then, for every $n \in \mathbb{N}$, g_n is bounded on $\mathcal{R}$ and $\mu(\operatorname{supp} f_n) < \infty$. Moreover,

$$|g - g_n| = \max\{|g| - |f_n|, 0\} \leq ||f| - |f_n|| \leq |f - f_n|,$$

whence

$$\|g - g_n\|_X \leq \|f - f_n\|_X \to 0.$$

By Theorem 6.3.15, we get $g \in X_b$. In other words, X_b is an ideal in X.

Finally, let $f \in X_a$ and let $\{R_k\}_{k=1}^\infty$ be the sequence of μ-measurable subsets of $\mathcal{R}$ of finite positive measure. Then, by Theorem 6.3.15, the functions f_k, defined for $k \in \mathbb{N}$ and $x \in \mathcal{R}$ by

$$f_k(x) := \operatorname{sign}(f(x)) \min\{|f(x)|, k\chi_{R_k}(x)\},$$

satisfy $f_k \in X_b$ for every $k \in \mathbb{N}$, and

$$\lim_{k\to\infty} f_k(x) = f(x) \quad \mu\text{-a.e. on } \mathcal{R}.$$

Theorem 6.3.9 then gives

$$\lim_{k\to\infty} \|f_k - f\|_X = 0.$$

But, X_b is a closed subspace of X, hence also $f \in X_b$. This establishes the inclusion

$$X_a \subset X_b,$$

finishing the proof. □

As we know from Example 6.3.5 (ii), the space X_a can be very small. That cannot happen with X_b, which is always a rather rich structure, as the following result shows.

Theorem 6.3.17. *Let X be a Banach function space. Then X_b is isometrically isomorphic to a norm-fundamental subspace of $(X')^*$.*

Proof. Applying the assertion of Theorem 6.2.13 to X' instead of X, and using the fact that $X'' = X$ (cf. Theorem 6.2.9), we get that X'' is isometrically isomorphic to a closed subspace of $(X')^*$. Since X_b is also a closed linear subspace of $X'' = X$, the same is true for X_b, that is, X_b is isometrically isomorphic to a closed subspace of $(X')^*$. We have to show that it is norm-fundamental.

Let $g \in X'$ and let $\varepsilon > 0$. Then, by the definition of the associate space, there exists a function $f \in X$, $\|f\|_X \leq 1$, such that

$$\|g\|_{X'} \leq \int_{\mathcal{R}} |fg|\,d\mu + \varepsilon.$$

For $k \in \mathbb{N}$, define

$$f_k := \min\{|f|, k\}\chi_{R_k}.$$

(Here, $\{R_k\}_{k=1}^\infty$ is as usual.) Let B be the unit ball in X_b, that is,

$$B := \{h \in X_b, \|h\|_X \leq 1\}.$$

By Theorem 6.3.15, we have $f_k \in B$ for every $k \in \mathbb{N}$. Since $0 \leq f_k \nearrow |f|$, the monotone convergence theorem yields

$$\|g\|_{X'} \leq \sup_{k\in\mathbb{N}} \int_{\mathcal{R}} |f_k g|\,d\mu + \varepsilon \leq \sup_{h\in B} \int_{\mathcal{R}} |hg|\,d\mu + \varepsilon.$$

We now send $\varepsilon \to 0_+$ to get

$$\|g\|_{X'} \leq \sup_{h\in B} \int_{\mathcal{R}} |hg|\,d\mu,$$

which, in fact, together with the Hölder inequality yields

$$\|g\|_{X'} = \sup_{h \in B} \int_{\mathcal{R}} |hg| \, d\mu.$$

Using Lemma 6.2.10, we get

$$\|g\|_{X'} = \sup_{h \in B} \left| \int_{\mathcal{R}} hg \, d\mu \right|.$$

Here we have to note that the only property used in the proof of Lemma 6.2.10 is the lattice property, which however is possessed by X_b since, by Theorem 6.3.16, it is an ideal. The proof is complete. □

We finish this section by characterizing when $X_a = X_b$. The key role is played by characteristic functions of sets of finite measure and the question whether they have absolutely continuous norms.

Theorem 6.3.18. *Let X be a Banach function space. Then $X_a = X_b$ if and only if $\chi_E \in X_a$ for every set $E \subset \mathcal{R}$ of finite measure.*

Proof. ⇐ Assume that $E \subset \mathcal{R}$ and $\mu(E) < \infty$. Then, by Theorem 6.3.16,

$$\chi_E \in X_b \subset X_a.$$

⇒ Assume that $\chi_E \in X_a$ for every set $E \subset \mathcal{R}$ of finite measure. This means that $S \subset X_a$. But X_a is a closed subspace of X by Theorem 6.3.12. Therefore, $X_b \subset X_a$. Together with Theorem 6.3.16, this shows that $X_b = X_a$.

The proof is complete. □

6.4 Reflexivity of Banach function spaces

Let X be a Banach function space and let $Y \subset X$ be a closed linear subspace of X such that $S \subset Y$, where S is the set of simple functions. Let $g \in X'$. Then the mapping

$$g \mapsto L_g \in X^* \subset Y^*,$$

where

$$L_g(f) := \int_{\mathcal{R}} fg \, d\mu$$

is injective as a map from X' to Y^*. It is also an isometry, since, by Theorem 6.3.17, $Y \supset X_b$. Consequently, through this mapping, X' is isometrically isomorphic to a closed linear subspace of Y^*. The following theorem shows what happens when Y is moreover an ideal in X.

Theorem 6.4.1. *Let X be a Banach function space and let Y be an ideal in X such that $S \subset Y$. Then $Y^* = X'$ if and only if $Y \subset X_a$.*

Proof. $\Leftarrow$

Step 1. Assume that $Y \subset X_a$. We know that Y contains S, hence also its closure. In other words, $X_b \supset X_a$. Together with Theorem 6.3.16, this implies

$$Y = X_a = X_b. \tag{6.4.1}$$

We already know that $X' \subset Y^*$, so we only have to prove that $Y^* \subset X'$.

Assume that $\Lambda \in Y^*$. We shall find a function $g \in X'$ such that $\Lambda(f) = \int_{\mathcal{R}} fg \, d\mu$ for every $f \in Y$.

Let $\{S_N\}_{N=1}^\infty$ be a disjoint system of measurable subsets of $\mathcal{R}$ such that $\bigcup_{N=1}^\infty S_N = \mathcal{R}$ and $\mu(S_N) < \infty$ for every $N \in \mathbb{N}$. For every fixed $N \in \mathbb{N}$, we denote by $\mathcal{A}_N$ the σ-algebra of all μ-measurable subsets of S_N. We then define the function

$$\lambda_N : \mathcal{A}_N \to [0, \infty)$$

by

$$\lambda_N(A) := \Lambda(\chi_A), \quad A \in \mathcal{A}_N.$$

The function λ_N is well-defined on every $A \in \mathcal{A}_N$ since $\chi_A \in X_b \subset Y$.

Step 2. We claim that λ is countably additive on $\mathcal{A}_N$. Let $\{A_i\}_{i=1}^\infty$ be a sequence of disjoint sets in $\mathcal{A}_N$. We then define

$$B_n := \bigcup_{i=1}^n A_i, \quad n \in \mathbb{N},$$

and

$$A := \bigcup_{n=1}^\infty A_n = \bigcup_{n=1}^\infty B_n. \tag{6.4.2}$$

Because $A \in \mathcal{A}_N$, one has $\chi_A \in X_b$, that is, by (6.4.1), $\chi_A \in X_a$. Next, the obvious inclusion $A \setminus B_n \subset A$ implies $\chi_{(A \setminus B_n)} \leq \chi_A$, and (6.4.2) implies that $(A \setminus B_n) \searrow \emptyset$. Hence, $\chi_{(A \setminus B_n)} \searrow 0$. By Theorem 6.3.7, this implies that

$$\lim_{n \to \infty} \|\chi_A - \chi_{B_n}\|_X = 0.$$

Now, the continuity of Λ on Y shows that

$$\lambda_N(A) = \Lambda(\chi_A) = \lim_{n \to \infty} \Lambda(\chi_{B_n}) = \lim_{n \to \infty} \sum_{i=1}^n \Lambda(\chi_{A_i}) = \sum_{i=1}^\infty \Lambda(\chi_{A_i}),$$

proving the claim.

Step 3. Now we claim that λ_N is bounded on $\mathcal{A}_N$. Indeed, for any $A \in \mathcal{A}_N$, we have

$$|\lambda_N(A)| = |\Lambda(\chi_A)| \le \|\Lambda\|_{Y*} \|\chi_A\|_Y = \|\Lambda\|_{Y*} \|\chi_A\|_X$$
$$\le \|\Lambda\|_{Y*} \|\chi_{S_N}\|_Y < \infty.$$

Step 4. Next, we shall prove that λ_N is absolutely continuous with respect to μ. To this end, let $A \in \mathcal{A}_N$ be such that $\mu(A) = 0$. Then

$$|\lambda_N(A)| \le \|\Lambda\|_{Y*} \|\chi_A\|_X,$$

whence also $|\lambda_N(A)| = 0$, establishing the claim.

Step 5. We have verified all the assumptions of the Radon–Nikodým theorem (Theorem 1.21.15). Thus, for each $N \in \mathbb{N}$, there exists a uniquely determined function g_N on S_N such that

$$\Lambda(\chi_A) = \lambda_N(\chi_A) = \int_\mathcal{R} \chi_A g_N \, d\mu, \quad A \in \mathcal{A}_N.$$

We set

$$g = \sum_{N=1}^\infty g_N.$$

Since S_N are mutually disjoint, this defines a function g on $\mathcal{R}$. Then

$$\Lambda(\chi_A) = \int_\mathcal{R} \chi_A g \, d\mu, \quad A \subset \bigcup_{N=1}^\infty \mathcal{A}_N. \qquad (6.4.3)$$

Step 6. We have to show that $g \in X'$. Assume that $h \in S$, $h \ge 0$ and $\operatorname{supp} h \subset G_n := \bigcup_{j=1}^n S_j$. We then have

$$\int_\mathcal{R} |hg| \, d\mu = \int_\mathcal{R} h \operatorname{sign}(g) g \, d\mu = \Lambda(h \operatorname{sign}(g))$$
$$\le \|\Lambda\|_{Y*} \|h \operatorname{sign}(g)\|_Y$$
$$= \|\Lambda\|_{Y*} \|h\|_X.$$

Now let $f \in X$ be arbitrary. Then we can find a sequence $\{h_n\}_{n=1}^\infty \subset S$ such that $\operatorname{supp} h_n \subset G_n$, satisfying $0 \le h_n \nearrow |f|$. By the monotone convergence theorem and the Fatou lemma, we obtain, for every $f \in X$,

$$\int_\mathcal{R} |fg| \, d\mu \le \|\Lambda\|_{Y*} \|f\|_X.$$

Step 7. We will prove (6.4.3) for a general case (so far we have it established only for a special case of characteristic functions). To this end, let $g \in X'$. Then the functional Λ_g, defined by

$$\Lambda_g(f) := \int_\mathcal{R} fg \, d\mu, \quad f \in Y,$$

is in Y^*. Let $f \in Y$. Again, then there exists a sequence of simple functions $\{h_n\}_{n=1}^\infty$ such that $\operatorname{supp} h_n \subset G_n$ and $0 \leq h_n \nearrow |f|$ μ-a.e. on $\mathcal{R}$. Define

$$f_n := h_n \operatorname{sign}(f) \quad n \in \mathbb{N}.$$

Then $f_n \in S$ for every $n \in \mathbb{N}$ and $\operatorname{supp} f_n \subset G_n$. Moreover, $f_n \to f$ μ-a.e. on $\mathcal{R}$. However, we know that $Y = X_a$. This implies $f \in X_a$, and, via Theorem 6.3.9, also

$$\lim_{n \to \infty} \|f_n - f\|_Y = 0.$$

Next, we have $\Lambda(f_n) = \Lambda_g(f_n)$, and since both Λ and Λ_g are continuous, we finally get $\Lambda(f) = \Lambda_g(f)$, as desired. It thus follows that $Y^* \subset X'$, whence, finally, $X' = Y^*$.

$\Rightarrow$ Assume now that $Y^* = X'$. We shall show that then $Y \subset X_a$ by contradiction. Suppose that there exists some $f \in Y \setminus X_a$. We can with no loss of generality assume that $f \geq 0$, because Y is an ideal. The fact that f does not have an absolutely continuous norm in X means that there is a sequence $\{E_n\}_{n=1}^\infty$ satisfying $E_n \searrow \emptyset$ μ-a.e., and an $\varepsilon > 0$ such that

$$\|f \chi_{E_n}\|_X \geq \varepsilon. \tag{6.4.4}$$

We define, for $n \in \mathbb{N}$, the sets

$$G_n := \{g \in X', \int_{\mathcal{R}} |f \chi_{E_n} g| \, d\mu < \frac{\varepsilon}{2}\}.$$

Then it follows from the dominated convergence theorem and the assumption of the theorem that

$$\bigcup_{n=1}^\infty G_n = X' = Y^*.$$

Again, since Y is an ideal, we have $f \chi_{E_n} \in Y$ for every $n \in \mathbb{N}$. Therefore, for every $n \in \mathbb{N}$, the set G_n is weakly* open in Y^*. By the Banach–Alaoglu theorem (Theorem 1.17.6), there is some $k \in \mathbb{N}$ and $n_1, \ldots, n_k$ such that

$$B_{Y^*} \subset \bigcup_{j=1}^k G_{n_j},$$

where B_{Y^*} is the unit ball in Y^*. Hence, for every $g \in X'$ such that $\|g\|_{X'} \leq 1$ there exists a $j \in \mathbb{N}$ such that $g \in G_{n_j}$. Now, because $f \geq 0$ and $E_n \searrow \emptyset$ μ-a.e., we have, for every $n \in \mathbb{N}$, $n \geq \max\{n_1, \ldots, n_k\}$. Hence, by the Lorentz–Luxemburg theorem (Theorem 6.2.9) and by Lemma 6.2.10, we finally obtain, for every $n \in \mathbb{N}$,

$$\int_{\mathcal{R}} |f \chi_{E_n} g| \, d\mu \leq \int_{\mathcal{R}} |f \chi_{n_j} g| \, d\mu < \frac{\varepsilon}{2},$$

which contradicts (6.4.4). The proof is complete. □

Section 6.4 Reflexivity of Banach function spaces

Corollary 6.4.2. *Let X be a Banach function space and let Y be an ideal in X such that $S \subset Y$. Then*
$$Y = X_a = X_b.$$

Proof. Since $S \subset Y$ and Y is closed, we have
$$X_b \subset Y \subset X_a \subset X_b,$$
and the assertion follows. □

Corollary 6.4.3. *Let X be a Banach function space such that $S \subset X_a$. Then*
$$(X_a)^* = X'.$$

Proof. Set $Y := X_a$. Then Y is an ideal in X by Corollary 6.3.13. Since $S \subset Y$, the assertion follows from Theorem 6.4.1. □

Example 6.4.4. Let Φ be a Young function. and let L^Φ be the corresponding Orlicz space. Then $X_a = E^\Phi$. Thus,
$$(E^\Phi)^* = L^\Psi,$$
where Ψ is the complementary function of Φ.

Corollary 6.4.5. *Let X be a Banach function space. Then $X^* = X'$ if and only if $X = X_a$.*

Proof. This follows by applying Theorem 6.4.1 to $Y = X$. □

Corollary 6.4.6. *Let X be a Banach function space. Then X is reflexive if and only if $X = X_a$ and $X' = (X_a)'$.*

Proof. $\Leftarrow$ Assume first that both X and X' have absolutely continuous norms. Then, by Theorem 6.4.1, we have
$$X^* = X' \quad \text{and also} \quad (X')^* = (X')' = X''.$$
Therefore, combining this with the Lorentz–Luxemburg theorem (Theorem 6.2.9), we obtain
$$X^{**} = (X^*)^* = (X')^* = X'' = X,$$
whence X is reflexive.

$\Rightarrow$ Assume now that X is reflexive. We know that $X' \subset X^*$. Suppose that this inclusion is proper, that is, $X' \neq X^*$. By the Hahn–Banach theorem, then there exists a nonzero functional $F \in X^{**}$ such that $F(g) = 0$ for every $g \in X'$. Since the space

X^{**} is reflexive, there exists a function $f \in X$ such that $F(\Lambda) = \Lambda(f)$ for every $\Lambda \in X^*$. In particular, for every $g \in X'$, one has

$$F(g) = \int_{\mathcal{R}} f(x)g(x)\,d\mu.$$

This, however, means that, for every $g \in X'$,

$$\int_{\mathcal{R}} f(x)g(x)\,d\mu = 0.$$

By Theorem 6.2.13, X' is a closed norm-fundamental subspace of X^*. Thus, necessarily, $f \equiv 0$ μ-a.e. on $\mathcal{R}$. This contradicts the fact that F is a nonzero functional, hence $X^* = X'$. By Corollary 6.4.5, $X = X_a$. Using the reflexivity of X and Lorentz–Luxemburg theorem (Theorem 6.2.9) once again, we get

$$X^{**} = X = X'' = (X')'.$$

Since X has absolutely continuous norm, we have

$$X' = X^*,$$

hence also

$$(X')^* = (X^*)^*.$$

Setting all this information together, we get

$$(X')^* = (X^*)^* = X^{**} = (X')'.$$

By Corollary 6.4.5, this implies that also $X' = (X')_a$, as desired. □

6.5 Separability in Banach function spaces

Our main aim in this section is to characterize when a given Banach function space is separable. The main tool will be the weak topology on X generated by certain subspace of X'.

Remark 6.5.1. Let X be a Banach function space and let $Z \subset X'$ be an ideal in X' satisfying $S \subset Z$. Then Z contains $(X')_b$. By Theorem 6.3.18, X_b is a norm-fundamental subspace of $(X')^*$ and applying the same theorem to X', we get that $(X_b)'$ is a norm-fundamental subspace of $(X'')^* = X^*$ by the Lorentz–Luxemburg theorem (Theorem 6.2.9). Therefore, also Z is a norm-fundamental subspace of X^*.

Definition 6.5.2. Let X be a Banach function space and let $Z \subset X'$ be an ideal in X' satisfying $S \subset Z$. Then the system of seminorms defined by

$$f \mapsto \left| \int fg \, d\mu \right|, \qquad g \in Z,$$

is separating and it defines on X a structure of Hausdorff locally convex topological linear space. This topology is called the *weak topology on X generated by Z* and is denoted by $\sigma(X, Z)$.

Remark 6.5.3. Let X be a Banach function space and let $Z \subset X'$ be an ideal in X' satisfying $S \subset Z$. Then, a sequence $\{f_n\}_{n=1}^{\infty} \subset X$ satisfies

$$f_n \to f \quad \text{in } \sigma(X, Z)$$

for some $f \in X$ if and only if

$$\int f_n g \, d\mu \to \int fg \, d\mu \quad \text{for every } g \in Z.$$

Definition 6.5.4. We say that a set $A \subset X$ is $\sigma(X, Z)$ bounded if, for every $g \in Z$,

$$\sup_{f \in A} \{|\int fg \, d\mu|\} < \infty. \tag{6.5.1}$$

Lemma 6.5.5. *Let X be a Banach function space and let $Z \subset X'$ be an ideal in X' satisfying $S \subset Z$. Let $A \subset X$. Then A is $\sigma(X, Z)$ bounded if and only if it is bounded in X.*

Proof. $\Leftarrow$ Let $A \subset X$ be norm bounded in X. By the Hölder inequality (Theorem 6.2.6), we obtain for every $g \in Z$ (since $Z \subset X'$)

$$\sup_{f \in A} \left\{ \left| \int_{\mathcal{R}} fg \, d\mu \right| \right\} \leq \sup_{f \in A} \{\|f\|_X \|g\|_{X'}\} \leq C \|g\|_{X'}$$

for an appropriate positive constant C. Thus, A is $\sigma(X, Z)$ bounded.
$\Rightarrow$ Conversely, assume that $A \subset X$ is $\sigma(X, Z)$ bounded and $g \in Z$. Then

$$\sup_{f \in A} \left\{ \left| \int_{\mathcal{R}} fg \, d\mu \right| \right\} < \infty.$$

Let $f \in A$. We define the functional Λ_f at a function $g \in Z$ by

$$\Lambda_f(g) := \int_{\mathcal{R}} fg \, d\mu.$$

Then the Hölder inequality (Theorem 6.2.6) guarantees that $\Lambda_f \in Z^*$ and

$$\|\Lambda_f\|_{Z^*} = \sup\left\{\left|\int_{\mathcal{R}} fg\,d\mu\right|, \ g \in Z^*, \ \|g\|_{X^*} \leq 1\right\} = \|f\|_X,$$

because Z is norm fundamental in X^*. For every $g \in Z$ one thus has, by (6.5.1),

$$\sup_{f \in A} |\Lambda_f(g)| =: C_g < \infty,$$

where C_g is a positive constant depending on g. Hence, by the principle of uniform boundedness (Theorem 1.18.1), we finally get

$$\sup_{f \in A} \|\Lambda_f\|_{Z^*} = \sup_{f \in A} \|f\|_X < \infty.$$

The proof is complete. □

Theorem 6.5.6. *Let X be a Banach function space and let $Z \subset X'$ be an ideal in X' satisfying $S \subset Z$. Then X is $\sigma(X, Z)$ complete.*

Proof. Step 1. Let $\{f_n\}_{n=1}^\infty$ be a $\sigma(X, Z)$-Cauchy sequence in X, that is, for every $g \in Z$, the sequence

$$\left\{\int_{\mathcal{R}} f_n g\,d\mu\right\}_{n=1}^\infty$$

is Cauchy in $\mathbb{R}$. In other words, $\{f_n\}_{n=1}^\infty$ is a $\sigma(X, Z)$ bounded sequence. By Lemma 6.5.5, it is also bounded in X, that is, there exists a positive constant M such that

$$\|f_n\|_X \leq M \quad \text{for every } n \in \mathbb{N}.$$

Step 2. We claim that there exists a function $f_0 \in X$ such that $\|f_0\|_X \leq M$, and, for every $g \in X'_b$,

$$\lim_{n \to \infty} \int_{\mathcal{R}} f_n g\,d\mu = \int_{\mathcal{R}} f_0 g\,d\mu.$$

To this end, we define, for $n \in \mathbb{N}$, the set functions

$$\nu_n : E \mapsto \int_E f_n\,d\mu, \quad E \subset \mathcal{R}.$$

Then each ν_n, $n \in \mathbb{N}$, is a measure, and it follows from the axioms of a Banach function space and the Hölder inequality that ν_n is for every $n \in \mathbb{N}$ absolutely continuous with respect to μ. We now fix $k \in \mathbb{N}$. Since $\{f_n\}_{n=1}^\infty$ is a Cauchy sequence and $S \subset X_b$, the sequence $\{\nu_n(E)\}_{n=1}^\infty$ is a Cauchy sequence in $\mathbb{R}$ for every μ-measurable subset E of R_k, since $E \subset R_k$ means in particular that $\mu(E) < \infty$ (here the sequence

Section 6.5 Separability in Banach function spaces

$\{R_k\}_{k=1}^{\infty}$ has the usual meaning). Therefore, for every such E, there exists the finite limit
$$\lim_{n \to \infty} \nu_n(E) =: \nu(E).$$
It follows from the Hahn–Saks theorem (Theorem 1.23.4) that the measures $\{\nu_n\}_{n=1}^{\infty}$, restricted to R_k, form a system of uniformly absolutely continuous measures (with respect to μ) and ν is a measure on R_k, which is absolutely continuous with respect to μ. It follows on letting $k \to \infty$ that there exists a function $f_0 \in L^1_{\text{loc}}(\mathcal{R})$ such that, for every $E \subset \mathcal{R}$, $\mu(E) < \infty$,
$$\lim_{n \to \infty} \nu_n(E) = \nu(E) = \int_{\mathcal{R}} f_0 \chi_E \, d\mu.$$
Let $g \in S$. Then, by linearity of S,
$$\lim_{n \to \infty} \int_{\mathcal{R}} f_n g \, d\mu = \int_{\mathcal{R}} f_0 g \, d\mu. \tag{6.5.2}$$
We have to prove $f_0 \in X$ (so far we know that $f_0 \in L^1_{\text{loc}}(\mathcal{R})$). If moreover $\|g\|_{X'} \leq 1$, then we have from the Hölder inequality
$$\left| \int_{\mathcal{R}} f_0 g \, d\mu \right| \leq \limsup_{n \to \infty} \|f_n\|_X \|g\|_{X'} \leq M.$$
Consequently, $f_0 \in X''$, and, by Theorem 6.2.9,
$$\|f_0\|_{X''} = \|f_0\|_X \leq M.$$
Now let g be bounded and such that supp $g \subset E$, where $\mu(E) < \infty$. Then we uniformly approximate g by functions from S, and (6.5.2) follows again.

Step 3. We now claim that (6.5.2) is true for every $g \in Z$. Thus, for any $g \in Z$ and $n \in \mathbb{N}$, define the set function
$$\omega_n(E) := \int_E f_n g \, d\mu.$$
Then every ω_n, $n \in \mathbb{N}$, is a finite measure, absolutely continuous with respect to μ. Since for every μ-measurable E, $g \chi_E \in Z$, the sequence $\{\omega_n(E)\}$ is convergent by the hypothesis. We use the Hahn–Saks theorem (Theorem 1.23.4) once again and get that $\{\omega_n(E)\}$ is uniformly absolutely continuous with respect to μ, in other words, $\omega_n(E) \to 0$ uniformly in $n \in \mathbb{N}$ as $\mu(E) \to 0$. We then define the measure ω_0 by
$$\omega_0(E) := \int_E f_0 g \, d\mu.$$
Then, again, ω_0 is absolutely continuous with respect to μ. We define the sequence of sets $\{E_k\}_{k=1}^{\infty}$ by
$$E_k := \{x \in \mathcal{R}; \ |g(x)| > k\} \cup (\mathcal{R} \setminus R_k).$$

Let $\varepsilon > 0$. Then there exists a k_0 such that for every $n \in \mathbb{N}$,
$$|\omega_n(E_{k_0})| < \frac{\varepsilon}{3}.$$
Set
$$F_{k_0} := \mathcal{R} \setminus E_{k_0}.$$
Then $g\chi_{F_{k_0}}$ is a bounded function and $\mu(\mathrm{supp}(g\chi_{F_{k_0}})) < \infty$. For such functions (6.5.2) was already established, so, there exists an $n_0 \in \mathbb{N}$ such that, for every $n \in \mathbb{N}, n \geq n_0$, one has
$$\left| \int_{F_{k_0}} f_n g \, d\mu - \int_{F_{k_0}} f_0 g \, d\mu \right| < \frac{\varepsilon}{3}.$$
Therefore,
$$\left| \int_{\mathcal{R}} f_n g \, d\mu - \int_{\mathcal{R}} f_0 g \, d\mu \right| < \frac{\varepsilon}{3} + |\omega_n(E_{k_0})| + |\omega_0(E_{k_0})| < \varepsilon.$$
Thus (6.5.2) holds for every $g \in Z$, proving the claim.

We have shown that $f_n \to f_0$ in the $\sigma(X, Z)$-topology. Hence, X is $\sigma(X, Z)$ complete, as desired. The proof is complete. □

Corollary 6.5.7. *Let X be a Banach function space. Then X is $\sigma(X, X')$ complete.*

Definition 6.5.8. The measure μ is called *separable* on $\mathcal{R}$ if the metric space $(\mathcal{A}, d)$ is separable, where
$$\mathcal{A} := \{E \subset \mathcal{R}, \, \mu(E) < \infty\}$$
and d is the metric defined on $\mathcal{A}$ by
$$d(E, F) := \int_{\mathcal{R}} |\chi_E - \chi_F| \, d\mu.$$

Theorem 6.5.9. *Let X be a Banach function space over $(\mathcal{R}, \mu)$ and let $Y \subset X$ be an ideal in X satisfying $S \subset Y$. Then Y is separable if and only if $Y = Y_a$ and μ is a separable measure.*

Proof. ⇒ Let Y be separable and let $\{f_n\}_{n=1}^{\infty}$ be a dense set in Y. Suppose that $f_0 \in Y \setminus X_a$. Then, by Proposition 6.3.4, there exists a sequence $\{E_n\}_{n=1}^{\infty}$ of measurable subsets of $\mathcal{R}$ such that $E_n \searrow \emptyset$ μ-a.e. and
$$\|f_0 \chi_{E_n}\|_X \geq \varepsilon \quad \text{for every } n \in \mathbb{N}.$$
Since X' is a norm-fundamental subspace of X^* (Theorem 6.2.13), there exists a sequence $\{g_n\}_{n=1}^{\infty}$ of functions $g_n \in X'$ satisfying $\|g_n\|_{X'} \leq 1$, and such that
$$\int_{\mathcal{R}} |f_0 g_n \chi_{E_n}| \, d\mu \geq \frac{\varepsilon}{2}. \tag{6.5.3}$$

We define now the functions

$$h_n := \operatorname{sign}(f_0)|g_n|\chi_{E_n}, \quad n \in \mathbb{N}.$$

Then also $\|h_n\|_{X'} \leq 1$, whence, by the Hölder inequality, for every $k, n \in \mathbb{N}$,

$$\left|\int_{\mathcal{R}} f_k h_n \, d\mu\right| \leq \|f_k\|_X \|h_n\|_{X'} \leq \|f_k\|_X.$$

So, for each fixed $k \in \mathbb{N}$, the sequence

$$\left|\int_{\mathcal{R}} f_k h_n \, d\mu\right|$$

is bounded in $\mathbb{R}$. By the Bolzano–Weierstrass theorem, this sequence contains a convergent subsequence. Using the diagonalization procedure, we get a sequence $\{h_{n_j}\}_{j=1}^{\infty}$ such that

$$\int_{\mathcal{R}} f_k h_{n_j} \, d\mu \text{ converges in } \mathbb{R} \text{ as } j \to \infty \text{ for every } k \in \mathbb{N}.$$

Now, $\{f_n\}_{n=1}^{\infty}$ is dense in Y, hence, by the Hölder inequality, the limit

$$\lim_{j \to \infty} \int_{\mathcal{R}} f h_{n_j} \, d\mu$$

exists for every $f \in Y$ (recall that $f \in Y \subset X = X''$). This means that the sequence $\{h_{n_j}\}_{j=1}^{\infty}$ is $\sigma(X', Y)$-Cauchy in X'. We now apply Theorem 6.5.6 to X' in place of X and Y in place of Z. It implies that there exists a function $h \in X'$ such that, for every $f \in Y$,

$$\lim_{j \to \infty} \int_{\mathcal{R}} f h_{n_j} \, d\mu = \int_{\mathcal{R}} f h \, d\mu. \tag{6.5.4}$$

We know that, for each $j \in \mathbb{N}$, the support of h_{n_j} is contained in E_{n_j}, a nonincreasing (with respect to set inclusion) sequence of sets. Fix $j \in \mathbb{N}$. Then, for any $E \subset \mathcal{R}$ satisfying $\mu(E) < \infty$ and $E \cap E_{n_j} = \emptyset$ we get by (6.5.4) (recall $\chi_E \in Y$) that

$$\int_E h \, d\mu = 0.$$

Keeping this $j \in \mathbb{N}$ still fixed and using the statement just proved for any such E, we obtain that $h = 0$ μ-a.e. on $\mathcal{R} \setminus E_{n_j}$. Letting $j \to \infty$, we get that, in fact, $h = 0$ μ-a.e. on the entire $\mathcal{R}$, since $E_{n_j} \searrow \emptyset$ as $j \to \infty$. Finally, (6.5.4) yields

$$0 = \lim_{j \to \infty} \int_{\mathcal{R}} f_0 h_{n_j} \, d\mu = \lim_{j \to \infty} \int_{\mathcal{R}} |f_0 g_{n_j}| \chi_{E_{n_j}} \, d\mu.$$

That, however, contradicts (6.5.3). Thus, $Y \subset X_a$.

We shall now prove that μ is a separable measure. We recall that the sequence $\{R_k\}_{k=1}^\infty$ will have its usual meaning. For $k \in \mathbb{N}$, we denote by $\mathcal{A}_k$ the σ-algebra of μ-measurable subsets of R_k.

Let $A \in \mathcal{A}_k$. Then, in particular, $\chi_A \in Y$. Since Y is separable, each its subset is separable, too. Let $k \in \mathbb{N}$. Then there exists a countable set $\{\chi_{E_{k,j}}\}_{j=1}^\infty$, where $E_{k,j} \in \mathcal{A}_k$, dense in the set $\{\chi_E;\ E \in \mathcal{A}_k\}$. We define the set

$$\mathcal{F} := \{E_{k,j};\ k, j \in \mathbb{N}\}$$

and we claim that the set $\mathcal{F}$ is a countable dense subset of the metric space $(\mathcal{A}, d)$.

Assume, thus, that $F \in \mathcal{A}$. Then $\mu(F) < \infty$ and $(F \cap R_k) \nearrow F$ as $k \to \infty$. Hence, by the dominated convergence theorem,

$$\mu(F \setminus R_k) \to 0, \quad k \to \infty.$$

Let $\varepsilon > 0$. Then there exists a $k_0 \in \mathbb{N}$ such that

$$\mu(F \setminus R_{k_0}) < \frac{\varepsilon}{2}.$$

By Remark 6.1.16(v), there exists a positive constant C_{k_0} such that, for every $f \in X$,

$$\int_{R_{k_0}} |f|\,d\mu \le C_{k_0} \|f\|_X.$$

Thus, the definition of $\mathcal{F}$ now guarantees that there is a set $E \in \mathcal{F}$, $E \subset R_{k_0}$, satisfying

$$\|\chi_E - \chi_{F \cap R_{k_0}}\|_X \le \frac{\varepsilon}{2C_{k_0}}.$$

Finally, altogether,

$$d(E, F) = \int_{\mathcal{R}} |\chi_F - \chi_E|\,d\mu = \int_{R_{k_0}} \left|\chi_F - \chi_{E \cap R_{k_0}}\right|\,d\mu + \int_{\mathcal{R} \setminus R_{k_0}} |0 - \chi_F|\,d\mu$$

$$= \int_{R_{k_0}} \left|\chi_F - \chi_{E \cap R_{k_0}}\right|\,d\mu + \mu(F \setminus R_{k_0})$$

$$\le \frac{\varepsilon}{2} + \frac{\varepsilon}{2} = \varepsilon.$$

Thus, $\mathcal{F}$ is dense in $(\mathcal{A}, d)$, that is, μ is a separable measure.

$\Leftarrow$ Assume that μ is separable and $Y = Y_a$. Then, by Theorem 6.4.1, also $Y = Y_a = Y_b$. Let $\mathcal{F}_1$ be a dense countable set in $(\mathcal{A}, d)$. We denote its elements by $E_1, E_2, \ldots$ Next, we define

$$\mathcal{F} := \{E_j \cap R_k,\ j, k \in \mathbb{N}\}$$

and

$$\mathcal{D} := \left\{ f \in S, \ f = \sum_{k=1}^{K} r_k \chi_{F_k}, \ K \in \mathbb{N}, \ r_k \in \mathbb{Q}, \ F_k \in \mathcal{F} \right\}.$$

Then $\mathcal{D}$ is obviously countable. We have to show that it is dense in Y. It will suffice to prove that it is dense in X_b. We know that

$$X_b = \overline{\left\{ g \in \mathcal{M}_0, \ g = \sum_{n=1}^{m} c_n \chi_{G_n}, \ m \in \mathbb{N} \right\}},$$

where the closure is in the norm of X, $c_n \in \mathbb{R}$ and $G_n \subset \mathcal{R}$ are such that $\mu(G_n) < \infty$. We can approximate c_n by rational numbers, hence it is enough to replace χ_{G_n} by χ_{F_k}.

Assume now that $\mu(G) < \infty$ and $\varepsilon > 0$. We want to find a sequence $\{F_j\}_{j=1}^{\infty}$ such that

$$\|\chi_{F_j} - \chi_G\|_X \to 0.$$

We know that $\chi_G \in X_b = X_a$ and that

$$\chi_{(G \cap R_k)} \nearrow \chi_G.$$

Thus, by Theorem 6.3.9,

$$\|\chi_G - \chi_{(G \cap R_k)}\|_X < \frac{\varepsilon}{2}.$$

It is thus enough to approximate $\chi_{G \cap R_k}$ by functions χ_{F_k}. Let $n \in \mathbb{N}$. Then there exists a function $F_n \in \mathcal{F}$ such that

$$\lim_{n \to \infty} d(G \cap R_k, F_n) = 0.$$

This means that the sequence $\{\chi_{F_n}\}_{n=1}^{\infty}$ converges to $\chi_{(G \cap R_k)}$ in the norm of the space $L^1(\mathcal{R}, \mu)$. Thus, there is a subsequence $\{\chi_{F_{n_j}}\}_{j=1}^{\infty}$, which converges pointwise μ-a.e. on $\mathcal{R}$ as $j \to \infty$. Now, for every $j \in \mathbb{N}$, the function $\chi_{F_{n_j}}$ is dominated by $\chi_{R_k} \in X_a$. Thus, by Theorem 6.3.9 again, we finally get

$$\lim_{j \to \infty} \|\chi_{F_{n_j}} - \chi_{(G \cap R_k)}\|_X = 0,$$

establishing our claim. The proof is complete. □

Corollary 6.5.10. *Let X be a Banach function space over $(\mathcal{R}, \mu)$. Then X is separable if and only if $X = X_a$ and μ is separable.*

Corollary 6.5.11. *Let X be a Banach function space and let $X_a = X_b$. Then X_a is separable if and only if μ is separable.*

Corollary 6.5.12. *Let X be a Banach function space such that its dual X^* is separable. Then X is reflexive.*

Proof. It is a classical fact (see, e.g., [66]) that when X^* is separable, then also X itself is separable. Since, then, every subset of X^* is separable, in particular, X' is separable. Now, by Theorem 6.5.9, the separability of X implies $X = X_a$ and likewise the separability of X' implies $X' = (X_a)'$. Finally, by Corollary 6.4.6, X is reflexive. □

Chapter 7

Rearrangement-invariant spaces

7.1 Distribution functions and nonincreasing rearrangements

As in the previous chapter, $(\mathcal{R}, \mu)$ denotes a σ-finite measure space, similarly for $(\mathcal{S}, \nu)$.

Definition 7.1.1. Let $f : (\mathcal{R}, \mu) \to \mathbb{R}$ be a measurable function. Then the function μ_f, defined at every $\lambda > 0$ by

$$\mu_f(\lambda) = \mu(\{x \in \mathcal{R} : |f(x)| > \lambda\}), \qquad (7.1.1)$$

is called the *distribution function* of f.

Remark 7.1.2. The distribution function μ_f of a function f depends only on $|f|$. Moreover, μ_f is allowed to attain the value ∞.

Definition 7.1.3. Let $f \in \mathcal{M}_0(\mathcal{R}, \mu)$ and $g \in \mathcal{M}_0(\mathcal{S}, \nu)$. We say that f and g are *equimeasurable* if they have the same distribution function, that is, if $\mu_f(\lambda) = \nu_g(\lambda)$ for all $\lambda \geq 0$. We write $f \sim g$.

Proposition 7.1.4. *Let* $f \in \mathcal{M}_0(\mathcal{R}, \mu)$. *Then the distribution function* μ_f *of* f *is a nonnegative, nonincreasing, and right-continuous function on* $[0, \infty)$.

Proof. It is clear that μ_f is nonnegative and nonincreasing. Let

$$E_\lambda := \{x : |f(x)| > \lambda\}, \quad \lambda \geq 0.$$

Then $E_{\lambda_2} \subset E_{\lambda_1}$ for $\lambda_1 < \lambda_2$. Moreover, for every $\lambda_0 \in \mathbb{R}$ we have

$$E_{\lambda_0} = \bigcup_{\lambda > \lambda_0} E_\lambda = \bigcup_{n=1}^{\infty} E_{\lambda_0 + \frac{1}{n}}.$$

Hence, by the monotone convergence theorem (Theorem 1.21.4),

$$\mu_f\left(\lambda_0 + \frac{1}{n}\right) = \mu\left(E_{\lambda_0 + \frac{1}{n}}\right) \uparrow \mu(E_{\lambda_0}) = \mu_f(\lambda_0),$$

whence μ_f is right-continuous. $\square$

Exercise 7.1.5. Let $f, g \in \mathcal{M}_0(\mathcal{R}, \mu)$, and let $\{f_n\}_{n=1}^{\infty}$ be a sequence of functions from $\mathcal{M}_0(\mathcal{R}, \mu)$. Let $a \in \mathbb{R}, a \neq 0$. Then

$$|g| \leq |f| \ \mu\text{-a.e.} \quad \Rightarrow \quad \mu_g \leq \mu_f, \tag{7.1.2}$$

$$\mu_{af}(\lambda) = \mu_f\left(\frac{\lambda}{|a|}\right), \quad \lambda \geq 0, \tag{7.1.3}$$

$$\mu_{f+g}(\lambda_1 + \lambda_2) \leq \mu_f(\lambda_1) + \mu_g(\lambda_2), \quad \lambda_1, \lambda_2 \geq 0, \tag{7.1.4}$$

$$|f| \leq \liminf_{n \to \infty} |f_n| \ \mu\text{-a.e.} \quad \Rightarrow \quad \mu_f \leq \liminf_{n \to \infty} \mu_{f_n}, \tag{7.1.5}$$

in particular,

$$|f_n| \uparrow |f| \ \mu\text{-a.e.} \quad \Rightarrow \quad \mu_{f_n} \uparrow \mu_f.$$

Definition 7.1.6. Let $f \in \mathcal{M}_0(\mathcal{R}, \mu)$. Then the function $f^* : [0, \infty) \to [0, \infty)$, defined by

$$f^*(t) = \inf\{\lambda : \mu_f(\lambda) \leq t\}, \quad t \in [0, \infty), \tag{7.1.6}$$

is called the *nonincreasing rearrangement* of f.

Convention 7.1.7. We use here the convention that $\inf \emptyset = \infty$. In particular, if $\mu_f(\lambda) > t$ for every $\lambda \geq 0$, then $f^*(t) = \infty$.

Convention 7.1.8. When $\mu(\mathcal{R}) < \infty$, then $\mu_f(\lambda) \leq \mu(\mathcal{R})$ for every $\lambda \in [0, \infty)$. Consequently, $f^*(t) = 0$ for every $t \geq \mu(\mathcal{R})$. In this case, we may regard f^* as a function defined on the interval $[0, \mu(\mathcal{R}))$.

Remark 7.1.9. If $f \in \mathcal{M}_0(\mathcal{R}, \mu)$ is such that μ_f is continuous and strictly decreasing on $(0, \infty)$, then f^* is just the ordinary inverse of μ_f on a suitable interval. Moreover, from (7.1.6) and the monotonicity of μ_f we have

$$f^*(t) = \sup\{\lambda : \mu_f(\lambda) > t\} = \mu_{\mu_f}(t), \quad t \geq 0. \tag{7.1.7}$$

Remark 7.1.10. Let

$$f(x) = 1 - \frac{1}{x+1}, \quad x \in [0, \infty).$$

Then it is not difficult to verify that $f^*(t) = 1$ for all $t \geq 0$. Thus, some information can be lost in passing to the nonincreasing rearrangement of a function.

Remark 7.1.11. Let $f \in \mathcal{M}_0(\mathcal{R}, \mu)$. Then the nonincreasing rearrangement and the distribution function of f are connected in the following way:

$$\begin{aligned} f^*(\mu_f(\lambda)) &\leq \lambda, \quad \lambda \in (0, \infty) \\ \mu_f(f^*(t)) &\leq t, \quad t \in (0, \mu(\mathcal{R})). \end{aligned} \tag{7.1.8}$$

Indeed, let $\lambda \in (0, \infty)$ and define $t := \mu_f(\lambda)$. Then (7.1.6) implies that

$$f^*(\mu_f(\lambda)) = f^*(t) = \inf\{\lambda' : \mu_f(\lambda') \leq t = \mu_f(\lambda)\} \leq \lambda,$$

and the first inequality in (7.1.8) follows. To prove the other one, let $t \in [0, \mu(\mathcal{R}))$. Then, by (7.1.6), there exists a sequence $\lambda_n \downarrow \lambda$ with $\mu_f(\lambda_n) \leq t$. Consequently, using the right-continuity of μ_f (Proposition 7.1.4), we arrive at

$$\mu_f(f^*(t)) = \mu_f(\lambda) = \lim_{n \to \infty} \mu_f(\lambda_n) \leq t,$$

as desired.

Proposition 7.1.12. *Let $f, g \in \mathcal{M}_0(\mathcal{R}, \mu)$ and let $\{f_n\}_{n=1}^\infty$ be a sequence in $\mathcal{M}_0(\mathcal{R}, \mu)$. Let $\lambda \in \mathbb{R}$. Then f^* is a nonnegative, nonincreasing, right-continuous function on $[0, \infty)$ such that*

$$|g| \leq |f| \; \mu\text{-a.e. on } \mathcal{R} \quad \Rightarrow \quad g^*(t) \leq f^*(t), \quad t \in [0, \mu(\mathcal{R})), \tag{7.1.9}$$

$$(af)^* = |a| f^*, \tag{7.1.10}$$

$$|f| \leq \liminf_{n \to \infty} |f_n| \; \mu\text{-a.e. on } \mathcal{R} \quad \Rightarrow \quad f^* \leq \liminf_{n \to \infty} f_n^*. \tag{7.1.11}$$

In particular, one has

$$|f_n| \uparrow |f| \; \mu\text{-a.e. on } \mathcal{R} \quad \Rightarrow \quad f_n^* \uparrow f^*.$$

Proof. All the assertions follow immediately from Proposition 7.1.4 or from their corresponding counterparts in Proposition 7.1.4 and from the definition of the nonincreasing rearrangement. □

The operation of taking the nonincreasing rearrangement is not subadditive. Instead, we have the following result.

Proposition 7.1.13. *Let $f, g \in \mathcal{M}_0(\mathcal{R}, \mu)$ and let $s, t \in [0, \mu(\mathcal{R}))$ be such that $s + t \in [0, \mu(\mathcal{R}))$. Then*

$$(f + g)^*(s + t) \leq f^*(s) + g^*(t). \tag{7.1.12}$$

Proof. Set $\lambda := f^*(s) + g^*(t)$ and $y := \mu_{f+g}(\lambda)$. Then, by the triangle inequality and the second inequality in (7.1.8), we obtain

$$\begin{aligned}
y &= \mu\{x \in \mathcal{R}; \; |f(x) + g(x)| > f^*(s) + g^*(t)\} \\
&\leq \mu\{x \in \mathcal{R}; \; |f(x)| > f^*(s)\} + \mu\{x \in \mathcal{R}; \; |g(x)| > g^*(t)\} \\
&= \mu_f(f^*(s)) + \mu_f(g^*(t)) \\
&\leq s + t.
\end{aligned}$$

Thus, by the first of the inequalities in (7.1.8) and the monotonicity of $(f+g)^*$, we get

$$(f+g)^*(s+t) \le (f+g)^*(y) = (f+g)^*(\mu_{f+g}(\lambda))$$
$$\le \lambda = f^*(s) + g^*(t),$$

and the assertion follows. The proof is complete. □

Remark 7.1.14. It is worth pointing out that probably the most useful special case of (7.1.12) reads as

$$(f+g)^*(t) \le f^*(\tfrac{t}{2}) + g^*(\tfrac{t}{2}), \quad f,g \in \mathcal{M}_0(\mathcal{R},\mu), \ t \in (0,\mu(\mathcal{R})). \quad (7.1.13)$$

Proposition 7.1.15. *Let $f \in \mathcal{M}_0(\mathcal{R},\mu)$. Then the functions f and f^* are equimeasurable.*

Proof. There exists a sequence $\{f_n\}_{n=1}^\infty$ of nonnegative simple functions $f_n \uparrow f$. It follows from definitions that for each n the functions f_n and f_n^* are equimeasurable, that is,

$$\mu_{f_n}(\lambda) = \mu_{f_n^*}(\lambda), \quad \lambda \in (0,\infty). \quad (7.1.14)$$

We however know that $f_n \uparrow |f|$ and $f_n^* \uparrow f^*$ (see (7.1.11)). Thus, the assertion follows from (7.1.5). □

Exercise 7.1.16. Let $n \in \mathbb{N}$ and let $\{E_j\}_{j=1}^n$ be pairwise disjoint measurable subsets of $\mathcal{R}$ of finite μ-measure. Let further $\{\alpha_j\}_{j=1}^n$ be positive real numbers satisfying $\alpha_1 > \alpha_2 > \cdots > \alpha_n > 0$. Define

$$f(x) := \sum_{j=1}^n \alpha_j \chi_{E_j}(x), \quad x \in \mathcal{R}. \quad (7.1.15)$$

For $j = 1, \ldots, n$, we denote

$$\beta_j := \sum_{i=1}^j \mu(E_i) \quad (7.1.16)$$

and $\beta_0 := 0$. Then

$$f^*(t) = \sum_{j=1}^n \alpha_j \chi_{[\beta_{j-1},\beta_j]}(t), \quad t \in (0,\infty). \quad (7.1.17)$$

Proposition 7.1.17. *Let $f \in \mathcal{M}_0$. If $0 < p < \infty$, then*

$$\int_\mathcal{R} |f|^p \, d\mu = p \int_0^\infty \lambda^{p-1} \mu_f(\lambda) \, d\lambda = \int_0^\infty f^*(t)^p \, dt. \quad (7.1.18)$$

Furthermore, in the case $p = \infty$,

$$\operatorname*{ess\,sup}_{x \in \mathcal{R}} |f(x)| = \inf\{\lambda;\ \mu_f(\lambda) = 0\} = f^*(0). \tag{7.1.19}$$

Proof. Let f be an arbitrary nonnegative simple function. Then there exist $n \in \mathbb{N}$, pairwise disjoint measurable subsets $\{E_j\}_{j=1}^n$ of $\mathcal{R}$ of finite μ-measure and positive real numbers $\{\alpha_j\}_{j=1}^n$ satisfying $\alpha_1 > \alpha_2 > \cdots > \alpha_n > 0$ such that f obeys (7.1.15). By Exercise 7.1.16, f^* satisfies (7.1.17). So, (7.1.16) yields

$$\int |f(x)|^p \, d\mu(x) = \sum_{j=1}^n \alpha_j^p \mu(E_j) = \sum_{j=1}^n \alpha_j^p \mu([\beta_{j-1}, \beta_j)) = \int_0^\infty f^*(t)^p \, dt.$$

Similarly, one gets by (7.1.16) and the summation by parts,

$$p \int_0^\infty \lambda^{p-1} \mu_f(\lambda) \, d\lambda = p \sum_{j=1}^n \int_{\alpha_{j+1}}^{\alpha_j} \lambda^{p-1} \mu_f(\lambda) \, d\lambda = \sum_{j=1}^n (\alpha_j^p - \alpha_{j+1}^p) \beta_j$$

$$= \sum_{j=1}^n \alpha_j^p \mu(E_j) = \int |f(x)|^p \, d\mu(x).$$

This shows the first part of the claim for an arbitrary nonnegative simple function f. The extension to a general f follows from (7.1.5), (7.1.11) and the monotone convergence theorem (Theorem 1.21.4). We have thus proved (7.1.18). The proof of (7.1.19) is even easier and hence omitted. □

7.2 Hardy–Littlewood inequality

Proposition 7.2.1. *Let $g \in \mathcal{M}_+(\mathcal{R}, \mu)$ and let $E \subset \mathcal{R}$ be μ-measurable. Then*

$$\int_E g \, d\mu \leq \int_0^{\mu(E)} g^*(s) \, ds. \tag{7.2.1}$$

Proof. Let $\{E_j\}_{j=1}^n$ be a finite sequence of μ-measurable subsets of $\mathcal{R}$ such that $E_1 \subset E_2 \subset \cdots \subset E_n$ and let $\{\alpha_j\}_{j=1}^n$ be real numbers. Let

$$g(x) = \sum_{j=1}^n \alpha_j \chi_{E_j}(x), \quad x \in \mathcal{R}.$$

Then it is not difficult to verify that

$$g^*(t) = \sum_{j=1}^n \alpha_j \chi_{[0, \mu(E_j))}(t), \quad t \in [0, \mu(\mathcal{R})).$$

Thus,

$$\int_E g \, d\mu = \sum_{j=1}^n \alpha_j \mu(E \cap E_j) \leq \sum_{j=1}^n \alpha_j \min(\mu(E), \mu(E_j))$$

$$= \sum_{j=1}^n \alpha_j \int_0^{\mu(E)} \chi_{(0,\mu(E_j))} \, ds = \int_0^{\mu(E)} g^*(s) \, ds,$$

establishing (7.2.1) for nonnegative simple functions. For arbitrary $g \in \mathcal{M}_+$ we apply the usual density argument. □

One of the most important theoretical tools in the theory of rearrangement-invariant spaces is the inequality of Hardy and Littlewood, which we now shall state and prove.

Theorem 7.2.2 (Hardy–Littlewood). *Let $f, g \in \mathcal{M}_0(\mathcal{R}, \mu)$. Then*

$$\int_{\mathcal{R}} |fg| \, d\mu \leq \int_0^\infty f^*(t) g^*(t) \, dt. \tag{7.2.2}$$

Proof. Let first f be a simple nonnegative function, that is, let $\{E_j\}_{j=1}^n$ be a finite sequence of μ-measurable subsets of $\mathcal{R}$ such that $E_1 \subset E_2 \subset \cdots \subset E_n$, let $\{\alpha_j\}_{j=1}^n$ be real numbers and let

$$f(x) = \sum_{j=1}^n \alpha_j \chi_{E_j}(x), \quad x \in \mathcal{R}.$$

Then, again,

$$f^*(t) = \sum_{j=1}^n \alpha_j \chi_{[0,\mu(E_j))}(t).$$

Let $g \in \mathcal{M}_+(\mathcal{R}, \mu)$. By Lemma 7.2.1, we obtain

$$\int_{\mathcal{R}} |fg| \, d\mu = \sum_{j=1}^n \alpha_j \int_{E_j} g \, d\mu \leq \sum_{j=1}^n \alpha_j \int_0^{\mu(E_j)} g^*(s) \, ds$$

$$= \int_0^\infty \sum_{j=1}^n \alpha_j \chi_{[0,\mu(E_j))} g^*(s) \, ds$$

$$= \int_0^\infty f^*(s) g^*(s) \, ds.$$

When $f \in \mathcal{M}_+(\mathcal{R}, \mu)$, we find a sequence $\{f_j\}_{j=1}^\infty$ such that $0 \leq f_j \nearrow f$ and use the monotone convergence theorem (Theorem 1.21.4). This shows the assertion for the case when f and g are nonnegative. The general case follows in turn because the right-hand side depends only of f^* and g^*, and hence on the absolute value of f and g. The proof is complete. □

Corollary 7.2.3. *Let $f, g \in \mathcal{M}_0(\mathcal{R}, \mu)$. Let $\tilde{g} \in \mathcal{M}_0(\mathcal{R}, \mu)$ be equimeasurable with g on $\mathcal{R}$. Then*

$$\int_{\mathcal{R}} |f\tilde{g}| \, d\mu \leq \int_0^\infty f^*(t) g^*(t) \, dt. \tag{7.2.3}$$

Proof. The assertion is an immediate consequence of Theorem 7.2.2 and the fact that $\tilde{g}^* = g^*$. □

7.3 Resonant measure spaces

Definition 7.3.1. A σ-finite measure space $(\mathcal{R}, \mu)$ is said to be *resonant* if, for each f and g in $\mathcal{M}_0(\mathcal{R}, \mu)$, one has

$$\int_0^\infty f^*(t) g^*(t) \, dt = \sup_{\tilde{g} \sim g} \int_{\mathcal{R}} |f\tilde{g}| \, d\mu, \tag{7.3.1}$$

and *strongly resonant* if for every $f, g \in \mathcal{M}_0(\mathcal{R}, \mu)$, there exists a function $\tilde{g}$ on $\mathcal{R}$ such that $\tilde{g} \sim g$ and

$$\int_{\mathcal{R}} |f\tilde{g}| \, d\mu = \int_0^\infty f^*(t) g^* \, dt. \tag{7.3.2}$$

Remark 7.3.2. Of course, strong resonance implies resonance. The converse is not true in general, as the following example shows.

Example 7.3.3. The measure space $([0, \infty), dx)$, endowed with the one-dimensional Lebesgue measure $d\mu(x) = dx$, is resonant. Let f be the function described in Remark 7.1.10, that is,

$$f(x) = 1 - \frac{1}{x+1}, \quad x \in [0, \infty).$$

Then, $f^*(t) \equiv 1$ for every $t \in [0, \infty)$. Let now $g = \chi_{[0,1)}$ and $\tilde{g} = \chi_{[1,2)}$. Then $g^* = \tilde{g}^* = \chi_{[0,1)}$. However, we obviously have

$$\int_0^\infty |f(x)\tilde{g}(x)| \, dx = \int_1^2 \left(1 - \tfrac{1}{x}\right) dx = 1 - \log 2 < 1,$$

while

$$\int_0^\infty f^*(t) g^*(t) \, dt = 1,$$

hence our measure space is not strongly resonant.

Remark 7.3.4. Not every measure space is resonant. A typical example of a nonresonant measure space is an atomic measure space having at least two atoms of nonequal measure. More precisely, let $a, b \in \mathcal{R}$ be such that $\mu(a) = \alpha$, $\mu(b) = \beta$ with

$0 < \alpha < \beta$. Let $f = \chi_{\{b\}}$ and $g = \chi_{\{a\}}$. Then it is clear that every function $\tilde{g}$ satisfying $\tilde{g} \sim g$ must obey $b \notin \mathrm{supp}\,\tilde{g}$. Therefore, for every such $\tilde{g}$, we have

$$\int_{\mathcal{R}} |f\tilde{g}|\, d\mu = 0.$$

On the other hand, $f^* = \chi_{[0,\beta)}$ and $g^* = \chi_{[0,\alpha)}$, hence

$$\int_0^\infty f^*(t)\, g^*(t) = \alpha.$$

Consequently, no such measure space can be resonant.

Notation 7.3.5. We shall denote by $\mathrm{Rng}(\mu)$ the range of the measure μ, that is,

$$\mathrm{Rng}(\mu) := \{t \in [0, \infty);\ \text{there exists } E \subset \mathcal{R} \text{ satisfying } \mu(E) = t\}.$$

We note that if μ is a nonatomic measure on $\mathcal{R}$, then

$$\mathrm{Rng}(\mu) = [0, \mu(\mathcal{R})]. \tag{7.3.3}$$

Lemma 7.3.6. *Let $(\mathcal{R}, \mu)$ be a nonatomic measure space such that $\mu(\mathcal{R}) < \infty$. Let $f \in \mathcal{M}_0(\mathcal{R}, \mu)$ and let $t \in [0, \mu(\mathcal{R})]$. Then there exists a μ-measurable set $E_t \subset \mathcal{R}$ satisfying $\mu(E_t) = t$ and such that*

$$\int_{E_t} |f|\, d\mu = \int_0^t f^*(s)\, ds. \tag{7.3.4}$$

Moreover,

$$0 \leq s \leq t \leq \mu(\mathcal{R}) \quad \Rightarrow \quad E_s \subset E_t. \tag{7.3.5}$$

Proof. First, assume that $t \in \mathrm{Rng}(\mu_f)$, that is, there exists $\alpha > 0$ such that $\mu_f(\alpha) = t$. Then

$$f^*(t) = \inf\{\lambda;\ \mu_f(\lambda) = t\}.$$

Since μ_f is right-continuous, this implies $\mu_f(f^*(t)) = t$. We then set

$$E_t := \{x;\ |f(x)| > f^*(t)\}.$$

Then $\mu(E_t) = t$. Moreover, one has

$$\mu_{f\chi_{E_t}}(\lambda) = \mu\{x \in E_t;\ |f(x)| > \lambda\} = \begin{cases} t & \text{if } \lambda \in [0, f^*(t)], \\ \mu_f(\lambda) & \text{if } \lambda \in (f^*(t), \infty). \end{cases}$$

On the other hand,

$$\mu_{f^*\chi_{[0,t)}}(\lambda) = \mu\{s \in [0,t];\ f^*(s) > \lambda\} = \begin{cases} t & \text{if } \lambda \in [0, f^*(t)], \\ \mu_{f^*}(\lambda) & \text{if } \lambda \in (f^*(t), \infty). \end{cases}$$

Section 7.3 Resonant measure spaces

By Proposition 7.1.15, $f \sim f^*$. Hence, $\mu_f = m_{f^*}$. In particular, we have $f_{\chi E_t} \sim f^* \chi_{[0,t)}$, and therefore also

$$\int_{E_t} |f| \, d\mu = \int_{\mathcal{R}} |f_{\chi E_t}| \, d\mu = \int_0^\infty f^*(s) \chi_{[0,t]}(s) \, ds = \int_0^t f^*(s) \, ds.$$

This establishes (7.3.4) and (7.3.5) in the case when $t \in \text{Rng}(\mu_f)$.

Now assume that $t \notin \text{Rng}(\mu(f))$. Since $\mu(\mathcal{R}) < \infty$ and $f \in \mathcal{M}_0$, the dominated convergence theorem (Theorem 1.21.5) and the monotonicity of the distribution function μ_f give

$$\lim_{\lambda \to \infty} \mu_f(\lambda) = \lim_{n \to \infty} \mu\{x;\ |f(x)| > n\} = \mu\{x;\ |f(x)| = \infty\} = 0. \quad (7.3.6)$$

We denote $\lambda_0 := f^*(t)$. Since $t \notin \text{Rng}(\mu_f)$, (7.1.8) and (7.1.6) imply that

$$t \in (\mu_f(\lambda_0), \mu_f(\lambda)) \quad \text{for every } \lambda \in (0, \lambda_0).$$

Next, let $t_1 := \lim_{s \to 0_-} \mu_f(\lambda_0 - s)$ (the limit exists due to the monotonicity of μ_f), and $t_0 := \mu_f(\lambda_0)$. Then we have

$$t_0 < t < t_1, \quad (7.3.7)$$

hence, also by (7.1.6),

$$f^*(s) = \lambda_0, \quad \text{for every } s \in (t_0, t_1). \quad (7.3.8)$$

We shall now prove that
$$t_1 = \mu\{x;\ |f(x)| \geq \lambda_0\}. \quad (7.3.9)$$

Define
$$F_n = \{x;\ |f(x)| < \lambda_0 - \frac{1}{n}\}, \ n \in \mathbb{N}.$$

Then $F_1 \supset F_2 \supset \ldots$ and, since $\mu(\mathcal{R})$ is finite, the dominated convergence theorem (Theorem 1.21.5) implies

$$\mu\{x \in \mathcal{R};\ |f(x)| \geq \lambda_0\} = \lim_{n \to \infty} \mu(F_n) = \lim_{n \to \infty} \mu_f\left(\lambda_0 - \frac{1}{n}\right) = \mu_f(\lambda_0-).$$

Again, the limit exists due to the monotonicity of μ_f.

We next denote $G := \{x \in \mathcal{R};\ |f(x)| = \lambda_0\}$. Then, by (7.3.7) and (7.3.9), one has $\mu(G_t) = t_1 - t_0$. Since $(\mathcal{R}, \mu)$ is nonatomic, there exists a set $F_t \subset G$ such that $\mu(F_t) = t - t_0$. We define

$$E_t := \{x \in \mathcal{R};\ |f(x)| < \lambda_0\} \cup F_t. \quad (7.3.10)$$

Then $\mu(E_t) = \mu_f(\lambda_0-) + (t - t_0) = t$. Next,

$$\int_{E_t} |f|\, d\mu = \int_{|f|>\lambda_0} |f|\, d\mu + \int_{F_t} |f|\, d\mu. \qquad (7.3.11)$$

It follows from (7.1.7) that $t_0 \in \mathrm{Rng}(\mu_f)$. Thus, by the result just proved, we have

$$\int_{|f|>\lambda_0} |f|\, d\mu = \int_0^{t_0} f^*(s)\, ds.$$

Furthermore, since $|f| = \lambda_0$ on F_t, one has, by (7.3.8),

$$\int_{F_t} |f|\, d\mu = \lambda_0 \mu(F_t) = \lambda_0(t - t_0) = \int_{t_0}^t f^*(s)\, ds.$$

This and (7.3.11) now imply (7.3.4).

Finally, assume that $\lambda_0 = 0$. Then we get, using an argument analogous to that from the proof of (7.3.7), that

$$\mu\{x \in \mathcal{R};\ |f(x)| > 0\} = t_0 < t.$$

There exists a set F_t satisfying $F_t \cap \mathrm{supp}\, f = \emptyset$ and $\mu(F_t) = t - t_0$. Let E_t be the set from (7.3.10). Since $f^*(t) = 0$ for $t \geq t_0$, we obtain

$$\int_{E_t} |f|\, d\mu = \int_0^{t_0} f^*(s)\, ds = \int_0^t f^*(s)\, ds,$$

and (7.3.4) follows, again.

It follows easily from the construction that it can be done so that (7.3.5) is satisfied. $\square$

Theorem 7.3.7. *A σ-finite measure space $(\mathcal{R}, \mu)$ is strongly resonant if and only if $\mu(\mathcal{R}) < \infty$ and $(\mathcal{R}, \mu)$ is either nonatomic or completely atomic with all atoms having the same measure.*

Proof. Let $f, g \in \mathcal{M}_0(\mathcal{R}, \mu)$ be nonnegative functions. It suffices to find a nonnegative function $\tilde{g}$ such that $\tilde{g} \sim g$ and

$$\int_{\mathcal{R}} f \tilde{g}\, d\mu = \int_0^\infty f^*(t) g^*(t)\, dt. \qquad (7.3.12)$$

This is an easy exercise when $(\mathcal{R}, \mu)$ is completely atomic with all atoms having the same measure, because then there are only a finite number of atoms.

Assume thus that $(\mathcal{R}, \mu)$ is nonatomic. Let $\{g_n\}_{n=1}^\infty \subset S$ be such that $0 \leq g_n \uparrow g$ μ-a.e. on $\mathcal{R}$.

Section 7.3 Resonant measure spaces

Let $n \in \mathbb{N}$, then there exist a $m \in \mathbb{N}$, measurable sets $F_1 \subset F_2 \subset \ldots \subset F_m$ and nonnegative numbers $\alpha_1, \alpha_2, \ldots, \alpha_m$ such that

$$g_n = \sum_{i=1}^{m} \alpha_i \chi_{F_i}.$$

By Lemma 7.3.6, there exist sets $E_1 \subset E_2 \subset \ldots \subset E_m$ satisfying $\mu(E_j) = \mu(F_j)$ and

$$\int_{E_j} f \, d\mu = \int_0^{\mu(F_j)} f^*(t) \, dt, \quad j = 1, 2, \ldots, m. \tag{7.3.13}$$

Set

$$\tilde{g}_n := \sum_{j=1}^{m} \alpha_j \chi_{E_j}.$$

Then,

$$g_n^* = \sum_{i=1}^{m} \alpha_i \chi_{[0,\mu(F_i))} = (\tilde{g}_n)^*,$$

that is, $g_n \sim \tilde{g}_n$. Furthermore, by (7.3.13),

$$\int_{\mathcal{R}} f \tilde{g}_n \, d\mu = \sum_{i=1}^{m} \alpha_i \int_{E_i} g \, d\mu = \sum_{i=1}^{m} \alpha_i \int_0^{\mu(F_i)} f^*(t) \, dt$$

$$= \int_0^{\infty} \sum_{i=1}^{m} \alpha_i \chi_{[0,\mu(F_i))} f^*(t) \, dt$$

$$= \int_0^{\infty} f^*(t) g_n^*(t) \, dt.$$

Since $g_n \uparrow$, it follows from Lemma 7.3.6 that the sets E_j in (7.3.13) can be constructed so that $E_n \uparrow$ in the sense of inclusion. Consequently, we get $\tilde{g}_n \uparrow$ and, in turn, by the monotone convergence theorem (Theorem 1.21.4), also (7.3.12). As mentioned above, this implies that $(\mathcal{R}, \mu)$ is strongly resonant.

As for the "only if" part of the assertion of the theorem, let $(\mathcal{R}, \mu)$ be resonant. Then the considerations given in Remark 7.3.4 show that if $(\mathcal{R}, \mu)$ has any atoms at all, then they must be of an identical measure and also that there cannot be a nontrivial atom and a nontrivial nonatomic part of $(\mathcal{R}, \mu)$ at the same time. Moreover, if $\mu(\mathcal{R}) = \infty$, then an example of the type pointed out in Remark 7.3.2 can be constructed ruling out strong resonance. □

Theorem 7.3.8. *A σ-finite measure space $(\mathcal{R}, \mu)$ is resonant if and only if it is either nonatomic or completely atomic with all atoms having equal measure.*

Proof. The "only if" part follows immediately from Remarks 7.3.2 and 7.3.4, as it was mentioned also in the proof of Theorem 7.3.7.

For the "if" part, let $f, g \in \mathcal{M}_+(\mathcal{R}, \mu)$. With no loss of generality assume that $\int_0^\infty f^*(t)g^*(t)\,dt > 0$, since otherwise (7.3.1) is trivially satisfied. Let $\{f_n\}_{n=1}^\infty$ and $\{g_n\}_{n=1}^\infty$ be two sequences of simple functions such that $\operatorname{supp} f_n \cup \operatorname{supp} g_n \subset R_n$, $n \in \mathbb{N}$, and $0 \le f_n \uparrow f$, $0 \le g_n \uparrow g$. Let, moreover, $\alpha \in \mathbb{R}$ be such that

$$0 < \alpha < \int_0^\infty f^*(t)g^*(t)\,dt. \tag{7.3.14}$$

Now, by (7.1.11), the monotone convergence theorem (Theorem 1.21.4) and (7.3.14), there exists $N \in \mathbb{N}$ such that

$$\alpha < \int_0^\infty f_N^*(t)g_N^*(t)\,dt. \tag{7.3.15}$$

The measure space (R_N, μ) is strongly resonant by Theorem 7.3.7. Thus, there exists a function $\overline{g} \ge 0$ on R_N such that $\overline{g} \sim g\chi_{R_N}$ and

$$\int_{R_N} f\overline{g}\,d\mu = \int_0^{\mu(R_N)} (f\chi_{R_N})^*(t)(g\chi_{R_N})^*(t)\,dt. \tag{7.3.16}$$

Set

$$\tilde{g}(x) := \begin{cases} \overline{g}(x) & \text{for } x \in R_N \\ 0 & \text{for } x \in \mathcal{R} \setminus R_N. \end{cases}$$

Since $f\chi_{R_N} \ge f_N$ and $g\chi_{R_N} \ge g_N$, (7.3.16) and (7.3.15) yield

$$\int_\mathcal{R} f\overline{g}\,d\mu > \alpha. \tag{7.3.17}$$

Hence $\tilde{g} \sim g$ and, by (7.3.17),

$$\int_\mathcal{R} f\tilde{g}\,d\mu \ge \int_\mathcal{R} fh\,d\mu > \alpha.$$

We have shown that, for each α satisfying (7.3.14) we can find a nonnegative function $\tilde{g}$ such that $\tilde{g} \sim g$ and

$$\int_\mathcal{R} f\tilde{g}\,d\mu > \alpha. \tag{7.3.18}$$

In other words, $(\mathcal{R}, \mu)$ is resonant. $\square$

The following result is an immediate consequence of Theorems 7.3.7 and 7.3.8.

Corollary 7.3.9. *A σ-finite measure space $(\mathcal{R}, \mu)$ is strongly resonant if and only if it is resonant and $\mu(\mathcal{R}) < \infty$.*

7.4 Maximal nonincreasing rearrangement

Let E be a μ-measurable subset of $\mathcal{R}$ satisfying $\mu(E) = t \in (0, \infty)$. Then, taking $g = \chi_E$ and inserting g into the Hardy–Littlewood inequality (7.2.2), we get

$$\frac{1}{\mu(E)} \int_E |f| \, d\mu \le \frac{1}{t} \int_0^t f^*(s) \, ds, \quad f \in \mathcal{M}_0(\mathcal{R}, \mu). \tag{7.4.1}$$

The last expression, that is, the integral average of f^* over the integral $(0, t)$, turns out to be of great importance. In this section we shall study its basic properties.

Definition 7.4.1. For $f \in \mathcal{M}_0(\mathcal{R}, \mu)$, we define the function f^{**} on $(0, \mu(\mathcal{R}))$ by

$$f^{**}(t) := \frac{1}{t} \int_0^t f^*(s) \, ds, \quad t \in (0, \mu(\mathcal{R})). \tag{7.4.2}$$

Remark 7.4.2. Let $f \in \mathcal{M}_0(\mathcal{R}, \mu)$. Then

$$f^*(t) \le f^{**}(t), \quad t \in (0, \mu(\mathcal{R})). \tag{7.4.3}$$

Indeed, let $t \in (0, \mu(\mathcal{R}))$. Then, by the monotonicity of f^*, we obtain

$$f^{**}(t) = \frac{1}{t} \int_0^t f^*(s) \, ds \ge \frac{f^*(t)}{t} \int_0^t ds = f^*(t).$$

Proposition 7.4.3. *Let $f \in \mathcal{M}_0(\mathcal{R}, \mu)$, then f^{**} is a nonnegative, nonincreasing and continuous function on $(0, \mu(\mathcal{R}))$.*

Proof. It follows from the definition of $f^{**}(t)$ and the monotonicity of f^* that either $f^{**}(t) < \infty$ for every $t \in (0, \mu(\mathcal{R}))$ or $f^{**}(t) = \infty$ for every $t \in (0, \mu(\mathcal{R}))$. Obviously, in any case, f^{**} is nonnegative and continuous. It is also nonincreasing, being an integral mean of a nonincreasing function. More precisely, let $0 < s < t < \mu(\mathcal{R})$. Then, by (7.4.3),

$$\begin{aligned} f^{**}(t) &= \frac{1}{t} \int_0^t f^*(y) \, dy \\ &= \frac{1}{t} \int_0^s f^*(y) \, dy + \frac{1}{t} \int_s^t f^*(y) \, dy \\ &\le \frac{s}{t} f^{**}(s) + \frac{f^*(s)}{t} \int_s^t dy \\ &= \frac{s}{t} f^{**}(s) + \left(1 - \frac{s}{t}\right) f^*(s) \\ &\le \frac{s}{t} f^{**}(s) + \left(1 - \frac{s}{t}\right) f^{**}(s) \\ &= f^{**}(s), \end{aligned}$$

as desired. The proof is complete. □

Proposition 7.4.4. *For every pair of functions $f, g \in \mathcal{M}_0(\mathcal{R}, \mu)$, every sequence $\{f_n\}_{n=1}^{\infty}$ of functions from $\mathcal{M}_0(\mathcal{R}, \mu)$ and every $a \in \mathbb{R}$, we have*

$$f^{**} \equiv 0 \iff f = 0 \ \mu\text{-a.e.} \tag{7.4.4}$$

$$0 \leq g \leq f \ \mu\text{-a.e.} \implies g^{**} \leq f^{**} \tag{7.4.5}$$

$$(af)^{**} = |a| f^{**} \tag{7.4.6}$$

$$0 \leq f_n \uparrow f \ \mu\text{-a.e.} \implies f_n^{**} \uparrow f^{**}. \tag{7.4.7}$$

Proof. All the properties of f^{**} easily follow from the corresponding ones of f^* (see Proposition 7.1.12). □

Proposition 7.4.5. *Let $t \in (0, \mu(\mathcal{R})) \cap \text{Rng}(\mu)$ and let $f \in \mathcal{M}_0(\mathcal{R}, \mu)$.*

(i) *If $(\mathcal{R}, \mu)$ is resonant, then*

$$f^{**}(t) = \frac{1}{t} \sup \left\{ \int_E |f| \, d\mu; \ \mu(E) = t \right\}. \tag{7.4.8}$$

(ii) *If $(\mathcal{R}, \mu)$ is strongly resonant, then there is a subset E of $\mathcal{R}$ with $\mu(E) = t$ such that*

$$f^{**}(t) = \frac{1}{t} \int_E |f| \, d\mu. \tag{7.4.9}$$

Proof. Let $F \subset \mathcal{R}$ be such that $\mu(F) = t$. Define $g = \chi_F$. Then $g^* = \chi_{[0,t)}$. Therefore, if a function $\tilde{g}$ is equimeasurable with g, then there exists a set $E \subset \mathcal{R}$ such that $\mu(E) = t$ and $|\tilde{g}| = \chi_E$. Both assertions thus follow from (7.3.1) and (7.3.2). □

Corollary 7.4.6. *Let $(\mathcal{R}, \mu)$ be a resonant measure space. Let $f, g \in \mathcal{M}_0(\mathcal{R}, \mu)$ and let $t \in (0, \mu(\mathcal{R})) \cap \text{Rng}(\mu)$. Then*

$$(f + g)^{**}(t) \leq f^{**}(t) + g^{**}(t). \tag{7.4.10}$$

Proof. The assertion follows immediately from (7.4.8) and the subadditivity of the supremum. □

Remark 7.4.7. The subadditivity of the functional $f \mapsto f^{**}$, claimed in Corollary 7.4.6, holds for arbitrary measure spaces, not only for the resonant ones. This can be proved by the so-called *method of retracts*. The details can be found in [14, p. 54].

Example 7.4.8. Unlikely the operation $f \mapsto f^{**}$, the functional $f \mapsto f^*$ is not subadditive, as simple examples show (for instance, take $\mathcal{R} = [0, \infty)$, let μ be the one-dimensional Lebesgue measure, and set $f = \chi_{[0,1)}$ and $g = \chi_{[1,2)}$). The best we can hope for in the case of $f \mapsto f^*$ is the considerably weaker property (7.1.12).

7.5 Hardy lemma

As we shall see in the following section, the relation $f^{**} \leq g^{**}$ between two given functions $f, g \in \mathcal{M}_0(\mathcal{R}, \mu)$, is of a crucial importance. Note that it is weaker than the pointwise estimate $f^* \leq g^*$. We shall now study this relation in detail.

Definition 7.5.1. We say that the functions $f, g \in \mathcal{M}_0(\mathcal{R}, \mu)$ are in the Hardy–Littlewood–Pólya relation, and write $f \prec g$, if, for every $t \in [0, \mu(\mathcal{R}))$, one has

$$\int_0^t f^*(s)\,ds \leq \int_0^t g^*(s)\,ds. \tag{7.5.1}$$

We shall now formulate the Hardy lemma, an important technical background result which will be very useful in what follows.

Proposition 7.5.2 (Hardy lemma). *Let $a \in (0, \infty]$, let f and g be two nonnegative measurable functions on $(0, a)$ and suppose that*

$$\int_0^t f(s)\,ds \leq \int_0^t g(s)\,ds \tag{7.5.2}$$

for all $t \in (0, a)$. Let h be a nonnegative nonincreasing function on $(0, a)$. Then

$$\int_0^a f(t)h(t)\,dt \leq \int_0^a g(t)h(t)\,dt. \tag{7.5.3}$$

Proof. Assume first that h is of the form

$$h(t) = \sum_{j=1}^n c_j \chi_{(0,t_j)}(t),$$

where $c_j \in (0, \infty)$ and $0 < t_1 < \ldots < t_n < a$. Then, by (7.5.2), we obtain

$$\int_0^a f(t)h(t)\,dt = \sum_{j=1}^n c_j \int_0^{t_j} f(t)\,dt \leq \sum_{j=1}^n c_j \int_0^{t_j} g(t)\,dt = \int_0^a g(t)h(t)\,dt,$$

proving the claim in the special case. For a general function, the assertion follows by using the monotone convergence theorem (Theorem 1.21.4) and the property (P3) from Definition 6.1.5. The proof is complete. □

Theorem 7.5.3. *Let $(\mathcal{R}, \mu)$ be a resonant measure space, let I be a countable index set and let $\{E_i\}_{i \in I}$ be a sequence of pairwise disjoint subsets of $\mathcal{R}$ satisfying $0 < \mu(E_i) < \infty$ for every $i \in I$. Define*

$$E := \mathcal{R} \setminus \bigcup_{i \in I} E_i.$$

Assume that $f \in \mathcal{M}_0(\mathcal{R}, \mu)$ and $\int_{E_i} f \, d\mu < \infty$ for every $i \in I$. We define the function Af by

$$Af := f\chi_E + \sum_{i \in I} \left(\frac{1}{\mu(E_i)} \int_{E_i} f \, d\mu \right) \chi_{E_i}. \tag{7.5.4}$$

Then

$$Af \prec f.$$

Proof. If I has only one element E_1, then

$$Af = f\chi_E + \left(\frac{1}{\mu(E_1)} \int_{E_1} f \, d\mu \right) \chi_{E_1}.$$

We want to show that

$$\int_0^t (Af)^*(s) \, ds \leq \int_0^t f^*(s) \, ds, \quad t \in (0, a). \tag{7.5.5}$$

Suppose first that $t \in (0, a) \cap \text{Rng}(\mu)$. Then there is some $F \subset \mathcal{R}$ such that $\mu(F) = t$. Define

$$t_0 := \mu(F \cap E_1).$$

Then

$$\int_F |Af| \, d\mu = \int_{F \cap E} |f| \, d\mu + t_0 \left| \frac{1}{\mu(E_1)} \int_{E_1} f \, d\mu \right|. \tag{7.5.6}$$

By (7.4.1) and the monotonicity of the function $(f\chi_{E_1})^{**}$, we get

$$t_0 \left| \frac{1}{\mu(E_1)} \int_{E_1} f \, d\mu \right| \leq t_0 (f_{E_1})^{**}(\mu(E_1)) \leq \int_0^{t_0} (f_{E_1})^*(s) \, ds. \tag{7.5.7}$$

Now, $(\mathcal{R}, \mu)$ is resonant and moreover $\mu(E_1) < \infty$. Thus, denoting the restriction of μ to E_1 by μ again, we conclude, using Corollary 7.3.9, that the measure space (E_1, μ) is strongly resonant. Next, $\mu(F \cap E_1) = t_0$, hence, in particular, $t_0 \in \text{Rng}(\mu)$. Proposition 7.4.5 (ii) now guarantees the existence of some $G \subset E_1$ such that $\mu(G) = t_0$ and

$$\int_0^{t_0} (f\chi_{E_1})^*(s) \, ds = \int_G |f\chi_{E_1}| \, d\mu = \int_G |f| \, d\mu.$$

Combining this with (7.5.7) and (7.5.6), we obtain

$$\int_F |Af| \, d\mu \leq \int_{F \cap E_1} |f| \, d\mu + \int_G |f| \, d\mu. \tag{7.5.8}$$

But $F \cap E_1$ and G are disjoint and

$$\mu((F \cap E_1) \cup G) = \mu(F \cap E_1) + \mu(G) = (t - t_0) + t_0 = t,$$

so applying (7.4.1) to the right-hand side of (7.5.8) we obtain

$$\int_F |Af|\,d\mu \le \int_0^t f^*(s)\,ds.$$

Finally, taking the supremum over all sets F of measure t and applying Proposition 7.4.5 (i), we obtain (7.5.5), at least for all t in the range of μ. Since $(\mathcal{R}, \mu)$ is of one of the types described in Theorem 7.3.8, however, it is clear that (7.5.5) must then hold for all $t > 0$.

Now let us consider the case when I is an arbitrary finite index set. In such cases, it suffices to iterate the result just proved.

If I is countable and infinite, we find a sequence $\{f_n\}_{n=1}^\infty$ satisfying $0 \le f_n \uparrow f$ μ-a.e. and $0 \le Af_n \uparrow Af$ μ-a.e. Then

$$Af_n = A_n f_n \prec f_n \prec f, \quad n \in \mathbb{N},$$

where

$$A_n f := f \chi_{R_n} + \sum_{m=1}^n \left(\frac{1}{\mu(E_m)} \int_{E_m} f\,d\mu\right) \chi_{E_m},$$

and

$$R_n := R \setminus \bigcup_{m=1}^n E_m.$$

and the result follows from (7.4.7). The proof is complete. □

7.6 Rearrangement-invariant spaces

We throughout this section assume that $(\mathcal{R}, \mu)$ is resonant.

Definition 7.6.1. Let ϱ be a Banach function norm over $(\mathcal{R}, \mu)$. We say that ϱ is *rearrangement-invariant* (r.i.) if

$$\varrho(f) = \varrho(g)$$

for every pair of equimeasurable functions $f, g \in \mathcal{M}_0^+(\mathcal{R}, \mu)$. In such cases, the corresponding Banach function space $X = X(\varrho)$ is then said to be a *rearrangement-invariant Banach function space*.

Remark 7.6.2. A Banach function space X is rearrangement-invariant if and only if, whenever f belongs to X and g is equimeasurable with f, then g also belongs to X and $\|g\|_X = \|f\|_X$.

Example 7.6.3. Assume that $p \in [1, \infty]$. Then the Lebesgue space $L^p(\mathcal{R}, \mu)$ is a rearrangement-invariant Banach function space. Indeed, this follows from Proposition 7.1.17.

Remark 7.6.4. In many situations, a natural choice of a function equimeasurable with a given function f is its nonincreasing rearrangement f^*.

Proposition 7.6.5. *Let ϱ be a rearrangement-invariant function norm over $(\mathcal{R}, \mu)$. Then the associate norm ϱ' is also rearrangement-invariant. Moreover, we have*

$$\varrho'(g) = \sup\left\{\int_0^{\mu(\mathcal{R})} f^*(t)g^*(t)\,dt;\ \varrho(f) \leq 1\right\}, \quad g \in \mathcal{M}_0^+, \tag{7.6.1}$$

and

$$\varrho(f) = \sup\left\{\int_0^{\mu(\mathcal{R})} f^*(t)g^*(t)\,dt;\ \varrho'(g) \leq 1\right\}, \quad f \in \mathcal{M}_0^+. \tag{7.6.2}$$

Proof. We know from Definition 6.2.1 that

$$\varrho'(g) = \sup\left\{\int_{\mathcal{R}} |fg|\,d\mu;\ \varrho(f) \leq 1\right\}, \quad g \in \mathcal{M}_0^+. \tag{7.6.3}$$

First, it follows immediately from the Hardy–Littlewood inequality (7.2.2) that

$$\varrho'(g) \leq \sup\left\{\int_0^{\mu(\mathcal{R})} f^*(t)g^*(t)\,dt;\ \varrho(f) \leq 1\right\}, \quad g \in \mathcal{M}_0^+.$$

To prove the converse inequality, note that the rearrangement-invariance of ϱ guarantees that $\varrho(\tilde{f}) = \varrho(f)$ for every $\tilde{f}$ equimeasurable with f. Thus, thanks to the resonance of $(\mathcal{R}, \mu)$, we have, for every $g \in \mathcal{M}_0(\mathcal{R}, \mu)$,

$$\sup\left\{\int_0^{\mu(\mathcal{R})} f^*(t)g^*(t)\,dt;\ \varrho(f) \leq 1\right\} = \sup\left\{\int_{\mathcal{R}} |\tilde{f}g|\,d\mu;\ \tilde{f} \sim f,\ \varrho(f) \leq 1\right\}$$

$$\geq \sup\left\{\int_{\mathcal{R}} |fg|\,d\mu;\ \varrho(f) \leq 1\right\}.$$

Altogether, this establishes (7.6.1). Finally, applying the already proved assertion to ϱ'' in place of ϱ' and using Theorem 6.2.9, we get (7.6.2). The proof is complete. □

Theorem 7.6.6 (Hölder inequality). *Let ϱ be a rearrangement-invariant norm over a resonant measure space $(\mathcal{R}, \mu)$. If f and g belong to $\mathcal{M}_0^+(\mathcal{R}, \mu)$, then*

$$\int_{\mathcal{R}} fg\,d\mu \leq \int_0^{\mu(\mathcal{R})} f^*(s)g^*(s)\,ds \leq \varrho(f)\varrho'(g). \tag{7.6.4}$$

Proof. The assertion follows immediately from Theorem 7.2.2 and Proposition 7.6.5. □

It will be useful to formulate also the "space-versions" of the results just obtained for norms.

Corollary 7.6.7. *Let X be a Banach function space over a resonant measure space. Then X is rearrangement-invariant if and only if the associate space X' is rearrangement-invariant. Furthermore, we have*

$$\|g\|_{X'} = \sup\left\{\int_0^{\mu(\mathcal{R})} f^*(t)g^*(t)\,dt;\ \|f\|_X \le 1\right\}, \quad g \in X', \tag{7.6.5}$$

and

$$\|f\|_X = \sup\left\{\int_0^{\mu(\mathcal{R})} f^*(t)g^*(t)\,dt;\ \|g\|_{X'} \le 1\right\}, \quad f \in X. \tag{7.6.6}$$

Corollary 7.6.8 (Hölder inequality). *Let X be a rearrangement-invariant space over a resonant measure space $(\mathcal{R}, \mu)$. If $f \in X$ and $g \in X'$, then*

$$\int_{\mathcal{R}} |fg|\,d\mu = \int_0^{\mu(\mathcal{R})} f^*(t)g^*(t)\,dt \le \|f\|_X \|g\|_{X'}. \tag{7.6.7}$$

Exercise 7.6.9. Let $\Omega \subset \mathbb{R}^N$ be a suitable domain.

(i) Let $1 \le p \le \infty$. Then the Lebesgue space $L^p(\Omega)$ is a rearrangement-invariant Banach function space.

(ii) Let $1 \le p \le \infty$ and let ϱ be a weight on Ω. Then the weighted Lebesgue space $L^p(\Omega, \varrho)$ (cf. Remark 3.19.8) is a Banach function space which in general is not rearrangement-invariant (unless the rearrangement is taken with respect to the measure $\varrho(x)\,dx$).

(iii) Let Φ be a Young function. Then the Orlicz space $L^\Phi(\Omega)$ is a rearrangement-invariant Banach function space.

(iv) Let $\lambda \ge 0$ and $p \in (1, \infty)$. Then the Morrey space $L_M^{p,\lambda}(\Omega)$ is a Banach function space which is not rearrangement-invariant.

We shall see more examples of which is not rearrangement-invariant Banach function spaces in subsequent chapters.

7.7 Hardy–Littlewood–Pólya principle

We shall now state and prove one of the key theoretical results of the theory of rearrangement-invariant Banach function spaces. It also illustrates the significance of the Hardy–Littlewood–Pólya relation, introduced in Definition 7.5.1.

Theorem 7.7.1 (Hardy–Littlewood–Pólya principle). *Let $(\mathcal{R}, \mu)$ be a resonant measure space. Suppose that the functions $f, g \in \mathcal{M}_0^+(\mathcal{R}, \mu)$ satisfy $f \prec g$. Let X be any rearrangement-invariant Banach function space over $(\mathcal{R}, \mu)$. Then*

$$\|f\|_X \leq \|g\|_X.$$

Proof. Let $h \in \mathcal{M}_0(\mathcal{R}, \mu)$ be such that $\|h\|_{X'} \leq 1$. Then, since h^* is nonincreasing on $(\mathcal{R}, \mu)$, we obtain from the Hardy lemma (Proposition 7.5.2)

$$\int_0^{\mu(\mathcal{R})} f^*(t) h^*(t)\, dt \leq \int_0^{\mu(\mathcal{R})} g^*(t) h^*(t)\, dt.$$

The assertion now follows from (7.6.2). □

Theorem 7.7.1 has many important consequences. One of them asserts that the averaging operator A introduced in (7.5.4) is bounded with norm equal to 1 on every rearrangement-invariant Banach function space.

Theorem 7.7.2. *Let $(\mathcal{R}, \mu)$ be a resonant measure space, let I be a countable index set and let $\{E_i\}_{i \in I}$ be a sequence of pairwise disjoint subsets of $\mathcal{R}$ such that $0 < \mu(E_i) < \infty$ for every $i \in I$. Let $E := \mathcal{R} \setminus \bigcup_{i \in I} E_i$. Let the operator A be defined at every $g \in \mathcal{M}_0(\mathcal{R}, \mu)$ by*

$$Ag := g\chi_E + \sum_{j \in I} \left(\frac{1}{\mu(E_i)} \int_{E_i} g\, d\mu \right) \chi_{E_i}.$$

Then

$$\|Ag\|_X \leq \|g\|_X, \quad g \in X.$$

Proof. Assume that $f \in X$. Then also $f \in L^1(E_i)$, $i \in I$, thanks to Remark 6.1.16 (vi). By Theorem 7.5.3, we have $Af \prec f$. Thus, the assertion follows from the Hardy–Littlewood–Pólya principle (Theorem 7.7.1). □

7.8 Luxemburg representation theorem

In this section it will be convenient to have a special symbol for the one-dimensional Lebesgue measure.

Notation 7.8.1. We shall denote by μ_1 the one-dimensional Lebesgue measure.

Theorem 7.8.2. *Assume that λ is a rearrangement-invariant norm over $([0, \infty), \mu_1)$. Let $(\mathcal{R}, \mu)$ be an arbitrary σ-finite measure space. Then the functional ϱ, defined by*

$$\varrho(g) := \lambda(g^*), \quad g \in \mathcal{M}_0^+(\mathcal{R}, \mu), \tag{7.8.1}$$

is a rearrangement-invariant Banach function norm over $(\mathcal{R}, \mu)$.

Section 7.8 Luxemburg representation theorem

Proof. Suppose that $f, g \in \mathcal{M}_0^+(\mathcal{R}, \mu)$. By (7.4.10) and the fact that $(f^*)^* = f^*$, we have
$$(f + g)^{**} \leq f^{**} + g^{**} = (f^* + g^*)^{**}.$$
Thus, $(f + g)^* \prec f^* + g^*$. Now, λ is a norm on $([0, \infty), \mu_1)$, which is resonant. Thus, by Theorem 7.7.1 and the triangle inequality for λ, we obtain
$$\lambda((f + g)^*) \leq \lambda(f^* + g^*) \leq \lambda(f)^* + \lambda(g^*).$$

By (7.8.1), we get the triangle inequality for ϱ. All the other norm properties of ϱ required by the axiom (P1) of Definition 6.1.5 are clearly satisfied. The axioms (P2), (P3), (P4) and (P5) for ϱ follow immediately from their corresponding counterparts for λ. The proof is complete. □

The preceding theorem describes a simple procedure in which a rearrangement-invariant Banach function norm arises from a given norm over the interval $[0, \infty)$ endowed with the one-dimensional Lebesgue measure μ_1. We shall now state and prove the Luxemburg representation theorem – a key result which shows that, in fact, *every* such norm is obtained this way as long as the underlying measure space is resonant.

Theorem 7.8.3 (Luxemburg representation theorem). *Let X be a rearrangement-invariant Banach function space over a resonant measure space $(\mathcal{R}, \mu)$.*

(i) *There exists a (not necessarily unique) rearrangement-invariant Banach function space $\overline{X}$ over $([0, \mu(\mathcal{R})), \mu_1)$, called the representation space of X, such that*
$$\|g\|_X = \|g^*\|_{\overline{X}}, \quad g \in \mathcal{M}_0^+(\mathcal{R}, \mu). \tag{7.8.2}$$

(ii) *If $\overline{X}$ is any rearrangement-invariant function space over $([0, \mu(\mathcal{R})), \mu_1)$ which represents X in the sense of (7.8.2), then*
$$\|g\|_{X'} = \|g^*\|_{\overline{X}'}, \quad g \in \mathcal{M}_0^+(\mathcal{R}, \mu). \tag{7.8.3}$$

Proof. (i) We define the norm $\|\cdot\|_{\overline{X}}$ at every $h \in \mathcal{M}_0([0, \mu(\mathcal{R})), \mu_1)$ by
$$\|h\|_{\overline{X}} := \sup \left\{ \int_0^{\mu(\mathcal{R})} g^*(t) h^*(t) \, dt; \ \|g\|_{X'} \leq 1 \right\}. \tag{7.8.4}$$

Then, by (7.6.2), we get (7.8.2).

Let $h_1, h_2 \in \mathcal{M}_0([0, \mu(\mathcal{R})), \mu_1)$. Then, as in the proof of the preceding theorem, we obtain
$$(h_1 + h_2)^* \prec h_1^* + h_2^*.$$

Hence, using the Hardy lemma (Proposition 7.5.2) and (7.8.4), we obtain

$$\|h_1 + h_2\|_{\overline{X}} \leq \|h_1\|_{\overline{X}} + \|h_2\|_{\overline{X}},$$

and the triangle inequality for $\|\cdot\|_{\overline{X}}$ follows. Other norm properties required by (P1) are easy to verify. The lattice property (P2) follows from (7.1.9) and (7.8.4). The Fatou property (P3) is guaranteed by (7.1.11), (7.8.4), and the monotone convergence theorem. The rearrangement-invariance of $\|\cdot\|_{\overline{X}}$ is obvious from the definition. Let $t \in [0, \mu(\mathcal{R})) \cap \mathrm{Rng}(\mu)$. Then there exists a set $E \subset \mathcal{R}$ such that $\mu(E) = t$. Thus, using (7.8.2), we get

$$\|\chi_{[0,t]}\|_{\overline{X}} = \|\chi_E\|_X < \infty$$

by (P4) for $\|\cdot\|_X$. Using the (already proved) triangle inequality for $\|\cdot\|_{\overline{X}}$ and the mathematical induction, we get

$$\|\chi_{[0,kt]}\|_{\overline{X}} < \infty, \quad k \in \mathbb{N}.$$

So, if now $t \in [0, \mu(\mathcal{R}))$ is arbitrary (not necessarily in $\mathrm{Rng}(\mu)$), it suffices to use (P2) to show that

$$\|\chi_{[0,t]}\|_{\overline{X}} < \infty.$$

This establishes (P4). In a similar way we can prove (P5). Altogether, the functional $\|\cdot\|_{\overline{X}}$ is a rearrangement-invariant function norm over $([0, \mu(\mathcal{R})), \mu_1)$.

(ii) Assume that $\overline{X}$ is any rearrangement-invariant Banach function space over $([0, \mu(\mathcal{R})), \mu_1)$ representing ϱ in the sense of (7.8.2). By Theorem 7.3.8, the measure space $([0, \mu(\mathcal{R})), \mu_1)$ is resonant. Thus, in view of (7.6.1), we have

$$\|g\|_{\overline{X}'} = \sup\left\{\int_0^{\mu(\mathcal{R})} f^*(t)g^*(t)\,dt;\ \|f\|_{\overline{X}} \leq 1\right\}, \quad g \in \mathcal{M}_0^+(\mathcal{R}, \mu).$$

In particular, with g replaced by g^*, we have

$$\|g^*\|_{\overline{X}'} = \sup\left\{\int_0^{\mu(\mathcal{R})} f^*(t)g^*(t)\,dt;\ \|f\|_{\overline{X}} \leq 1\right\}, \quad g \in \mathcal{M}_0^+(\mathcal{R}, \mu). \tag{7.8.5}$$

Hence, it follows from (7.8.2) and (7.8.3) that

$$\|g\|_{X'} \leq \|g^*\|_{\overline{X}'}, \quad g \in \mathcal{M}_0^+(\mathcal{R}, \mu). \tag{7.8.6}$$

Assume that

$$f(x) = \sum_{j=1}^k c_j \chi_{(0,t_j)}, \tag{7.8.7}$$

where $k \in \mathbb{N}$, $c_j \in (0, \infty)$, $j \in \mathbb{N}$ and $0 < t_1 < \cdots < t_k$ (in other words, f is a nonincreasing simple function). It follows from the Fatou property (P3) for the norm $\|\cdot\|_{\overline{X}}$ that, for every $g \in \mathcal{M}_0^+(\mathcal{R}, \mu)$, one has

$$\|g^*\|_{\overline{X}'} = \sup\left\{\int_0^{\mu(\mathcal{R})} f^*(t) g^*(t)\, dt;\ f \text{ is of the form (7.8.7)},\ \|f\|_{\overline{X}} \le 1\right\}.$$

Now, if $(\mathcal{R}, \mu)$ is nonatomic and f is of the form (7.8.7), then it is the nonincreasing rearrangement of some simple function on $\mathcal{R}$. We thus get

$$\|g^*\|_{\overline{X}'} \le \|g\|_{X'},\quad g \in \mathcal{M}_0^+(\mathcal{R}, \mu),$$

which together with (7.8.6) establishes (7.8.3).

In the remaining case when $(\mathcal{R}, \mu)$ is completely atomic measure space with all atoms having the same positive measure, say α (cf. Theorem 7.3.8), the functions f of the form (7.8.7) may not necessarily be constant on the intervals of the form $I_k := [(k-1)\alpha, k\alpha)$, where $k \in \mathbb{N}$, so they may not be representable as nonincreasing rearrangements of simple functions on $(\mathcal{R}, \mu)$. We then replace every such f by

$$\tilde{f} = \sum_{k \in \mathbb{N}} \chi_{I_k} \frac{1}{\mu(I_k)} \int_{I_k} f.$$

Then

$$\int_0^{\mu(\mathcal{R})} f^*(t) g^*(t)\, dt = \int_0^{\mu(\mathcal{R})} \tilde{f}^*(t) g^*(t)\, dt$$

and, by Theorem 7.7.2,

$$\|\tilde{f}\|_{\overline{X}} \le \|f\|_{\overline{X}}.$$

The rest of the proof (with f replaced by $\tilde{f}$) is the same as in the nonatomic case. □

Remark 7.8.4. The Luxemburg representation theorem reduces rearrangement-invariant spaces to their representations on an interval and thereby it enables us to study one-dimensional analogues of rearrangement-invariant spaces without having to deal with intrinsic difficulties connected with the underlying measure spaces. This representation is not always unique. The uniqueness of the representation is however guaranteed in the cases when $(\mathcal{R}, \mu)$ is nonatomic and such that $\mu(\mathcal{R}) = \infty$.

7.9 Fundamental function

In this section we shall study a very important characteristic of a given r.i. space, namely the fundamental function.

Convention 7.9.1. All r.i. spaces in this section are supposed to be over a resonant measure space $(\mathcal{R}, \mu)$.

Definition 7.9.2. Let X be a rearrangement-invariant Banach function space. Assume that $t \in \mathrm{Rng}(\mu)$. We then define the function

$$\varphi_X : \mathrm{Rng}(\mu) \to [0, \infty)$$

by

$$\varphi_X(t) := \|\chi_E\|_X, \quad t \in \mathrm{Rng}(\mu). \tag{7.9.1}$$

We say that the function φ_X is the *fundamental function* of the space X.

Remark 7.9.3. We have to make sure that the fundamental function is well-defined by (7.9.1) for every rearrangement-invariant Banach function space X. To this end, note that the value of $\varphi_X(t)$ does not depend on the choice of the set E. Indeed, if $\tilde{E}$ is another such set, then

$$(\chi_E)^* = (\chi_{\tilde{E}})^* = \chi_{[0,\mu(E))},$$

hence

$$\|\chi_E\|_X = \|\chi_{\tilde{E}}\|_X.$$

Example 7.9.4. Assume that $(\mathcal{R}, \mu)$ is nonatomic.

(i) Let $X = L^p(\mathcal{R}, \mu)$ with $p \in [1, \infty)$. Then

$$\varphi_X(t) = t^{\frac{1}{p}}, \quad t \in [0, \mu(\mathcal{R})). \tag{7.9.2}$$

(ii) Let $X = L^\infty(\mathcal{R}, \mu)$. Then

$$\varphi_X(t) = \begin{cases} 1 & \text{if } t \in (0, \mu(\mathcal{R})), \\ 0 & \text{if } t = 0. \end{cases} \tag{7.9.3}$$

(iii) Let Φ be a Young function, let $L^\Phi(\mathcal{R}, \mu)$ be the corresponding Orlicz space endowed with the Orlicz norm and let $X = L^\Phi$. Let Ψ denote the complementary Young function to Φ. Then

$$\varphi_X(t) = \begin{cases} t\Psi^{-1}\left(\frac{1}{t}\right) & \text{if } t \in (0, \mu(\mathcal{R})), \\ 0 & \text{if } t = 0. \end{cases}$$

(iv) Let Φ be a Young function, let $L^\Phi(\mathcal{R}, \mu)$ be the corresponding Orlicz space endowed with the Luxemburg norm and let $X = L^\Phi$. Then

$$\varphi_X(t) = \begin{cases} \frac{1}{\Phi^{-1}(\frac{1}{t})} & \text{if } t \in (0, \mu(\mathcal{R})), \\ 0 & \text{if } t = 0. \end{cases} \tag{7.9.4}$$

Section 7.9 Fundamental function

Example 7.9.5. Assume that $(\mathcal{R}, \mu) = (\mathbb{N} \cup \{0\}, m)$, where m is the counting measure.

(i) Let $X = \ell^p$ with $p \in [1, \infty)$. Then
$$\varphi_X(n) = n^{\frac{1}{p}}, \quad n \in \mathbb{N}. \tag{7.9.5}$$

(ii) Let $X = \ell^\infty$. Then
$$\varphi_X(n) = \begin{cases} 0 & \text{if } n = 0, \\ 1 & \text{if } n \in \mathbb{N}. \end{cases} \tag{7.9.6}$$

One of the most important properties of fundamental functions is how they reflect the associate spaces. We have the following general result.

Theorem 7.9.6. *Let X be a rearrangement-invariant Banach function space and let X' be the associate space of X. Then*
$$\varphi_X(t)\varphi_{X'}(t) = t \tag{7.9.7}$$

for every $t \in [0, \infty) \cap \text{Rng}(\mu)$.

Proof. Every fundamental function satisfies $\varphi_X(0) = 0$, so the assertion is obvious for $t = 0$. Let $t \in (0, \infty) \cap \text{Rng}(\mu)$. Then there exists a measurable set $E \subset \mathcal{R}$ satisfying $\mu(E) = t$. By the Hölder inequality (7.6.7), we get
$$t = \int_E d\mu \leq \|\chi_E\|_X \|\chi_E\|_{X'} = \varphi_X(t)\varphi_{X'}(t).$$

Conversely, we have
$$\varphi_X(t) = \|\chi_E\|_X = \sup\left\{\int_E g\,d\mu;\ g \in X',\ g \geq 0,\ \|g\|_{X'} \leq 1\right\}. \tag{7.9.8}$$

Let $g \in X'$ be a nonnegative function satisfying $\|g\|_{X'} \leq 1$ and set
$$h(x) := \left(\frac{1}{t}\int_E g\,d\mu\right)\chi_E(x), \quad x \in \mathcal{R}.$$

Now, using Theorem 7.8.1, we obtain
$$\left(\frac{1}{t}\int_E g\,d\mu\right)\varphi_{X'}(t) = \|h\|_{X'} \leq \|g\|_{X'} \leq 1. \tag{7.9.9}$$

Finally, we take the supremum over all such g on the left and use (7.9.8). We arrive at
$$\frac{\varphi_X(t)\varphi_{X'}(t)}{t} \leq 1,$$

the desired converse inequality.
The proof is complete. □

Remark 7.9.7. Let X be a rearrangement-invariant Banach function space and let φ_X be its fundamental function. Then φ_X is nondecreasing on $[0, \mu(\mathcal{R}))$, vanishing at zero and only at zero, continuous on $(0, \mu(\mathcal{R}))$ (with the only possible discontinuity at the origin – see Example 7.9.4 (ii)), and

$$\frac{t}{\varphi_X(t)} \text{ is nondecreasing on } (0, \mu(\mathcal{R})). \tag{7.9.10}$$

Indeed, the monotonicity of φ_X on $[0, \mu(\mathcal{R}))$ is a direct consequence of the lattice property of Banach function spaces (Remark 6.1.16 (i)). Since X' is also a rearrangement-invariant Banach function space, $\varphi_{X'}$ is nondecreasing on $[0, \mu(\mathcal{R}))$, too. However, by (7.9.7), we have

$$\varphi_{X'}(t) = \frac{t}{\varphi_X(t)}, \quad t \in (0, \mu(\mathcal{R})).$$

As a consequence, we get (7.9.10). When $(\mathcal{R}, \mu)$ is atomic, then φ_X is automatically continuous. If $(\mathcal{R}, \mu)$ is nonatomic, then the monotonicity of φ_X implies that there are only countably many points of discontinuity of it on $(0, \infty)$. That, however, is impossible thanks to (7.9.10).

Theorem 7.9.8. *Let $(\mathcal{R}, \mu)$ be nonatomic and let X be a rearrangement-invariant space over $(\mathcal{R}, \mu)$. Then the following conditions on X are equivalent:*

(i) $\lim_{t \to 0+} \varphi_X(t) = 0$;

(ii) $X_a = X_b$;

(iii) $(X_b)^* = X'$.

If, in addition, μ is separable, then each of the properties (i), (ii), *and* (iii) *is equivalent to*

(iv) X_b *is separable.*

Proof. Theorem 6.4.1 shows that (ii)$\Leftrightarrow$(iii). When μ is a separable measure, then (ii)$\Leftrightarrow$(iv) by Theorem 6.5.9. Assume that (i) holds and $E \subset \mathcal{R}$ satisfies $\mu(E) < \infty$. Then we claim that $\chi_E \in X_a$. Indeed, let $E_n \downarrow \emptyset$ a.e. Then $\mu(E \cap E_n) \downarrow 0$ by the dominated convergence theorem, whence

$$\|\chi_E \chi_{E_n}\|_X = \|\chi_{E \cap E_n}\|_X \leq \varphi_X(\mu(E \cap E_n)) \downarrow 0.$$

By Theorem 6.3.18, this establishes (ii). We have thus shown (i)$\Rightarrow$(ii).

Conversely, let (ii) hold, that is, $X_a = X_b$. Let $E \subset \mathcal{R}$ with $0 < \mu(\mathcal{R}) < \infty$. Since $(\mathcal{R}, \mu)$ is nonatomic, there exists a sequence $\{E_n\}_{n=1}^\infty$ of subsets of E such that $\mu(E_n) = 2^{-n}\mu(E)$ and $E_{n+1} \subset E_n$ for every $n \in \mathbb{N}$. Then $E_n \downarrow \emptyset$ μ-a.e. and it follows from (ii) that $\chi_E \in X_a$. Therefore,

$$\varphi_X(2^{-n}\mu(E)) = \|\chi_{E_n}\|_X = \|\chi_E \chi_{E_n}\|_X \downarrow 0.$$

Because φ_X is nondecreasing on $(0, \mu(\mathcal{R}))$, this shows (i).

The proof is complete. $\square$

Remark 7.9.9. If $\lim_{t \to 0+} \varphi_X(t) > 0$, then $X_a = \{0\}$.

To see this, suppose that $0 \neq f \in X_a$ is a nonnegative function. Then there is $\varepsilon > 0$ and a set $E \subset \mathcal{R}$, $\mu(E) < \infty$, such that $\varepsilon \chi_E \leq f$. Hence $\chi_E \in X_a$, since, by Theorem 6.3.12, X_a is an ideal. Then (cf. the argument used in the proof of Theorem 7.9.8) $\varphi_X(t) \downarrow 0$ as $t \downarrow 0$.

Should a function φ, defined on an interval of the form $[0, a)$ for $a \in (0, \infty]$, be a fundamental function of some rearrangement-invariant Banach function space, then it has to satisfy the properties mentioned in Remark 7.9.7. It will be useful to single out the class of functions obeying these requirements.

Definition 7.9.10. Let $a \in (0, \infty]$. A nondecreasing function $\varphi : [0, a) \to [0, \infty)$ is called *quasi-concave* on $[0, a)$ if

$$\varphi(t) = 0 \iff t = 0, \qquad (7.9.11)$$

$$\frac{t}{\varphi(t)} \text{ is nondecreasing on } (0, a). \qquad (7.9.12)$$

Remark 7.9.11. Every nonnegative concave function on $[0, a)$ with $a \in (0, \infty]$ that vanishes only at the origin is automatically quasi-concave on $[0, a)$. The converse is not true. For example, take $a = \infty$, then the function

$$\varphi(t) := \begin{cases} \max\{1, t\} & \text{if } t \in (0, \infty), \\ 0 & \text{if } t = 0, \end{cases}$$

is quasi-concave but not concave on $[0, a)$.

Remark 7.9.12. It follows from Remark 7.9.7 that if X is a rearrangement-invariant Banach function space, and φ_X is its fundamental function, then φ_X is quasi-concave on $[0, \mu(\mathcal{R}))$. The converse is in some sense true, too, namely, if φ is a quasi-concave function on $[0, \infty)$, then there exists at least one rearrangement-invariant Banach function space X such that $\varphi = \varphi_X$.

We will finish this section with a simple but useful inequality.

Lemma 7.9.13. *Let X be a rearrangement-invariant Banach function space and let φ_X be its fundamental function. Then, for every $t \in (0, \infty) \cap \mathrm{Rng}(\mu)$ and every $f \in \mathcal{M}(\mathcal{R}, \mu)$, one has*

$$\int_0^t f^*(s) \, ds \leq \varphi_X(t) \|f\|_{X'}. \qquad (7.9.13)$$

Proof. Since $t \in (0, \infty) \cap \mathrm{Rng}(\mu)$, there is a μ-measurable subset E of $\mathcal{R}$ such that $\mu(E) = t$. Thus, by the Hölder inequality (Theorem 6.2.6),

$$\int_0^t f^*(s) \, ds = \int_0^\infty \chi_E^*(s) f^*(s) \, ds \leq \|\chi_E\|_X \|f\|_{X'} = \varphi_X(t) \|f\|_{X'},$$

as desired. The proof is complete. $\square$

7.10 Endpoint spaces

Definition 7.10.1. Let φ be a quasi-concave function on $[0, \mu(\mathcal{R}))$. Then, the collection $M_\varphi = M_\varphi(\mathcal{R}, \mu)$ is defined as

$$M_\varphi := \left\{ g \in \mathcal{M}_0(\mathcal{R}, \mu); \sup_{t \in (0, \mu(\mathcal{R}))} \varphi(t) g^{**}(t) < \infty \right\}$$

is called the *Marcinkiewicz endpoint space*. (We shall throughout use the shorter name *Marcinkiewicz space* as there is no confusion likely to arise. In fact, the Marcinkiewicz space $L^{p,\infty}$ which was introduced in Definition 3.18.3 is (at least for $p \in (1, \infty)$) a special case of an endpoint Marcinkiewicz space).

Proposition 7.10.2. *If φ is quasi-concave on $[0, \mu(\mathcal{R}))$, then the functional $\|\cdot\|_{M_\varphi M_\varphi = M_\varphi(\mathcal{R}, \mu)}$, defined by*

$$\|g\|_{M_\varphi(\mathcal{R}, \mu)} := \sup_{t \in (0, \mu(\mathcal{R}))} \varphi(t) g^{**}(t), \quad g \in \mathcal{M}_0(\mathcal{R}, \mu), \tag{7.10.1}$$

is a Banach function norm and the corresponding Marcinkiewicz space $M_\varphi(\mathcal{R}, \mu)$ is a rearrangement-invariant Banach function space. Moreover,

$$\varphi_{M_\varphi}(t) = \varphi(t), \quad t \in [0, \mu(\mathcal{R})). \tag{7.10.2}$$

Proof. The functional

$$g \mapsto \sup_{0 < t < \mu(\mathcal{R})} \{g^{**}(t) \varphi(t)\}, \quad g \in \mathcal{M}_0(\mathcal{R}, \mu)$$

satisfies all properties of a norm; in particular, the triangle inequality is guaranteed by Corollary 7.4.6 and by the subadditivity of supremum. Other properties of (P1) as well as (P2) and (P3) follow from the elementary properties of the functional $g \mapsto g^{**}$. As for (P4), let $E \subset \mathcal{R}$ such that $\mu(\mathcal{R}) < \varepsilon$ and denote $t := \mu(E)$. Then $\chi_E^* = \chi_{[0,t]}$ and so

$$\|\chi_E\|_{M_\varphi} = \sup_{s \in (0, \mu(\mathcal{R}))} \left\{ \chi_{(0,t)}^{**}(s) \varphi(s) \right\} = \sup_{s \in (0, \mu(\mathcal{R}))} \left\{ \min\left(1, \tfrac{t}{s}\right) \varphi(s) \right\}$$

$$= \max \left\{ \sup_{s \in (0,t]} \varphi(s), t \sup_{s \in [t, \mu(\mathcal{R}))} \frac{\varphi(s)}{s} \right\} = \varphi(t) < \infty, \tag{7.10.3}$$

since $\varphi(s)$ is nondecreasing on $[0, t]$ while $\frac{\varphi(s)}{s}$ is nonincreasing on $[t, \mu(\mathcal{R}))$, and (P4) follows. Let $g \in M_\varphi$ and let $E \subset \mathcal{R}$ be such that $0 < \mu(E) < \infty$. Denote

Section 7.10 Endpoint spaces

$t := \mu(E)$ and $C_E := \frac{t}{\varphi(t)}$. Note that $C_E < \infty$, since the function φ is strictly positive everywhere outside zero. Thus, by (7.4.1),

$$\left|\int_E g(x)\,dx\right| \leq \int_0^t g^*(s)\,ds = \frac{t}{\varphi(t)} g^{**}(t)\varphi(t)$$

$$\leq \frac{t}{\varphi(t)} \sup_{s\in(0,\mu(\mathcal{R}))} g^{**}(s)\varphi(s) = C_E \, \|g\|_{M_\varphi},$$

which yields (P5). The rearrangement-invariance of the norm $\|\cdot\|_{M_\varphi}$ is obvious from its definition. This shows that M_φ is a rearrangement-invariant Banach function space. Finally, (7.10.2) follows from (7.10.3). $\square$

Definition 7.10.3. Let φ be a quasi-concave function on $[0, \mu(\mathcal{R}))$. We then denote by $\|\cdot\|_{m_\varphi(\mathcal{R},\mu)}$ the functional defined by

$$\|g\|_{m_\varphi(\mathcal{R},\mu)} := \sup_{t\in(0,\mu(\mathcal{R}))} \varphi(t)g^*(t), \quad g \in \mathcal{M}_0(\mathcal{R}, \mu),$$

and by $m_\varphi = m_\varphi(\mathcal{R}, \mu)$ the collection

$$m_\varphi := \{g \in \mathcal{M}_0(\mathcal{R}, \mu);\ \|g\|_{m_\varphi(\mathcal{R},\mu)} < \infty\}.$$

Remark 7.10.4. The functional $\|\cdot\|_{m_\varphi(\mathcal{R},\mu)}$ is not necessarily a norm on $\mathcal{M}_0(\mathcal{R}, \mu)$.

Proposition 7.10.5. *Let φ be a quasi-concave function on $[0, \mu(\mathcal{R}))$. Assume that there is a positive constant C such that*

$$\frac{1}{t}\int_0^t \frac{ds}{\varphi(s)} \leq \frac{C}{\varphi(t)} \quad \text{for every } t \in (0, \mu(\mathcal{R})). \tag{7.10.4}$$

Then, for every $g \in \mathcal{M}_0(\mathcal{R}, \mu)$, one has

$$\sup_{t\in(0,\mu(\mathcal{R}))} g^{**}(t)\varphi(t) \leq C \sup_{t\in(0,\mu(\mathcal{R}))} g^*(t)\varphi(t). \tag{7.10.5}$$

Consequently, in particular,

$$M_\varphi(\mathcal{R}, \mu) = m_\varphi(\mathcal{R}, \mu). \tag{7.10.6}$$

Proof. Assume that $g \in m_\varphi(\mathcal{R}, \mu)$ and denote

$$K := \|g\|_{m_\varphi(\mathcal{R},\mu)} = \sup_{t\in(0,\mu(\mathcal{R}))} g^*(t)\varphi(t).$$

Then, for every fixed $t \in (0, \mu(\mathcal{R}))$, one has

$$g^*(t) \leq \frac{K}{\varphi(t)},$$

whence, by (7.10.4),

$$g^{**}(t) = \frac{1}{t}\int_0^t g^*(s)\,ds \leq \frac{K}{t}\int_0^t \frac{ds}{\varphi(s)} \leq \frac{CK}{\varphi(t)}.$$

Multiplying through by $\varphi(t)$ and then passing to the supremum over all $t \in (0, \mu(\mathcal{R}))$, we obtain

$$\sup_{t\in(0,\mu(\mathcal{R}))} \varphi(t) g^{**}(t) \leq CK = C \sup_{t\in(0,\mu(\mathcal{R}))} \varphi(t) g^*(t),$$

establishing (7.10.5).

It remains to verify (7.10.6). The embedding

$$M_\varphi(\mathcal{R}, \mu) \hookrightarrow m_\varphi(\mathcal{R}, \mu)$$

follows from the inequality

$$\|g\|_{m_\varphi(\mathcal{R},\mu)} \leq \|g\|_{M_\varphi(\mathcal{R},\mu)}, \quad g \in \mathcal{M}_0(\mathcal{R}, \mu),$$

which is an immediate consequence of the definitions and (7.4.3). On the other hand, assume that $g \in m_\varphi(\mathcal{R}, \mu)$. Then (7.10.5), just proved, in particular implies that $g \in M_\varphi(\mathcal{R}, \mu)$ and

$$\|g\|_{M_\varphi(\mathcal{R},\mu)} \leq C\|g\|_{m_\varphi(\mathcal{R},\mu)}.$$

Therefore, the converse embedding

$$m_\varphi(\mathcal{R}, \mu) \hookrightarrow M_\varphi(\mathcal{R}, \mu)$$

follows, and hence so does (7.10.6). The proof is complete. $\square$

It turns out that the Marcinkiewicz space has a special role among all the r.i. spaces sharing the same fundamental function; indeed, it is the largest such space. This interesting and important fact will be noted in the following result.

Proposition 7.10.6. *Let X be a rearrangement-invariant Banach function space. Let φ be its fundamental function and let M_φ be the corresponding Marcinkiewicz space, introduced in Definition 7.10.1. Then*

$$X \hookrightarrow M_\varphi,$$

with the norm of the embedding equal to 1, more precisely,

$$\|g\|_{M_\varphi} \leq \|g\|_X \quad \text{for every } g \in \mathcal{M}_0(\mathcal{R}, \mu). \tag{7.10.7}$$

Section 7.10 Endpoint spaces

Proof. Let $t \in (0, \mu(\mathcal{R}))$. Then we get, by the Hölder inequality and (7.9.7), for every $g \in X$,

$$\int_0^t g^*(s)\,ds = \int_0^1 g^*(s)\chi_{(0,t)}(s)\,ds \leq \|g\|_X \,\|\chi_{(0,t)}\|_{X'}$$

$$= \varphi_{X'}(t)\|g\|_X = \frac{t}{\varphi(t)}\|g\|_X.$$

Hence,

$$g^{**}(t)\varphi(t) \leq \|g\|_X,$$

and (7.10.7) follows on taking the supremum over $t \in (0, \mu(\mathcal{R}))$ on the left. □

The preceding result raises a natural question whether for every given rearrangement-invariant Banach function space X there exists the smallest rearrangement-invariant space with the same fundamental function. The answer is affirmative under an extra condition that the fundamental function of X is concave (so far we needed only quasi-concavity). Before we introduce this space we need some preliminary work on quasi-concave and concave functions on an interval.

Remark 7.10.7. Let $a \in (0, \infty]$ and let $\{\varphi_\alpha\}_{\alpha \in I}$, where I is an index set of any cardinality, be a collection of concave functions on $[0, a)$. Define

$$\varphi(t) := \inf_{\alpha \in I} \varphi_\alpha(t), \quad t \in [0, a).$$

Then it is an easy exercise to observe that the function φ is concave on $[0, a)$.

Proposition 7.10.8. *Let $a \in (0, \infty]$ and let φ be a quasi-concave function on $[0, a)$. Then there exists a function $\tilde{\varphi}$ with the following properties:*

(i) *$\tilde{\varphi}$ is concave on $[0, a)$;*

(ii) *$\varphi(t) \leq \tilde{\varphi}(t)$ for every $t \in [0, a)$;*

(iii) *if ψ is a concave function on $[0, a)$ such that $\varphi(t) \leq \psi(t)$ for every $t \in [0, a)$, then $\tilde{\varphi}(t) \leq \psi(t)$ for every $t \in [0, a)$.*

Proof. We define the function $\tilde{\varphi}$ as the pointwise infimum of all concave functions that majorize φ on $[0, a)$. By (7.9.11) and (7.9.12),

$$\varphi(t) \leq \varphi(1)\max\{1, t\}, \quad t \in [0, a).$$

Consequently, in particular,

$$\varphi(t) \leq \varphi(1)(1 + t), \quad t \in [0, a).$$

Now, the function $\varphi(1)(1 + t)$ is concave on $[0, a)$. Thus, there exists a concave majorant of φ; hence, $\tilde{\varphi}$ is well-defined on $[0, a)$. By Remark 7.10.7, $\tilde{\varphi}$ is concave on $[0, a)$. All the other required properties of $\tilde{\varphi}$ are obviously satisfied. The proof is complete. □

Definition 7.10.9. Let $a \in (0, \infty]$ and let φ be a quasi-concave function on $[0, a)$. Then the function $\tilde{\varphi}$ from Proposition 7.10.8 is called the *least concave majorant* of φ on $[0, a)$.

Proposition 7.10.10. *Let $a \in (0, \infty]$ and let φ be quasi-concave on $[0, a)$ and let $\tilde{\varphi}$ be its least concave majorant on $[0, a)$. Then*

$$\tfrac{1}{2}\tilde{\varphi}(t) \leq \varphi(t) \leq \tilde{\varphi}(t), \quad t \in [0, a). \tag{7.10.8}$$

Proof. The latter inequality has been already observed in Proposition 7.10.8. To prove the former one, fix $s \in (0, a)$. Then it follows from (7.9.11) and (7.9.12) that

$$\varphi(t) \leq \left(1 + \tfrac{t}{s}\right) \varphi(s)$$

for all $t \in (0, a)$. Thus, $\varphi(t)$ is dominated on $[0, a)$ by the concave function

$$\psi(t) = \left(1 + \tfrac{t}{s}\right) \varphi(s), \quad t \in [0, a).$$

Consequently, it follows from the definition of the least concave majorant that, for every $t \in [0, a)$, one has

$$\tilde{\varphi}(t) \leq \psi(t).$$

In particular, on taking $t = s$, we obtain

$$\tilde{\varphi}(s) \leq \psi(s) = 2\varphi(s),$$

as desired. The proof is complete. $\square$

Definition 7.10.11. We say that two Banach spaces X and Y are *equivalent* if $X = Y$ in the set-theoretic sense and there exists a positive constant C such that for every $x \in X$, one has
$$C^{-1}\|x\|_X \leq \|x\|_Y \leq C\|x\|_X.$$

Theorem 7.10.12. *Let X be a rearrangement-invariant Banach function space over $(\mathcal{R}, \mu)$. Then there exists another rearrangement-invariant Banach function space over $(\mathcal{R}, \mu)$, say, Y, having the following two properties:*

(i) *Y is equivalent to X;*

(ii) *the fundamental function of Y is concave on $[0, \mu(\mathcal{R}))$.*

Proof. Let φ be the fundamental function of X and let $\tilde{\varphi}$ be a concave function on $[0, \mu(\mathcal{R}))$ satisfying (7.10.8). Let $M_{\tilde{\varphi}}$ be the Marcinkiewicz space associated with $\tilde{\varphi}$. We define

$$\nu(f) := \max\{\|g\|_X, \|g\|_{M_{\tilde{\varphi}}}\}, \quad g \in \mathcal{M}_0^+(\mathcal{R}, \mu).$$

Since both X and $M_{\tilde{\varphi}}$ are rearrangement-invariant spaces, it is simple matter to check that ν is a rearrangement-invariant function norm. Furthermore, it follows from (7.10.7) and (7.10.8) that

$$\|f\|_X \leq \nu(f) \leq \max(\|f\|_X, 2\|f\|_{M_\varphi}) \leq 2\|f\|_X,$$

so ν is equivalent to the norm of X. Finally, since

$$\nu(\chi_{(0,t)}) \max(\varphi(t), \tilde{\varphi}(t)) = \tilde{\varphi}(t),$$

we see that $X = X(\nu)$ has concave fundamental function $\tilde{\varphi}$. □

Definition 7.10.13. Let φ be a concave function on $[0, \mu(\mathcal{R}))$. Then the collection

$$\Lambda_\varphi(\mathcal{R}, \mu) := \left\{ g \in \mathcal{M}_0(\mathcal{R}, \mu); \int_0^{\mu(\mathcal{R})} g^*(t)\, d\varphi(t) < \infty \right\} \quad (7.10.9)$$

is called the *Lorentz endpoint space*.

Remark 7.10.14. Note that the Lebesgue–Stieltjes integral in (7.10.9) is well-defined because φ is nondecreasing. Furthermore, since φ is nonnegative and concave, it may be represented as the integral of a nonnegative, nonincreasing function, say ϕ, on $(0, \infty)$. Hence the integral in (7.10.9) may be rewritten in the form

$$\int_0^{\mu(\mathcal{R})} f^*(t)\, d\varphi(t) = \|g\|_{L^\infty} \varphi(0+) + \int_0^\infty g^*(t)\phi(t)\, dt. \quad (7.10.10)$$

Proposition 7.10.15. *Let X be a rearrangement-invariant Banach function space over $(\mathcal{R}, \mu)$ and assume that its fundamental function $\varphi = \varphi_X$ is concave on $[0, \mu(\mathcal{R}))$. Then*

$$\|g\|_X \leq \int_0^{\mu(\mathcal{R})} g^*(t)\, d\varphi(t). \quad (7.10.11)$$

Proof. Assume first that g is such that g^* is a nonincreasing simple function, that is, it can be written in the form

$$g^*(t) = \sum_{k=1}^n c_k \chi_{(0,t_k)},$$

for some $c_k \in (0, \infty)$ and $0 < t_1 < t_2 < \ldots < t_n$ (cf. Remark 7.1.10). Then

$$\|g\|_X \leq \sum_{k=1}^n c_k \|\chi_{(0,t_k)}\|_X = \sum_{k=1}^n c_k \varphi(t_k)$$

$$= \int_0^\infty g^*(s)\, d\varphi(s) = \|f\|_{\Lambda_\varphi}.$$

When $g \in \mathcal{M}_0(\mathcal{R}, \mu)$ is arbitrary, the assertion follows from rearrangement-invariance of the spaces X and Λ_φ and their Fatou property. The proof is complete. □

Proposition 7.10.16. *If φ is concave on $[0, \mu(\mathcal{R}))$, then the functional $\|\cdot\|_{\Lambda_\varphi(\mathcal{R},\mu)}$, defined by*

$$\|g\|_{\Lambda_\varphi(\mathcal{R},\mu)} := \int_0^{\mu(\mathcal{R})} f^*(s)\,d\varphi(s), \quad g \in \mathcal{M}_0(\mathcal{R},\mu), \qquad (7.10.12)$$

is a Banach function norm and the corresponding Lorentz endpoint space $\Lambda_\varphi(\mathcal{R},\mu)$ is a rearrangement-invariant Banach function space. Moreover,

$$\varphi_{\Lambda_\varphi}(t) = \varphi(t), \quad t \in [0, \mu(\mathcal{R})). \qquad (7.10.13)$$

Proof. We recall that

$$(f + g)^* \prec f^* + g^*, \quad f, g \in \mathcal{M}_0(\mathcal{R},\mu).$$

Thus, by the Hardy lemma (Proposition 7.5.2), we have, for every $f, g \in \mathcal{M}_0(\mathcal{R},\mu)$,

$$\int_0^{\mu(\mathcal{R})} (f+g)^*(t)\phi(t)\,dt \le \int_0^{\mu(\mathcal{R})} f^*(t)\phi(t)\,dt + \int_0^{\mu(\mathcal{R})} g^*(t)\phi(t)\,dt,$$

because ϕ is nonincreasing on $(0, \mu(\mathcal{R}))$. This and (7.10.10) gives the triangle inequality

$$\|f + g\|_{\Lambda_\varphi(\mathcal{R},\mu)} \le \|f\|_{\Lambda_\varphi(\mathcal{R},\mu)} + \|g\|_{\Lambda_\varphi(\mathcal{R},\mu)}.$$

All the other properties in (P1) as well as (P2) and (P3) follow easily from the corresponding properties of the functional $f \mapsto f^*$.

Assume that $E \subset \mathcal{R}$ is such that $0 < \mu(E) < \infty$ and denote $t := \mu(E)$. Then

$$\|\chi_E\|_{\Lambda_\varphi} = \int_0^{\mu(\mathcal{R})} \chi_{(0,t)}(s)\,d\varphi(s) = \varphi(t), \quad t \in (0, \mu(\mathcal{R})). \qquad (7.10.14)$$

From this identity we immediately obtain both (P4) and (7.10.13). We now note that the particular case of (7.10.11) with $X = M_\varphi$ (M_φ is admissible for (7.10.11) thanks to (7.10.2)) gives

$$\|g\|_{M_\varphi} \le \|g\|_{\Lambda_\varphi}.$$

This and (P5) for the space M_φ now clearly gives (P5) for the space Λ_φ. The rearrangement-invariance of Λ_φ is obvious from its definition. The proof is complete. □

It will be useful to introduce a symbol for Lorentz and Marcinkiewicz endpoint spaces that correspond to a given rearrangement-invariant Banach function space.

Definition 7.10.17. Let X be a rearrangement-invariant Banach function space over $(\mathcal{R}, \mu)$ with the fundamental function φ_X. Then the *Marcinkiewicz endpoint space* $M_X = M_X(\mathcal{R}, \mu)$, corresponding to X, is defined as

$$M_X(\mathcal{R}, \mu) := M_{\varphi_X}(\mathcal{R}, \mu).$$

If, moreover, φ_X is concave on $[0, \mu(\mathcal{R}))$, then the *Lorentz endpoint space* $\Lambda_X = \Lambda_X(\mathcal{R}, \mu)$, corresponding to X, is defined as

$$\Lambda_X(\mathcal{R}, \mu) = \Lambda_{\varphi_X}(\mathcal{R}, \mu).$$

Section 7.10 Endpoint spaces

The following assertion is a summary of the results obtained in this section.

Theorem 7.10.18. *Let X be a rearrangement-invariant Banach function space over a nonatomic resonant measure space $(\mathcal{R}, \mu)$. Assume that φ_X, the fundamental function of X, is concave on $[0, \mu(\mathcal{R}))$. Then the endpoint spaces Λ_X and M_X are rearrangement-invariant Banach function spaces having fundamental function equal to φ_X. Furthermore,*

$$\Lambda_X \hookrightarrow X \hookrightarrow M_X \tag{7.10.15}$$

with norm constants equal to 1.

Corollary 7.10.19. *Let X be a rearrangement-invariant Banach function space over a nonatomic resonant measure space $(\mathcal{R}, \mu)$. Assume that φ_X, the fundamental function of X, is concave on $[0, \mu(\mathcal{R}))$. Then the Lorentz endpoint space Λ_X is the smallest rearrangement-invariant Banach function spaces having fundamental function φ_X. Similarly, the Marcinkiewicz endpoint space M_X is the largest such rearrangement-invariant Banach function space.*

Example 7.10.20. Let $p \in (1, \infty)$ and let $\varphi(t) := t^{\frac{1}{p}}$. Then

$$\Lambda_\varphi = L^{p,1} \quad \text{and} \quad M_\varphi = L^{p,\infty}.$$

Examples 7.10.21. Let $(\mathcal{R}, \mu)$ be a resonant measure space. Let us for simplicity assume that $\mu(\mathcal{R}) = \infty$.

(i) Define

$$\varphi(t) := t, \quad t \in [0, \infty).$$

Then φ is the fundamental function of the space L^1. Moreover,

$$\Lambda_{L^1} = M_{L^1} = L^1. \tag{7.10.16}$$

Indeed, this follows from the identities

$$\|f\|_{\Lambda_{L^1}} = \int_0^\infty f^*(t)\,dt = \|f\|_{L^1}$$

and

$$\|f\|_{M_{L^1}} = \sup_{t \in (0,\infty)} t f^{**}(t) = \sup_{t \in (0,\infty)} \int_0^t f^*(s)\,ds = \int_0^\infty f^*(t)\,dt = \|f\|_{L^1},$$

which are valid for any $f \in \mathcal{M}_0(\mathcal{R}, \mu)$.

(ii) Define
$$\varphi(t) := \begin{cases} 0 & \text{if } t = 0, \\ 1 & \text{if } t \in (0, \infty). \end{cases}$$

Then φ is the fundamental function of the space L^∞. Moreover,
$$\Lambda_{L^\infty} = M_{L^\infty} = L^\infty. \tag{7.10.17}$$

Indeed, this follows from the identities
$$\|f\|_{\Lambda_{L^\infty}} = \int_0^\infty f^*(t)\, d\varphi(t) = \lim_{s \to 0+} f^*(s) = \|f\|_{L^\infty}$$

and
$$\|f\|_{M_{L^\infty}} = \sup_{t \in (0,\infty)} f^{**}(t) = \lim_{s \to 0+} f^{**}(s) = \|f\|_{L^\infty},$$

which are again valid for any $f \in \mathcal{M}_0(\mathcal{R}, \mu)$.

Remark 7.10.22. The results mentioned in the preceding examples can be also summarized as follows. Let, for $t \in [0, \infty)$, either $\varphi(t) = t$ or
$$\varphi(t) := \begin{cases} 0, & \text{if } t = 0, \\ 1 & \text{if } t \in (0, \infty). \end{cases}$$

Then, in each case, we have
$$\Lambda_\varphi = M_\varphi. \tag{7.10.18}$$

We finish this section with characterization of the set of functions having absolutely continuous norm in a Marcinkiewicz endpoint space. We follow the argument of [205].

Theorem 7.10.23. Let $(\mathcal{R}, \mu)$ be a nonatomic σ-finite measure space with $\mu(\mathcal{R}) = \infty$. Suppose that φ is a quasi-convave function on $(0, \infty)$.

(i) If $\lim_{t \to \infty} \frac{\varphi(t)}{t} = 0$, then
$$(M_\varphi)_a = \left\{ f \in \mathcal{M};\ \lim_{t \to 0+} \varphi(t) f^{**}(t) = \lim_{t \to \infty} \varphi(t) f^{**}(t) = 0 \right\}. \tag{7.10.19}$$

(ii) If $\lim_{t \to \infty} \frac{\varphi(t)}{t} > 0$, then
$$(M_\varphi)_a = \left\{ f \in \mathcal{M};\ \lim_{t \to 0+} \varphi(t) f^{**}(t) = 0 \text{ and } f \in L^1 \right\}.$$

Section 7.10 Endpoint spaces

Proof. (i) Suppose that a function $f \in \mathcal{M}$ belongs to $(M_\varphi)_a$. Choose $\varepsilon > 0$ arbitrarily. According to Lemma 6.3.6, there exists $\delta > 0$ such that $\|f\chi_E\|_{M_\varphi} < \varepsilon$, whenever $\mu(E) < \delta$. Thus, using that the measure space $(\mathcal{R}, \mu)$ is resonant, for every $t \in (0, \delta)$ we have

$$\frac{\varphi(t)}{t} \int_0^t f^*(s)\,ds = \sup_{\mu(E)=t} \frac{\varphi(t)}{t} \int_\mathcal{R} |f\chi_E|\,d\mu$$

$$= \sup_{\mu(E)=t} \frac{\varphi(t)}{t} \int_0^t (f\chi_E)^*(s)\,ds$$

$$\leq \sup_{\mu(E)=t} \sup_{r>0} \frac{\varphi(r)}{r} \int_0^r (f\chi_E)^*(s)\,ds$$

$$= \sup_{\mu(E)=t} \|f\chi_E\|_{M_\varphi} \leq \varepsilon,$$

which yields that $\lim_{t \to 0_+} \varphi(t) f^{**}(t) = 0$.

To prove an analogous result near infinity, consider an increasing sequence R_n of sets of finite measure whose union is $\mathcal{R}$ (which can be found because $(\mathcal{R}, \mu)$ is σ-finite). Denote $E_n = \mathcal{R} \setminus R_n$. Then $E_n \downarrow \emptyset$, so for every $\varepsilon > 0$ there is $k \in \mathbb{N}$ such that

$$\sup_{t>0} \varphi(t)(f\chi_{E_k})^{**}(t) < \frac{\varepsilon}{2}.$$

For every $t > 0$ we have

$$\varphi(t) f^{**}(t) \leq \varphi(t)(f\chi_{E_k})^{**}(t) + \varphi(t)(f\chi_{R_k})^{**}(t)$$

$$= \varphi(t)(f\chi_{E_k})^{**}(t) + \frac{\varphi(t)}{t} \int_0^t (f\chi_{R_k})^*(s)\,ds$$

$$\leq \varphi(t)(f\chi_{E_k})^{**}(t) + \frac{\varphi(t)}{t} \int_{R_k} |f|\,d\mu.$$

By the property (P5) from Definition 6.1.5, every Banach function space is locally (i.e. on sets of finite measure) embedded to L^1. We thus have $\int_{R_k} |f|\,d\mu < \infty$. Combining this with our assumption $\lim_{t \to \infty} \frac{\varphi(t)}{t} = 0$, we can find $t_0 > 0$ such that

$$\frac{\varphi(t)}{t} \int_{R_k} |f|\,d\mu < \frac{\varepsilon}{2},$$

whenever $t > t_0$. Thus, for $t > t_0$ we have $\varphi(t) f^{**}(t) < \varepsilon$, as required.

Now assume that a function $f \in \mathcal{M}$ satisfies $\lim_{t \to 0_+} \varphi(t) f^{**}(t) = 0$ and $\lim_{t \to \infty} \varphi(t) f^{**}(t) = 0$. Let $(E_n)_{n=1}^\infty$ be any sequence of subsets of $\mathcal{R}$ such that $E_n \downarrow \emptyset$ μ-a.e. Choose $\varepsilon > 0$ arbitrarily. Then we can find δ_1, δ_2 such that

$0 < \delta_1 < \delta_2$ and for every $t \in (0, \delta_1) \cup (\delta_2, \infty)$ we have $\varphi(t) f^{**}(t) < \varepsilon$. So, for every $n \in \mathbb{N}$,

$$\sup_{t \in (0,\delta_1) \cup (\delta_2,\infty)} \varphi(t)(f\chi_{E_n})^{**}(t) \leq \sup_{t \in (0,\delta_1) \cup (\delta_2,\infty)} \varphi(t) f^{**}(t) \leq \varepsilon.$$

Furthermore, we observe that $\lim_{t \to \infty} f^*(t) = 0$. Indeed, if $\lim_{t \to \infty} f^*(t) = c > 0$, we have $f^{**}(t) \geq c$ for every $t > 0$, which implies $\varphi(t) f^{**}(t) \geq c\varphi(t)$, $t > 0$, contradicting the assumption $\lim_{t \to \infty} \varphi(t) f^{**}(t) = 0$.

Denote

$$A = \left\{ x \in \mathcal{R};\ |f(x)| \geq \frac{\varepsilon}{2\varphi(\delta_2)} \right\}.$$

Then $\mu(A) < \infty$, which together with the fact that $E_n \downarrow \emptyset$ μ-a.e. implies $\lim_{n \to \infty} \mu(E_n \cap A) = 0$. The assumptions given on f obviously ensure that $\int_0^1 f^*(s)\,ds < \infty$, so we can find $n_0 \in \mathbb{N}$ such that

$$\int_0^{\mu(E_n \cap A)} f^*(s)\,ds < \frac{\varepsilon \delta_1}{2\varphi(\delta_2)}$$

whenever $n > n_0$. Thus, for $n > n_0$ we have

$$\sup_{t \in [\delta_1,\delta_2]} \varphi(t)(f\chi_{E_n})^{**}(t) \leq \varphi(\delta_2)(f\chi_{E_n})^{**}(\delta_1)$$

$$\leq \varphi(\delta_2)(f\chi_{E_n \cap A})^{**}(\delta_1) + \varphi(\delta_2)(f\chi_{E_n \cap A^c})^{**}(\delta_1)$$

$$\leq \frac{\varphi(\delta_2)}{\delta_1} \int_0^{\mu(E_n \cap A)} f^*(s)\,ds + \varphi(\delta_2)(f\chi_{E_n \cap A^c})^{**}(\delta_1) \leq \varepsilon.$$

We have just shown that whenever $n > n_0$,

$$\|f\chi_{E_n}\|_{M_\varphi} = \sup_{t>0} \varphi(t)(f\chi_{E_n})^{**}(t) \leq \varepsilon,$$

which concludes the proof.

(ii) Suppose that $f \in (M_\varphi)_a$. Then $f \in M_\varphi$ and $M_\varphi \hookrightarrow L^1$ (because $\lim_{t \to \infty} \frac{\varphi(t)}{t} > 0$), i.e. $f \in L^1$. Moreover, from the proof of part (i) it follows that $\lim_{t \to 0_+} \varphi(t) f^{**}(t) = 0$.

Conversely, assume that a function $f \in L^1$ satisfies $\lim_{t \to 0_+} \varphi(t) f^{**}(t) = 0$. Let $(E_n)_{n=1}^\infty$ be a sequence of subsets of $\mathcal{R}$ such that $E_n \downarrow \emptyset$ μ-a.e. A similar proof as in the part (i) yields

$$\lim_{n \to \infty} \sup_{t \in (0,1]} \varphi(t)(f\chi_{E_n})^{**}(t) = 0.$$

Thus, to get $f \in (M_\varphi)_a$ it only remains to show that

$$\lim_{n \to \infty} \sup_{t > 1} \varphi(t)(f\chi_{E_n})^{**}(t) = 0,$$

which is, due to the assumption $\lim_{t\to\infty} \frac{\varphi(t)}{t} > 0$, the same as

$$\lim_{n\to\infty} \sup_{t>1} t(f\chi_{E_n})^{**}(t) = 0.$$

Using that $f \in L^1 = (L^1)_a$, we obtain

$$\lim_{n\to\infty} \sup_{t>1} t(f\chi_{E_n})^{**}(t) = \lim_{n\to\infty} \sup_{t>1} \int_0^t (f\chi_{E_n})^*(s)\,ds$$
$$= \lim_{n\to\infty} \int_0^\infty (f\chi_{E_n})^*(s)\,ds = 0,$$

as required. $\square$

7.11 Almost-compact embeddings

In this section we shall study an important relation between function spaces which is stronger than an ordinary embedding but weaker than a compact embedding. We have already once briefly met this relation, namely in connection with inclusions of type $L^{\Phi_1}(\Omega) \hookrightarrow E^{\Phi_2}(\Omega)$ between Orlicz spaces, where Φ_1 and Φ_2 are Young functions (see Remark 4.17.8). The almost-compact embeddings, called in literature also *absolutely continuous embeddings*, have recently proved its importance and usefulness and have been studied by several authors (let us mention, for instance, [54, 80, 206]). Our exposition is based on a mixture of results from [80, 204, 206].

Definition 7.11.1. Suppose that X and Y are Banach function spaces over a measure space $(\mathcal{R}, \mu)$. We say that X is *almost-compactly embedded into* Y if for every sequence $\{E_n\}_{n=1}^\infty$ of μ-measurable subsets of $\mathcal{R}$ satisfying $E_n \to \emptyset$ μ-a.e., we have

$$\lim_{n\to\infty} \sup_{\|f\|_X \le 1} \|f\chi_{E_n}\|_Y = 0.$$

We denote the almost-compact embedding of X into Y by

$$X \stackrel{*}{\hookrightarrow} Y.$$

We first observe that, in the definition of an almost-compact embedding, the sequence $\{E_n\}$ can be taken nonincreasing. This assertion easily follows by replacing $\{E_n\}$ by $\{\bigcup_{k\ge n} E_k\}$ so we omit the proof.

Theorem 7.11.2. *Let X and Y be Banach function spaces over a σ-finite measure space $(\mathcal{R}, \mu)$. Then $X \stackrel{*}{\hookrightarrow} Y$ if and only if*

$$\lim_{n\to\infty} \sup_{\|f\|_X \le 1} \|f\chi_{E_n}\|_Y = 0$$

holds for every sequence $\{E_n\}_{n=1}^\infty$ satisfying $E_n \downarrow \emptyset$ μ-a.e.

We next note that an almost-compact embedding is always preserved between swapped associate spaces.

Theorem 7.11.3. *Let X and Y be Banach function spaces over a σ-finite measure space $(\mathcal{R}, \mu)$. Then $X \stackrel{*}{\hookrightarrow} Y$ if and only if $Y' \stackrel{*}{\hookrightarrow} X'$.*

Proof. Suppose that $X \stackrel{*}{\hookrightarrow} Y$. Let $\{E_n\}_{n=1}^\infty$ be an arbitrary sequence of sets in $\mathcal{R}$ satisfying $E_n \downarrow \emptyset$ μ-a.e. Using the definition of the associate norm and the fact that $Y'' = Y$, we get

$$\lim_{n \to \infty} \sup_{\|g\|_{Y'} \leq 1} \|g \chi_{E_n}\|_{X'} = \lim_{n \to \infty} \sup_{\|g\|_{Y'} \leq 1} \left(\sup_{\|f\|_X \leq 1} \int_\mathcal{R} |fg\chi_{E_n}| \, d\mu \right)$$

$$= \lim_{n \to \infty} \sup_{\|f\|_X \leq 1} \left(\sup_{\|g\|_{Y'} \leq 1} \int_\mathcal{R} |fg\chi_{E_n}| \, d\mu \right)$$

$$= \lim_{n \to \infty} \sup_{\|f\|_X \leq 1} \|f \chi_{E_n}\|_{Y''}$$

$$= \lim_{n \to \infty} \sup_{\|f\|_X \leq 1} \|f \chi_{E_n}\|_Y = 0,$$

i.e. $Y' \stackrel{*}{\hookrightarrow} X'$, as required.

Now suppose that $Y' \stackrel{*}{\hookrightarrow} X'$. From the first part of the proof we obtain $X'' \stackrel{*}{\hookrightarrow} Y''$, that is, by the Lorentz–Luxemburg theorem (Theorem 6.2.9), $X \stackrel{*}{\hookrightarrow} Y$, as desired. The proof is complete. □

The following theorem provides a characterization of $X \stackrel{*}{\hookrightarrow} Y$ in terms of convergence μ-a.e.

Theorem 7.11.4. *Let X and Y be Banach function spaces over a σ-finite measure space $(\mathcal{R}, \mu)$. Then $X \stackrel{*}{\hookrightarrow} Y$ if and only if for every sequence $\{f_n\}_{n=1}^\infty$ of μ-measurable functions on $\mathcal{R}$ satisfying $\|f_n\|_X \leq 1$ and $f_n \to 0$ μ-a.e., one has $\|f_n\|_Y \to 0$.*

Proof. Suppose that $X \stackrel{*}{\hookrightarrow} Y$. First, we will construct a μ-measurable function g such that $g > 0$ on $\mathcal{R}$ and $\|g\|_Y < \infty$. Let $\{R_n\}_{n=1}^\infty$ be the sequence of sets of finite measure satisfying $R_n \uparrow \mathcal{R}$. For every positive integer n, consider a function g_n given by

$$g_n = \frac{1}{2^n} \cdot \frac{1}{1 + \|\chi_{R_n}\|_Y} \cdot \chi_{R_n}.$$

Let us also define the function g by $g = \sum_{n=1}^\infty g_n$. We have

$$\|g_n\|_Y = \frac{1}{2^n} \cdot \frac{1}{1 + \|\chi_{R_n}\|_Y} \cdot \|\chi_{R_n}\|_Y < \frac{1}{2^n}.$$

Thus
$$\|g\|_Y = \lim_{n\to\infty} \left\|\sum_{k=1}^n g_k\right\|_Y \le \lim_{n\to\infty} \sum_{k=1}^n \|g_k\|_Y \le \sum_{k=1}^\infty \frac{1}{2^k} = 1.$$

Because, obviously, $g > 0$ on $\mathcal{R}$, g has the required properties.

Let $\{f_n\}_{n=1}^\infty$ be a sequence of μ-measurable functions on $\mathcal{R}$ satisfying $\|f_n\|_X \le 1$ and $f_n \to 0$ μ-a.e. Choose $\varepsilon > 0$ arbitrarily. Let $E_n = \{x \in \mathcal{R} : |f_n(x)| \ge \varepsilon g(x)\}$. Because $f_n \to 0$ μ-a.e. and $\varepsilon g > 0$ on $\mathcal{R}$, for μ-a.e. $x \in \mathcal{R}$ we have that $x \in E_n$ holds only for a finite number of positive integers n. This implies $E_n \to \emptyset$ μ-a.e.

Observe that
$$\|f_n\|_Y = \|f_n \chi_{E_n} + f_n \chi_{E_n^c}\|_Y \le \|f_n \chi_{E_n}\|_Y + \|f_n \chi_{E_n^c}\|_Y.$$

The assumptions $X \overset{*}{\hookrightarrow} Y$ and $\|f_n\|_X \le 1$ give
$$\lim_{n\to\infty} \|f_n \chi_{E_n}\|_Y \le \lim_{n\to\infty} \sup_{\|h\|_X \le 1} \|h \chi_{E_n}\|_Y = 0.$$

Moreover,
$$\|f_n \chi_{E_n^c}\|_Y \le \|\varepsilon g\|_Y = \varepsilon \|g\|_Y \le \varepsilon.$$

Altogether, we have
$$\limsup_{n\to\infty} \|f_n\|_Y \le \varepsilon,$$
which holds for every $\varepsilon > 0$. So, $\lim_{n\to\infty} \|f_n\|_Y = 0$.

Conversely, suppose that for every sequence $\{f_n\}_{n=1}^\infty$ of μ-measurable functions on $\mathcal{R}$ satisfying $\|f_n\|_X \le 1$ and $f_n \to 0$ μ-a.e., it holds $\|f_n\|_Y \to 0$. Let $\{E_n\}_{n=1}^\infty$ be a sequence of subsets of $\mathcal{R}$ satisfying $E_n \to \emptyset$ μ-a.e. Then we can find a sequence of functions $\{f_n\}_{n=1}^\infty$ such that $\|f_n\|_X \le 1$ and
$$\|f_n \chi_{E_n}\|_Y + \frac{1}{n} > \sup_{\|f\|_X \le 1} \|f \chi_{E_n}\|_Y.$$

Because $E_n \to \emptyset$ μ-a.e., we have $f_n \chi_{E_n} \to 0$ μ-a.e. Due to the assumption, $\|f_n \chi_{E_n}\|_Y \to 0$. Thus
$$\lim_{n\to\infty} \sup_{\|f\|_X \le 1} \|f \chi_{E_n}\|_Y \le \lim_{n\to\infty} \left(\|f_n \chi_{E_n}\|_Y + \frac{1}{n}\right) = 0. \qquad \square$$

Our aim now is to study relations of an almost-compact embedding to other types of embeddings. In particular, in the following two theorems we will show that an almost-compact embedding is in general stronger than a usual (continuous) embedding but weaker than a compact embedding.

Theorem 7.11.5. *Suppose that $(\mathcal{R}, \mu)$ is a σ-finite measure space and X and Y are Banach function spaces over $(\mathcal{R}, \mu)$ satisfying $X \overset{*}{\hookrightarrow} Y$. Then $X \hookrightarrow Y$.*

Proof. Let $\{f_n\}_{n=1}^{\infty}$ be a sequence in X such that $\|f_n - f\|_X \to 0$ for some $f \in X$. To get a contradiction, assume that $\|f_n - f\|_Y \not\to 0$. Then we can find $\varepsilon > 0$ and a subsequence $\{g_k\}_{k=1}^{\infty}$ of $\{f_n\}_{n=1}^{\infty}$ satisfying $\|g_k - f\|_Y \geq \varepsilon$ for every $k \in \mathbb{N}$. Because $g_k \to f$ in X, there is a subsequence $\{h_l\}_{l=1}^{\infty}$ of $\{g_k\}_{k=1}^{\infty}$ such that $h_l \to f$ μ-a.e. Using that $X \overset{*}{\hookrightarrow} Y$, by Theorem 7.11.4 we obtain $\|h_l - f\|_Y \to 0$, which gives a contradiction. So, $X \hookrightarrow Y$. □

Theorem 7.11.6. *Suppose that $(\mathcal{R}, \mu)$ is a σ-finite measure space and X and Y are Banach function spaces over $(\mathcal{R}, \mu)$ satisfying $X \hookrightarrow\hookrightarrow Y$. Then $X \overset{*}{\hookrightarrow} Y$.*

Proof. Let $\{f_n\}_{n=1}^{\infty}$ be a sequence in X such that $\|f_n\|_X \leq 1$ for every $n \in \mathbb{N}$ and $f_n \to 0$ μ-a.e. To get a contradiction, assume that $\|f_n\|_Y \not\to 0$. Then there is $\varepsilon > 0$ and a subsequence $\{g_k\}_{k=1}^{\infty}$ of $\{f_n\}_{n=1}^{\infty}$ satisfying $\|g_k\|_Y \geq \varepsilon$ for every $k \in \mathbb{N}$. Because $\{g_k\}_{k=1}^{\infty}$ is bounded in X and $X \hookrightarrow\hookrightarrow Y$, we can find a subsequence $\{h_l\}_{l=1}^{\infty}$ of $\{g_k\}_{k=1}^{\infty}$ such that $\{h_l\}_{l=1}^{\infty}$ is convergent in Y. But $h_l \to 0$ μ-a.e., so the limit must be 0. So, $\|h_l\|_Y \to 0$, which contradicts the assumption. Thus, $X \overset{*}{\hookrightarrow} Y$. □

Now we will use the results obtained so far to observe that in the cases that might be a possible interest, a Banach function space cannot be almost-compactly embedded into itself.

Definition 7.11.7. We say that a σ-finite measure space $(\mathcal{R}, \mu)$ is *friendly* if there exists a sequence $\{E_n\}_{n=1}^{\infty}$ of μ-measurable subsets of $\mathcal{R}$ such that $E_n \downarrow \emptyset$ μ-a.e. and $\mu(E_n) > 0$ for every $n \in \mathbb{N}$. Every sequence $\{E_n\}_{n=1}^{\infty}$ having such properties will be called a *friendly sequence*.

It is not hard to observe that a measure space is not friendly if and only if its nonatomic part has measure 0 and its atomic part consists of at most finitely many atoms.

Remark 7.11.8. Measure spaces, which are not friendly, are exactly those on which Banach function spaces are almost-compactly embedded into itself. Indeed, an easy observation shows that in the definition of an almost-compact embedding, one can consider only friendly sequences. Thus, if the measure space $(\mathcal{R}, \mu)$ is not friendly, then for every pair of Banach function spaces X and Y we have $X \overset{*}{\hookrightarrow} Y$ and also $Y \overset{*}{\hookrightarrow} X$. So, $X \hookrightarrow Y$ and $Y \hookrightarrow X$, i.e. every two Banach function spaces X and Y coincide and $X \overset{*}{\hookrightarrow} X$.

Conversely, assume that $(\mathcal{R}, \mu)$ is friendly and fix a friendly sequence $\{E_n\}_{n=1}^{\infty}$ in $\mathcal{R}$. The sets $\{E_n\}$ can be with no loss of generality taken in such a way that $\mu(E_n) < \infty$ for every $n \in \mathbb{N}$. If X is a Banach function space over $(\mathcal{R}, \mu)$, one can consider a sequence $\{f_n\}_{n=1}^{\infty}$ of functions in X defined by $f_n = \frac{1}{\|\chi_{E_n}\|_X} \chi_{E_n}$. For every $n \in \mathbb{N}$, we have

$$\sup_{\|f\|_X \leq 1} \|f \chi_{E_n}\|_X \geq \|f_n\|_X = 1,$$

which shows that $X \stackrel{*}{\hookrightarrow} X$ cannot hold in this situation.

Next theorem shows that, on atomic measure spaces, an almost-compact embedding coincides with a compact one.

Theorem 7.11.9. *Suppose that $(\mathcal{R}, \mu)$ is a σ-finite completely atomic measure space. Then for any pair of Banach function spaces X and Y over $(\mathcal{R}, \mu)$, $X \stackrel{*}{\hookrightarrow} Y$ holds if and only if $X \hookrightarrow\hookrightarrow Y$.*

Proof. We have already observed that a compact embedding always implies the almost-compact one.

Assume that $X \stackrel{*}{\hookrightarrow} Y$ for some pair of Banach function spaces X and Y. Let $\{f_n\}_{n=1}^{\infty}$ be a bounded sequence in X. We will show that there is a subsequence $\{f_{n_k}\}_{k=1}^{\infty}$ which converges pointwise to some function f (by the Fatou lemma, $f \in X$). From Theorem 7.11.4 it follows that then $\{f_{n_k}\}_{k=1}^{\infty}$ converges to f in the norm of the space Y, so $X \hookrightarrow\hookrightarrow Y$, as required.

Because the measure space $(\mathcal{R}, \mu)$ is σ-finite, it has at most countably many atoms. We will consider only the case of infinitely many atoms, in the other case the proof proceeds similarly.

Denote by $\{a_k\}_{k=1}^{\infty}$ the sequence containing all atoms of $(\mathcal{R}, \mu)$. Because $\{f_n\}_{n=1}^{\infty}$ is bounded in X, the sequence $\{f_n(a_k)\}_{n=1}^{\infty}$ must be bounded for every $k \in \mathbb{N}$. We will construct a pointwise converging subsequence of $\{f_n\}_{n=1}^{\infty}$ by induction as follows:

We formally set $f_n^0 = f_n$. Suppose that, for some $m \in \mathbb{N} \cup \{0\}$, we have already constructed the sequence $\{f_n^m\}_{n=1}^{\infty}$. Because $(\{f_n^m(a_{m+1})\}_{n=1}^{\infty}$ is bounded, we can find a subsequence $\{f_n^{m+1}\}_{n=1}^{\infty}$ of $\{f_n^m\}_{n=1}^{\infty}$ such that $\{f_n^{m+1}(a_{m+1})\}_{n=1}^{\infty}$ is convergent. Then the diagonal sequence $\{f_n^n\}_{n=1}^{\infty}$ converges pointwise on the entire $\mathcal{R}$, as required. □

Thus, in view of Theorem 7.11.9, in what follows we especially focus on almost-compact embeddings between Banach function spaces over a nonatomic measure space. The case when the measure of the space is 0 is trivial and it was discussed above (note that such a space is not friendly). Furthermore, in the following theorem we observe that there are no almost-compact embeddings between Banach function spaces over nonatomic infinite measure space.

Theorem 7.11.10. *Suppose that $(\mathcal{R}, \mu)$ is a nonatomic σ-finite measure space with $\mu(\mathcal{R}) = \infty$. Then there is no pair of Banach function spaces X and Y over $(\mathcal{R}, \mu)$ such that $X \overset{*}{\hookrightarrow} Y$.*

To prove the theorem we need the following auxiliary assertion.

Lemma 7.11.11. *Suppose that X is a Banach function space over a nonatomic σ-finite measure space $(\mathcal{R}, \mu)$ with $\mu(\mathcal{R}) = \infty$. Let $c \in (0, \infty)$. Then there are constants $C_1, C_2 > 0$ such that for every $A \subseteq \mathcal{R}$ satisfying $\mu(A) = c$, we have*

$$C_1 \leq \|\chi_A\|_X \leq C_2.$$

Proof. To get a contradiction, suppose that

$$\sup\{\|\chi_A\|_X : A \subseteq \mathcal{R}, \mu(A) = c\} = \infty.$$

Then we can find a sequence $\{A_n\}_{n=1}^\infty$ of subsets of $\mathcal{R}$ such that $\mu(A_n) = c$ and $\|\chi_{A_n}\|_X > n^3$ for every $n \in \mathbb{N}$. Because the measure space $(\mathcal{R}, \mu)$ is nonatomic, each A_n can be written as a union of n^2 disjoint sets $A_n^1, A_n^2, \ldots, A_n^{n^2}$ of measure $\frac{c}{n^2}$. We have

$$\sum_{i=1}^{n^2} \|\chi_{A_n^i}\|_X \geq \|\chi_{A_n}\|_X > n^3,$$

so for every $n \in \mathbb{N}$ there is $i(n) \in \{1, 2, \ldots, n^2\}$ such that $\|\chi_{A_n^{i(n)}}\|_X > n$. We denote $B_n = A_n^{i(n)}$.

Consider the set $B = \bigcup_{n=1}^\infty B_n$. Then

$$\mu(B) \leq \sum_{n=1}^\infty \mu(B_n) = c \sum_{n=1}^\infty \frac{1}{n^2} < \infty,$$

so, due to the property (P4) of Banach function spaces, $\|\chi_B\|_X < \infty$. On the other hand, for every $n \in \mathbb{N}$ we have

$$\|\chi_B\|_X \geq \|\chi_{B_n}\|_X > n,$$

which is a contradiction. This justifies the existence of the constant C_2.

Consider now the associate space X'. From the first part of the proof we know that there is a constant $D > 0$ such that $\|\chi_A\|_{X'} \leq D$, whenever A is a subset of $\mathcal{R}$ of measure c. Fix such a set A. Because the function $\frac{\chi_A}{\|\chi_A\|_{X'}}$ belongs to the unit ball of X', we have

$$\|\chi_A\|_X = \sup_{\|f\|_{X'} \leq 1} \int_A |f|\, d\mu \geq \frac{\mu(A)}{\|\chi_A\|_{X'}} \geq \frac{c}{D}.$$

By setting $C_1 = \frac{c}{D}$, we get the result. □

Proof of Theorem 7.11.10. Denote $E_n = \mathcal{R} \setminus R_n$. Then $E_n \downarrow \emptyset$. Moreover, $\mu(E_n) = \infty$ for every $n \in \mathbb{N}$. Combining this with the fact that $(\mathcal{R}, \mu)$ is nonatomic, we get that for every $n \in \mathbb{N}$ there is a set $A_n \subseteq E_n$ with $\mu(A_n) = 1$.

Let X and Y be any pair of Banach function spaces over $(\mathcal{R}, \mu)$. According to Lemma 7.11.11, we can find positive constants C, D such that for every $n \in \mathbb{N}$, $\|\chi_{A_n}\|_X \leq C$ and $\|\chi_{A_n}\|_Y \geq D$. The first inequality implies that $\|\frac{1}{C}\chi_{A_n}\|_X \leq 1$, so

$$\sup_{\|f\|_X \leq 1} \|f \chi_{E_n}\|_Y \geq \left\|\frac{1}{C}\chi_{A_n}\right\|_Y \geq \frac{D}{C} > 0.$$

Therefore $X \overset{*}{\hookrightarrow} Y$ cannot be true. □

We shall now present a characterization of almost-compact embeddings on nonatomic measure spaces. It follows from this lemma that our definition of an almost-compact embedding is a generalization of the definition of absolutely continuous embedding, given in [80] for the case of rearrangement-invariant spaces over $((0, 1), dx)$.

Lemma 7.11.12. *Suppose that $(\mathcal{R}, \mu)$ is a nonatomic measure space with $0 < \mu(\mathcal{R}) < \infty$ and X and Y are Banach function spaces over $(\mathcal{R}, \mu)$. Then the following two statements are equivalent:*

(i) $X \overset{*}{\hookrightarrow} Y$;

(ii) $\lim_{t \to 0_+} \sup_{\|f\|_X \leq 1} \sup_{\mu(E) \leq t} \|f \chi_E\|_Y = 0.$

Proof. (i) $\Rightarrow$ (ii) Consider a function H defined by

$$H(t) = \sup_{\|f\|_X \leq 1} \sup_{\mu(E) \leq t} \|f \chi_E\|_Y, \quad t \in (0, \mu(\mathcal{R})].$$

Clearly, H is nondecreasing on $(0, \mu(\mathcal{R})]$. Thus, it will be enough to prove

$$\lim_{n \to \infty} \sup_{\|f\|_X \leq 1} \sup_{\mu(E) \leq a_n} \|f \chi_E\|_Y = 0 \tag{7.11.1}$$

for some sequence $\{a_n\}_{n=1}^\infty$ with $a_n \downarrow 0$. Our proof will work for any such sequence satisfying, moreover, that $\sum_{n=1}^\infty a_n < \infty$.

For every $n \in \mathbb{N}$ we can find $f_n \in X$, $E_n \subseteq \mathcal{R}$ such that $\|f_n\|_X \leq 1$, $\mu(E_n) \leq a_n$ and

$$\sup_{\|f\|_X \leq 1} \sup_{\mu(E) \leq a_n} \|f \chi_E\|_Y < \|f_n \chi_{E_n}\|_Y + \frac{1}{n}. \tag{7.11.2}$$

Denote $F_n = \bigcup_{k=n}^\infty E_k$. Then $F_1 \supseteq F_2 \supseteq \ldots$ and

$$\mu\left(\bigcap_{n=1}^\infty F_n\right) = \lim_{n \to \infty} \mu(F_n) \leq \lim_{n \to \infty} \sum_{k=n}^\infty \mu(E_k) \leq \lim_{n \to \infty} \sum_{k=n}^\infty a_k = 0.$$

This implies $F_n \downarrow \emptyset$ μ-a.e. Because $E_n \subseteq F_n$ for every $n \in \mathbb{N}$ and $X \overset{*}{\hookrightarrow} Y$, we have

$$\lim_{n\to\infty} \|f_n \chi_{E_n}\|_Y \le \lim_{n\to\infty} \sup_{\|f\|_X \le 1} \|f \chi_{E_n}\|_Y \le \lim_{n\to\infty} \sup_{\|f\|_X \le 1} \|f \chi_{F_n}\|_Y = 0.$$

Using the inequality (7.11.2), we obtain (7.11.1).

(ii) $\Rightarrow$ (i) Choose an arbitrary sequence $\{E_n\}_{n=1}^\infty$ of subsets of $\mathcal{R}$ such that $E_n \downarrow \emptyset$ μ-a.e. Because $\mu(\mathcal{R}) < \infty$ we have

$$\lim_{n\to\infty} \mu(E_n) = \mu\left(\bigcap_{n=1}^\infty E_n\right) = 0.$$

Thus

$$\lim_{n\to\infty} \sup_{\|f\|_X \le 1} \|f \chi_{E_n}\|_Y \le \lim_{n\to\infty} \sup_{\|f\|_X \le 1} \sup_{\mu(E) = \mu(E_n)} \|f \chi_E\|_Y = 0,$$

so $X \overset{*}{\hookrightarrow} Y$. $\square$

We shall now point out that L^∞ is almost-compactly embedded into every rearrangement-invariant Banach function space essentially different from itself.

Theorem 7.11.13. *Let $(\mathcal{R}, \mu)$ be a nonatomic measure space satisfying $0 < \mu(\mathcal{R}) < \infty$ and let X be a rearrangement-invariant Banach function space over $(\mathcal{R}, \mu)$. Denote by φ the fundamental function of X. Then the following statements are equivalent:*

(i) $X \ne L^\infty$;

(ii) $L^\infty \overset{*}{\hookrightarrow} X$;

(iii) $\lim_{t \to 0+} \varphi(t) = 0$.

Proof. (i) $\Rightarrow$ (iii) According to (i), there is a function $f \in X$ such that ess $\sup_\mathcal{R} |f| = \infty$. Consider a sequence of sets $(E_n)_{n=1}^\infty$ defined by

$$E_n = \{x \in \mathcal{R} : |f(x)| \ge n\}.$$

Because ess $\sup_\mathcal{R} |f| = \infty$, we have $\mu(E_n) > 0$, $n \in \mathbb{N}$. From the inequality $n \chi_{E_n} \le |f|$ we obtain $\|\chi_{E_n}\|_X \le \frac{1}{n} \|f\|_X$. Hence, $0 = \inf_{\mu(E)>0} \|\chi_E\|_X = \inf_{t>0} \varphi(t)$. Since the function φ is nondecreasing on $(0, 1)$, we get (iii).

(iii) $\Rightarrow$ (ii) The constant function 1 belongs to the unit ball of L^∞ and for every function f from the unit ball of L^∞ we have $|f| \le 1$ μ-a.e. This implies

$$\lim_{t\to 0+} \sup_{\|f\|_{L^\infty} \le 1} \sup_{\mu(E) \le t} \|f \chi_E\|_X = \lim_{t\to 0+} \sup_{\mu(E) \le t} \|\chi_E\|_X = \lim_{t\to 0+} \varphi(t) = 0.$$

Using Lemma 7.11.12 we get the result.

(ii) ⇒ (i) Because every nonatomic measure space $(\mathcal{R}, \mu)$ with $\mu(\mathcal{R}) > 0$ is friendly, $L^\infty \overset{*}{\hookrightarrow} L^\infty$ cannot be true. □

Near the other endpoint space, L^1, we have an analogous result. Namely, L^1 is an almost-compact target for every rearrangement-invariant Banach function space essentially different from itself. This assertion can be obtained from Theorem 7.11.13 by duality arguments.

Theorem 7.11.14. *Let $(\mathcal{R}, \mu)$ be a nonatomic measure space satisfying $0 < \mu(\mathcal{R}) < \infty$ and let X be a rearrangement-invariant Banach function space over $(\mathcal{R}, \mu)$. Denote by φ the fundamental function of X. Then the following statements are equivalent:*

(i) $X \neq L^1$;

(ii) $X \overset{*}{\hookrightarrow} L^1$;

(iii) $\lim_{t \to 0+} \frac{t}{\varphi(t)} = 0$.

We shall present an important necessary condition for an almost-compact embedding between two rearrangement-invariant spaces in terms of their fundamental functions.

Lemma 7.11.15. *Suppose that $(\mathcal{R}, \mu)$ is a nonatomic measure space satisfying $0 < \mu(\mathcal{R}) < \infty$ and let X and Y be rearrangement-invariant Banach function spaces over $(\mathcal{R}, \mu)$. Let S denote the set of nonnegative nonzero simple functions on $\mathcal{R}$. Then the following conditions are equivalent.*

(i) $X \overset{*}{\hookrightarrow} Y$;

(ii) $\lim_{t \to 0+} \sup_{\|f\|_X \leq 1} \|f^* \chi_{[0,t)}\|_{\bar{Y}} = 0$;

(iii) $\lim_{t \to 0+} \sup_{u \in S} \frac{\|u^* \chi_{[0,t)}\|_{\bar{Y}}}{\|u^* \chi_{[0,t)}\|_{\bar{X}}} = 0$.

Proof. (i) ⇔ (ii) Due to Lemma 7.11.12, $X \overset{*}{\hookrightarrow} Y$ is equivalent to

$$\lim_{t \to 0+} \sup_{\|f\|_X \leq 1} \sup_{\mu(E) \leq t} \|f \chi_E\|_Y = 0.$$

Thus it is sufficient to show that, for every $f \in X$ and for every $t \in (0, \mu(\mathcal{R}))$,

$$\sup_{\mu(E) \leq t} \|f \chi_E\|_Y = \|f^* \chi_{[0,t)}\|_{\bar{Y}}.$$

Fix $f \in X$ and $t \in (0, \mu(\mathcal{R}))$. Whenever E is a measurable subset of $\mathcal{R}$ with $\mu(E) \leq t$, we have

$$\|f \chi_E\|_Y = \|(f \chi_E)^*\|_{\bar{Y}} = \|(f \chi_E)^* \chi_{[0,t)}\|_{\bar{Y}} \leq \|f^* \chi_{[0,t)}\|_{\bar{Y}},$$

and therefore
$$\sup_{\mu(E)\leq t} \|f\chi_E\|_Y \leq \|f^*\chi_{[0,t)}\|_{\bar Y}.$$

For $f \in X$ and $t \in (0, \mu(\mathcal{R}))$, we can find a measurable set $F \subseteq \mathcal{R}$ with $\mu(F) = t$ such that $f^*\chi_{[0,t)} = (f\chi_F)^*$ (this follows from the proof of Lemma 7.3.6). Thus, we can write
$$\sup_{\mu(E)\leq t} \|f\chi_E\|_Y = \sup_{\mu(E)\leq t} \|(f\chi_E)^*\|_{\bar Y} \geq \|(f\chi_F)^*\|_{\bar Y} = \|f^*\chi_{[0,t)}\|_{\bar Y},$$
which gives the reverse inequality.

(ii) $\Leftrightarrow$ (iii) We will show that for every $t \in (0, \mu(\mathcal{R}))$
$$\sup_{\|f\|_X \leq 1} \|f^*\chi_{[0,t)}\|_{\bar Y} = \sup_{u \in S} \frac{\|u^*\chi_{[0,t)}\|_{\bar Y}}{\|u^*\chi_{[0,t)}\|_{\bar X}},$$
which is obviously enough for the proof.

Suppose that $f \in X$, $\|f\|_X \leq 1$. Then we have
$$f^*\chi_{[0,t)} \leq \frac{f^*}{\|f\|_X}\chi_{[0,t)} = \left(\frac{f}{\|f\|_X}\right)^*\chi_{[0,t)},$$
which gives
$$\|f^*\chi_{[0,t)}\|_{\bar Y} \leq \left\|\left(\frac{f}{\|f\|_X}\right)^*\chi_{[0,t)}\right\|_{\bar Y} \leq \sup_{\|g\|_X=1} \|g^*\chi_{[0,t)}\|_{\bar Y},$$
so
$$\sup_{\|f\|_X\leq 1} \|f^*\chi_{[0,t)}\|_{\bar Y} \leq \sup_{\|f\|_X=1} \|f^*\chi_{[0,t)}\|_{\bar Y}.$$
The reverse inequality is obvious, thus
$$\sup_{\|f\|_X\leq 1} \|f^*\chi_{[0,t)}\|_{\bar Y} = \sup_{\|f\|_X=1} \|f^*\chi_{[0,t)}\|_{\bar Y}.$$
Furthermore,
$$\sup_{\|f\|_X=1} \|f^*\chi_{[0,t)}\|_{\bar Y} = \sup_{0\neq f\in X} \left\|\left(\frac{f}{\|f\|_X}\right)^*\chi_{[0,t)}\right\|_{\bar Y}$$
$$= \sup_{0\neq f\in X} \frac{\|f^*\chi_{[0,t)}\|_{\bar Y}}{\|f\|_X}$$
$$= \sup_{0\neq f\in X} \frac{\|f^*\chi_{[0,t)}\|_{\bar Y}}{\|f^*\|_{\bar X}}.$$

Section 7.11 Almost-compact embeddings

We need to show that

$$\sup_{0\neq f\in X} \frac{\|f^*\chi_{[0,t)}\|_{\tilde{Y}}}{\|f^*\|_{\tilde{X}}} = \sup_{0\neq f\in X} \frac{\|f^*\chi_{[0,t)}\|_{\tilde{Y}}}{\|f^*\chi_{[0,t)}\|_{\tilde{X}}}.$$

Because $\|f^*\|_{\tilde{X}} \geq \|f^*\chi_{[0,t)}\|_{\tilde{X}}$, it must be

$$\sup_{0\neq f\in X} \frac{\|f^*\chi_{[0,t)}\|_{\tilde{Y}}}{\|f^*\|_{\tilde{X}}} \leq \sup_{0\neq f\in X} \frac{\|f^*\chi_{[0,t)}\|_{\tilde{Y}}}{\|f^*\chi_{[0,t)}\|_{\tilde{X}}}.$$

On the other hand, whenever $f \in X$, $f \neq 0$, we can find a measurable set F such that $\mu(F) = t$ and $f^*\chi_{[0,t)} = (f\chi_F)^* = (f\chi_F)^*\chi_{[0,t)}$. Then

$$\frac{\|f^*\chi_{[0,t)}\|_{\tilde{Y}}}{\|f^*\chi_{[0,t)}\|_{\tilde{X}}} = \frac{\|(f\chi_F)^*\chi_{[0,t)}\|_{\tilde{Y}}}{\|(f\chi_F)^*\|_{\tilde{X}}} \leq \sup_{0\neq g\in X} \frac{\|g^*\chi_{[0,t)}\|_{\tilde{Y}}}{\|g^*\|_{\tilde{X}}},$$

which gives the reverse inequality.

Finally, we observe that the supremum can be taken over the (smaller) set S instead of $X \setminus \{0\}$. Indeed, for every $f \in X$, $f \neq 0$, we can find a sequence $\{u_n\}_{n=1}^\infty$ with $u_n \in S$, $(n \in \mathbb{N})$, and $u_n \uparrow |f|$. This implies $u_n^*\chi_{[0,t)} \uparrow f^*\chi_{[0,t)}$, and thus

$$\lim_{n\to\infty} \frac{\|u_n^*\chi_{[0,t)}\|_{\tilde{Y}}}{\|u_n^*\chi_{[0,t)}\|_{\tilde{X}}} = \frac{\|f^*\chi_{[0,t)}\|_{\tilde{Y}}}{\|f^*\chi_{[0,t)}\|_{\tilde{X}}},$$

which gives the result. □

The next result yields a necessary condition for an almost-compact embedding in terms of fundamental functions.

Theorem 7.11.16. *If $(\mathcal{R}, \mu)$ is a nonatomic measure space satisfying $0 < \mu(\mathcal{R}) < \infty$ and X and Y are rearrangement-invariant Banach function spaces over $(\mathcal{R}, \mu)$ such that $X \stackrel{*}{\hookrightarrow} Y$, then*

$$\lim_{t\to 0+} \frac{\varphi_Y(t)}{\varphi_X(t)} = 0,$$

where φ_X, φ_Y are fundamental functions of X, Y respectively.

Proof. The function $f = 1$ on $\mathcal{R}$ belongs to S defined in the Lemma 7.11.15. Thus, for $t \in (0, \mu(\mathcal{R})]$,

$$\frac{\varphi_Y(t)}{\varphi_X(t)} = \frac{\|f^*\chi_{[0,t)}\|_Y}{\|f^*\chi_{[0,t)}\|_X} \leq \sup_{u\in S} \frac{\|u^*\chi_{[0,t)}\|_Y}{\|u^*\chi_{[0,t)}\|_X}.$$

According to the lemma, we have

$$\lim_{t\to 0+} \frac{\varphi_Y(t)}{\varphi_X(t)} = 0. \quad \Box$$

Corollary 7.11.17. *Let $(\mathcal{R}, \mu)$ be a nonatomic measure space satisfying $0 < \mu(\mathcal{R}) < \infty$ and let φ be a positive nondecreasing concave function on $(0, \mu(\mathcal{R}))$. Then it does not hold that $\Lambda_\varphi \stackrel{*}{\hookrightarrow} M_\varphi$.*

Proof. Λ_φ and M_φ have the same fundamental function φ so the necessary condition for almost-compact embedding from Theorem 7.11.16 cannot hold. □

Our next example shows that the necessary condition from Theorem 7.11.16 is not sufficient for an almost-compact embedding.

Example 7.11.18. Suppose that X, Y are rearrangement-invariant Banach function spaces with fundamental functions φ_X, φ_Y, respectively. We will show that the condition $\lim_{t \to 0+} \frac{\varphi_Y(t)}{\varphi_X(t)} = 0$ does not imply $X \hookrightarrow Y$. In particular, it does not imply $X \stackrel{*}{\hookrightarrow} Y$.

Let $p \in (1, \infty)$. Denote by p' the conjugate index satisfying $\frac{1}{p} + \frac{1}{p'} = 1$. We will consider for X the Marcinkiewicz space $L^{p,\infty}$ and for Y the *Lorentz-Zygmund space* $L^{p,1;-1}$ over $((0,1), \mu_1)$, consisting of all measurable functions f such that

$$\|f\|_{p,\infty} = \sup_{t \in (0,1)} f^*(t) t^{\frac{1}{p}} < \infty$$

and

$$\|f\|_{p,1;-1} = \int_0^1 \frac{f^*(s) s^{\frac{1}{p}-1}}{e - \log s} \, ds = \int_0^1 \frac{f^*(s)}{s^{\frac{1}{p'}}(e - \log s)} \, ds < \infty,$$

respectively. We note that the functionals $\|\cdot\|_{p,\infty}$ and $\|\cdot\|_{p,1;-1}$ are equivalent to rearrangement-invariant Banach function norms.

Next, we have $L^{p,\infty} \not\hookrightarrow L^{p,1;-1}$ (see Theorem 9.5.14). Moreover, if we denote by φ the fundamental function of $L^{p,\infty}$ and by ψ the fundamental function of $L^{p,1;-1}$, then, for every $t \in (0,1)$, we have

$$\varphi(t) = \sup_{s \in (0,1)} \chi_{(0,t)}(s) s^{\frac{1}{p}} = t^{\frac{1}{p}},$$

while

$$\psi(t) = \int_0^t \frac{1}{s^{\frac{1}{p'}}(e - \log s)} \, ds.$$

Because

$$\lim_{t \to 0+} \varphi(t) = \lim_{t \to 0+} \psi(t) = 0,$$

we can use the L'Hospital rule to get

$$\lim_{t \to 0+} \frac{\psi(t)}{\varphi(t)} = \lim_{t \to 0+} \frac{\psi'(t)}{\varphi'(t)} = \lim_{t \to 0+} \frac{p}{e - \log t} = 0.$$

Example 7.11.19. Let $(\mathcal{R}, \mu)$ be a nonatomic measure space such that $\mu(\mathcal{R}) = 1$. Suppose that Φ_1, Φ_2 are Young functions. Then the almost-compact embedding $L^{\Phi_1}(\mathcal{R}, \mu) \overset{*}{\hookrightarrow} L^{\Phi_2}(\mathcal{R}, \mu)$ between the corresponding Orlicz spaces holds if and only if, for every $\lambda > 0$,

$$\lim_{t \to \infty} \frac{\Phi_2(\lambda t)}{\Phi_1(t)} = 0 \tag{7.11.3}$$

(cf. Theorems 4.17.7 and 4.17.9).

Remark 7.11.20. Suppose that $p \in (1, \infty)$ and consider the Young function $\Phi(t) = t^p$. Then the Orlicz space L^Φ coincides with the Lebesgue space L^p. The characterization of almost-compact embedding between Orlicz spaces from the previous example together with Theorems 7.11.13 and 7.11.14 shows that for $1 \leq p, q \leq \infty$, $L^p \overset{*}{\hookrightarrow} L^q$ holds if and only if $q < p$. Note that $L^p \hookrightarrow L^q$ if and only if $q \leq p$, while $L^p \hookrightarrow\hookrightarrow L^q$ is never true.

In the rest of section we will present a complete characterization of all possible mutual almost-compact embeddings among the Lorentz and Marcinkiewicz endpoint spaces. We shall work for our typographical convenience on the measure space $((0, 1), dx)$. This of course can be done with no loss of generality and the results of this section can be easily extended to all nonatomic finite measure spaces.

Suppose that φ is a quasi-concave function on $(0, 1)$. In the following text, $\tilde{\varphi}$ denotes the quasi-concave function satisfying $\tilde{\varphi}(t) = \frac{t}{\varphi(t)}$ for every $t \in (0, 1)$.

Lemma 7.11.21. *Let φ and ψ be quasi-concave functions on $(0, 1)$. Suppose that there exist positive constants C_1, C_2 such that*

$$C_1 \varphi(t) \leq \psi(t) \leq C_2 \varphi(t)$$

for every $t \in (0, 1)$. Then $M_\varphi = M_\psi$.

Proof. Assume that $f \in M_\varphi$. Then $f \in M_\psi$, because

$$\|f\|_{M_\psi} = \sup_{t \in (0,1)} \psi(t) f^{**}(t) \leq C_2 \sup_{t \in (0,1)} \varphi(t) f^{**}(t) = C_2 \|f\|_{M_\varphi} < \infty.$$

The converse embedding follows from symmetry. $\square$

Lemma 7.11.22. *Suppose that φ is a quasi-concave function on $(0, 1)$. Let α be the least nondecreasing concave majorant of $\tilde{\varphi}$. Then $M'_\varphi = \Lambda_\alpha$.*

Proof. The function α satisfies $\frac{\alpha(t)}{2} \leq \frac{t}{\varphi(t)} \leq \alpha(t)$ for $t \in (0, 1)$. Thus also $\frac{\varphi(t)}{2} \leq \frac{t}{\alpha(t)} \leq \varphi(t)$ on $(0, 1)$. Due to Lemma 7.11.21, $M_\varphi = M_{\tilde{\tilde{\alpha}}}$.

First, we will show that $M'_\varphi \subseteq \Lambda_\alpha$. This is equivalent to $\Lambda'_\alpha \subseteq M''_\varphi = M_\varphi = M_{\tilde{\tilde{\alpha}}}$. But Λ'_α has the fundamental function $\tilde{\alpha}$ and $M_{\tilde{\tilde{\alpha}}}$ is the largest rearrangement-invariant space with this fundamental function, so $\Lambda'_\alpha \subseteq M_{\tilde{\tilde{\alpha}}}$.

On the other hand, because $M_\varphi = M_{\tilde{\alpha}}$, we have $M'_\varphi = M'_{\tilde{\alpha}}$. Using that $M'_{\tilde{\alpha}}$ has the fundamental function α and Λ_α is the smallest rearrangement-invariant space with fundamental function α, we obtain $\Lambda_\alpha \subseteq M'_{\tilde{\alpha}} = M'_\varphi$. □

Lemma 7.11.23. *Let φ and ψ be positive nondecreasing concave functions on $(0, 1)$. Suppose that there exist positive constants C_1, C_2 such that*

$$C_1 \varphi(t) \leq \psi(t) \leq C_2 \varphi(t) \tag{7.11.4}$$

for every $t \in (0, 1)$. Then $\Lambda_\varphi = \Lambda_\psi$.

Proof. According to Lemma 7.11.22, $\Lambda_\varphi = M'_{\tilde{\varphi}}$ and $\Lambda_\psi = M'_{\tilde{\psi}}$. The assumption (7.11.4) gives

$$C_1 \tilde{\psi}(t) \leq \tilde{\varphi}(t) \leq C_2 \tilde{\psi}(t)$$

for every $t \in (0, 1)$. So, due to Lemma 7.11.21, $M_{\tilde{\varphi}} = M_{\tilde{\psi}}$, thus also $\Lambda_\varphi = M'_{\tilde{\varphi}} = M'_{\tilde{\psi}} = \Lambda_\psi$. □

Theorem 7.11.24. *Suppose that φ and ψ are positive nondecreasing concave functions on $(0, 1)$. Then the following four statements are equivalent.*

(i) $\Lambda_\varphi \overset{*}{\hookrightarrow} \Lambda_\psi$;

(ii) $M_\varphi \overset{*}{\hookrightarrow} M_\psi$;

(iii) $\Lambda_\varphi \overset{*}{\hookrightarrow} M_\psi$;

(iv) $\lim_{t \to 0+} \frac{\psi(t)}{\varphi(t)} = 0$.

Proof. According to Theorem 7.11.16, each of the conditions (i), (ii), (iii) implies (iv).

(iv) $\Rightarrow$ (i) Due to Lemma 7.11.15, we only need to prove that

$$\lim_{t \to 0+} \sup_{u \in S} \frac{\|u^* \chi_{(0,t)}\|_{\Lambda_\psi}}{\|u^* \chi_{(0,t)}\|_{\Lambda_\varphi}} = 0,$$

where S denotes the set of nonnegative nonzero simple functions on $(0, 1)$.

Suppose that $u \in S$. Given $t \in (0, 1)$, we have

$$u^* \chi_{(0,t)} = \sum_{i=1}^{n} c_i \chi_{(0,t_i)},$$

where $c_i > 0$, $i = 1, 2, \ldots, n$ and $0 < t_1 < \cdots < t_n \leq t$. Because

$$\frac{\psi(t_i)}{\varphi(t_i)} \leq \sup_{0 < s \leq t} \frac{\psi(s)}{\varphi(s)},$$

Section 7.11 Almost-compact embeddings

we have

$$\frac{\|u^*\chi_{(0,t)}\|_{\Lambda_\psi}}{\|u^*\chi_{(0,t)}\|_{\Lambda_\varphi}} = \frac{\sum_{i=1}^n c_i \psi(t_i)}{\sum_{i=1}^n c_i \varphi(t_i)} \leq \frac{\sum_{i=1}^n c_i \varphi(t_i) \sup_{0<s\leq t} \frac{\psi(s)}{\varphi(s)}}{\sum_{i=1}^n c_i \varphi(t_i)} = \sup_{0<s\leq t} \frac{\psi(s)}{\varphi(s)}.$$

Thus,

$$\lim_{t\to 0+} \sup_{u\in S} \frac{\|u^*\chi_{(0,t)}\|_{\Lambda_\psi}}{\|u^*\chi_{(0,t)}\|_{\Lambda_\varphi}} \leq \lim_{t\to 0+} \sup_{0<s\leq t} \frac{\psi(s)}{\varphi(s)} = 0.$$

(iv) $\Rightarrow$ (ii) Denote by α, β the least nondecreasing concave majorant of $\tilde\varphi, \tilde\psi$, respectively. Then

$$\frac{\alpha(t)}{\beta(t)} \leq 2\frac{\tilde\varphi(t)}{\tilde\psi(t)} = 2\frac{\psi(t)}{\varphi(t)}$$

for every $t \in (0, 1)$. The assumption (iv) gives

$$\lim_{t\to 0+} \frac{\alpha(t)}{\beta(t)} \leq \lim_{t\to 0+} 2\frac{\psi(t)}{\varphi(t)} = 0.$$

Using the implication (iv) $\Rightarrow$ (i), which was just proved, for functions α, β, we obtain $\Lambda_\beta \overset{*}{\hookrightarrow} \Lambda_\alpha$. Due to Lemma 7.11.22, $\Lambda_\alpha = M'_\varphi$ and $\Lambda_\beta = M'_\psi$. Thus we have $M'_\psi \overset{*}{\hookrightarrow} M'_\varphi$, which (by Theorem 7.11.3) implies $M_\varphi \overset{*}{\hookrightarrow} M_\psi$, as required.

(ii) $\Rightarrow$ (iii) This is a consequence of the facts that $\Lambda_\varphi \hookrightarrow M_\varphi$ and $M_\varphi \overset{*}{\hookrightarrow} M_\psi$. $\square$

It will be useful to study inclusions into the subset of an r.i. space containing the functions having absolutely continuous norms.

If X and Y are Banach function spaces, then an inclusion of the space X into the subspace of functions having absolutely continuous norm in Y, denoted by $X \subseteq Y_a$, is a necessary condition for $X \overset{*}{\hookrightarrow} Y$.

Remark 7.11.25. Assume that Λ_φ is a Lorentz endpoint space different from L^∞ (i.e. $\lim_{t\to 0+} \varphi(t) = 0$). Whenever X is a rearrangement-invariant space such that $\Lambda_\varphi \hookrightarrow X$, then $\Lambda_\varphi \subseteq X_a$. Indeed, for every $f \in \Lambda_\varphi$ we have

$$\lim_{t\to 0+} \|f^*\chi_{(0,t)}\|_X \leq \lim_{t\to 0+} C\|f^*\chi_{(0,t)}\|_{\Lambda_\varphi} = C \lim_{t\to 0+} \int_0^t f^*(s)\varphi'(s)\,ds = 0,$$

where the existence of the constant C is due to $\Lambda_\varphi \hookrightarrow X$. Thus, in particular, $\Lambda_\varphi \subseteq (M_\varphi)_a$.

While for Lorentz endpoint spaces the inclusion into the subspace of functions having absolutely continuous norm coincides with the regular inclusion, the situation is quite different when one deals with Marcinkiewicz spaces.

Theorem 7.11.26. *Let X be a rearrangement-invariant Banach function space over $((0, 1), \mathrm{d}x)$ and let φ be a positive nondecreasing concave function on $(0, 1)$ such that $\lim_{t \to 0+} \frac{t}{\varphi(t)} = 0$. Denote by α the least nondecreasing concave majorant of $\frac{t}{\varphi(t)}$. Then the following conditions are equivalent:*

(i) $M_\varphi \overset{*}{\hookrightarrow} X$;

(ii) $M_\varphi \subseteq X_a$;

(iii) $\alpha' \in X_a$.

Proof. The implication (i) $\Rightarrow$ (ii) obviously holds. To prove that (ii) implies (iii) we will show that $\alpha' \in M_\varphi$. Recall that the function α satisfies $\frac{\alpha(t)}{2} \leq \frac{t}{\varphi(t)} \leq \alpha(t)$ for every $t \in (0, 1)$. We have

$$\lim_{t \to 0+} \alpha(t) \leq \lim_{t \to 0+} \frac{2t}{\varphi(t)} = 0.$$

We also have

$$\|\alpha'\|_{M_\varphi} = \sup_{t \in (0,1)} \frac{\varphi(t)}{t} \int_0^t \alpha'(s) \, \mathrm{d}s = \sup_{t \in (0,1)} \frac{\varphi(t)}{t} \cdot \alpha(t) \leq \sup_{t \in (0,1)} \frac{\varphi(t)}{t} \cdot \frac{2t}{\varphi(t)} = 2 < \infty,$$

so $\alpha' \in M_\varphi$.

Finally, suppose that (iii) holds. To prove (i) it is enough to show that $X' \overset{*}{\hookrightarrow} M'_\varphi = \Lambda_\alpha$, which is equivalent to

$$\lim_{t \to 0+} \sup_{\|f\|_{X'} \leq 1} \|f^* \chi_{(0,t)}\|_{\Lambda_\alpha} = 0.$$

Due to the Hölder inequality, for $t \in (0, 1)$

$$\|f^* \chi_{(0,t)}\|_{\Lambda_\alpha} = \int_0^1 f^*(s) \alpha'(s) \chi_{(0,t)} \, \mathrm{d}s \leq \|f\|_{X'} \|\alpha' \chi_{(0,t)}\|_X.$$

So, using that $\alpha' \in X_a$,

$$\lim_{t \to 0+} \sup_{\|f\|_{X'} \leq 1} \|f^* \chi_{(0,t)}\|_{\Lambda_\alpha} \leq \lim_{t \to 0+} \|\alpha' \chi_{(0,t)}\|_X = 0.$$

This completes the proof. □

When we use the previous theorem with $X = \Lambda_\psi$, we get the following corollary, in which we characterize almost-compact embeddings of Marcinkiewicz endpoint spaces into Lorentz endpoint spaces.

Corollary 7.11.27. *Suppose that φ and ψ are positive nondecreasing concave functions on $(0, 1)$ and α is the least nondecreasing concave majorant of $\frac{t}{\varphi(t)}$. Assume that $\lim_{t \to 0+} \psi(t) = 0$ and $\lim_{t \to 0+} \frac{t}{\varphi(t)} = 0$. Then the following conditions are equivalent:*

(i) $M_\varphi \stackrel{*}{\hookrightarrow} \Lambda_\psi$;

(ii) $M_\varphi \hookrightarrow \Lambda_\psi$;

(iii) $\int_0^1 \alpha'(s)\psi'(s)\,ds < \infty.$

Moreover, if ψ' is locally absolutely continuous on $(0, 1)$, then the conditions (i), (ii), (iii) are equivalent to

(iv) $\int_0^1 \frac{s(-\psi''(s))}{\varphi(s)}\,ds < \infty.$

Proof. Only the equivalence of (iii) and (iv) requires proof.

Because ψ' is locally absolutely continuous on $(0, 1)$, we have $\psi'(s) - \psi'(t) = \int_s^t (-\psi''(r))\,dr$, whenever $0 < s, t < 1$. Passing to the limit when t tends to 1, we get $\psi'(s) - \psi'(1_-) = \int_s^1 (-\psi''(r))\,dr$, $0 < s < 1$. Using the Fubini theorem, we obtain

$$\int_0^1 \alpha'(s)\psi'(s)\,ds = \int_0^1 \alpha'(s) \int_s^1 (-\psi''(r))\,dr\,ds + \int_0^1 \psi'(1_-)\alpha'(s)\,ds$$

$$= \int_0^1 (-\psi''(r)) \int_0^r \alpha'(s)\,ds\,dr + \alpha(1_-)\psi'(1_-)$$

$$= \int_0^1 \alpha(r)(-\psi''(r))\,dr + \alpha(1_-)\psi'(1_-).$$

The product $\alpha(1_-)\psi'(1_-)$ is always finite, so

$$\int_0^1 \alpha'(s)\psi'(s)\,ds < \infty \text{ iff } \int_0^1 \alpha(s)(-\psi''(s))\,ds < \infty.$$

Using the inequality $\frac{\alpha(s)}{2} \leq \frac{s}{\varphi(s)} \leq \alpha(s)$, $0 < s < 1$, the assertion follows. $\square$

Our next example shows that the condition (iv) is not sufficient for an almost-compact embedding when ψ' is not locally absolutely continuous.

Example 7.11.28. Suppose that $p \in (0, 1)$. Let $\varphi(t) = t^p$. Define a function ψ_0 to be linear on each of the intervals $[\frac{1}{n+1}, \frac{1}{n}]$ ($n \in \mathbb{N}$) in such a way that $\psi_0(\frac{1}{n}) = \frac{1}{n^p}$ holds for every $n \in \mathbb{N}$. Set $\psi = \psi_0 \upharpoonright (0, 1)$. It is easy to see that ψ is positive, concave and nondecreasing on $(0, 1)$. Observe that $\psi'' = 0$ a.e., which gives

$$\int_0^1 \frac{s(-\psi''(s))}{\varphi(s)}\,ds = 0 < \infty.$$

We will show that $\Lambda_\psi = \Lambda_\varphi$. Proving this, we will get $M_\varphi \not\hookrightarrow \Lambda_\psi$ because $M_\varphi \not\hookrightarrow \Lambda_\varphi$ (the function $f(t) = \frac{1}{t^p}$ belongs to M_φ but not to Λ_φ).

Fix $n \in \mathbb{N}$ and suppose that t belongs to $[\frac{1}{n+1}, \frac{1}{n})$. Then $\varphi(t) \in [\frac{1}{(n+1)^p}, \frac{1}{n^p})$, $\psi(t) \in [\frac{1}{(n+1)^p}, \frac{1}{n^p})$. Thus,

$$\left(\frac{n}{n+1}\right)^p \le \frac{\psi(t)}{\varphi(t)} \le \left(\frac{n+1}{n}\right)^p.$$

But for every $n \in \mathbb{N}$, we have $\frac{1}{2^p} \le (\frac{n}{n+1})^p$ and $(\frac{n+1}{n})^p \le 2^p$. So,

$$\frac{1}{2^p} \le \frac{\psi(t)}{\varphi(t)} \le 2^p$$

holds for every $t \in (0, 1)$. It is easy to check that this implies $\Lambda_\psi = \Lambda_\varphi$.

The following corollary summarizes the almost-compact embeddings between endpoint spaces.

Corollary 7.11.29. *Let X and Y be Banach function spaces of type M or Λ (not necessarily both of the same type) over $((0, 1), \lambda)$. Denote by φ_X, φ_Y the fundamental functions of X and Y, respectively. Then $X \overset{*}{\hookrightarrow} Y$ if and only if $X \hookrightarrow Y$ and $\lim_{t \to 0+} \frac{\varphi_Y(t)}{\varphi_X(t)} = 0$.*

7.12 Gould space

In this section we shall introduce an important example of a rearrangement-invariant Banach function space. We start with a result of fundamental importance which also has applications in the interpolation theory.

Theorem 7.12.1. *Let $(\mathcal{R}, \mu)$ be a σ-finite measure space and let $f \in \mathcal{M}_0(\mathcal{R}, \mu)$. Then*

$$\inf_{f=g+h} \{\|g\|_{L^1} + t\|h\|_{L^\infty}\} = \int_0^t f^*(s)\, ds, \quad t \in (0, \mu(\mathcal{R})). \tag{7.12.1}$$

Proof. Fix $t \in (0, \mu(\mathcal{R}))$ and denote

$$\alpha_t := \inf_{f=g+h} \{\|g\|_{L^1} + t\|h\|_{L^\infty}\}.$$

We shall first prove that

$$\alpha_t \ge \int_0^t f^*(s)\, ds.$$

Section 7.12 Gould space

We can assume that there exists a decomposition $f = g + h$ such that $g \in L^1$ and $h \in L^\infty$, since otherwise the infimum on the left of (7.12.1) is not finite and there is nothing to prove. By Corollary 7.4.6, we have

$$\int_0^t f^*(s)\,ds \leq \int_0^t g^*(s)\,ds + \int_0^t h^*(s)\,ds.$$

By Proposition 7.1.17, this means that

$$\int_0^t f^*(s)\,ds \leq \|g\|_{L^1} + t\|h\|_{L^\infty},$$

and hence the claim follows on taking the infimum over all such decompositions.

Conversely, we claim that

$$\alpha_t \leq \int_0^t f^*(s)\,ds.$$

Define

$$E := \{x \in \mathcal{R};\ |f(x)| > f^*(t)\}$$

and

$$t_0 := \mu(E).$$

Then the first inequality in (7.1.8) guarantees that $t_0 \leq t$, whence f is integrable over E. We now define the functions g and h by

$$g(x) := \max\{|f(x)| - f^*(t); 0\}\operatorname{sign} f(x),$$
$$h(x) := \min\{|f(x)|, f^*(t)\}\operatorname{sign} f(x).$$

Then, obviously, $g \in L^1(\mathcal{R}, \mu)$ and $h \in L^\infty(\mathcal{R}, \mu)$. Moreover, $\|h\|_{L^\infty} \leq t$. We thus have

$$\|g\|_{L^1} = \int_E |f(x)|\,d\mu(x) - \mu(E)f^*(t) \leq \int_0^{t_0} f^*(s)\,ds - t_0 f^*(t),$$

which implies

$$\|g\|_{L^1} + t\|h\|_{L^\infty} \leq \int_0^{t_0} f^*(s)\,ds + (t - t_0)f^*(t) = \int_0^t f^*(s)\,ds,$$

since f is constant on $[t_0, t]$. This proves the claim.

The proof is complete. □

Making obvious modifications in the proof of Theorem 7.12.1, one can obtain the following result, again of independent interest, in which the space L^1 is replaced by the space $L^{1,\infty}$. We omit the details.

Theorem 7.12.2. *Let $(\mathcal{R}, \mu)$ be a σ-finite measure space and let $f \in \mathcal{M}_0(\mathcal{R}, \mu)$. Then*

$$\inf_{f=g+h} \{\|g\|_{L^{1,\infty}} + t\|h\|_{L^\infty}\} = \sup_{s\in(0,t)} sf^*(s), \quad t \in (0, \mu(\mathcal{R})). \tag{7.12.2}$$

Definition 7.12.3. Let $(\mathcal{R}, \mu)$ be a σ-finite measure space. Then the space

$$(L^1 + L^\infty) = (L^1 + L^\infty)(\mathcal{R}, \mu)$$

is the collection of all functions $f \in \mathcal{M}_0(\mathcal{R}, \mu)$ for which there exist functions $g \in L^1(\mathcal{R}, \mu)$ and $h \in L^\infty(\mathcal{R}, \mu)$ such that $f = g + h$ on $\mathcal{R}$. This space, called also the *Gould space*, is endowed with the functional

$$\|f\|_{L^1+L^\infty} := \inf_{f=g+h} \{\|g\|_{L^1} + \|h\|_{L^\infty}\},$$

where the infimum is extended over all such representations.

Remark 7.12.4. It follows from Theorem 7.12.1 that

$$\|f\|_{L^1+L^\infty} = \int_0^1 f^*(t)\,dt = f^{**}(1). \tag{7.12.3}$$

As an easy consequence we can derive from here that the functional $\|f\|_{L^1+L^\infty}$ is a rearrangement-invariant norm on $\mathcal{M}_0(\mathcal{R}, \mu)$.

We shall now study the "duality" properties of the space $(L^1 + L^\infty)(\mathcal{R}, \mu)$.

Definition 7.12.5. Let $(\mathcal{R}, \mu)$ be a σ-finite measure space. Then, for each $f \in L^1(\mathcal{R}, \mu) \cap L^\infty(\mathcal{R}, \mu)$, we define the functional

$$\|f\|_{L^1\cap L^\infty} := \max\{\|f\|_{L^1}, \|f\|_{L^\infty}\}.$$

Then $\|f\|_{L^1\cap L^\infty}$ is a rearrangement-invariant norm on $\mathcal{M}_0(\mathcal{R}, \mu)$, turning the set $L^1(\mathcal{R}, \mu) \cap L^\infty(\mathcal{R}, \mu)$, which we shall denote briefly as

$$(L^1 \cap L^\infty) = (L^1 \cap L^\infty)(\mathcal{R}, \mu),$$

into a rearrangement-invariant Banach function space over $(\mathcal{R}, \mu)$.

Remark 7.12.6. Let $(\mathcal{R}, \mu)$ be a σ-finite measure space. Note that, by (7.1.18) and (7.1.19),

$$\|f\|_{L^1\cap L^\infty} = \max\left\{\sup_{t\in(1,\infty)} \int_0^t f^*(s)\,ds;\ \sup_{t\in(0,1)} f^{**}(t)\right\},$$

Section 7.12 Gould space

hence

$$\|f\|_{L^1 \cap L^\infty} = \sup_{t \in (0,\infty)} \frac{\int_0^t f^*(s)\,ds}{\min\{1;t\}} = \sup_{t \in (0,\infty)} f^{**}(t) \max\{1;t\}. \quad (7.12.4)$$

It thus follows from Proposition 7.10.2 that

$$L^1 \cap L^\infty = M_\varphi,$$

where

$$\varphi(t) := \begin{cases} 0, & \text{if } t = 0, \\ \max\{1,t\} & \text{if } t \in (0, \mu(\mathcal{R})). \end{cases}$$

We shall now state and prove a useful estimate that will eventually provide us with a suitable version of the Hölder inequality.

Lemma 7.12.7. *Let f and g be nonnegative nonincreasing functions on $(0, \infty)$. Then*

$$\int_0^\infty f(t)g(t)\,dt \leq \left(\int_0^1 g(t)\,dt\right) \cdot \max\left\{\int_0^\infty f(t)\,dt;\ \sup_{s \in (0,\infty)} f(s)\right\}. \quad (7.12.5)$$

Proof. We know that f is nonincreasing, hence it follows from an observation analogous to (7.12.4) that

$$\max\left\{\int_0^\infty f(t)\,dt;\ \sup_{s \in (0,\infty)} f(s)\right\} = \sup_{t \in (0,\infty)} \frac{\int_0^t f^*(s)\,ds}{\min\{1;t\}}.$$

This implies

$$\int_0^t f^*(s)\,ds \leq \max\left\{\int_0^\infty f(t)\,dt;\ \sup_{s \in (0,\infty)} f(s)\right\} \cdot \min\{1;t\}$$

$$= \int_0^t \max\left\{\int_0^\infty f(y)\,dy;\ \sup_{y \in (0,\infty)} f(y)\right\} \chi_{(0,1)}(s)\,ds.$$

By the Hardy lemma (Proposition 7.5.2), applied to the nonincreasing function g, we then get

$$\int_0^\infty f(s)g(s)\,ds \leq \int_0^\infty \max\left\{\int_0^\infty f(y)\,dy;\ \sup_{y \in (0,\infty)} f(y)\right\} \chi_{(0,1)}(s)\,ds$$

$$\leq \max\left\{\int_0^\infty f(t)\,dt;\ \sup_{s \in (0,\infty)} f(s)\right\} \int_0^1 f(s)\,ds.$$

The proof is complete. □

Remark 7.12.8. Let $(\mathcal{R}, \mu)$ be a resonant σ-finite measure space. Then the norms in both the spaces $(L^1 + L^\infty)(\mathcal{R}, \mu)$ and $(L^1 \cap L^\infty)(\mathcal{R}, \mu)$ can be expressed in terms of the maximal nonincreasing rearrangement $f \mapsto f^{**}$ in the following manner. For every function $f \in \mathcal{M}_0(\mathcal{R}, \mu)$, we have

$$\|f\|_{(L^1+L^\infty)} = \sup_{t\in(0,\infty)} f^{**}(t)\min\{1;t\},$$

$$\|f\|_{(L^1\cap L^\infty)} = \sup_{t\in(0,\infty)} f^{**}(t)\max\{1;t\}. \tag{7.12.6}$$

Theorem 7.12.9. *Let $(\mathcal{R}, \mu)$ be a resonant σ-finite measure space. Then the spaces $(L^1 + L^\infty)(\mathcal{R}, \mu)$ and $(L^1 \cap L^\infty)(\mathcal{R}, \mu)$ are mutually associate rearrangement-invariant Banach function spaces, that is,*

$$(L^1 + L^\infty)' = (L^1 \cap L^\infty) \quad \text{and} \quad (L^1 \cap L^\infty)' = (L^1 + L^\infty).$$

Proof. By Remark 7.12.4, both the spaces in question are rearrangement-invariant Banach function spaces. It follows from Remark 7.12.8 that, denoting $\varphi(t) := \min\{1, t\}$ and $\psi(t) := \max\{1, t\}$, we have the following relation between our spaces and Marcinkiewicz endpoint spaces (cf. Section 7.10):

$$(L^1 + L^\infty) = M_\varphi \quad \text{and} \quad (L^1 \cap L^\infty) = M_\psi.$$

Incidentally, this shows, again, via Proposition 7.10.2, that our spaces are rearrangement-invariant Banach function spaces. We thus only need to verify that each is the associate space of the other. Using the Hölder inequality from Lemma 7.12.7, we obtain

$$\int_\mathcal{R} f(x)g(x)\,d\mu(x) \le \int_0^\infty f^*(t)g^*(t)\,dt \le \|f\|_{L^1+L^\infty}\|g\|_{L^1\cap L^\infty}.$$

By the definition of the associate space (Definition 6.2.5), we immediately see that

$$\|g\|_{(L^1+L^\infty)'} \le \|g\|_{L^1\cap L^\infty}, \quad g \in \mathcal{M}_0(\mathcal{R}, \mu),$$

that is,

$$(L^1 \cap L^\infty) \hookrightarrow (L^1 + L^\infty)'.$$

In order to prove the converse inequality, the Hölder inequality (7.6.6) ensures that, for every $t \in (0, \mu(\mathcal{R})$ and $g \in \mathcal{M}_0(\mathcal{R}, \mu)$,

$$\int_0^t g^*(s)\,ds \le \|\chi_{(0,t)}\|_{L^1+L^\infty}\|g\|_{(L^1+L^\infty)'}$$

$$= \min\{1;t\}\|g\|_{(L^1+L^\infty)'}.$$

Thus, dividing by $\min\{1;t\}$, taking the supremum over all g and using (7.12.6), we finally arrive at
$$(L^1 + L^\infty)' \hookrightarrow (L^1 \cap L^\infty).$$
This establishes the desired identity
$$(L^1 + L^\infty)' = (L^1 \cap L^\infty).$$
The proof is complete. □

Corollary 7.12.10. *Let $(\mathcal{R}, \mu)$ be a resonant σ-finite measure space. Then fundamental functions of the spaces $(L^1 + L^\infty)(\mathcal{R}, \mu)$ and $(L^1 \cap L^\infty)(\mathcal{R}, \mu)$ are given by*
$$\varphi_{(L^1+L^\infty)}(t) = \min\{1, t\} \quad \text{and} \quad \varphi_{(L^1 \cap L^\infty)}(t) = \max\{1, t\}, \quad t \in (0, \mu(\mathcal{R})).$$

The facts just established about the Gould space and its associate space lead in particular to the following assertion, which is of a key importance in the theory of rearrangement-invariant Banach function spaces. It shows that the spaces $(L^1 + L^\infty)$ and $(L^1 \cap L^\infty)$ are in some sense cornerstones of the entire class.

Theorem 7.12.11. *Let X be a rearrangement-invariant Banach function space over a resonant measure space $(\mathcal{R}, \mu)$. Then*
$$(L^1 \cap L^\infty) \hookrightarrow X \hookrightarrow (L^1 + L^\infty).$$

Proof. Assume first that $(\mathcal{R}, \mu) = ((0, \infty), \mu_1)$. Then, by the Hölder inequality (7.6.7),
$$\int_0^1 f^*(s)\,ds \leq \|f\|_X \|\chi_{(0,1)}\|_{X'} = \|f\|_X \varphi_{X'}(1).$$
This and (7.12.3) shows the embedding
$$X \hookrightarrow (L^1 + L^\infty)$$
(with embedding constant equal to $\varphi_{X'}(1) < \infty$). By symmetry, we also obtain
$$X' \hookrightarrow (L^1 + L^\infty)$$
(this time with constant $\varphi_X(1) < \infty$). Proposition 6.2.14 and Theorem 7.12.9 now imply that, in fact,
$$(L^1 \cap L^\infty) \hookrightarrow X$$
and
$$\|f\|_X \leq \varphi_X(1)\|f\|_{L^1 \cap L^\infty}, \quad f \in \mathcal{M}_0(\mathcal{R}, \mu).$$
Altogether, this establishes the desired sandwich
$$(L^1 \cap L^\infty) \hookrightarrow X \hookrightarrow (L^1 + L^\infty).$$
The proof is complete. □

We note that if $\mu(X) < \infty$, then we obviously have (cf. also Proposition 6.2.14)

$$\|f\|_{L^1} \leq \mu(\mathcal{R})\|f\|_{L^\infty},$$

that is,
$$(L^1 \cap L^\infty) = L^\infty \quad \text{and} \quad (L^1 + L^\infty) = L^1.$$

Likewise, if $(\mathcal{R}, \mu)$ is completely atomic, consisting of countably many atoms all of which have the same positive measure (we recall that in the discrete case we denote the corresponding spaces by ℓ^1 and ℓ^∞), then

$$\|f\|_{\ell^\infty} \leq \mu(\mathcal{R})\|f\|_{\ell^1},$$

that is,
$$(L^1 \cap L^\infty) = \ell^1 \quad \text{and} \quad (L^1 + L^\infty) = \ell^\infty.$$

This observation leads to the following two simplifications of Theorem 7.12.11.

Corollary 7.12.12. *Let X be a rearrangement-invariant Banach function space over a finite measure resonant measure space $(\mathcal{R}, \mu)$. Then*

$$L^\infty \hookrightarrow X \hookrightarrow L^1.$$

Corollary 7.12.13. *Let X be a rearrangement-invariant Banach function space over a resonant measure space $(\mathcal{R}, \mu)$ which is completely atomic, consisting of countably many atoms all of which have the same positive measure. Then*

$$\ell^1 \hookrightarrow X \hookrightarrow \ell^\infty.$$

Remark 7.12.14. Let $(\mathcal{R}, \mu)$ be a resonant σ-finite measure space. In order to avoid technical complications, we will assume that $\mu(\mathcal{R}) = \infty$. As we have seen in Corollary 7.12.10, the fundamental function of the space $(L^1 + L^\infty)(\mathcal{R}, \mu)$ is equal to $\varphi(t) := \min\{1, t\}$. It is not difficult to verify that this function is concave on $(0, \infty)$. Therefore, in accord with Definition 7.10.13, we can define the corresponding endpoint Lorentz space $\Lambda_\varphi(\mathcal{R}, \mu)$ with the help of the norm

$$\int_0^\infty f^*(t)\, d\varphi(t), \quad g \in \mathcal{M}_0(\mathcal{R}, \mu).$$

We note that, using also (7.12.3),

$$\|f\|_{\Lambda_{(L^1+L^\infty)}} = \int_0^\infty f^*(t)\, d(\min\{1, t\}) = \int_0^1 f^*(t)\, dt = \|f\|_{L^1+L^\infty}.$$

Section 7.12 Gould space

Similarly, using (7.12.3) again, we get

$$\|f\|_{M_{(L^1+L^\infty)}} = \sup_{t\in(0,\infty)} f^{**}(t)\min\{1,t\}$$

$$= \max\{t \sup_{t\in(0,1)} f^{**}(t); \sup_{t\in(1,\infty)} f^{**}(t)\}$$

$$= \int_0^1 f^*(t)\,dt = \|f\|_{L^1+L^\infty}.$$

The last two estimates show that

$$\Lambda_{(L^1+L^\infty)} = M_{(L^1+L^\infty)} = (L^1+L^\infty). \tag{7.12.7}$$

Similar facts are true also for the space $(L^1 \cap L^\infty)$ but the situation is not so straightforward because the fundamental function of this space, namely the function $\max\{1;t\}$, is not concave. In order to carry out the similar argument as above, we have to replace it with an equivalent concave function. In this case, an appropriate such function is $1+t$, which is, in fact, the least concave majorant of $\max\{1,t\}$. We will first observe that this idea indeed leads to an equivalent norm on the space $(L^1 \cap L^\infty)$. We define

$$\|f\|'_{(L^1\cap L^\infty)} := \|f\|_{L^1} + \|f\|_{L^\infty}.$$

Then, since, for every $t \in (0,\infty)$, one has $\max\{1;t\} \leq 1+t \leq 2\max\{1;t\}$, we immediately see that, for every $f \in \mathcal{M}_0(\mathcal{R},\mu)$,

$$\|f\|_{(L^1\cap L^\infty)} \leq \|f\|'_{(L^1\cap L^\infty)} \leq 2\|f\|_{(L^1\cap L^\infty)},$$

so the norms are indeed equivalent. Moreover, if the space $(L^1 \cap L^\infty)$ is normed by $\|\cdot\|'_{(L^1\cap L^\infty)}$, then the corresponding fundamental function satisfies $\varphi_{(L^1\cap L^\infty)} = 1+t$. This function is convex, hence we can define the Lorentz endpoint space $\Lambda_{(L^1\cap L^\infty)}$. Moreover, then, for a function $f \in \mathcal{M}_0(\mathcal{R},\mu)$, one has

$$\|f\|_{\Lambda_{(L^1\cap L^\infty)}} = \int_0^\infty f^*(t)\,d(1+t) = \lim_{t\to 0+} f^*(t) + \int_0^\infty f^*(s)\,ds$$

$$= \|f\|'_{(L^1\cap L^\infty)} \approx \|f\|_{(L^1\cap L^\infty)}.$$

Similarly,

$$\|f\|_{M_{(L^1\cap L^\infty)}} = \sup_{t\in(0,\infty)} f^{**}(t)(1+t) \geq \sup_{t\in(0,\infty)} f^{**}(t) + \sup_{t\in(0,\infty)} tf^{**}(t)$$

$$= \lim_{t\to 0+} f^*(t) + \int_0^\infty f^*(s)\,ds = \|f\|'_{(L^1\cap L^\infty)} \approx \|f\|_{(L^1\cap L^\infty)}.$$

We thus obtain

$$\Lambda_{(L^1\cap L^\infty)} = M_{(L^1\cap L^\infty)} = (L^1 \cap L^\infty). \tag{7.12.8}$$

Remark 7.12.15. It follows from (7.10.16), (7.10.17), (7.12.7) and (7.12.8) that for each of the following four fundamental functions

$$\varphi(t) := t,$$

$$\varphi(t) := \begin{cases} 0, & \text{if } t = 0, \\ 1 & \text{if } t \in (0, \infty), \end{cases}$$

$$\varphi(t) := \min\{1; t\},$$

$$\varphi(t) := 1 + t,$$

we have

$$\Lambda_\varphi = M_\varphi. \tag{7.12.9}$$

This interesting fact, combined with the embeddings in (7.10.15), shows that, for each of these four special choices of φ, there is *only one* rearrangement-invariant Banach function space whose fundamental function is equivalent to φ. These solitary spaces are, respectively, L^1, L^∞, $L^1 + L^\infty$ and $L^1 \cap L^\infty$.

It is of interest to notice that these four choices of a fundamental function φ, in fact, happen to be *the only* such cases. That is, they present themselves as the only possible fundamental functions for which the equality (7.12.9) holds. This observation will follow from certain deeper weighted inequalities which we will derive in Chapter 10. See, in particular, Corollary 10.3.30.

Chapter 8

Lorentz spaces

We have seen that Lebesgue spaces as well as more general Orlicz spaces happen to be rearrangement-invariant Banach function spaces. As a consequence, their norms can be introduced solely through the representation norms applied to the nonincreasing rearrangement of a function. On the other hand, in order to introduce Lebesgue or Orlicz spaces, one does not need even to know about the existence of the operation of taking the nonincreasing rearrangement.

In this chapter we shall treat a different family of function spaces, which also presents a very important generalization of Lebesgue spaces, called the Lorentz spaces. However, unlikely the Lebesgue and Orlicz spaces, the Lorentz spaces can only be introduced with the help of the nonincreasing rearrangement of a given function.

Throughout this chapter we assume that $(\mathcal{R}, \mu)$ is a totally σ-finite measure space.

We shall often appeal to Convention 7.1.8, that is, we shall consider the functions f^* defined on entire $(0, \infty)$, extended by zero outside $(0, \mu(\mathcal{R}))$ when necessary. As a result, we often work on the interval $(0, \infty)$ rather than $(0, \mu(\mathcal{R}))$. We shall do this only in the case when it is clear that no confusion can arise.

We start with a two-parameter family of Lorentz spaces, denoted by $L^{p,q}$, where, in general, $0 < p, q \leq \infty$.

8.1 Definition and basic properties

Definition 8.1.1. Assume that $0 < p, q \leq \infty$. The *Lorentz space* $L^{p,q} = L^{p,q}(\mathcal{R}, \mu)$ is the collection of all $f \in \mathcal{M}(\mathcal{R}, \mu)$ such that $\|f\|_{L^{p,q}} < \infty$, where

$$\|f\|_{L^{p,q}} = \|t^{\frac{1}{p}-\frac{1}{q}} f^*(t)\|_{L^q(0,\infty)}. \tag{8.1.1}$$

In other words, we have

$$\|f\|_{L^{p,q}} = \begin{cases} \left(\int_0^\infty [t^{\frac{1}{p}} f^*(t)]^q \frac{dt}{t}\right)^{\frac{1}{q}} & \text{if } 0 < q \leq \infty, \\ \sup_{0 < t < \infty} t^{\frac{1}{p}} f^*(t) & \text{if } q = \infty. \end{cases} \tag{8.1.2}$$

We also define the space $L^{(p,q)} = L^{(p,q)}(\mathcal{R}, \mu)$ as the collection of all $f \in \mathcal{M}(\mathcal{R}, \mu)$ such that $\|f\|_{L^{(p,q)}} < \infty$, where

$$\|f\|_{L^{(p,q)}} = \|t^{\frac{1}{p}-\frac{1}{q}} f^{**}(t)\|_{L^q(0,\infty)}, \tag{8.1.3}$$

that is,

$$\|f\|_{L^{(p,q)}} = \begin{cases} \left(\int_0^\infty [t^{\frac{1}{p}} f^{**}(t)]^q \dfrac{dt}{t} \right)^{\frac{1}{q}} & \text{if } 0 < q \leq \infty, \\ \sup_{0<t<\infty} t^{\frac{1}{p}} f^{**}(t) & \text{if } q = \infty. \end{cases} \quad (8.1.4)$$

The functional $\|\cdot\|_{L^{p,q}}$ is not always a norm on $\mathcal{M}(\mathcal{R}, \mu)$. We shall characterize the cases when it is so below. First let us make some trivial observations.

Remark 8.1.2. The two-parameter Lorentz spaces generalize the Lebesgue spaces in the following way: if $0 < p \leq \infty$, then $L^{p,p} = L^p$.

Convention 8.1.3. We will often write $\|\cdot\|_{p,q}$ and $\|\cdot\|_{(p,q)}$ instead of $\|\cdot\|_{L^{p,q}}$ and $\|\cdot\|_{L^{(p,q)}}$.

Remark 8.1.4. Since $f^* \leq f^{**}$ for every $f \in \mathcal{M}(\mathcal{R}, \mu)$, we immediately have

$$L^{(p,q)} \hookrightarrow L^{p,q} \quad (8.1.5)$$

for every choice of $p, q \in (0, \infty]$ (with the embedding constant equal to 1).

For certain, rather rare, particular cases of the parameters p and q, the space $L^{p,q}$ happens to be trivial in the sense that it contains only the function which is equal to zero μ-a.e. on $\mathcal{R}$. We shall single out these cases in the following proposition.

Proposition 8.1.5. *Assume that $(\mathcal{R}, \mu)$ is a nonatomic resonant measure space. Let $p, q \in (0, \infty]$. Then the space $L^{p,q}$ is trivial in the sense that $L^{p,q} = \{0\}$ if and only if $p = \infty$ and $q < \infty$.*

Proof. Let $p = \infty$ and $q < \infty$ and let $f \in \mathcal{M}(\mathcal{R}, \mu)$ be such that $f \not\equiv 0$. Then $f^* > 0$ on certain right neighborhood of zero, whence

$$\|f\|_{L^{p,q}}^q = \int_0^{\mu(\mathcal{R})} f^*(t)^q \frac{dt}{t} = \infty.$$

Therefore, no such function is allowed into the space.

Conversely, if either $p < \infty$ or $p = q = \infty$, then characteristic functions of sets of finite measure belong to the Lorentz space $L^{p,q}$, hence the space is nontrivial. □

Example 8.1.6. We have

$$L^{(1,\infty)} = L^1.$$

It follows immediately from Proposition 8.1.5 and (8.1.5) that also the space $L^{(p,q)}$ is trivial when $p = \infty$ and $q < \infty$. However, for these spaces, this is not the only case when they contain only the zero function. When the underlying measure space is of infinite measure, then the small values of p become troublesome, too. We shall study this question in more detail now.

Proposition 8.1.7. *Assume that $(\mathcal{R}, \mu)$ is a nonatomic resonant measure space, satisfying $\mu(\mathcal{R}) = \infty$. Let $p, q \in (0, \infty]$. Assume that either $p < 1$ or $p = 1$ and $q < \infty$. Then the space $L^{(p,q)}$ is trivial in the sense that $L^{(p,q)} = \{0\}$.*

Proof. Let $f \in \mathcal{M}(\mathcal{R}, \mu)$ be such that $f \not\equiv 0$. Then there exist constants $c > 0$ and $\delta > 0$ such that $f^*(t) > c$ for every $t \in (0, \delta)$. Then

$$f^{**}(t) \geq c\chi_{(0,\delta)}(t) + \frac{c\delta}{t}\chi_{(\delta,\infty)}(t), \quad t \in (0, \infty).$$

Thus,

$$\|f\|_{L^{(p,q)}} = \left(\int_0^\infty f^{**}(t)^q t^{\frac{q}{p}-1}\,dt\right)^{\frac{1}{q}} \geq c\delta \left(\int_\delta^\infty t^{\frac{q}{p}-q-1}\,dt\right)^{\frac{1}{q}} = \infty,$$

since $\frac{q}{p} - q - 1 \geq 0$. The proof is complete. $\square$

It is important to know when the spaces $L^{p,q}$ and $L^{(p,q)}$ coincide. For this reason, we shall now characterize those parameters p, q, for which the converse inclusion to (8.1.5) holds.

Proposition 8.1.8. *Let $1 < p \leq \infty$ and $0 < q \leq \infty$. Then*

$$L^{(p,q)} = L^{p,q}. \tag{8.1.6}$$

Proof. The inclusion

$$L^{(p,q)} \hookrightarrow L^{p,q}$$

is true for every possible choice of p, q, as was noted in Remark 8.1.4.
We will prove the converse embedding

$$L^{p,q} \hookrightarrow L^{(p,q)}. \tag{8.1.7}$$

If $p, q < \infty$, then (8.1.7) follows from the Hardy inequality (Theorem 3.11.8). Assume that $p = \infty$ and $q < \infty$. Then both the spaces $L^{p,q}$ and $L^{(p,q)}$ are trivial by Propositions 8.1.5 and 8.1.7, hence so is (8.1.7). In the remaining case when $1 < p \leq \infty$ and $q = \infty$, we note that the function $\varphi(t) := t^{\frac{1}{p}}$ satisfies (7.10.4) with $C = p'$. Thus, by Proposition 7.10.5, we get

$$\sup_{0 < t \leq \mu(\mathcal{R})} t^{\frac{1}{p}} g^{**}(t) \leq p' \sup_{0 < t \leq \mu(\mathcal{R})} t^{\frac{1}{p}} g^*(t) \quad \text{for every } g \in \mathcal{M}(\mathcal{R}, \mu), \tag{8.1.8}$$

and

$$m_\varphi \hookrightarrow M_\varphi.$$

Since, by definitions, $M_\varphi = L^{(p,\infty)}$ and $m_\varphi = L^{p,\infty}$, we are done. $\square$

We shall now point out that, for $p = 1$, the relation (8.1.6) is not true (hence $L^{p,q}$ coincides with $L^{(p,q)}$ if and only if $p > 1$).

Proposition 8.1.9. *Assume $(\mathcal{R}, \mu)$ is a nonatomic resonant measure space. Then,*

$$L^{(1,q)} \neq L^{1,q}.$$

Proof. Assume first that $\mu(\mathcal{R}) = \infty$. Then, by Propositions 8.1.5 and 8.1.7, we have

$$L^{(1,q)} = \{0\} \neq L^{1,q}.$$

Suppose, then, that $\mu(\mathcal{R}) < \infty$ and let $a \in (0, \mu(\mathcal{R}))$. Since the underlying measure space is nonatomic, there exists a μ-measurable set $E_a \subset \mathcal{R}$ such that $\mu(E_a) = a$. We set $f_a := \chi_{E_a}$. Then $f_a^* = \chi_{(0,a)}$ and

$$f_a^{**}(t) = \chi_{(0,a)}(t) + \chi_{(a,\mu(\mathcal{R}))}(t)\frac{a}{t}, \quad t \in (0, \mu(\mathcal{R})).$$

Thus,

$$\|f_a\|_{L^{(1,q)}} = \left(\int_0^{\mu(\mathcal{R})} f_a^{**}(t)^q t^{q-1}\,dt\right)^{\frac{1}{q}}$$

$$\geq a\left(\int_a^{\mu(\mathcal{R})} t^{-1}\,dt\right)^{\frac{1}{q}} = a\left(\log\frac{\mu(\mathcal{R})}{a}\right)^{\frac{1}{q}},$$

while

$$\|f_a\|_{L^{1,q}} = \left(\int_0^{\mu(\mathcal{R})} f_a^*(t)^q t^{q-1}\,dt\right)^{\frac{1}{q}} = \left(\int_0^a t^{q-1}\,dt\right)^{\frac{1}{q}} = Ca$$

with $C = q^{-\frac{1}{q}}$. Therefore,

$$\sup_{f \neq 0} \frac{\|f\|_{L^{(1,q)}}}{\|f\|_{L^{1,q}}} \geq \sup_{a \in (0,\mu(\mathcal{R}))} \frac{\|f_a\|_{L^{(1,q)}}}{\|f_a\|_{L^{1,q}}} \geq \lim_{a \to 0+} \frac{\|f_a\|_{L^{(1,q)}}}{\|f_a\|_{L^{1,q}}} = \infty.$$

This rules out the embedding $L^{(1,q)} \hookrightarrow L^{1,q}$. The proof is complete. □

Remark 8.1.10. It follows in particular from the proof of Proposition 8.1.9 that, on finite measure nonatomic spaces, the Lorentz space $L^{(1,1)}$ coincides with the Orlicz space $L \log L$. In general, relations between Lorentz type spaces and Orlicz spaces are of great interest, and we shall return to this question several times in the subsequent text.

8.2 Embeddings between Lorentz spaces

Our next aim is to study embeddings between various Lorentz spaces. We start with certain nesting property of two-parameter Lorentz spaces $L^{p,q}$ according to the second parameter q. This relation is independent of the underlying measure space $(\mathcal{R}, \mu)$.

Proposition 8.2.1. *Suppose that $p, q, r \in (0, \infty]$ and*
$$0 < q \leq r \leq \infty.$$
Then
$$L^{p,q} \hookrightarrow L^{p,r}. \tag{8.2.1}$$
If moreover $q < r$ and $L^{p,r}$ is nontrivial (that is, $L^{p,r} \neq \{0\}$), then $L^{p,q} \neq L^{p,r}$.

Proof. We start with establishing the embedding (8.2.1).

Assume first that $p = \infty$. If $q < \infty$, then the space on the left is trivial (cf. Proposition 8.1.5), hence so is the embedding. If $q = \infty$, then, necessarily, also $r = \infty$, but then the spaces coincide.

Suppose that $p < \infty$. Let first $r = \infty$. If $q = \infty$, there is nothing to prove. Let $q < \infty$. Assume that $t \in (0, \infty)$ and $f \in \mathcal{M}_0(\mathcal{R}, \mu)$. Then, by the monotonicity of f^*, we have

$$t^{\frac{1}{p}} f^*(t) = f^*(t) \left(\frac{p}{q} \int_0^t s^{\frac{q}{p}-1} ds \right)^{\frac{1}{q}} \leq \left(\frac{p}{q} \int_0^t f^*(s)^q s^{\frac{q}{p}-1} ds \right)^{\frac{1}{q}}$$

$$\leq \left(\frac{p}{q} \int_0^\infty f^*(s)^q s^{\frac{q}{p}-1} ds \right)^{\frac{1}{q}} = \left(\frac{p}{q} \right)^{\frac{1}{q}} \|f\|_{L^{p,q}}.$$

Taking the supremum over all $t \in (0, \infty)$ on the left-hand side, we get

$$\|f\|_{L^{p,\infty}} \leq \left(\frac{p}{q} \right)^{\frac{1}{q}} \|f\|_{L^{p,q}}. \tag{8.2.2}$$

This shows (8.2.1) for the case $r = \infty$. Now let $r < \infty$. Then,

$$\|f\|_{L^{p,r}} = \left(\int_0^\infty f^*(s)^r s^{\frac{r}{p}-1} ds \right)^{\frac{1}{r}} = \left(\int_0^\infty \left[f^*(s) s^{\frac{1}{p}} \right]^r \frac{ds}{s} \right)^{\frac{1}{r}}$$

$$= \left(\int_0^\infty \left[f^*(s) s^{\frac{1}{p}} \right]^{r-q+q} \frac{ds}{s} \right)^{\frac{1}{r}}$$

$$\leq \left(\left[\sup_{0 < s < \infty} f^*(s) s^{\frac{1}{p}} \right]^{r-q} \int_0^\infty \left[f^*(s) s^{\frac{1}{p}} \right]^q \frac{ds}{s} \right)^{\frac{1}{r}}$$

$$= \|f\|_{L^{p,\infty}}^{1-\frac{q}{r}} \|f\|_{L^{p,q}}^{\frac{q}{r}}.$$

By the inequality (8.2.2) just proved, we get

$$\|f\|_{L^{p,r}} \le C\|f\|_{L^{p,q}}$$

with

$$C = \left(\frac{p}{q}\right)^{\frac{1}{q}-\frac{1}{r}},$$

and (8.2.1) follows. The proof is complete. □

As mentioned above, the quantity $\|\cdot\|_{L^{p,q}}$ is not always a norm, not even when $p, q \ge 1$. We shall now pursue this question in detail.

Theorem 8.2.2. *Assume that $1 \le q \le p < \infty$ or $p = q = \infty$. Then the quantity $\|\cdot\|_{L^{p,q}}$ is a rearrangement-invariant norm on $\mathcal{M}(\mathcal{R}, \mu)$.*

Proof. When $p = q = 1$ or $p = q = \infty$, then the space $L^{p,q}$ coincides with the appropriate Lebesgue space, hence the result is clear. Suppose now that $1 < p < \infty$ and $1 \le q \le p$. Then

$$\|f + g\|_{L^{p,q}} = \left(\int_0^\infty ([f+g]^*(t))^q t^{\frac{q}{p}-1}\, dt\right)^{\frac{1}{q}}$$

$$= \sup_{\|h\|_{L^{q'}(0,\infty)} \le 1} \int_0^\infty [f+g]^*(t) t^{\frac{1}{p}-\frac{1}{q}} h^*(t)\, dt.$$

Our assumption $q \le p$ guarantees that the function

$$t \mapsto t^{\frac{1}{p}-\frac{1}{q}}$$

is nonincreasing on $(0, \infty)$. Therefore, also the function

$$t \mapsto t^{\frac{1}{p}-\frac{1}{q}} h^*(t)$$

is nonincreasing on $(0, \infty)$ for every $h \in L^{q'}(0, \infty)$. We next know that

$$(f+g)^* \prec f^* + g^*.$$

Thus, by the Hardy lemma and the Hölder inequality, we get

$$\int_0^\infty [f+g]^*(t) t^{\frac{1}{p}-\frac{1}{q}} h^*(t)\, dt$$

$$\le \int_0^\infty f^*(t) t^{\frac{1}{p}-\frac{1}{q}} h^*(t)\, dt + \int_0^\infty g^*(t) t^{\frac{1}{p}-\frac{1}{q}} h^*(t)\, dt$$

$$\le \|f^*(t) t^{\frac{1}{p}-\frac{1}{q}}\|_{L^q(0,\infty)} \|h\|_{L^{q'}(0,\infty)} + \|g^*(t) t^{\frac{1}{p}-\frac{1}{q}}\|_{L^q(0,\infty)} \|h\|_{L^{q'}(0,\infty)}$$

$$= \|f\|_{L^{p,q}} + \|g\|_{L^{p,q}},$$

hence the triangle inequality holds for $\|\cdot\|_{L^{p,q}}$. The other properties of the Banach function norm are readily verified and the rearrangement-invariance is obvious. □

The restriction $1 \le q \le p$ in the previous theorem is indispensable. However, when the functional $\|\cdot\|_{L^{p,q}}$ is replaced by $\|\cdot\|_{L^{(p,q)}}$, no such restriction is needed.

Theorem 8.2.3. *Assume that $1 \le p < \infty$ and $1 \le q \le \infty$ or $p = q = \infty$. Then the quantity $\|\cdot\|_{L^{(p,q)}}$ is a rearrangement-invariant norm on $\mathcal{M}(\mathcal{R}, \mu)$.*

Proof. The triangle inequality follows immediately from the fact that the mapping $g \mapsto g^{**}$ is subadditive (cf. (7.4.10)). Other properties of rearrangement-invariant Banach function norm are clear. □

Corollary 8.2.4. *Assume that either $p = q = 1$ or $1 < p < \infty$ and $1 \le q \le \infty$ or $p = q = \infty$. Then the two-parameter Lorentz space $L^{p,q}$ can be equivalently renormed so that the resulting space is a rearrangement-invariant Banach function space.*

Proof. For $p = q = 1$ we have $L^{p,q} = L^1$. For all the other cases, the functional $\|\cdot\|_{L^{(p,q)}}$ is a rearrangement-invariant Banach function norm (by Theorem 8.2.3), which is equivalent to $\|\cdot\|_{L^{p,q}}$ (by Proposition 8.1.8). The proof is complete. □

8.3 The associate space

Our next aim is to characterize the associate space of a given two-parameter Lorentz space. We restrict ourselves only to those cases of p, q when $L^{p,q}$ is a rearrangement-invariant Banach function space.

Theorem 8.3.1. *Assume that either $p = q = 1$ or $1 < p < \infty$ and $1 \le q \le \infty$ or $p = q = \infty$. Then the associate space of the space $L^{p,q}$ is equivalent to the space $L^{p',q'}$.*

Proof. In the cases when either $p = q = 1$ or $p = q = \infty$ the result follows from the corresponding theorems on Lebesgue spaces. Assume that $1 < p < \infty$ and $1 \le q \le \infty$. Then, by the Hardy–Littlewood inequality (Theorem 7.2.2) and the Hölder inequality for Lebesgue spaces (Theorem 3.1.6 and Remark 3.10.5),

$$\left| \int_{\mathcal{R}} f(x)g(x) \, d\mu \right| \le \int_0^\infty f^*(t)g^*(t) \, dt = \int_0^\infty f^*(t) t^{\frac{1}{p}-\frac{1}{q}} g^*(t) t^{\frac{1}{q}-\frac{1}{p}} \, dt$$

$$= \int_0^\infty f^*(t) t^{\frac{1}{p}-\frac{1}{q}} g^*(t) t^{\frac{1}{p'}-\frac{1}{q'}} \, dt$$

$$\le \| f^*(t) t^{\frac{1}{p}-\frac{1}{q}} \|_{L^q(0,\infty)} \| g^*(t) t^{\frac{1}{p'}-\frac{1}{q'}} \|_{L^{q'}(0,\infty)}$$

$$= \|f\|_{p,q} \|g\|_{p',q'}.$$

Thus, passing on the left to the supremum over the unit ball of $L^{p,q}$, we get

$$\|g\|_{(L^{p,q})'} = \sup_{\|f\|_{L^{p,q}}\leq 1} \left|\int_{\mathcal{R}} f(x)g(x)\,d\mu\right| \leq \|g\|_{p',q'}.$$

Conversely, we have to saturate the Hölder inequality by an appropriate function. We shall restrict ourselves to the situation when $(\mu, \mathcal{R})$ is resonant, nonatomic and such that $\mu(\mathcal{R}) = \infty$. Other cases can be handled in a similar manner. Suppose first that $g \in L^{p,q}$ is given. Assume $1 < q < \infty$. Let $f_0 \in \mathcal{M}_0(\mathcal{R},\mu)$ be such that

$$f_0^*(t) := \int_{\frac{t}{2}}^{\infty} s^{\frac{q'}{p'}-2} g^*(s)^{q'-1}\,ds, \quad t \in (0,\infty).$$

Then, by the monotonicity of g^*, we have

$$f_0^*(t) \geq \int_{\frac{t}{2}}^{t} s^{\frac{q'}{p'}-2} g^*(s)^{q'-1}\,ds$$

$$\geq g^*(t)^{q'-1} \int_{\frac{t}{2}}^{t} s^{\frac{q'}{p'}-2}\,ds = C_{p,q} g^*(t)^{q'-1} t^{\frac{q'}{p'}-1}, \quad t \in (0,\infty).$$

Thus,

$$\int_0^{\infty} f_0^*(t) g^*(t)\,dt = \int_0^{\infty} f_0^*(t)(t)\,dt$$

$$\geq C_{p,q} \int_0^{\infty} g^*(t)^{q'} t^{\frac{q'}{p'}-1}\,dt = C_{p,q} \|g\|_{p',q'}^{q'}.$$

On the other hand, by change of variables and, again, the monotonicity of g^*,

$$\|f_0\|_{p,q} = \left(\int_0^{\infty} t^{\frac{q}{p}-1} \left(\int_{\frac{t}{2}}^{\infty} s^{\frac{q'}{p'}-2} g^*(s)^{q'-1}\,ds\right)^q dt\right)^{\frac{1}{q}}$$

$$= C'_{p,q} \left(\int_0^{\infty} t^{\frac{q}{p}-1} \left(\int_t^{\infty} s^{\frac{q'}{p'}-2} g^*(s)^{q'-1}\,ds\right)^q dt\right)^{\frac{1}{q}}$$

$$\leq C'_{p,q} \left(\int_0^{\infty} t^{\frac{q}{p}-1} g^*(t)^{q'} \left(\int_t^{\infty} s^{\frac{q'}{p'}-2}\,ds\right)^q dt\right)^{\frac{1}{q}}$$

$$= C''_{p,q} \left(\int_0^{\infty} t^{\frac{q'}{p'}-1} g^*(t)^{q'}\,dt\right)^{\frac{1}{q}} = C''_{p,q} \|g\|_{p',q'}^{q'-1}.$$

Thus, combining all the estimates, we get

$$\|g\|_{(L^{p,q})'} = \sup_{f\neq 0} \frac{\int_0^{\infty} f^*(t) g^*(t)\,dt}{\|f\|_{p,q}} \geq \frac{\int_0^{\infty} f_0^*(t) g^*(t)\,dt}{\|f_0\|_{p,q}}$$

$$\geq \frac{\int_0^{\infty} f_0^*(t) g^*(t)\,dt}{C''_{p,q} \|g\|_{p',q'}^{q'-1}} \geq \frac{C_{p,q}}{C''_{p,q}} \|g\|_{p',q'},$$

as desired. The proof is complete. $\square$

8.4 The fundamental function

In this section we shall characterize the fundamental function and the endpoint spaces of Lorentz spaces. We shall restrict ourselves to the case when μ is nonatomic, hence $\operatorname{Rng}(\mu) = [0, \mu(\mathcal{R})]$. The proof of the first proposition is an easy exercise and is therefore omitted.

Proposition 8.4.1. *Assume that $1 < p < \infty$ and $1 \le q \le \infty$. Then the fundamental function of the space $L^{p,q}$ satisfies*

$$\varphi_{L^{p,q}}(t) = t^{\frac{1}{p}}, \quad t \in [0, \mu(\mathcal{R})).$$

The spaces $L^{p,1}$ and $L^{p,\infty}$ are equal to the Lorentz and Marcinkiewicz endpoint spaces Λ_φ and M_φ, respectively, with $\varphi(t) = t^{\frac{1}{p}}$.

Remark 8.4.2. Assume that $p = q = 1$. Then the Lorentz space $L^{p,q} = L^{1,1}$ coincides with the Lebesgue space L^1 and its fundamental function satisfies

$$\varphi_{L^{p,q}}(t) = t, \quad t \in [0, \mu(\mathcal{R})).$$

In this case the corresponding Lorentz and Marcinkiewicz endpoint spaces satisfy

$$\Lambda_\varphi = M_\varphi = L^1.$$

Remark 8.4.3. Assume that $p = q = \infty$. Then the Lorentz space $L^{p,q} = L^{\infty,\infty}$ coincides with the Lebesgue space L^∞ and its fundamental function satisfies

$$\varphi_{L^{p,q}}(t) = \begin{cases} 0 & t = 0, \\ 1 & t \in (0, \mu(\mathcal{R})). \end{cases}$$

In this case the corresponding Lorentz and Marcinkiewicz endpoint spaces satisfy

$$\Lambda_\varphi = M_\varphi = L^\infty.$$

8.5 Absolute continuity of norm

In this section we shall characterize the subsets $L^{p,q}_a$ of Lorentz spaces $L^{p,q}$ containing the functions with absolutely continuous norms. We shall restrict ourselves to those spaces which are (equivalent to) rearrangement-invariant Banach function spaces, that is, by Corollary 8.2.4, to the cases when either $p = q = 1$ or $p = q = \infty$ or $1 < p < \infty$ and $1 \le q \le \infty$. Of these, we can also leave out the case $p = q = \infty$, since then $L^{p,q} = L^\infty$, in which case we know that $L^{p,q}_a = \{0\}$.

Theorem 8.5.1. *Assume that either $p = q = 1$ or $1 < p < \infty$ and $1 \leq q < \infty$. Then every function in the space $L^{p,q}$ has absolutely continuous norm, that is, $L_a^{p,q} = L^{p,q}$.*

Proof. Let $f \in L^{p,q}$ and let E be a measurable subset of $\mathcal{R}$. We first recall that

$$(f\chi_E)^*(t) \leq f^*(t)\chi_{(0,\mu(E))}(t). \tag{8.5.1}$$

Indeed, if $t > \mu(E)$, then $(f\chi_E)^*(t) = 0$, hence the inequality is trivial, while, when $t \leq \mu(E)$, the pointwise inequality $|f|\chi_E \leq |f|$ yields $(f\chi_E)^*(t) \leq f^*(t)$. This establishes (8.5.1). Using (8.5.1), we get

$$\|f\chi_E\|_{L^{p,q}}^q = \int_0^\infty (f\chi_E)^*(t)^q t^{\frac{q}{p}-1}\,dt \leq \int_0^{\mu(E)} f^*(t)^q t^{\frac{q}{p}-1}\,dt.$$

Now, $f \in L^{p,q}$ implies $f^*(t)^q t^{\frac{q}{p}-1} \in L^1(0,\infty)$. Thus, by Theorem 1.21.13, for a given $\varepsilon > 0$ there exists a $\delta > 0$ such that if $\mu(E) < \delta$, then

$$\int_0^{\mu(E)} f^*(t)^q t^{\frac{q}{p}-1}\,dt < \varepsilon.$$

Since $q < \infty$, this implies that, again, to a given $\varepsilon > 0$, there exists a $\delta' > 0$ such that if $\mu(E) < \delta'$, then

$$\|f\chi_E\|_{L^{p,q}} < \varepsilon.$$

But then, Lemma 6.3.6 guarantees $f \in L_a^{p,q}$. The proof is complete. □

The following characterization of the dual space of a Lorentz space is an immediate consequence of Theorem 8.5.1 and Corollary 6.4.3.

Corollary 8.5.2. *Assume that either $p = q = 1$ or $1 < p < \infty$ and $1 \leq q < \infty$. Then the dual space $(L^{p,q})^*$ of $L^{p,q}$ is equivalent to the space $L^{p',q'}$.*

It remains to consider the case when $1 < p < \infty$ and $q = \infty$. In that case, the situation is quite different as the set $L_a^{p,\infty}$ is neither equal to $\{0\}$ nor to $L^{p,\infty}$.

Theorem 8.5.3. *Assume that $1 < p < \infty$. Then*

$$L_a^{p,\infty} = \left\{ f \in \mathcal{M}(\mathcal{R},\mu);\ \lim_{t \to 0+} t^{\frac{1}{p}} f^{**}(t) = \lim_{t \to \infty} t^{\frac{1}{p}} f^{**}(t) = 0 \right\}.$$

Proof. By Proposition 8.1.8, we have $L^{p,\infty} = L^{(p,\infty)}$. By Theorem 8.2.3, this space is a rearrangement-invariant Banach function space. By Proposition 8.4.1, the fundamental function of the space $L^{p,\infty}$ reads as $\varphi_{L^{p,\infty}}(t) = t^{\frac{1}{p}}$ and the space $L^{p,\infty}$ is the Marcinkiewicz space corresponding to this fundamental function, that is, $L^{p,\infty} = M_{t^{\frac{1}{p}}}$. Therefore, the assertion follows from Theorem 7.10.23. □

We finish this chapter by characterizing when a Lorentz space $L^{p,q}$ is separable and reflexive. Both the following assertions are immediate consequences of Theorem 6.5.9, Corollary 6.4.6 and Theorem 8.5.1.

Corollary 8.5.4. *Assume that μ is a separable measure. Let either $p = q = 1$ or $1 < p < \infty$ and $1 \le q \le \infty$. Then the Lorentz space $L^{p,q}$ is separable.*

Corollary 8.5.5. *Let $1 < p < \infty$ and $1 \le q \le \infty$. Then the Lorentz space $L^{p,q}$ is reflexive.*

8.6 Remarks on $\|\cdot\|_{1,\infty}$

We shall note that although the functional $\|\cdot\|_{1,\infty}$ is not equivalent to any norm, it still has some interesting properties. We first need to introduce the notion of an α-norm.

Definition 8.6.1. Let X be a vector space and let $\alpha \in (0, 1]$. A mapping $\|\cdot\| : X \longrightarrow [0, \infty)$ is called an α-*norm* on X if for every $a \in \mathbb{R}$ and $x, y \in X$ one has

$$\|x\| = 0 \Leftrightarrow x = 0,$$
$$\|ax\| = |a|\|x\|,$$
$$\|x + y\|^\alpha \le \|x\|^\alpha + \|y\|^\alpha.$$

We shall now point out a simple connection between an α-norm and a quasinorm.

Exercise 8.6.2. Let X be a vector space and let $0 < \alpha < 1$. Assume that $\|\cdot\|$ is an α-norm on X. Then $\|\cdot\|$ is also a quasinorm on X. More precisely, one has, for every $f, g \in X$,

$$\|f + g\| \le 2^{\frac{1}{\alpha}-1}(\|f\| + \|g\|).$$

The converse implication does not hold in the sense that there exist a vector space and a quasinorm on it which is not an α-norm for any $\alpha \in (0, 1]$.

The following assertion is probably folklore. We have taken it from [129], it also appears in [232].

Proposition 8.6.3. *The functional $\|\cdot\|_{1,\infty}$ is a $\frac{1}{2}$-norm. That is, for every $f, g \in L^{1,\infty}$, one has*

$$\|f + g\|_{1,\infty}^{\frac{1}{2}} \le \|f\|_{1,\infty}^{\frac{1}{2}} + \|g\|_{1,\infty}^{\frac{1}{2}}.$$

Proof. We want to establish

$$(\sup t(f + g)^*(t))^{\frac{1}{2}} \le (\sup t f^*(t))^{\frac{1}{2}} + (\sup t g^*(t))^{\frac{1}{2}}.$$

Denoting $a := \sup \sqrt{tf^*(t)}$ and $b := \sup \sqrt{tg^*(t)}$, we see that it suffices to show that, for every $t \in (0, \infty)$,

$$(f + g)^*(t) \leq \frac{1}{t}(a + b)^2.$$

By (7.1.12), we get, for every $\lambda \in [0, 1]$,

$$(f + g)^*(t) \leq f^*(\lambda t) + g^*((1 - \lambda)t).$$

We thus will be done if we prove that

$$f^*(\lambda t) + g^*((1 - \lambda)t) \leq \frac{1}{t}(a + b)^2. \tag{8.6.1}$$

Since $a^2 = \sup \lambda t f^*(\lambda t)$, we have, for every $t \in (0, \infty)$,

$$f^*(\lambda t) \leq \frac{a^2}{\lambda t}.$$

Similarly, we get

$$g^*((1 - \lambda)t) \leq \frac{b^2}{(1 - \lambda)t}.$$

Altogether, thus,

$$f^*(\lambda t) + g^*((1 - \lambda)t) \leq \frac{a^2}{\lambda t} + \frac{b^2}{(1 - \lambda)t}.$$

Setting $\lambda := \frac{a}{a+b}$, we get $\lambda \in (0, 1)$ and

$$\frac{a^2}{\lambda} + \frac{b^2}{(1 - \lambda)} = (a + b)^2.$$

Combining the last two estimates we get (8.6.1), as desired. The proof is complete. □

Corollary 8.6.4. *The functional $\|\cdot\|_{1,\infty}$ is a quasinorm with constant 2. That is, for every $f, g \in L^{1,\infty}$, one has*

$$\|f + g\|_{1,\infty} \leq 2(\|f\|_{1,\infty} + \|g\|_{1,\infty}).$$

Proof. This is an immediate consequence of Proposition 8.6.3 and Exercise 8.6.2. □

Exercise 8.6.5. The exponent $\alpha = \frac{1}{2}$ is optimal in Proposition 8.6.3 in the sense that $\|\cdot\|_{1,\infty}$ is not an α-norm for any $\alpha > \frac{1}{2}$.

We will finish this chapter by quoting an interesting relation between Orlicz spaces and Lorentz spaces. The proof can be found in [144].

Theorem 8.6.6. *Let $p \in (1, \infty)$. Then*

$$L^{p,1} = \bigcup \left\{ L^\Phi; \ \Phi \text{ is a Young function}, \ \int_0^\infty (\Phi'(t))^{\frac{1}{1-p}} \, dt = 1 \right\}.$$

Chapter 9

Generalized Lorentz–Zygmund spaces

In this chapter we shall study the so-called *generalized Lorentz–Zygmund spaces*, sometimes abbreviated as *GLZ spaces*. These spaces have been widely studied in connection with some specific limiting problems concerning double-exponential integrability and related topics. The norms of these spaces are defined with the help of logarithmic functions raised to different powers near 0 and near infinity. Therefore, such functions are in a sense "broken" in 1. For this reason they are called *broken logarithmic functions*. Our explanation basically follows that of [172], see also [73].

9.1 Measure-preserving transformations

In this chapter we shall need some theoretical background concerning the so-called measure-preserving transformations.

Definition 9.1.1. Assume that $(\mathcal{R}, \mu)$ and $(\mathcal{S}, \nu)$ are two σ-finite measure spaces and let σ be a mapping from $\mathcal{R}$ to $\mathcal{S}$. We say that τ is a *measure-preserving transformation* if, for every ν-measurable set $E \subset \mathcal{S}$, the set

$$\tau^{-1}(E) = \{x \in \mathcal{R}; \ \tau(x) \in E\}$$

is a μ-measurable subset of $\mathcal{R}$ and

$$\mu(\tau^{-1}E) = \nu(E). \tag{9.1.1}$$

We shall now recall that

Proposition 9.1.2. *Let $(\mathcal{R}, \mu)$ and $(\mathcal{S}, \nu)$ be two σ-finite measure spaces and let τ be a measure-preserving transformation from $\mathcal{R}$ to $\mathcal{S}$. Let g be a nonnegative ν-measurable function on $\mathcal{S}$. Then the function $f := g \circ \tau$ is a nonnegative ν-measurable function on $\mathcal{R}$. Moreover, $f \sim g$.*

Proof. For $\lambda \in \mathbb{R}$, define the level sets $E_\lambda := \{x \in \mathcal{R}; \ f(x) > \lambda\}$ and $F_\lambda := \{y \in \mathcal{S}; \ g(y) > \lambda\}$. Since g is ν-measurable and τ is measure preserving, we get that f is μ-measurable. By (9.1.1), for every $\lambda \in \mathbb{R}$, we get $\mu(E_\lambda) = \nu(F_\lambda)$. Therefore, f and g are equimeasurable. □

We shall need two important assertions concerning measure-preserving transformations. The proofs of these results are based on a rather deep theory whose details go beyond the scope of this book. We will therefore just state the theorems and for the proofs we refer the reader to [14, Chapter 2, Section 7] where all the details are given.

Theorem 9.1.3. *Let $(\mathcal{R}, \mu)$ be a resonant measure space and let f be a nonnegative μ-measurable function on $\mathcal{R}$. Suppose that $\lim_{t \to \infty} f^*(t) = 0$. Then there exists a measure-preserving transformation τ from the support of f onto the support of f^* such that $f = f^* \circ \tau$ μ-a.e. on the support of f.*

The most important result concerning measure-preserving transformations is the following theorem.

Theorem 9.1.4. *Let $(\mathcal{R}, \mu)$ be a nonatomic resonant measure space and let h be a nonnegative nonincreasing right-continuous function on $(0, \infty)$. Then there exists a μ-measurable function f on $\mathcal{R}$ such that $h = f^*$ on $(0, \infty)$.*

9.2 Definition and basic properties

Convention 9.2.1. We will assume everywhere in the rest of this chapter that $(\mathcal{R}, \mu)$ is a nonatomic σ-finite measure space. We shall again often use Convention 7.1.8 with no further explicit reference.

We shall use the letters $\mathbb{A}, \mathbb{B}, \mathbb{D}, \mathbb{L}, \mathbb{E}, \mathbb{S}$ and $\mathbb{W}$ for two-dimensional real vectors, that is, $\mathbb{A} = (\alpha_0, \alpha_\infty)$, $\mathbb{B} = (\beta_0, \beta_\infty)$, $\mathbb{D} = (\delta_0, \delta_\infty)$, $\mathbb{L} = (\lambda_0, \lambda_\infty)$, $\mathbb{E} = (\varepsilon_0, \varepsilon_\infty)$, $\mathbb{S} = (\sigma_0, \sigma_\infty)$, and $\mathbb{W} = (\omega_0, \omega_\infty) \in \mathbb{R}^2$. Given $\sigma \in \mathbb{R}$, we shall use the convention $\mathbb{A} + \sigma = (\alpha_0 + \sigma, \alpha_\infty + \sigma)$ and $\sigma \mathbb{A} = (\sigma \alpha_0, \sigma \alpha_\infty)$. We also write $\mathbb{A} < \mathbb{B}$ and $\mathbb{A} \leq \mathbb{B}$ when $\alpha_i < \beta_i$ and $\alpha_i \leq \beta_i$, respectively, $i = 0, \infty$. If $\mathbb{A} = (\alpha_0, \alpha_\infty) \in \mathbb{R}^2$, we set $\widetilde{\mathbb{A}} = (\alpha_\infty, \alpha_0)$.

We shall use the abbreviations

$$\ell(t) = 1 + |\log t|, \quad \ell\ell(t) = 1 + \log(\ell(t)), \quad \ell\ell\ell(t) = 1 + \log(\ell\ell(t)), \quad t > 0.$$

If $\mathbb{A} = (\alpha_0, \alpha_\infty) \in \mathbb{R}^2$, we define

$$\ell^{\mathbb{A}}(t) = \begin{cases} \ell^{\alpha_0}(t), & 0 < t \leq 1 \\ \ell^{\alpha_\infty}(t), & 1 < t < \infty, \end{cases}$$

and analogously for $\ell\ell^{\mathbb{A}}(t)$ and $\ell\ell\ell^{\mathbb{A}}(t)$. We call the functions $\ell^{\mathbb{A}}(t)$, $\ell\ell^{\mathbb{A}}(t)$ and $\ell\ell\ell^{\mathbb{A}}(t)$, etc., the *broken logarithmic functions*.

Notation 9.2.2. For any measurable function g acting on an interval $(a, b) \subset (0, \infty)$ and every $q \in (0, \infty]$, we will use the symbol $\|g\|_{q,(a,b)}$ for $\|g \chi_{(a,b)}\|_{L^q(0,\infty)}$.

We shall now define generalized Lorentz–Zygmund spaces.

Definition 9.2.3. Let $0 < p, q \leq \infty$ and $\mathbb{A}, \mathbb{B} \in \mathbb{R}^2$. The *generalized Lorentz–Zygmund* (GLZ) space $L_{p,q;\mathbb{A},\mathbb{B}}$ is given by

$$L_{p,q;\mathbb{A},\mathbb{B}} := \left\{ f \in \mathcal{M}(\mathcal{R}, \mu); \|f\|_{p,q;\mathbb{A},\mathbb{B}} = \|t^{\frac{1}{p}-\frac{1}{q}} \ell^{\mathbb{A}}(t) \ell\ell^{\mathbb{B}}(t) f^*(t)\|_{q,(0,\infty)} < \infty \right\}.$$

Remarks 9.2.4. (i) Generalized Lorentz–Zygmund spaces $L_{p,q;\mathbb{A},\mathbb{B}}$ include many familiar ones: When $\mathbb{A} = \mathbb{B} = (0,0)$, we obtain just the Lorentz space $L^{p,q}$. If, moreover, $p = q$, then $L_{p,q;\mathbb{A},\mathbb{B}} = L^p$ is the classical Lebesgue space, and the (quasi) norms coincide. If $\mu(\mathcal{R}) < \infty$, $\alpha \in \mathbb{R}$, and $\mathbb{A} = (\alpha,\alpha)$, $\mathbb{B} = (0,0)$, then $L_{p,q;\mathbb{A},\mathbb{B}}$ is the Lorentz–Zygmund space $L^{p,q}(\log L)^{\alpha}$, considered by Bennett and Rudnick in [12], which coincides with the *Zygmund class* $L^p(\log L)^{\alpha}$ when $p = q$.

(ii) Note that the use of different powers near 0 and near ∞ is reasonable only if $\mu(\mathcal{R}) = \infty$.

When $\mu(\mathcal{R}) < \infty$, the space $L_{p,q;\mathbb{A},\mathbb{B}}$ coincides with $L_{p,q;\alpha_0,\beta_0}$ introduced in [67]. Below we shall use the following slight modification of these spaces.

Let $0 < p, q \leq \infty$, $\alpha, \beta \in \mathbb{R}$, and $T \in (0, \mu(\mathcal{R})]$. Then we set

$$L_{p,q;\alpha,\beta}(0,T) := \{ f \in \mathcal{M}(\mathcal{R}, \mu); \|f\|_{p,q;\alpha,\beta,(0,T)} < \infty \},$$

where, for $0 \leq t < T \leq \mu(\mathcal{R})$,

$$\|f\|_{p,q;\alpha,\beta,(t,T)} = \|s^{\frac{1}{p}-\frac{1}{q}} \ell^{\alpha}(s) \ell\ell^{\beta}(s) f^*(s)\|_{q,(t,T)}.$$

If $0 < T < \infty$, then it is easy to see that

$$f \in L_{p,q;\alpha,\beta}(0,T) \quad \text{if and only if} \quad f \in L_{p,q;\alpha,\beta}(0,1),$$

and for all $f \in \mathcal{M}(\mathcal{R}, \mu)$,

$$\|f\|_{p,q;\alpha,\beta,(0,T)} \approx \|f\|_{p,q;\alpha,\beta,(0,1)}.$$

We set

$$L_{p,q;\alpha,\beta} := L_{p,q;\alpha,\beta}(0, \mu(\mathcal{R})) = L_{p,q;(\alpha,\alpha),(\beta,\beta)}$$

and

$$\|\cdot\|_{p,q;\alpha,\beta} := \|\cdot\|_{p,q;(\alpha,\alpha),(\beta,\beta)}.$$

In addition to the above notation we write for $g \in \mathcal{M}(\mathcal{R}, \mu)$, $p, q \in (0, \infty]$, $\mathbb{A}, \mathbb{B} \in \mathbb{R}^2$, and $0 \leq t < T \leq \mu(\mathcal{R})$,

$$\|g\|_{p,q;\mathbb{A},\mathbb{B},(t,T)} = \|s^{\frac{1}{p}-\frac{1}{q}} \ell^{\mathbb{A}}(s) \ell\ell^{\mathbb{B}}(s) g^*(s)\|_{q,(t,T)}.$$

Besides the spaces $L_{p,q;\mathbb{A},\mathbb{B}}$ we also introduce their analogues $L_{(p,q;\mathbb{A},\mathbb{B})}$ by replacing the nonincreasing rearrangement f^* by the maximal function f^{**}. Let us be more precise.

Definition 9.2.5. Let $0 < p, q \le \infty$ and $\mathbb{A}, \mathbb{B} \in \mathbb{R}^2$. The *generalized Lorentz–Zygmund* (GLZ) space $L_{(p,q;\mathbb{A},\mathbb{B})}$ is given by

$$L_{(p,q;\mathbb{A},\mathbb{B})} := \{f \in \mathcal{M}(\mathcal{R}, \mu); \|f\|_{(p,q;\mathbb{A},\mathbb{B})} < \infty\},$$

where

$$\|f\|_{(p,q;\mathbb{A},\mathbb{B})} := \|t^{\frac{1}{p}-\frac{1}{q}} \ell^{\mathbb{A}}(t) \ell\ell^{\mathbb{B}}(t) f^{**}(t)\|_{q,(0,\mu(\mathcal{R}))}.$$

Remarks 9.2.6. (i) Let $0 < p, q \le \infty$, $\alpha, \beta \in \mathcal{R}$, $\mathbb{A}, \mathbb{B} \in \mathbb{R}^2$, and $0 \le t < T \le \infty$. The spaces $L_{(p,q;\alpha,\beta)}(0,T)$, $L_{(p,q;\alpha,\beta)}$, and the quantities $\|f\|_{(p,q;\alpha,\beta)(t,T)}$, $\|f\|_{(p,q;\alpha,\beta)}$ and $\|f\|_{(p,q;\mathbb{A},\mathbb{B})(t,T)}$, are defined in an obvious way (cf. Remark 9.2.4 (ii)).

(ii) Occasionally we shall use a third tier of logarithms. In such cases we work with the spaces

$$L_{p,q;\mathbb{A},\mathbb{B},\mathbb{D}} := \{f \in \mathcal{M}(\mathcal{R}, \mu); \|f\|_{p,q;\mathbb{A},\mathbb{B},\mathbb{D}} < \infty\},$$

and

$$L_{(p,q;\mathbb{A},\mathbb{B},\mathbb{D})} := \{f \in \mathcal{M}(\mathcal{R}, \mu); \|f\|_{(p,q;\mathbb{A},\mathbb{B},\mathbb{D})} < \infty\},$$

where

$$\|f\|_{p,q;\mathbb{A},\mathbb{B},\mathbb{D}} := \|t^{\frac{1}{p}-\frac{1}{q}} \ell^{\mathbb{A}}(t) \ell\ell^{\mathbb{B}}(t) \ell\ell\ell^{\mathbb{D}}(t) f^*(t)\|_{q,(0,\mu(\mathcal{R}))},$$

and

$$\|f\|_{p,q;\mathbb{A},\mathbb{B},\mathbb{D}} := \|t^{\frac{1}{p}-\frac{1}{q}} \ell^{\mathbb{A}}(t) \ell\ell^{\mathbb{B}}(t) \ell\ell\ell^{\mathbb{D}}(t) f^{**}(t)\|_{q,(0,\mu(\mathcal{R}))}.$$

(iii) Since $f^* \le f^{**}$, we have

$$L_{(p,q;\mathbb{A},\mathbb{B})} \hookrightarrow L_{p,q;\mathbb{A},\mathbb{B}}, \quad L_{(p,q;\mathbb{A},\mathbb{B},\mathbb{D})} \hookrightarrow L_{p,q;\mathbb{A},\mathbb{B},\mathbb{D}}, \tag{9.2.1}$$

and so on.

Notation 9.2.7. Let φ be a nonnegative function defined on $[0, \infty)$. We will write $\varphi \in \mathcal{F}$ provided that

(i) $\varphi(t) = 0$ if and only if $t = 0$;

(ii) φ is continuous except perhaps at 0;

(iii) φ is equivalent to a nondecreasing concave function on $(0, \mu(R))$.

9.3 Nontriviality of GLZ spaces

Let us first clarify when the spaces $L_{p,q,\mathbb{A},\mathbb{B}}$ and $L_{(p,q,\mathbb{A},\mathbb{B})}$ are nontrivial.

Lemma 9.3.1. *Let $0 < p, q \leq \infty$, $\mathbb{A} = (\alpha_0, \alpha_\infty)$, and $\mathbb{B} = (\beta_0, \beta_\infty)$.*

(i) *The space $L_{p,q;\mathbb{A},\mathbb{B}}$ is nontrivial, that is, not equal to $\{0\}$, if and only if one of the following conditions holds:*

$$\begin{cases} p < \infty, \\ p = \infty, \ \alpha_0 + \frac{1}{q} < 0, \\ p = \infty, \ \alpha_0 + \frac{1}{q} = 0, \ \beta_0 + \frac{1}{q} < 0, \\ p = \infty, \ q = \infty, \ \alpha_0 = 0, \ \beta_0 = 0. \end{cases} \quad (9.3.1)$$

(ii) *The space $L_{(p,q;\mathbb{A},\mathbb{B})}$ is nontrivial, that is, not equal to $\{0\}$, if and only if one of the following conditions holds:*

$$\begin{cases} 1 < p < \infty, \\ p = \infty, \ \alpha_0 + \frac{1}{q} < 0, \\ p = \infty, \ \alpha_0 + \frac{1}{q} = 0, \ \beta_0 + \frac{1}{q} < 0, \\ p = \infty, \ q = \infty, \ \alpha_0 = 0, \ \beta_0 = 0, \\ p = 1, \ \alpha_\infty + \frac{1}{q} < 0, \\ p = 1, \ \alpha_\infty + \frac{1}{q} = 0, \ \beta_\infty + \frac{1}{q} < 0, \\ p = 1, \ q = \infty, \ \alpha_\infty = 0, \ \beta_\infty = 0. \end{cases} \quad (9.3.2)$$

Proof. Set $w(t) := t^{\frac{1}{p} - \frac{1}{q}} \ell^{\mathbb{A}}(t) \ell \ell^{\mathbb{B}}(t)$, $t \in (0, \infty)$. Let $E \subset \mathcal{R}$ be such that $\mu(E) \in (0, \infty)$ and denote $t = \mu(E)$. Note that $\chi_E^* = \chi_{(0,t)}$ and

$$\chi_E^{**}(s) = \chi_{(0,t)}(s) + t s^{-1} \chi_{(t,\infty)}(s), \quad s \in (0, \infty). \quad (9.3.3)$$

Thus, $\chi_E \in L_{p,q;\mathbb{A},\mathbb{B}}$ if and only if

$$\|w\|_{q,(0,t)} < \infty, \quad (9.3.4)$$

while $\chi_E \in L_{(p,q;\mathbb{A},\mathbb{B})}$ if and only if

$$\|w(s)\|_{q,(0,t)} < \infty \quad \text{and} \quad \|s^{-1} w(s)\|_{q,(t,\infty)} < \infty. \quad (9.3.5)$$

It is not difficult to verify that (9.3.4) is satisfied if and only if one of the conditions in (9.3.1) holds and, analogously (9.3.5) is satisfied if and only if one of the conditions in (9.3.2) holds. The proof is complete. □

Remark 9.3.2. Let $0 < p \leq \infty$, $1 \leq q \leq \infty$, $\mathbb{A} = (\alpha_0, \alpha_\infty)$, and $\mathbb{B} = (\beta_0, \beta_\infty) \in \mathbb{R}^2$. Then $L_{(p,q;\mathbb{A},\mathbb{B})}$ is a rearrangement-invariant Banach function space if and only if it is nontrivial.

Indeed, if $L_{(p,q;\mathbb{A},\mathbb{B})}$ is nontrivial, then, by Lemma 9.3.1, one of the conditions in (9.3.2) holds. The axioms (P1)–(P3) readily follow from elementary properties of rearrangements. The axiom (P4) is a consequence of (9.3.5), which follows from (9.3.2), as mentioned in Lemma 9.3.1. As for (P5), note that for a set E with $\mu(E) = t$, one has

$$\int_E |f| \, d\mu \leq \int_0^t f^*(y) \, dy = C_t \, \|s^{-1} w(s)\|_{q,(t,\infty)} \int_0^t f^*(y) \, dy$$

$$\leq C_t \, \|s^{-1} w(s) \int_0^t f^*(y) \, dy\|_{q,(t,\infty)}$$

$$\leq C_t \, \|s^{-1} w(s) \int_0^s f^*(y) \, dy\|_{q,(t,\infty)}$$

$$= C_t \, \|f\|_{L_{(p,q;\mathbb{A},\mathbb{B})}},$$

where $w(t) := t^{\frac{1}{p}-\frac{1}{q}} \ell^{\mathbb{A}}(t) \ell\ell^{\mathbb{B}}(t)$ and $C_t = (\|s^{-1} w(s)\|_{q,(t,\infty)})^{-1} < \infty$. If, conversely, (9.3.5) is not satisfied, then, by Lemma 9.3.1 again, (P4) does not hold and therefore $L_{(p,q;\mathbb{A},\mathbb{B})}$ is not a Banach function space.

9.4 Fundamental function

We shall now list the fundamental functions of GLZ spaces.

Lemma 9.4.1. *Let $0 < p, q \leq \infty$, $\mathbb{A} = (\alpha_0, \alpha_\infty)$, and $\mathbb{B} = (\beta_0, \beta_\infty)$.*

(i) *Assume that the space $X = L_{p,q;\mathbb{A},\mathbb{B}}$ is nontrivial (cf. Lemma 9.3.1). Then, for $0 < t \leq 1$,*

$$\varphi_X(t) \approx \begin{cases} t^{\frac{1}{p}} \ell^{\alpha_0}(t) \ell\ell^{\beta_0}(t) & \text{if } 0 < p < \infty, \\ \ell^{\alpha_0 + \frac{1}{q}}(t) \ell\ell^{\beta_0}(t) & \text{if } p = \infty, \; \alpha_0 + \frac{1}{q} < 0, \\ \ell\ell^{\beta_0 + \frac{1}{q}}(t) & \text{if } p = \infty, \; \alpha_0 + \frac{1}{q} = 0, \; \beta_0 + \frac{1}{q} < 0, \\ 1 & \text{if } p = \infty, \; q = \infty, \; \alpha_0 = 0, \; \beta_0 = 0, \end{cases}$$

Section 9.4 Fundamental function

whereas, for $1 < t < \infty$,

$$\varphi_X(t) \approx \begin{cases} t^{\frac{1}{p}} \ell^{\alpha_\infty}(t) \ell\ell^{\beta_\infty}(t) & \text{if } 0 < p < \infty, \\ 1 & \text{if } p = \infty, \text{ either } \alpha_\infty + \frac{1}{q} < 0, \\ & \quad \text{or } \alpha_\infty + \frac{1}{q} = 0, \beta_\infty + \frac{1}{q} < 0, \\ & \quad \text{or } q = \infty, \alpha_\infty = 0, \beta_\infty = 0, \\ \ell^{\alpha_\infty + \frac{1}{q}}(t) \ell\ell^{\beta_\infty}(t) & \text{if } p = \infty, \alpha_\infty + \frac{1}{q} > 0, \\ \ell\ell^{\beta_\infty + \frac{1}{q}}(t) & \text{if } p = \infty, \alpha_\infty + \frac{1}{q} = 0, \\ & \quad \beta_\infty + \frac{1}{q} > 0, \\ \ell\ell\ell^{\frac{1}{q}}(t) & \text{if } p = \infty, \alpha_\infty + \frac{1}{q} = 0, \\ & \quad \beta_\infty + \frac{1}{q} = 0. \end{cases}$$

(ii) *Assume that the space* $Y = L_{(p,q;\mathbb{A},\mathbb{B})}$ *is nontrivial. Then, for* $0 < t \leq 1$,

$$\varphi_Y(t) \approx \begin{cases} t^{\frac{1}{p}} \ell^{\alpha_0}(t) \ell\ell^{\beta_0}(t) & \text{if } 1 < p < \infty, \\ \ell^{\alpha_0 + \frac{1}{q}}(t) \ell\ell^{\beta_0}(t) & \text{if } p = \infty, \alpha_0 + \frac{1}{q} < 0, \\ \ell\ell^{\beta_0 + \frac{1}{q}}(t) & \text{if } p = \infty, \alpha_0 + \frac{1}{q} = 0, \beta_0 + \frac{1}{q} < 0, \\ 1 & \text{if } p = \infty, q = \infty, \alpha_0 = 0, \beta_0 = 0, \\ t\ell^{\alpha_0 + \frac{1}{q}}(t) \ell\ell^{\beta_0}(t) & \text{if } p = 1, \alpha_0 + \frac{1}{q} > 0, \\ t\ell\ell^{\beta_0 + \frac{1}{q}}(t) & \text{if } p = 1, \alpha_0 + \frac{1}{q} = 0, \beta_0 + \frac{1}{q} > 0, \\ t\ell\ell\ell^{\frac{1}{q}}(t) & \text{if } p = 1, \alpha_0 + \frac{1}{q} = 0, \beta_0 + \frac{1}{q} = 0, \\ t & \text{if } p = 1, \text{ either } \alpha_0 + \frac{1}{q} < 0, \\ & \quad \text{or } \alpha_0 + \frac{1}{q} = 0, \beta_0 + \frac{1}{q} < 0, \end{cases}$$

whereas, for $1 < t < \infty$,

$$\varphi_Y(t) \approx \begin{cases} t^{\frac{1}{p}} \ell^{\alpha_\infty}(t) \ell\ell^{\beta_\infty}(t) & \text{if } 1 < p < \infty, \\ 1 & \text{if } p = \infty, \text{ either } \alpha_\infty + \frac{1}{q} < 0, \\ & \quad \text{or } \alpha_\infty + \frac{1}{q} = 0, \beta_\infty + \frac{1}{q} < 0, \\ & \quad \text{or } q = \infty, \alpha_\infty = 0, \beta_\infty = 0, \\ \ell^{\alpha_\infty + \frac{1}{q}}(t) \ell\ell^{\beta_\infty}(t) & \text{if } p = \infty, \alpha_\infty + \frac{1}{q} > 0, \\ \ell\ell^{\beta_\infty + \frac{1}{q}}(t) & \text{if } p = \infty, \alpha_\infty + \frac{1}{q} = 0, \beta_\infty + \frac{1}{q} > 0, \\ \ell\ell\ell^{\frac{1}{q}}(t) & \text{if } p = \infty, \alpha_\infty + \frac{1}{q} = 0, \beta_\infty + \frac{1}{q} = 0, \\ t\ell^{\alpha_\infty + \frac{1}{q}}(t) \ell\ell^{\beta_\infty}(t) & \text{if } p = 1, \alpha_\infty + \frac{1}{q} < 0, \\ t\ell\ell^{\beta_\infty + \frac{1}{q}}(t) & \text{if } p = 1, \\ & \quad \text{either } \alpha_\infty + \frac{1}{q} = 0, \beta_\infty + \frac{1}{q} < 0, \\ & \quad \text{or } q = \infty, \alpha_\infty = 0, \beta_\infty = 0. \end{cases}$$

Proof. The assertion follows by an elementary calculation and is left to the reader. In the case (ii) one has to use (9.3.3). □

9.5 Embeddings between GLZ spaces

Our next aim is to study relations between the spaces $L_{p,q,\mathbb{A},\mathbb{B}}$ and $L_{(p,q,\mathbb{A},\mathbb{B})}$.

Theorem 9.5.1. *Let* $1 \leq p \leq \infty$, $0 < q \leq \infty$, $\mathbb{A} = (\alpha_0, \alpha_\infty)$, $\mathbb{B} = (\beta_0, \beta_\infty) \in \mathbb{R}^2$, *and assume that one of the conditions in (9.3.2) is satisfied.*

(i) *If* $1 < p \leq \infty$, *then*
$$L_{(p,q;\mathbb{A},\mathbb{B})} = L_{p,q;\mathbb{A},\mathbb{B}}. \tag{9.5.1}$$

(ii) *The space* $L_{(1,1;\mathbb{A},\mathbb{B})}$ *coincides with the space*

$L_{1,1;(0,\alpha_\infty+1),(0,\beta_\infty)}$ if $\alpha_\infty + 1 < 0$, either $\alpha_0 + 1 < 0$,
or $\alpha_0 + 1 = 0$, $\beta_0 + 1 < 0$,

$L_{1,1;\mathbb{A}+1,\mathbb{B}}$ if $\alpha_\infty + 1 < 0$, $\alpha_0 + 1 > 0$,

$L_{1,1;(0,\alpha_\infty+1),(\beta_0+1,\beta_\infty)}$ if $\alpha_\infty + 1 < 0$, $\alpha_0 + 1 = 0$, $\beta_0 + 1 > 0$,

$L_{1,1;(0,\alpha_\infty+1),(0,\beta_\infty),(1,0)}$ if $\alpha_\infty + 1 < 0$, $\alpha_0 + 1 = 0$, $\beta_0 + 1 = 0$,

$L_{1,1;(0,0),(0,\beta_\infty+1)}$ if $\alpha_\infty + 1 = 0$, $\beta_\infty + 1 < 0$,
and either $\alpha_0 + 1 < 0$,
or $\alpha_0 + 1 = 0$, $\beta_0 + 1 < 0$,

$L_{1,1;(\alpha_0+1,0),(\beta_0,\beta_\infty+1)}$ if $\alpha_\infty + 1 = 0$, $\beta_\infty + 1 < 0$, $\alpha_0 + 1 > 0$,

$L_{1,1;(0,0),\mathbb{B}+1}$ if $\alpha_\infty + 1 = 0$, $\beta_\infty + 1 < 0$, $\alpha_0 + 1 = 0$,
$\beta_0 + 1 > 0$,

$L_{1,1;(0,0),(0,\beta_\infty+1),(1,0)}$ if $\alpha_\infty + 1 = 0$, $\beta_\infty + 1 < 0$, $\alpha_0 + 1 = 0$,
$\beta_0 + 1 = 0$.

(iii) *Let* $1 < q \leq \infty$. *Then*
$$L_{1,q;\mathbb{A}+1,\mathbb{B}} \subsetneq L_{(1,q;\mathbb{A},\mathbb{B})} \quad \text{if } \alpha_0 + \frac{1}{q} > 0, \; \alpha_\infty + \frac{1}{q} < 0, \tag{9.5.2}$$

and

$$L_{1,q;(\frac{1}{q'},\frac{1}{q'}),\mathbb{B}+1} \subsetneq L_{(1,q;(-\frac{1}{q},-\frac{1}{q}),\mathbb{B})} \quad \text{if } \beta_0 + \frac{1}{q} > 0, \; \beta_\infty + \frac{1}{q} < 0. \tag{9.5.3}$$

(iv) *Let* $0 < q < 1$. *Then*
$$L_{1,q;\mathbb{A}+\frac{1}{q},\mathbb{B}} \subsetneq L_{(1,q;\mathbb{A},\mathbb{B})} \quad \text{if } \alpha_0 + \frac{1}{q} > 0, \; \alpha_\infty + \frac{1}{q} < 0$$

Section 9.5 Embeddings between Generalized Lorentz–Zygmund spaces

and

$$L_{1,q;(0,0),\mathbb{B}+\frac{1}{q}} \subset L_{(1,q;(-\frac{1}{q},-\frac{1}{q}),\mathbb{B})} \quad \text{if } \beta_0 + \frac{1}{q} > 0, \ \beta_\infty + \frac{1}{q} < 0.$$

Proof. (i) Assume first that $1 \leq q \leq \infty$. Since $p > 1$, the Hardy inequality

$$\left\| t^{\frac{1}{p}-\frac{1}{q}} \ell^{\mathbb{A}}(t) \ell\ell^{\mathbb{B}}(t) t^{-1} \int_0^t g(s)\,ds \right\|_q \lesssim \left\| t^{\frac{1}{p}-\frac{1}{q}} \ell^{\mathbb{A}}(t) \ell\ell^{\mathbb{B}}(t) g(t) \right\|_q \quad (9.5.4)$$

holds for every $g \in \mathcal{M}^+(0,\infty)$ (Theorem 3.11.8). Applied to $g = f^*$, (9.5.4) implies $L_{p,q;\mathbb{A},\mathbb{B}} \hookrightarrow L_{(p,q;\mathbb{A},\mathbb{B})}$. Combined with (9.2.1), this yields (9.5.1). If $0 < q < 1$, we use an analogous argument, applying [131, Theorem 2.2].

(ii) By the Fubini theorem,

$$\|f\|_{(1,1;\mathbb{A},\mathbb{B})} = \int_0^\infty f^*(s) \left(\int_s^\infty t^{-1} \ell^{\mathbb{A}}(t) \ell\ell^{\mathbb{B}}(t)\,dt \right) ds.$$

Calculating the inner integral, we obtain the assertion.

(iii) Both embeddings in (9.5.2) and (9.5.3) follow from the corresponding Hardy inequality (cf. [74, Lemmas 4.2 and 4.3]). The distinction of the spaces follows by comparing their fundamental functions (cf. Lemma 9.4.1).

(iv) This follows from [131, Theorem 2.2]. □

Now we shall prove some auxiliary results which will be needed later.

Lemma 9.5.2. *Let $0 < q \leq \infty$, and $\mathbb{A}, \mathbb{B} \in \mathbb{R}^2$. Assume that for each $i \in \{0,\infty\}$ one of the following conditions holds:*

$$\alpha_i + \frac{1}{q} < 0, \quad \alpha_i + \frac{1}{q} = 0, \ \beta_i + \frac{1}{q} < 0, \quad q = \infty, \ \alpha_i = 0, \ \beta_i = 0.$$

Then for all $f \in L_{\infty,q;\mathbb{A},\mathbb{B}}$,

$$\left\| t^{-\frac{1}{q}} \ell^{\alpha_\infty}(t) \ell\ell^{\beta_\infty}(t) f^*(t) \right\|_{q,(1,\infty)} \lesssim \left\| t^{-\frac{1}{q}} \ell^{\alpha_0}(t) \ell\ell^{\beta_0}(t) f^*(t) \right\|_{q,(0,1)}.$$

Proof. Our assumptions imply that

$$\left\| t^{-\frac{1}{q}} \ell^{\alpha_\infty}(t) \ell\ell^{\beta_\infty}(t) \right\|_{q,(1,\infty)} \approx 1 \approx \left\| t^{-\frac{1}{q}} \ell^{\alpha_0}(t) \ell\ell^{\beta_0}(t) \right\|_{q,(0,1)}. \quad (9.5.5)$$

Consequently, for all $f \in L_{\infty,q;\mathbb{A},\mathbb{B}}$,

$$\left\| t^{-\frac{1}{q}} \ell^{\alpha_\infty}(t) \ell\ell^{\beta_\infty}(t) f^*(t) \right\|_{q,(1,\infty)} \leq f^*(1) \left\| t^{-\frac{1}{q}} \ell^{\alpha_\infty}(t) \ell\ell^{\beta_\infty}(t) \right\|_{q,(1,\infty)}$$

$$\approx f^*(1) \left\| t^{-\frac{1}{q}} \ell^{\alpha_0}(t) \ell\ell^{\beta_0}(t) \right\|_{q,(0,1)} \leq \left\| t^{-\frac{1}{q}} \ell^{\alpha_0}(t) \ell\ell^{\beta_0}(t) f^*(t) \right\|_{q,(0,1)}. \quad □$$

Corollary 9.5.3. *Let all the assumptions of Lemma 9.5.2 be satisfied. Then*

$$L_{\infty,q;\mathbb{A},\mathbb{B}} = L_{\infty,q;\alpha_0,\beta_0}(0,1).$$

If moreover $q = \infty$, then

$$L_{\infty,\infty;\mathbb{A},\mathbb{B}} = L_{\infty,\infty;\alpha_0,\beta_0}(0,1) = L_{\infty,\infty;(\alpha_0,0),(\beta_0,0)}. \quad (9.5.6)$$

The following result is a dual version of Lemma 9.5.2.

Lemma 9.5.4. *Let $0 < q \le \infty$, and $\mathbb{A}, \mathbb{B} \in \mathbb{R}^2$. Assume that for each $i \in \{0,\infty\}$ one of the following conditions holds:*

$$\alpha_i + \frac{1}{q} < 0,$$

$$\alpha_i + \frac{1}{q} = 0, \ \beta_i + \frac{1}{q} < 0,$$

$$q = \infty, \ \alpha_i = 0, \ \beta_i = 0.$$

Then for all $f \in L_{(1,q;\mathbb{A},\mathbb{B})}$,

$$\|t^{1-\frac{1}{q}}\ell^{\alpha_0}(t)\ell\ell^{\beta_0}(t)f^{**}(t)\|_{q,(0,1)} \lesssim \|t^{1-\frac{1}{q}}\ell^{\alpha_\infty}(t)\ell\ell^{\beta_\infty}(t)f^{**}(t)\|_{q,(1,\infty)}.$$

Proof. Using (9.5.5), we have for all $f \in L_{(1,q;\mathbb{A},\mathbb{B})}$,

$$\|t^{1-\frac{1}{q}}\ell^{\alpha_0}(t)\ell\ell^{\beta_0}(t)f^{**}(t)\|_{q,(0,1)}$$

$$= \|t^{-\frac{1}{q}}\ell^{\alpha_0}(t)\ell\ell^{\beta_0}(t)\int_0^t f^*(s)\,ds\|_{q,(0,1)}$$

$$\le \|t^{-\frac{1}{q}}\ell^{\alpha_0}(t)\ell\ell^{\beta_0}(t)\|_{q,(0,1)}\int_0^1 f^*(s)\,ds$$

$$\approx \|t^{-\frac{1}{q}}\ell^{\alpha_\infty}(t)\ell\ell^{\beta_\infty}(t)\|_{q,(1,\infty)}\int_0^1 f^*(s)\,ds$$

$$\le \|t^{-\frac{1}{q}}\ell^{\alpha_\infty}(t)\ell\ell^{\beta_\infty}(t)\int_0^t f^*(s)\,ds\|_{q,(1,\infty)}$$

$$= \|t^{1-\frac{1}{q}}\ell^{\alpha_\infty}(t)\ell\ell^{\beta_\infty}(t)f^{**}(t)\|_{q,(1,\infty)}. \quad \square$$

Corollary 9.5.5. *Let $\mathbb{A}, \mathbb{B} \in \mathbb{R}^2$. Assume that for each $i \in \{0,\infty\}$ one of the following conditions holds:*

$$\alpha_i < 0,$$
$$\alpha_i = 0, \ \beta_i < 0,$$
$$\alpha_i = 0, \ \beta_i = 0.$$

Then

$$L_{(1,\infty;\mathbb{A},\mathbb{B})} = L_{(1,\infty;(0,\alpha_\infty),(0,\beta_\infty))}.$$

Section 9.5 Embeddings between Generalized Lorentz–Zygmund spaces

We shall now formulate modifications of Lemmas 9.3.1 and 9.4.1 and of Theorem 9.5.1 for the case when $\mu(\mathcal{R}) < \infty$. The proofs are analogous to the corresponding ones above and therefore omitted.

Lemma 9.5.6. *Let $\mu(\mathcal{R}) < \infty$, $0 < p, q \le \infty$, and $\alpha, \beta \in \mathbb{R}$. Let X be one of the spaces $L_{p,q;\alpha,\beta}$, $L_{(p,q;\alpha,\beta)}$. Then X is nontrivial if and only if one of the following conditions holds:*

$$\begin{cases} p < \infty, \\ p = \infty, \ \alpha + \frac{1}{q} < 0, \\ p = \infty, \ \alpha + \frac{1}{q} = 0, \ \beta + \frac{1}{q} < 0, \\ p = \infty, \ q = \infty, \ \alpha = 0, \ \beta = 0. \end{cases} \qquad (9.5.7)$$

Lemma 9.5.7. *Let $R = \mu(\mathcal{R}) < \infty$, $0 < p, q \le \infty$, and $\alpha, \beta \in \mathbb{R}$. Assume that one of the conditions in (9.5.7) is satisfied.*

(i) *Let $X = L_{p,q;\alpha,\beta}$. Then, for all $t \in (0, R]$,*

$$\varphi_X(t) \approx \begin{cases} t^{\frac{1}{p}} \ell^\alpha(t) \ell\ell^\beta(t) & \text{if } 0 < p < \infty, \\ \ell^{\alpha+\frac{1}{q}}(t) \ell\ell^\beta(t) & \text{if } p = \infty, \ \alpha + \frac{1}{q} < 0, \\ \ell\ell^{\beta+\frac{1}{q}}(t) & \text{if } p = \infty, \ \alpha + \frac{1}{q} = 0, \ \beta + \frac{1}{q} < 0, \\ 1 & \text{if } p = \infty, \ q = \infty, \ \alpha = 0, \ \beta = 0. \end{cases}$$

(ii) *Let $Y = L_{(p,q;\alpha,\beta)}$. Then, for all $t \in (0, R]$,*

$$\varphi_Y(t) \approx \begin{cases} t^{\frac{1}{p}} \ell^\alpha(t) \ell\ell^\beta(t) & \text{if } 1 < p < \infty, \\ \ell^{\alpha+\frac{1}{q}}(t) \ell\ell^\beta(t) & \text{if } p = \infty, \ \alpha + \frac{1}{q} < 0, \\ \ell\ell^{\beta+\frac{1}{q}}(t) & \text{if } p = \infty, \ \alpha + \frac{1}{q} = 0, \ \beta + \frac{1}{q} < 0, \\ 1 & \text{if } p = \infty, \ q = \infty, \ \alpha = 0, \ \beta = 0, \\ t \ell^{\alpha+\frac{1}{q}}(t) \ell\ell^\beta(t) & \text{if } p = 1, \ \alpha + \frac{1}{q} > 0, \\ t \ell\ell^{\beta+\frac{1}{q}}(t) & \text{if } p = 1, \ \alpha + \frac{1}{q} = 0, \ \beta + \frac{1}{q} > 0, \\ t \ell\ell\ell^{\frac{1}{q}}(t) & \text{if } p = 1, \ \alpha + \frac{1}{q} = 0, \ \beta + \frac{1}{q} = 0, \\ t & \text{if either } 0 < p < 1, \\ & \quad \text{or } p = 1, \ \alpha + \frac{1}{q} < 0, \\ & \quad \text{or } p = 1, \ \alpha + \frac{1}{q} = 0, \ \beta + \frac{1}{q} < 0. \end{cases}$$

If $\mu(\mathcal{R}) < \infty$, we see from Lemma 9.5.7 that, in certain cases, $L_{(p,q;\alpha,\beta)}$ has the same fundamental function as L^1. The next assertion states that in fact in such cases these spaces coincide.

Lemma 9.5.8. *Let $R = \mu(\mathcal{R}) < \infty$, $0 < q \leq \infty$, and $\alpha, \beta \in \mathbb{R}$, and let one of the following conditions be satisfied:*

$$\begin{cases} 0 < p < 1, \\ p = 1, \ \alpha + \frac{1}{q} < 0, \\ p = 1, \ \alpha + \frac{1}{q} = 0, \ \beta + \frac{1}{q} < 0, \\ p = 1, \ q = \infty, \ \alpha = 0, \ \beta = 0. \end{cases} \tag{9.5.8}$$

Then
$$L_{(p,q;\alpha,\beta)} = L^1.$$

Proof. Let $R = \mu(\mathcal{R})$. Our assumptions imply that

$$\|t^{\frac{1}{p}-1-\frac{1}{q}} \ell^\alpha(t) \ell\ell^\beta(t)\|_{q,(0,R)} \approx 1 \approx \|t^{\frac{1}{p}-\frac{1}{q}} \ell^\alpha(t) \ell\ell^\beta(t)\|_{q,(0,R)}.$$

Consequently, for all $f \in \mathcal{M}(\mathcal{R}, \mu)$,

$$\|f\|_{(p,q;\alpha,\beta)} \leq \left(\int_0^R f^*(s) \, ds\right) \|t^{\frac{1}{p}-1-\frac{1}{q}} \ell^\alpha(t) \ell\ell^\beta(t)\|_{q,(0,R)}$$
$$\approx \int_0^R f^*(s) \, ds \approx f^{**}(R) \|t^{\frac{1}{p}-\frac{1}{q}} \ell^\alpha(t) \ell\ell^\beta(t)\|_{q,(0,R)} \leq \|f\|_{(p,q;\alpha,\beta)},$$

and the result follows. □

Theorem 9.5.9. *Let $\mu(\mathcal{R}) < \infty$, $1 \leq p \leq \infty$, $0 < q \leq \infty$, and $\alpha, \beta \in \mathbb{R}$.*

(i) *Let $1 < p \leq \infty$ and let one of the conditions in (9.5.7) be satisfied. Then*
$$L_{(p,q;\alpha,\beta)} = L_{p,q;\alpha,\beta}.$$

(ii) *The space $L_{(1,1;\alpha,\beta)}$ coincides with the space*

$$\begin{array}{ll} L^1 & \text{if either } \alpha + 1 < 0, \\ & \text{or } \alpha + 1 = 0, \ \beta + 1 < 0, \\ L_{1,1;\alpha+1,\beta} & \text{if } \alpha + 1 > 0, \\ L_{1,1;0,\beta+1} & \text{if } \alpha + 1 = 0, \ \beta + 1 > 0; \\ L_{1,1;0,0,1} & \text{if } \alpha + 1 = 0, \ \beta + 1 = 0. \end{array}$$

(iii) *Let $1 < q \leq \infty$. Then*

$$L_{1,q;\alpha+1,\beta} \subsetneq L_{(1,q;\alpha,\beta)} \quad \text{if} \quad \alpha + \frac{1}{q} > 0,$$

and

$$L_{1,q;\frac{1}{q'},\beta+1} \subsetneq L_{(1,q;-\frac{1}{q},\beta)} \quad \text{if} \quad \beta + \frac{1}{q} > 0.$$

(iv) *Let* $0 < q < 1$. *Then*

$$L_{1,q;\alpha+\frac{1}{q},\beta} \subset L_{(1,q;\alpha,\beta)} \quad \text{if} \quad \alpha + \frac{1}{q} > 0,$$

and

$$L_{1,q;0,\beta+\frac{1}{q}} \subset L_{(1,q;-\frac{1}{q},\beta)} \quad \text{if} \quad \beta + \frac{1}{q} > 0.$$

Our next objective will be to characterize the embeddings of the form

$$L_{P_1,Q_1;\mathbb{L},\mathbb{E}} \hookrightarrow L_{P_2,Q_2;\mathbb{S},\mathbb{W}} \qquad (9.5.9)$$

with $0 < P_1, P_2, Q_1, Q_2 \le \infty$ and $\mathbb{L} = (\lambda_0, \lambda_\infty)$, $\mathbb{E} = (\varepsilon_0, \varepsilon_\infty)$, $\mathbb{S} = (\sigma_0, \sigma_\infty)$, $\mathbb{W} = (\omega_0, \omega_\infty) \in \mathbb{R}^2$.

First we shall investigate the embedding (9.5.9) with $P_1 = P_2 = P$. In the case when $\mu(\mathcal{R}) < \infty$ such an embedding is completely characterized in terms of inequalities involving the first components of vector exponents of logarithmic functions (cf. [73]). If $\mu(\mathcal{R}) = \infty$, the second components of these exponents will take place in the corresponding conditions as well.

For the case of brevity, we present only statements of the main results. If $0 < P_1, P_2 < \infty$, these follow from the more general theorems in [216]. If $P_1 = P_2 = \infty$, one can use the results of [208, Proposition 2.7] and [35, Section 3] to prove them under certain additional assumptions on weights involved. Proofs in the remaining cases are left to the reader as an exercise.

Our first theorem characterizes the embedding $L_{P,Q_1;\mathbb{L},\mathbb{E}} \hookrightarrow L_{P,Q_2;\mathbb{S},\mathbb{W}}$ provided that $0 < Q_1 \le Q_2 \le \infty$ and $0 < P < \infty$.

Theorem 9.5.10. *Let* $0 < Q_1 \le Q_2 \le \infty$, $0 < P < \infty$, $\mu(\mathcal{R}) = \infty$, *and* $L_{P,Q;\mathbb{L},\mathbb{E}} \ne \{0\}$. *Then*

$$L_{P,Q_1;\mathbb{L},\mathbb{E}} \hookrightarrow L_{P,Q_2;\mathbb{S},\mathbb{W}}$$

if and only if

$$\mathbb{L} \ge \mathbb{S}$$

and

$$\text{if} \quad \lambda_i = \sigma_i \quad \text{for some} \quad i \in \{0, \infty\}, \quad \text{then} \quad \varepsilon_i \ge \omega_i.$$

Next, we shall characterize the embedding

$$L_{P,Q_1;\mathbb{L},\mathbb{E}} \hookrightarrow L_{P,Q_2;\mathbb{S},\mathbb{W}}, \qquad (9.5.10)$$

provided that $0 < Q_1 \le Q_2 \le \infty$ and $P = \infty$.

Theorem 9.5.11. *Let* $0 < Q_1 \le Q_2 \le \infty$, $P = \infty$, $\mu(\mathcal{R}) = \infty$, *and* $L_{P,Q_1;\mathbb{L},\mathbb{E}} \ne \{0\}$. *Then*

$$L_{P,Q_1;\mathbb{L},\mathbb{E}} \hookrightarrow L_{P,Q_2;\mathbb{S},\mathbb{W}}$$

if and only if
$$\lambda_0 + \frac{1}{Q_1} > \sigma_0 + \frac{1}{Q_2},$$
or
$$0 = \lambda_0 + \frac{1}{Q_1} = \sigma_0 + \frac{1}{Q_2}, \qquad \varepsilon_0 + \frac{1}{Q_1} \geq \omega_0 + \frac{1}{Q_2},$$
or
$$0 > \lambda_0 + \frac{1}{Q_1} = \sigma_0 + \frac{1}{Q_2}, \qquad \varepsilon_0 \geq \omega_0$$

and simultaneously one of the following conditions is satisfied:

$$\lambda_\infty + \frac{1}{Q_1} < 0, \qquad \sigma_\infty + \frac{1}{Q_2} < 0,$$

$$\lambda_\infty + \frac{1}{Q_1} < 0, \qquad \sigma_\infty + \frac{1}{Q_2} = 0, \qquad \omega_\infty + \frac{1}{Q_2} < 0,$$

$$\lambda_\infty + \frac{1}{Q_1} < 0, \qquad Q_2 = \infty, \qquad \sigma_\infty = 0, \qquad \omega_\infty = 0,$$

$$0 \leq \lambda_\infty + \frac{1}{Q_1}, \qquad \lambda_\infty + \frac{1}{Q_1} > \sigma_\infty + \frac{1}{Q_2},$$

$$0 < \lambda_\infty + \frac{1}{Q_1} = \sigma_\infty + \frac{1}{Q_2}, \qquad \varepsilon_\infty \geq \omega_\infty,$$

$$0 = \lambda_\infty + \frac{1}{Q_1} = \sigma_\infty + \frac{1}{Q_2}, \qquad \varepsilon_\infty + \frac{1}{Q_1} \geq 0, \qquad \varepsilon_\infty + \frac{1}{Q_1} \geq \omega_\infty + \frac{1}{Q_2},$$

$$0 = \lambda_\infty + \frac{1}{Q_1} = \sigma_\infty + \frac{1}{Q_2}, \qquad \varepsilon_\infty + \frac{1}{Q_1} < 0, \qquad \omega_\infty + \frac{1}{Q_2} < 0.$$

Now, we shall characterize the embedding (9.5.10) provided that $0 < Q_2 < Q_1 \leq \infty$ and $0 < P \leq \infty$. We shall start with the case $0 < P < \infty$.

Theorem 9.5.12. *Let* $0 < Q_2 < Q_1 \leq \infty$, $0 < P < \infty$, $\mu(\mathcal{R}) = \infty$, *and* $L_{P,Q_1;\mathbb{L},\mathbb{E}} \neq \{0\}$. *Then*
$$L_{P,Q_1;\mathbb{L},\mathbb{E}} \hookrightarrow L_{P,Q_2;\mathbb{S},\mathbb{W}}$$
if and only if
$$\mathbb{L} + \frac{1}{Q_1} \geq \mathbb{S} + \frac{1}{Q_2}$$
and
$$\textit{if} \quad \lambda_i + \frac{1}{Q_1} = \sigma_i + \frac{1}{Q_2} \textit{ for some } i \in \{0, \infty\}, \textit{ then } \varepsilon_i + \frac{1}{Q_1} > \omega_i + \frac{1}{Q_2}.$$

Section 9.5 Embeddings between Generalized Lorentz–Zygmund spaces

In the next theorem we consider the embedding (9.5.10) in the case $0 < Q_2 < Q_1 \leq \infty$ and $P = \infty$.

Theorem 9.5.13. *Let $0 < Q_2 < Q_1 \leq \infty$, $P = \infty$, $\mu(\mathcal{R}) = \infty$, and $L_{P,Q_1;\mathbb{L},\mathbb{E}} \neq \{0\}$. Then*

$$L_{P,Q_1;\mathbb{L},\mathbb{E}} \hookrightarrow L_{P,Q_2;\mathbb{S},\mathbb{W}}$$

if and only if either

$$\lambda_0 + \frac{1}{Q_1} > \sigma_0 + \frac{1}{Q_2},$$

or

$$\lambda_0 + \frac{1}{Q_1} = \sigma_0 + \frac{1}{Q_2}, \qquad \varepsilon_0 + \frac{1}{Q_1} > \omega_0 + \frac{1}{Q_2}$$

and simultaneously one of the following conditions is satisfied:

$$\lambda_\infty + \frac{1}{Q_1} < 0, \qquad\qquad \sigma_\infty + \frac{1}{Q_2} < 0,$$

$$\lambda_\infty + \frac{1}{Q_1} < 0, \qquad\qquad \sigma_\infty + \frac{1}{Q_2} = 0, \qquad \omega_\infty + \frac{1}{Q_2} < 0,$$

$$0 \leq \lambda_\infty + \frac{1}{Q_1}, \qquad\qquad \lambda_\infty + \frac{1}{Q_1} > \sigma_\infty + \frac{1}{Q_2},$$

$$0 < \lambda_\infty + \frac{1}{Q_1} = \sigma_\infty + \frac{1}{Q_2}, \qquad \varepsilon_\infty + \frac{1}{Q_1} > \omega_\infty + \frac{1}{Q_2},$$

$$0 = \lambda_\infty + \frac{1}{Q_1} = \sigma_\infty + \frac{1}{Q_2}, \qquad \varepsilon_\infty + \frac{1}{Q_1} \geq 0, \qquad \varepsilon_\infty + \frac{1}{Q_1} > \omega_\infty + \frac{1}{Q_2},$$

$$0 = \lambda_\infty + \frac{1}{Q_1} = \sigma_\infty + \frac{1}{Q_2}, \qquad \varepsilon_\infty + \frac{1}{Q_1} < 0, \qquad \omega_\infty + \frac{1}{Q_2} < 0.$$

In all the preceding theorems we have assumed that $\mu(R) = \infty$. When $\mu(R) < \infty$, then the results remain valid if we omit all the assumptions on the second components of vectors $\mathbb{L}, \mathbb{E}, \mathbb{S}$, and $\mathbb{W}$ (cf. [73, Theorem 6.3] and remarks on GLZ spaces with $\mu(\mathcal{R}) < \infty$ above). We thus have the following result.

Theorem 9.5.14. *Assume that $\mu(\mathcal{R}) < \infty$ and $L_{P,Q_1;\mathbb{L},\mathbb{E}} \neq \{0\}$. Then*

$$L_{P,Q_1;\mathbb{L},\mathbb{E}} \hookrightarrow L_{P,Q_2;\mathbb{S},\mathbb{W}}$$

if and only if one of the following conditions is satisfied:

(i) $0 < Q_1 \leq Q_2 \leq \infty$, $\quad 0 < P < \infty$, $\quad \lambda_0 > \sigma_0$;

(ii) $0 < Q_1 \leq Q_2 \leq \infty$, $\quad 0 < P < \infty$, $\quad \lambda_0 = \sigma_0$, $\quad \varepsilon_0 \geq \omega_0$;

(iii) $0 < Q_1 \leq Q_2 \leq \infty$, $P = \infty$, $\lambda_0 + \frac{1}{Q_1} > \sigma_0 + \frac{1}{Q_2}$;

(iv) $0 < Q_1 \leq Q_2 \leq \infty$, $P = \infty$, $\lambda_0 + \frac{1}{Q_1} = \sigma_0 + \frac{1}{Q_2} = 0$,
$\varepsilon_0 + \frac{1}{Q_1} \geq \omega_0 + \frac{1}{Q_2}$;

(v) $0 < Q_1 \leq Q_2 \leq \infty$, $P = \infty$, $\lambda_0 + \frac{1}{Q_1} = \sigma_0 + \frac{1}{Q_2} < 0$, $\varepsilon_0 \geq \omega_0$;

(vi) $0 < Q_2 < Q_1 \leq \infty$, $0 < P \leq \infty$, $\lambda_0 + \frac{1}{Q_1} > \sigma_0 + \frac{1}{Q_2}$;

(vii) $0 < Q_2 < Q_1 \leq \infty$, $0 < P \leq \infty$, $\lambda_0 + \frac{1}{Q_1} = \sigma_0 + \frac{1}{Q_2}$,
$\varepsilon_0 + \frac{1}{Q_1} > \omega_0 + \frac{1}{Q_2}$.

So far we have investigated embeddings among $L_{p,q;\mathbb{A},\mathbb{B}}$ spaces provided that the first index p was fixed. Embeddings with p varying are similar to those for Lebesgue spaces L^p.

Theorem 9.5.15. *Let* $0 < P_1, P_2, Q_1, Q_2 \leq \infty$, $P_1 \neq P_2$, *and* $L_{P_1,Q_1;\mathbb{L},\mathbb{E}} \neq \{0\}$. *Then*
$$L_{P_1,Q_1;\mathbb{L},\mathbb{E}} \hookrightarrow L_{P_2,Q_2;\mathbb{S},\mathbb{W}}$$
if and only if $\mu(\mathcal{R}) < \infty$ *and* $P_1 > P_2$.

We shall turn our attention now to the characterization of embeddings of the form
$$L_{(P_1,Q_1;\mathbb{L},\mathbb{E})} \hookrightarrow L_{(P_2,Q_2;\mathbb{S},\mathbb{W})} \tag{9.5.11}$$
with $0 < P_1, P_2, Q_1, Q_2 \leq \infty$, and $\mathbb{L} = (\lambda_0, \lambda_\infty)$, $\mathbb{E} = (\varepsilon_0, \varepsilon_\infty)$, $\mathbb{S} = (\sigma_0, \sigma_\infty)$, $\mathbb{W} = (\omega_0, \omega_\infty) \in \mathbb{R}^2$.

To this end one can use the approach of [95, Theorem 5.2] where embeddings among classical Lorentz spaces are characterized (cf. also Chapter 10). However, the characterization is described in terms of discretizing sequences and thus it is not explicit. We shall point out a simple characterization of (9.5.11). As above, we present only the statements of results. Detailed proofs can be found in [172, Appendix].

First, we consider the embedding (9.5.11) with $P_1 = P_2 = P$, that is
$$L_{(P,Q_1;\mathbb{L},\mathbb{E})} \hookrightarrow L_{(P,Q_2;\mathbb{S},\mathbb{W})}. \tag{9.5.12}$$

We already know the necessary and sufficient conditions for the embedding
$$L_{P,Q_1;\mathbb{L},\mathbb{E}} \hookrightarrow L_{P,Q_2;\mathbb{S},\mathbb{W}}.$$

Because, for $0 < q \leq \infty$ and $\mathbb{A}, \mathbb{B} \in \mathbb{R}^2$, one has

$$\begin{aligned} L_{(p,q;\mathbb{A},\mathbb{B})} &= L_{p,q;\mathbb{A},\mathbb{B}} & \text{if } 1 < p \leq \infty, \\ L_{(p,q;\mathbb{A},\mathbb{B})} &= \{0\} & \text{if } 0 < p < 1 \text{ and } \mu(R) = \infty, \\ L_{(p,q;\mathbb{A},\mathbb{B})} &= L^1 & \text{if } 0 < p < 1 \text{ and } \mu(\mathcal{R}) < \infty, \end{aligned} \tag{9.5.13}$$

Section 9.5 Embeddings between Generalized Lorentz–Zygmund spaces

it remains to characterize the embedding (9.5.12) for $P = 1$. Such a characterization is given in Theorems (9.5.16)–(9.5.19) while Theorem 5.5 characterizes the embedding (9.5.11) for $P_1 \neq P_2$.

First, we consider the case when $0 < Q_1 \leq Q_2 \leq \infty$.

Theorem 9.5.16. *Let* $0 < Q_1 \leq Q_2 \leq \infty$, $\mu(\mathcal{R}) = \infty$, *and* $L_{(1,Q_1;\mathbb{L},\mathbb{E})} \neq \{0\}$. *Then*

$$L_{(1,Q_1;\mathbb{L},\mathbb{E})} \hookrightarrow L_{(1,Q_2;\mathbb{S},\mathbb{W})}$$

if and only if either

$$\lambda_\infty + \frac{1}{Q_1} > \sigma_\infty + \frac{1}{Q_2},$$

or

$$0 > \lambda_\infty + \frac{1}{Q_1} = \sigma_\infty + \frac{1}{Q_2}, \qquad \varepsilon_\infty \geq \omega_\infty,$$

or

$$0 = \lambda_\infty + \frac{1}{Q_1} = \sigma_\infty + \frac{1}{Q_2}, \qquad \varepsilon_\infty + \frac{1}{Q_1} \geq \omega_\infty + \frac{1}{Q_2},$$

and simultaneously one of the following conditions is satisfied:

$$\lambda_0 + \frac{1}{Q_1} < 0, \qquad \sigma_0 + \frac{1}{Q_2} < 0,$$

$$\lambda_0 + \frac{1}{Q_1} < 0, \qquad \sigma_0 + \frac{1}{Q_2} = 0, \qquad \omega_0 + \frac{1}{Q_2} < 0,$$

$$\lambda_0 + \frac{1}{Q_1} < 0, \qquad Q_2 = \infty, \qquad \sigma_0 = \omega_0 = 0,$$

$$0 \leq \lambda_0 + \frac{1}{Q_1}, \qquad \lambda_0 + \frac{1}{Q_1} > \sigma_0 + \frac{1}{Q_2},$$

$$0 < \lambda_0 + \frac{1}{Q_1} = \sigma_0 + \frac{1}{Q_2}, \qquad \varepsilon_0 \geq \omega_0,$$

$$0 = \lambda_0 + \frac{1}{Q_1} = \sigma_0 + \frac{1}{Q_2}, \qquad \varepsilon_0 + \frac{1}{Q_1} \geq 0, \qquad \varepsilon_0 + \frac{1}{Q_1} \geq \omega_0 + \frac{1}{Q_2},$$

$$0 = \lambda_0 + \frac{1}{Q_1} = \sigma_0 + \frac{1}{Q_2}, \qquad \varepsilon_0 + \frac{1}{Q_1} < 0, \qquad \omega_0 + \frac{1}{Q_2} < 0.$$

The following theorem characterizes the embedding

$$L_{(1,Q_1;\mathbb{L},\mathbb{E})} \hookrightarrow L_{(1,Q_2;\mathbb{S},\mathbb{W})},$$

in the case when $0 < Q_2 < Q_1 \leq \infty$.

Theorem 9.5.17. *Let $0 < Q_2 < Q_1 \leq \infty$, $\mu(R) = \infty$, and $L_{(1,Q_1;\mathbb{L},\mathbb{E})} \neq \{0\}$. Then*
$$L_{(1,Q_1;\mathbb{L},\mathbb{E})} \hookrightarrow L_{(1,Q_2;\mathbb{S},\mathbb{W})}$$
if and only if either
$$\lambda_\infty + \frac{1}{Q_1} > \sigma_\infty + \frac{1}{Q_2},$$
or
$$\lambda_\infty + \frac{1}{Q_1} = \sigma_\infty + \frac{1}{Q_2}, \quad \varepsilon_\infty + \frac{1}{Q_1} > \omega_\infty + \frac{1}{Q_2}$$
and simultaneously one of the following conditions is satisfied:

$$\lambda_0 + \frac{1}{Q_1} < 0, \qquad \sigma_0 + \frac{1}{Q_2} < 0,$$

$$\lambda_0 + \frac{1}{Q_1} < 0, \qquad \sigma_0 + \frac{1}{Q_2} = 0, \quad \omega_0 + \frac{1}{Q_2} < 0,$$

$$0 \leq \lambda_0 + \frac{1}{Q_1}, \qquad \lambda_0 + \frac{1}{Q_1} > \sigma_0 + \frac{1}{Q_2},$$

$$0 < \lambda_0 + \frac{1}{Q_1} = \sigma_0 + \frac{1}{Q_2}, \qquad \varepsilon_0 + \frac{1}{Q_1} > \omega_0 + \frac{1}{Q_2},$$

$$0 = \lambda_0 + \frac{1}{Q_1} = \sigma_0 + \frac{1}{Q_2}, \qquad \varepsilon_0 + \frac{1}{Q_1} \geq 0, \qquad \varepsilon_0 + \frac{1}{Q_1} > \omega_0 + \frac{1}{Q_2},$$

$$0 = \lambda_0 + \frac{1}{Q_1} = \sigma_0 + \frac{1}{Q_2}, \qquad \varepsilon_0 + \frac{1}{Q_1} < 0, \qquad \omega_0 + \frac{1}{Q_2} < 0.$$

In Theorems 9.5.16 and 9.5.16 we have assumed that $\mu(\mathcal{R}) = \infty$. Now, we shall characterize the embedding (9.5.12) provided that $P = 1$ and $\mu(\mathcal{R}) < \infty$. In this case Theorems 9.5.16 and 9.5.16 remain true if we omit all the assumptions on the second components of vectors $\mathbb{L}, \mathbb{E}, \mathbb{S}$ and $\mathbb{W}$ (cf. [73, Theorem 6.3] and remarks on the GLZ spaces with $\mu(\mathcal{R}) < \infty$ above). Since the condition $\mu(\mathcal{R}) < \infty$ implies that the spaces $L_{(1,q;\mathbb{A},\mathbb{B})}$ (with $0 < q \leq \infty$, $\mathbb{A} = (\alpha_0, \alpha_\infty)$, $\mathbb{B} = (\beta_0, \beta_\infty) \in \mathbb{R}^2$) and $L_{(1,q;\alpha_0,\beta_0)}$ coincide, we can consider, instead of (9.5.12), the embedding

$$L_{(1,Q_1;\lambda,\varepsilon)} \hookrightarrow L_{(1,Q_2;\sigma,\omega)},$$

where $0 < Q_1, Q_2 \leq \infty$ and $\lambda, \varepsilon, \sigma, \omega \in \mathbb{R}$. Using the same method as in the case $P = 1$ and $\mu(\mathcal{R}) = \infty$, one can prove the following two theorems.

Theorem 9.5.18. *Let $0 < \mu(\mathcal{R}) < \infty$, $0 < Q_1 \leq Q_2 \leq \infty$, and $\lambda, \varepsilon, \sigma, \omega \in \mathbb{R}$. Then*
$$L_{(1,Q_1;\lambda,\varepsilon)} \hookrightarrow L_{(1,Q_2;\sigma,\omega)}$$

Section 9.5 Embeddings between Generalized Lorentz–Zygmund spaces   331

if and only if one of the following conditions is satisfied:

$$\lambda + \frac{1}{Q_1} < 0, \qquad \sigma + \frac{1}{Q_2} < 0,$$

$$\lambda + \frac{1}{Q_1} < 0, \qquad \sigma + \frac{1}{Q_2} = 0, \qquad \omega + \frac{1}{Q_2} < 0,$$

$$\lambda + \frac{1}{Q_1} < 0, \qquad Q_2 = \infty, \qquad \sigma = \omega = 0,$$

$$0 \leq \lambda + \frac{1}{Q_1}, \qquad \lambda + \frac{1}{Q_1} > \sigma + \frac{1}{Q_2},$$

$$0 < \lambda + \frac{1}{Q_1} = \sigma + \frac{1}{Q_2}, \qquad \varepsilon \geq \omega,$$

$$0 = \lambda + \frac{1}{Q_1} = \sigma + \frac{1}{Q_2}, \qquad \varepsilon + \frac{1}{Q_1} \geq 0, \qquad \varepsilon + \frac{1}{Q_1} \geq \omega + \frac{1}{Q_2},$$

$$0 = \lambda + \frac{1}{Q_1} = \sigma + \frac{1}{Q_2}, \qquad \varepsilon + \frac{1}{Q_1} < 0, \qquad \omega + \frac{1}{Q_2} < 0.$$

Theorem 9.5.19. *Let* $0 < \mu(R) < \infty$, $0 < Q_2 < Q_1 \leq \infty$, *and* $\lambda, \varepsilon, \sigma, \omega \in \mathbb{R}$. *Then*

$$L_{(1,Q_1;\lambda,\varepsilon)} \hookrightarrow L_{(1,Q_2;\sigma,\omega)}$$

if and only if one of the following conditions is satisfied:

$$\lambda + \frac{1}{Q_1} < 0, \qquad \sigma + \frac{1}{Q_2} < 0,$$

$$\lambda + \frac{1}{Q_1} < 0, \qquad \sigma + \frac{1}{Q_2} = 0, \qquad \omega + \frac{1}{Q_2} < 0,$$

$$0 \leq \lambda + \frac{1}{Q_1}, \qquad \lambda + \frac{1}{Q_1} > \sigma + \frac{1}{Q_2},$$

$$0 < \lambda + \frac{1}{Q_1} = \sigma + \frac{1}{Q_2}, \qquad \varepsilon + \frac{1}{Q_1} > \omega + \frac{1}{Q_2},$$

$$0 = \lambda + \frac{1}{Q_1} = \sigma + \frac{1}{Q_2}, \qquad \varepsilon + \frac{1}{Q_1} \geq 0, \qquad \varepsilon + \frac{1}{Q_1} > \omega + \frac{1}{Q_2},$$

$$0 = \lambda + \frac{1}{Q_1} = \sigma + \frac{1}{Q_2}, \qquad \varepsilon + \frac{1}{Q_1} < 0, \qquad \omega + \frac{1}{Q_2} < 0.$$

The next theorem describes embedding among $L_{(p,q;\mathbb{A},\mathbb{B})}$ spaces with p varying.

Theorem 9.5.20. *Let* $0 < P_1, P_2, Q_1, Q_2 \leq \infty$, $P_1 \neq P_2$, $\mathbb{L}, \mathbb{E}, \mathbb{S}, \mathbb{W} \in \mathbb{R}^2$ *and*

$$L_{(P_1,Q_1;\mathbb{L},\mathbb{E})} \neq \{0\}. \tag{9.5.14}$$

Then

$$L_{(P_1,Q_1;\mathbb{L},\mathbb{E})} \hookrightarrow L_{(P_2,Q_2;\mathbb{S},\mathbb{W})} \tag{9.5.15}$$

if and only if $\mu(\mathcal{R}) < \infty$ and one of the following conditions is satisfied:

$$1 \leq P_2 < P_1 \leq \infty, \tag{9.5.16}$$
$$0 < P_2 < 1 \leq P_1 \leq \infty, \tag{9.5.17}$$
$$0 < P_1, P_2 < 1, \tag{9.5.18}$$
$$0 < P_1 < 1, \quad P_2 = 1, \quad \sigma_0 + \frac{1}{Q_2} < 0, \tag{9.5.19}$$
$$0 < P_1 < 1, \quad P_2 = 1, \quad \sigma + \frac{1}{Q_2} = 0, \quad \omega_0 + \frac{1}{Q_2} < 0, \tag{9.5.20}$$
$$0 < P_1 < 1, \quad P_2 = 1, \quad Q_2 = \infty, \quad \sigma_0 = \omega_0 = 0. \tag{9.5.21}$$

9.6 The associate space

Our next objective is to give a complete description of the associate space of a non-trivial GLZ space. To begin, we single out the GLZ spaces whose associate space is trivial.

Theorem 9.6.1. *Let $0 < p, q \leq \infty$ and $\mathbb{A} = (\alpha_0, \alpha_\infty)$, $\mathbb{B} = (\beta_0, \beta_\infty) \in \mathbb{R}^2$. Set $X = L_{p,q;\mathbb{A},\mathbb{B}}$.*

(i) *Assume that one of the following conditions holds:*

$$0 < p < 1,$$
$$p = 1, \quad 0 < q \leq 1, \quad \alpha_0 < 0,$$
$$p = 1, \quad 0 < q \leq 1, \quad \alpha_0 = 0, \quad \beta_0 < 0.$$

Then
$$X' = \{0\}.$$

(ii) *Assume that one of the following conditions holds:*

$$p = 1, \quad 1 < q \leq \infty, \quad \alpha_0 < \frac{1}{q'},$$
$$p = 1, \quad 1 < q \leq \infty, \quad \alpha_0 = \frac{1}{q'}, \quad \beta_0 \leq \frac{1}{q'}.$$

Then
$$X' = \{0\}.$$

Proof. (i) By Theorem 9.1.4, for every $t \in (0, 1)$, there exists a function $g_t \in \mathcal{M}(\mathcal{R}, \mu)$ such that $g_t^* = \chi_{(0,t)}$. By Lemma 9.4.1 (i),

$$\|g_t\|_X \approx t^{\frac{1}{p}} \ell^{\alpha_0}(t) \ell\ell^{\beta_0}(t), \quad 0 < t < 1.$$

Assume that $f \in \mathcal{M}(\mathcal{R}, \mu)$ and $f \not\equiv 0$. Then there exist two positive constants, ε, δ, such that $f^*(s) \geq \delta$ for $s \in (0, \varepsilon)$. We claim that $f \notin X'$, that is, $\|f\|_{X'} = \infty$. Indeed,

$$\|f\|_{X'} = \sup_{\|g\|_X \leq 1} \int_0^\infty f^*(s)g^*(s)\,ds \geq \sup_{0 < t < \varepsilon} \int_0^\infty f^*(s)\frac{g_t^*(s)}{\|g_t\|_X}\,ds$$

$$\gtrsim \delta \sup_{0 < t < \varepsilon} t^{-\frac{1}{p}} \ell^{-\alpha_0}(t) \ell\ell^{-\beta_0}(t) \int_0^t ds = \infty.$$

(ii) The function $h(t) = t^{-1}\ell^{-1}(t)\ell\ell^{-1}(t)\chi_{(0,\varepsilon)}(t)$ is, for some ε small enough, nonincreasing on $(0, \infty)$. Moreover,

$$\|t^{\frac{1}{p}-\frac{1}{q}}\ell^{\mathbb{A}}(t)\ell\ell^{\mathbb{B}}(t)h(t)\|_{q,(0,\infty)} < \infty, \tag{9.6.1}$$

but

$$\int_0^\delta h(t)\,dt = \infty \quad \text{for every} \quad \delta > 0. \tag{9.6.2}$$

By Theorem 9.1.4, there is a $g \in \mathcal{M}(\mathcal{R}, \mu)$ such that $g^* = h$, and, by (9.6.1), $g \in X$. Now, let $f \in X'$. Then $\int_0^\infty f^*(t)g^*(t)\,dt < \infty$. However, by (9.6.2), this is possible only if $f \equiv 0$.

The proof is complete. □

In the next theorem we describe associated spaces of $X = L_{p,q;\mathbb{A},\mathbb{B}}$ provided that $1 \leq p \leq \infty$ and $1 < q \leq \infty$. We restrict ourselves to the cases when $X \neq \{0\}$.

Theorem 9.6.2. *Let* $1 \leq p \leq \infty$, $1 < q \leq \infty$, $\mathbb{A} = (\alpha_0, \alpha_\infty)$, $\mathbb{B} = (\beta_0, \beta_\infty) \in \mathbb{R}^2$, *and assume that the space* $L_{p,q;\mathbb{A},\mathbb{B}}$ *is nontrivial (cf. (9.3.1)). Then* $(L_{p,q;\mathbb{A},\mathbb{B}})' = \mathfrak{L}$, *the space described below:*

(i) *Let* $1 \leq p < \infty$, $1 < q \leq \infty$. *Then*

$$\mathfrak{L} = L_{(p',q';-\mathbb{A},-\mathbb{B})} = L_{p',q';-\mathbb{A},-\mathbb{B}}. \tag{9.6.3}$$

(ii) *Let* $p = \infty$, $1 < q < \infty$, $\alpha_0 + \frac{1}{q} < 0$, *and either* $\alpha_\infty + \frac{1}{q} > 0$ *or* $\alpha_\infty + \frac{1}{q} = 0$, $\beta_\infty + \frac{1}{q} \geq 0$. *Then*

$$\mathfrak{L} = \begin{cases} L_{(1,q';-\mathbb{A}-1,-\mathbb{B})} & \text{if } \alpha_\infty + \frac{1}{q} > 0, \\ L_{(1,q';(-\alpha_0-1,-\frac{1}{q'}),(-\beta_0,-\beta_\infty-1))} & \text{if } \alpha_\infty + \frac{1}{q} = 0, \ \beta_\infty + \frac{1}{q} > 0, \\ L_{(1,q';(-\alpha_0-1,-\frac{1}{q'}),(-\beta_0,-\frac{1}{q'}),(0,-1))} & \text{if } \alpha_\infty + \frac{1}{q} = 0, \ \beta_\infty + \frac{1}{q} = 0. \end{cases}$$

(iii) Let $p = \infty$, $1 < q < \infty$, $\alpha_0 + \frac{1}{q} = 0$, $\beta_0 + \frac{1}{q} < 0$, and either $\alpha_\infty + \frac{1}{q} > 0$ or $\alpha_\infty + \frac{1}{q} = 0$, $\beta_\infty + \frac{1}{q} \geq 0$. Then

$$\mathfrak{L} = \begin{cases} L_{(1,q';(-\frac{1}{q'},-\alpha_\infty-1),(-\beta_0-1,-\beta_\infty))} & \text{if } \alpha_\infty + \frac{1}{q} > 0, \\ L_{(1,q';(-\frac{1}{q'},-\frac{1}{q'}),-\mathbb{B}-1)} & \text{if } \alpha_\infty + \frac{1}{q} = 0,\ \beta_\infty + \frac{1}{q} > 0, \\ L_{(1,q';(-\frac{1}{q'},-\frac{1}{q'}),(-\beta_0-1,-\frac{1}{q'}),(0,-1))} & \text{if } \alpha_\infty + \frac{1}{q} = 0,\ \beta_\infty + \frac{1}{q} = 0. \end{cases}$$

(iv) Let $p = \infty$, $q = \infty$, and either $\alpha_\infty > 0$ or $\alpha_\infty = 0$, $\beta_\infty \geq 0$. Then

$$\mathfrak{L} = L_{1,1;-\mathbb{A},-\mathbb{B}}.$$

(v) Let $p = \infty$, $1 < q \leq \infty$, and either $\alpha_\infty + \frac{1}{q} < 0$ or $\alpha_\infty + \frac{1}{q} = 0$ and $\beta_\infty + \frac{1}{q} < 0$. Then

$$\mathfrak{L} = \left\{ f \in \mathcal{M}(\mathcal{R}, \mu);\ \|f\|_{\mathfrak{L}} := \|f\|_{X(0,1)} + \int_0^\infty f^*(t)\,dt < \infty \right\},$$

where

$$X(0,1) = \begin{cases} L_{(1,q';-\alpha_0-1,-\beta_0)}(0,1) & \text{if } 1 < q < \infty,\ \alpha_0 + \frac{1}{q} < 0, \\ L_{(1,q';-\frac{1}{q'},-\beta_0-1)}(0,1) & \text{if } 1 < q < \infty,\ \alpha_0 + \frac{1}{q} = 0, \\ & \phantom{\text{if}} \beta_0 + \frac{1}{q} < 0, \\ L_{1,1;-\alpha_0,-\beta_0}(0,1) & \text{if } q = \infty. \end{cases}$$

Remark 9.6.3. Since $L_{(1,r;\sigma,\omega)}(0,1) \hookrightarrow L^1(0,1)$ for every $1 \leq r \leq \infty$, $\sigma, \omega \in \mathbb{R}$, and $L_{1,1;-\alpha_0,-\beta_0}(0,1) \hookrightarrow L^1(0,1)$ if either $\alpha_0 < 0$, or $\alpha_0 = 0$ and $\beta_0 \leq 0$, we can write in Theorem 9.6.2 (v),

$$\mathfrak{L} = \left\{ f \in \mathcal{M}(\mathcal{R}, \mu);\ \|f\|_{\mathfrak{L}} := \|f\|_{X(0,1)} + \int_1^\infty f^*(t)\,dt < \infty \right\}.$$

Proof of Theorem 9.6.2. We first prove the assertion in the cases (i)–(iii) for $1 < q < \infty$, since the technique of the proof is common. We shall start with proving the inclusion

$$(L_{p,q;\mathbb{A},\mathbb{B}})' \hookrightarrow \mathfrak{L}. \tag{9.6.4}$$

For this purpose it is enough to verify the inequality

$$\|f\|_{\mathfrak{L}} \lesssim \|f\|_{(L_{p,q;\mathbb{A},\mathbb{B}})'} \tag{9.6.5}$$

for all step functions f. For convenience, let us denote by $b(t)$ the function defined by

$$\|f\|_{\mathfrak{L}} = \|t^{\frac{1}{p'}-\frac{1}{q'}} b(t) f^{**}(t)\|_{q'}. \tag{9.6.6}$$

Section 9.6 The associate space

We further set
$$\varrho(t) = (f^{**}(t))^{q'-1} t^{\frac{q'}{p'}-1} b^{q'}(t), \qquad (9.6.7)$$

and
$$g(t) = \int_t^\infty \frac{\varrho(s)}{s}\, ds. \qquad (9.6.8)$$

Then $g^* = g$ and there exists $\widetilde{g} \in \mathcal{M}(\mathcal{R}, \mu)$ such that $\widetilde{g}^* = g$. By the Fubini theorem and the Hölder inequality,

$$\|f\|_{\mathcal{L}}^{q'} = \int_0^\infty \varrho(t) f^{**}(t)\, dt = \int_0^\infty g(t) f^*(t)\, dt \le \|\widetilde{g}\|_{p,q;\mathbb{A},\mathbb{B}} \|f\|_{(L_{p,q;\mathbb{A},\mathbb{B}})'}.$$

Now, in order to obtain (9.6.5), it is enough to show that

$$\|\widetilde{g}\|_{p,q;\mathbb{A},\mathbb{B}} \lesssim \|f\|_{\mathcal{L}}^{q'-1}. \qquad (9.6.9)$$

Rewriting (9.6.9) with the help of Definition 9.2.3 and (9.6.6) we get, using also (9.6.8) and (9.6.7),

$$\left\| t^{\frac{1}{p}-\frac{1}{q}} \ell^{\mathbb{A}}(t) \ell\ell^{\mathbb{B}}(t) \int_t^\infty (f^{**}(s))^{q'-1} s^{\frac{q'}{p'}-2} b^{q'}(s)\, ds \right\|_q$$
$$\lesssim \left\| t^{\frac{q'-1}{p'}-\frac{1}{q}} b^{q'-1}(t) (f^{**}(t))^{q'-1} \right\|_q.$$

Using an appropriate substitution, this amounts to the Hardy inequality

$$\left\| t^{\frac{1}{p}-\frac{1}{q}} \ell^{\mathbb{A}}(t) \ell\ell^{\mathbb{B}}(t) \int_t^\infty h(s)\, ds \right\|_q \lesssim \left\| t^{\frac{1}{p}+\frac{1}{q'}} (b(t))^{-1} h(t) \right\|_q. \qquad (9.6.10)$$

A sufficient condition for (9.6.10) is given by (cf. Theorem 3.11.8)

$$\sup_{0<x<\infty} \|t^{\frac{1}{p}-\frac{1}{q}} \ell^{\mathbb{A}}(t) \ell\ell^{\mathbb{B}}(t)\|_{q,(0,x)} \|t^{-\frac{1}{p}-\frac{1}{q'}} b(t)\|_{q',(x,\infty)} < \infty. \qquad (9.6.11)$$

Now it is a matter of a tedious but elementary calculation to verify (9.6.11) for appropriate $b(t)$, given by (9.6.6), in all the cases of $\mathcal{L}$. This yields the inclusion (9.6.4) for the cases (i)–(iii) restricted to $1 < q < \infty$.

Now let us prove the converse inclusion $\mathcal{L} \hookrightarrow (L_{p,q;\mathbb{A},\mathbb{B}})'$, that is,

$$\|f\|_{(L_{p,q;\mathbb{A},\mathbb{B}})'} \lesssim \|f\|_{\mathcal{L}} \qquad \text{for all} \quad f \in \mathcal{L} \qquad (9.6.12)$$

in the cases (i)–(iii) with $1 < q < \infty$. Since for every $f \in \mathcal{M}(\mathcal{R}, \mu)$,

$$\|f\|_{(L_{p,q;\mathbb{A},\mathbb{B}})'} = \sup_{\|g\|_{p,q;\mathbb{A},\mathbb{B}} \le 1} \int_{\mathcal{R}} f(x) g(x)\, d\mu, \qquad (9.6.13)$$

we see that, in view of the Hardy–Littlewood inequality (7.2.2), in order to prove (9.6.12) it is enough to show that

$$\int_0^\infty f^*(t) g^*(t)\, dt \lesssim \|f\|_{\mathfrak{L}}\, \|g\|_{p,q;\mathbb{A},\mathbb{B}} \qquad (9.6.14)$$

for all $f \in \mathfrak{L}$ and $g \in L_{p,q;\mathbb{A},\mathbb{B}}$. We use (9.6.6) and rewrite (9.6.14) as

$$\int_0^\infty f^*(t) g^*(t)\, dt \le \|t^{\frac{1}{p'}-\frac{1}{q'}} b(t) f^{**}(t)\|_{q'}\, \|t^{\frac{1}{p}-\frac{1}{q}} \ell^{\mathbb{A}}(t) \ell\ell^{\mathbb{B}}(t) g^*(t)\|_q. \qquad (9.6.15)$$

To get (9.6.15) we shall use the result of Sawyer [194, (1.7)]:

$$\int_0^\infty f^*(t) g^*(t)\, dt \le \|g^*\|_{q(v)} \left(\|f^{**}\|_{q'(\widetilde{v})} + \frac{1}{v(0,\infty)^{\frac{1}{q}}} \int_0^\infty f^*(t)\, dt \right), \qquad (9.6.16)$$

where $1 < q < \infty$, v is a positive weight, $v(a,b) = \int_a^b v(t) dt$ if $0 \le a < b \le \infty$, $\widetilde{v}$ is given by

$$\widetilde{v}(t) = \frac{t^{q'} v(t)}{(v(0,t))^{q'}}, \qquad t \in (0,\infty), \qquad (9.6.17)$$

and $\|h\|_{q(v)} = \left(\int_0^\infty h^q v \right)^{\frac{1}{q}}$. The suitable choice of v in our situation reads as

$$v(t) = t^{\frac{q}{p}-1} \ell^{\mathbb{A}q}(t) \ell\ell^{\mathbb{B}q}(t), t \in (0,\infty). \qquad (9.6.18)$$

Now, in the cases (i)–(iii) with $1 < q < \infty$, we have $v(0,\infty) = \infty$. Together with the convention $\frac{\infty}{\infty} = 0$ (used also in [194]) this implies that the second summand at the right-hand side of (9.6.16) disappears, and (9.6.14) will follows once we show that

$$t^{\frac{q'}{p'}-1} b^{q'}(t) = \widetilde{v}(t), \qquad (9.6.19)$$

where b is the function from (9.6.6) that corresponds to the space $\mathfrak{L}$ from (i)–(iii), and $v, \widetilde{v}$ are from (9.6.18), (9.6.17). Since (9.6.19) follows by a calculation, we have proved (i)–(iii) for $1 < q < \infty$.

Our next step will be to prove (v) for $1 < q < \infty$. Assume that $\alpha_0 + \frac{1}{q} < 0$. (The proof in the case $\alpha_0 + \frac{1}{q} = 0$ and $\beta_0 + \frac{1}{q} < 0$ is entirely analogous and therefore omitted.) By our definition of $\mathfrak{L}$,

$$\|f\|_{\mathfrak{L}} = \|t^{\frac{1}{q}} \ell^{-\alpha_0-1}(t) \ell\ell^{-\beta_0}(t) f^{**}(t)\|_{q',(0,1)} + \int_0^\infty f^*(t)\, dt.$$

Exactly in the same way as above we can show that

$$\|t^{\frac{1}{q}} \ell^{-\alpha_0-1}(t) \ell\ell^{-\beta_0}(t) f^{**}(t)\|_{q',(0,1)} \lesssim \|f\|_{(L_{\infty,q;\mathbb{A},\mathbb{B}})'}.$$

Section 9.6 The associate space

Further, by the Hölder inequality,

$$\int_0^\infty f^*(t)\, dt \leq \|f\|_{(L_{\infty,q;\mathbb{A},\mathbb{B}})'} \|1\|_{\infty,q;\mathbb{A},\mathbb{B}},$$

and, since $1 \in L_{\infty,q;\mathbb{A},\mathbb{B}}$, this yields $(L_{\infty,q;\mathbb{A},\mathbb{B}})' \hookrightarrow \mathfrak{L}$.

To prove the converse embedding, we shall use the Sawyer inequality (9.6.16) again, but this time the last summand does not disappear, as $v(0,\infty) < \infty$, where v is from (9.6.18), that is (recall $p = \infty$),

$$v(t) = t^{-1}\ell^{\mathbb{A}q}(t)\ell\ell^{\mathbb{B}q}(t), \qquad t \in (0,\infty).$$

We get

$$v(0,t) \approx \ell^{\alpha_0 q + 1}(t)\ell\ell^{\beta_0 q}(t)\chi_{(0,1)}(t) + \chi_{(1,\infty)}(t),$$

and, by (9.6.17),

$$\widetilde{v}(t) \approx t^{q'-1}\ell^{(-\alpha_0-1)q'}(t)\ell\ell^{-\beta_0 q'}(t)\chi_{(0,1)}(t) + t^{q'-1}\ell^{\alpha_\infty q}(t)\ell\ell^{\beta_\infty q}(t)\chi_{(1,\infty)}(t).$$

Therefore, by (9.6.16),

$$\int_0^\infty f^*(t)g^*(t)\, dt \lesssim \|g\|_{\infty,q;\mathbb{A},\mathbb{B}} \left(\left(\int_0^\infty (f^{**}(t))^{q'} \widetilde{v}(t)\, dt \right)^{\frac{1}{q'}} + \int_0^\infty f^*(t)\, dt \right).$$

Moreover,

$$\left(\int_0^\infty (f^{**}(t))^{q'} \widetilde{v}(t)\, dt \right)^{\frac{1}{q'}} = \|t^{\frac{1}{q'}} \ell^{-\alpha_0 - 1}(t)\ell\ell^{-\beta_0}(t) f^{**}(t)\|_{q',(0,1)}$$

$$+ \|t^{\frac{1}{q'}} \ell^{\alpha_\infty(q-1)}(t)\ell\ell^{\beta_\infty(q-1)}(t) f^{**}(t)\|_{q',(1,\infty)} =: I_1 + I_2.$$

It is clear that

$$I_1 = \|f^*\|_{X(0,1)}.$$

To estimate I_2, we insert for f^{**} and obtain thereby

$$I_2 \approx \int_0^1 f^*(t)\, dt\, \|t^{-\frac{1}{q'}} \ell^{\alpha_\infty(q-1)}(t)\ell\ell^{\beta_\infty(q-1)}(t)\|_{q',(1,\infty)}$$

$$+ \left\| t^{-\frac{1}{q'}} \ell^{\alpha_\infty(q-1)}(t)\ell\ell^{\beta_\infty(q-1)}(t) \int_1^t f^* \right\|_{q',(1,\infty)} =: I_3 + I_4.$$

Observe that

$$\|t^{-\frac{1}{q'}} \ell^{\alpha_\infty(q-1)}(t)\ell\ell^{\beta_\infty(q-1)}(t)\|_{q',(1,\infty)} < \infty,$$

and this yields

$$I_3 \lesssim \int_0^1 f^*(t)\, dt.$$

Further, by Theorem 3.11.8
$$I_4 \lesssim \int_1^\infty f^*(t)\,dt.$$

Consequently,
$$\int_0^\infty f^*(t)g^*(t)\,dt \lesssim \|g\|_{\infty,q;\mathbb{A},\mathbb{B}} \left(\|f\|_{X(0,1)} + \int_0^\infty f^*(t)\,dt \right).$$

Together with (9.6.13) and the Hardy–Littlewood inequality (7.2.2), this yields the embedding $\mathfrak{L} \hookrightarrow (L_{\infty,q;\mathbb{A},\mathbb{B}})'$, which completes the proof of (v) for $1 < q < \infty$.

Finally, let $q = \infty$. Then we have
$$\|f\|_{p,\infty;\mathbb{A},\mathbb{B}} = \sup_{0<t<\infty} t^{\frac{1}{p}} \ell^{\mathbb{A}}(t) \ell\ell^{\mathbb{B}}(t) f^*(t).$$

Assume that $1 < p \leq \infty$. Then, by Theorem 9.5.1 (i), also
$$\|f\|_{p,\infty;\mathbb{A},\mathbb{B}} \approx \sup_{0<t<\infty} t^{\frac{1}{p}} \ell^{\mathbb{A}}(t) \ell\ell^{\mathbb{B}}(t) f^{**}(t).$$

Moreover, our assumptions on $p, \mathbb{A}, \mathbb{B}$ imply
$$\|f\|_{p,\infty;\mathbb{A},\mathbb{B}} \approx \sup_{0<t<\infty} \varphi(t) f^{**}(t),$$

where
$$\varphi(t) \approx \begin{cases} t^{\frac{1}{p}} \ell^{\mathbb{A}}(t) \ell\ell^{\mathbb{B}}(t), & \text{if } 1 < p < \infty, \\ & \text{or } p = \infty,\ \alpha_\infty > 0, \\ & \text{or } p = \infty,\ \alpha_\infty = 0,\ \beta_\infty \geq 0, \\ t^{\frac{1}{p}} \ell^{\alpha_0}(t) \ell\ell^{\mathbb{B}_0}(t)\chi_{(0,1]}(t) + \chi_{(1,\infty)}(t) & \text{if } p = \infty,\ \alpha_\infty < 0, \\ & \text{or } p = \infty,\ \alpha_\infty = 0,\ \beta_\infty \leq 0. \end{cases}$$

By (9.3.1), $\varphi \in \mathcal{F}$ in all cases. Hence, by (7.10.1), $L_{p,\infty;\mathbb{A},\mathbb{B}} = M_\varphi$. Moreover, $L_{p,\infty;\mathbb{A},\mathbb{B}} = \Lambda^{p,\infty}(w)$ for an appropriate weight w. Thus, applying Lemma 7.11.22 (see also Theorem 10.4.1 (ii)), calculating $\frac{d\widetilde{\varphi}}{dt}$, where $\widetilde{\varphi}(t) = \frac{t}{\varphi(t)}$, and using (7.10.12), we get (i) for $q = \infty$ and $p > 1$, (iv), and (v) for $q = \infty$.

To finish the proof, we have only to verify (i) for $p = 1$ and $q = \infty$. The inequality
$$\int_0^\infty f^*(t)g^*(t)\,dt \leq \|g\|_{1,\infty;\mathbb{A},\mathbb{B}} \|f\|_{\infty,1;-\mathbb{A},-\mathbb{B}}$$

yields $L_{\infty,1;-\mathbb{A},-\mathbb{B}} \hookrightarrow (L_{(1,\infty;\mathbb{A},\mathbb{B})})'$. For the converse, set
$$\varrho(t) = t^{-1} \ell^{-\mathbb{A}}(t) \ell\ell^{-\mathbb{B}}(t).$$

Section 9.6 The associate space

Then $\|\varrho\|_{1,\infty;\mathbb{A},\mathbb{B}} \approx 1$, whence, by the Hölder inequality,

$$\|f\|_{\infty,1;-\mathbb{A},-\mathbb{B}} = \int_0^\infty \varrho(t) f^*(t)\, dt \leq \|\varrho\|_{1,\infty;\mathbb{A},\mathbb{B}} \|f\|_{(L_{1,\infty;\mathbb{A},\mathbb{B}})'}.$$

The proof is complete. (Note that the second equality in (9.6.3) follows from Theorem 9.5.1 (i).) □

It remains to describe associate spaces of $X = L_{p,q;\mathbb{A},\mathbb{B}}$ provided that $X \neq \{0\}$, $0 < q \leq 1$ and either $1 < p \leq \infty$, or $p = 1$, $\alpha_0 > 0$, or $p = 1$, $\alpha_0 = 0$, and $\beta_0 \geq 0$. Since in this case X is not necessarily a Banach space, we need some extension of the appropriate definitions.

Definition 9.6.4. Let Z be a nontrivial quasi-normed space over $(0, \infty)$ endowed with the quasinorm $\|\cdot\|_Z$ satisfying the rearrangement-invariance property $\|f\|_Z = \|f^*\|_Z$ for every $f \in Z$. We then say that Z is a *rearrangement-invariant quasi-Banach space*. We also define the functional

$$\|g\|_{Z'} := \sup_{f \neq 0} \frac{\int_0^\infty f^*(t) g^*(t)\, dt}{\|f\|_Z}$$

and the set

$$Z' := \{g \in \mathcal{M}(0, \infty); \|g\|_{Z'} < \infty\}.$$

We shall use the following two lemmas of independent interest.

Lemma 9.6.5. *Let Z be a nontrivial quasi-normed space over $(0, \infty)$ endowed with the quasinorm $\|\cdot\|_Z$ satisfying the rearrangement-invariance property $\|f\|_Z = \|f^*\|_Z$. Let φ_X be an absolutely continuous function satisfying $\varphi_X \in \mathcal{F}$. Assume that $Z \hookrightarrow \Lambda_{\varphi_X}$. Then $M_{\widetilde{\varphi}_X} \hookrightarrow Z'$, where $\widetilde{\varphi}_X(t) = \frac{t}{\varphi_X(t)}$.*

Proof. Note that $Z \hookrightarrow \Lambda_{\varphi_X}$ implies

$$\sup_{f \neq 0} \frac{\|f\|_{\Lambda_{\varphi_X}}}{\|f\|_Z} < \infty.$$

But, by the definition of Z' and by Lemma 7.11.22 (see also Corollary 10.4.2), we get

$$\sup_{f \neq 0} \frac{\|f\|_{\Lambda_{\varphi_X}}}{\|f\|_Z} = \sup_{f \neq 0} \sup_{g \neq 0} \frac{\int_0^\infty f^*(t) g^*(t)\, dt}{\|f\|_Z \|g\|_{M_{\widetilde{\varphi}_X}}} = \sup_{g \neq 0} \frac{\|g\|_{Z'}}{\|g\|_{M_{\widetilde{\varphi}_X}}} < \infty.$$

In other words,

$$M_{\widetilde{\varphi}_X} \hookrightarrow Z',$$

as desired. The proof is complete. □

Lemma 9.6.6. *Let X be a rearrangement-invariant quasi-Banach space. Let φ_X be an absolutely continuous function satisfying $\varphi_X \in \mathcal{F}$. Assume that $X \hookrightarrow \Lambda_{\varphi_X}$. Then $X' = M_{\widetilde{\varphi}_X}$, where $\widetilde{\varphi}_X(t) = \frac{t}{\varphi_X(t)}$.*

Proof. The inclusion $M_{\widetilde{\varphi}_X} \hookrightarrow X'$ follows as a particular case $X = Z$ from Lemma 9.6.5.

For the converse, noting that the Hölder inequality, and hence also (7.9.13) hold when X is just a rearrangement-invariant quasi-Banach space, one gets by (7.10.1), (7.9.13) and (7.9.7), we get

$$\|f\|_{M_{\widetilde{\varphi}_X}} = \sup_{0<t<\mu(\mathcal{R})} \frac{\widetilde{\varphi}_X(t)}{t} \int_0^t f^*(s)\,ds \leq \sup_{0<t<\mu(\mathcal{R})} \frac{\widetilde{\varphi}_X(t)}{t} \varphi_X(t) \|f\|_{X'}$$
$$= \|f\|_{X'},$$

as desired. The proof is complete. □

To check the condition $\varphi_X \in \mathcal{F}$ of Lemma 9.6.6, the next lemma will be useful. The proof can be obtained via a simple calculation and is omitted.

Lemma 9.6.7. *Let $0 < p \leq \infty$ and $\mathbb{L} = (\lambda_0, \lambda_\infty)$, $\mathbb{E} = (\varepsilon_0, \varepsilon_\infty) \in \mathbb{R}^2$. Then the function φ, given by $\varphi(0) = 0$ and $\varphi(t) = t^{\frac{1}{p}} \ell^{\mathbb{L}}(t) \ell\ell^{\mathbb{E}}(t)$ for $t \in (0,\infty)$, is equivalent on $(0,\infty)$ to a nondecreasing concave function if and only if one of the following conditions holds:*

$$\begin{cases} 1 < p < \infty; \\ p = 1,\ \lambda_0 > 0,\ \lambda_\infty < 0, \\ p = 1,\ \lambda_0 > 0,\ \lambda_\infty = 0,\ \varepsilon_\infty \leq 0, \\ p = 1,\ \lambda_0 = 0,\ \varepsilon_0 \geq 0,\ \lambda_\infty < 0, \\ p = 1,\ \lambda_0 = 0,\ \varepsilon_0 \geq 0,\ \lambda_\infty = 0,\ \varepsilon_\infty \leq 0, \\ p = \infty,\ \lambda_0 < 0,\ \lambda_\infty > 0, \\ p = \infty,\ \lambda_0 < 0,\ \lambda_\infty = 0,\ \varepsilon_\infty \geq 0, \\ p = \infty,\ \lambda_0 = 0,\ \varepsilon_0 \leq 0,\ \lambda_\infty > 0, \\ p = \infty,\ \lambda_0 = 0,\ \varepsilon_0 \leq 0,\ \lambda_\infty = 0,\ \varepsilon_\infty \geq 0. \end{cases} \quad (9.6.20)$$

Theorem 9.6.8. *Let $0 < q \leq 1$, $\mathbb{A} = (\alpha_0, \alpha_\infty)$, $\mathbb{B} = (\beta_0, \beta_\infty) \in \mathbb{R}^2$, and assume that the space $X = L_{p,q;\mathbb{A},\mathbb{B}}$ is nontrivial. Let one of the following conditions hold:*

$$1 < p \leq \infty,$$
$$p = 1, \quad \alpha_0 > 0, \qquad (9.6.21)$$
$$p = 1, \quad \alpha_0 = 0, \quad \beta_0 > 0, \qquad (9.6.22)$$
$$p = 1, \quad q = 1, \quad \alpha_0 = 0, \quad \beta_0 = 0. \qquad (9.6.23)$$

Then $X' = \mathfrak{L}$, a space described below:

Section 9.6 The associate space

(i) *Let* $1 \leq p < \infty$. *Then*
$$\mathfrak{L} = L_{(p',\infty;-\mathbb{A},-\mathbb{B})} = L_{p',\infty;-\mathbb{A},-\mathbb{B}}.$$

(ii) *Let* $p = \infty$, $\alpha_0 + \frac{1}{q} < 0$, *and either* $\alpha_\infty + \frac{1}{q} > 0$ *or* $\alpha_\infty + \frac{1}{q} = 0$, $\beta_\infty + \frac{1}{q} \geq 0$. *Then*

$$\mathfrak{L} = \begin{cases} L_{(1,\infty;-\mathbb{A}-\frac{1}{q},-\mathbb{B})} & \text{if } \alpha_\infty + \frac{1}{q} > 0, \\ L_{(1,\infty;(-\alpha_0-\frac{1}{q},0),(-\beta_0,-\beta_\infty-\frac{1}{q}))} & \text{if } \alpha_\infty + \frac{1}{q} = 0, \beta_\infty + \frac{1}{q} > 0, \\ L_{(1,\infty;(-\alpha_0-\frac{1}{q},0),(-\beta_0,0),(0,-\frac{1}{q}))} & \text{if } \alpha_\infty + \frac{1}{q} = 0, \beta_\infty + \frac{1}{q} = 0. \end{cases}$$

(iii) *Let* $p = \infty$, $\alpha_0 + \frac{1}{q} = 0$, $\beta_0 + \frac{1}{q} < 0$, *and either* $\alpha_\infty + \frac{1}{q} > 0$ *or* $\alpha_\infty + \frac{1}{q} = 0$, $\beta_\infty + \frac{1}{q} \geq 0$. *Then*

$$\mathfrak{L} = \begin{cases} L_{(1,\infty;(0,-\alpha_\infty-\frac{1}{q}),(-\beta_0-\frac{1}{q},-\beta_\infty))} & \text{if } \alpha_\infty + \frac{1}{q} > 0, \\ L_{(1,\infty;(0,0),-\mathbb{B}-\frac{1}{q})} & \text{if } \alpha_\infty + \frac{1}{q} = 0, \beta_\infty + \frac{1}{q} > 0, \\ L_{(1,\infty;(0,0),(-\beta_0-\frac{1}{q},0),(0,-\frac{1}{q}))} & \text{if } \alpha_\infty + \frac{1}{q} = 0, \beta_\infty + \frac{1}{q} = 0. \end{cases}$$

(iv) *Let* $p = \infty$, *and either* $\alpha_\infty + \frac{1}{q} < 0$ *or* $\alpha_\infty + \frac{1}{q} = 0$ *and* $\beta_\infty + \frac{1}{q} < 0$. *Then*

$$\mathfrak{L} = \left\{ f \in \mathcal{M}(\mathcal{R},\mu); \ \|f\|_{\mathfrak{L}} := \|f\|_{Y(0,1)} + \int_0^\infty f^*(t)\,dt < \infty \right\},$$

where

$$Y(0,1) = \begin{cases} L_{(1,\infty;-\alpha_0-\frac{1}{q},-\beta_0)}(0,1) & \text{if } \alpha_0 + \frac{1}{q} < 0 \\ L_{(1,\infty;0,-\beta_0-\frac{1}{q})}(0,1) & \text{if } \alpha_0 + \frac{1}{q} = 0, \beta_0 + \frac{1}{q} < 0. \end{cases}$$

Proof. (I) Assume first that $1 < p < \infty$. By Lemma 9.4.1 (i),

$$\varphi_X(t) \approx t^{\frac{1}{p}} \ell^{\mathbb{A}}(t) \ell \ell^{\mathbb{B}}(t), \qquad t \in (0,\infty). \tag{9.6.24}$$

Together with Lemma 9.6.7, this implies that $\varphi_X \in \mathcal{F}$. Moreover, by (9.6.24), $\Lambda_{\varphi_X} = L_{p,1;\mathbb{A},\mathbb{B}}$, and

$$\widetilde{\varphi}_X(t) \approx t^{\frac{1}{p'}} \ell^{-\mathbb{A}}(t) \ell \ell^{-\mathbb{B}}(t), \qquad t \in (0,\infty).$$

Consequently,
$$M_{\widetilde{\varphi}_X} = L_{(p',\infty;-\mathbb{A},-\mathbb{B})} = L_{p',\infty;-\mathbb{A},-\mathbb{B}}.$$

Thus, by Lemma 9.6.6,
$$X' = M_{\widetilde{\varphi}_X} = L_{p',\infty;-\mathbb{A},-\mathbb{B}},$$

provided that
$$X = L_{p,q;\mathbb{A},\mathbb{B}} \hookrightarrow L_{p,1;\mathbb{A},\mathbb{B}} = \Lambda_{\varphi_X}.$$
This however follows from Theorem 9.5.10.

(II) In the case when $p = \infty$ the proof is quite analogous to the one above (instead of Theorem 9.5.10 we use Theorem 9.5.11).

(III) Assume that either (9.6.21) or (9.6.22) or (9.6.23) holds. Then (cf. (9.6.20)), $\varphi_X \in \mathcal{F}$ if and only if

$$\text{either } \alpha_\infty < 0 \quad \text{or } \alpha_\infty = 0 \text{ and } \beta_\infty \leq 0. \tag{9.6.25}$$

In each of these cases we can again apply the argument from part (I) to get the result.

Assume now that (9.6.25) is not satisfied, that is, either $\alpha_\infty > 0$ or $\alpha_\infty = 0$ and $\beta_\infty > 0$. We set $Z = L_{1,q;(\alpha_0,0),(\beta_0,0)}$. Then clearly $X \hookrightarrow Z$, whence

$$Z' \hookrightarrow X'. \tag{9.6.26}$$

Now the parameters of the space Z' fit in the situation described by (9.6.25), and we thus obtain $Z' = L_{\infty,\infty;(-\alpha_0,0),(-\beta_0,0)}$. Moreover, by Corollary 9.5.3, $Z' = L_{\infty,\infty;-\mathbb{A},-\mathbb{B}}$. Consequently, $Z' = \mathfrak{L}$. Together with (9.6.26), this yields $\mathfrak{L} \hookrightarrow X'$.

Using (7.9.13) and Lemma 9.4.1 (i), we get

$$\|f\|_{\mathfrak{L}} = \|\ell^{-\mathbb{A}}(t)\ell\ell^{-\mathbb{B}}(t) t^{-1} \int_0^t f^*(s)\,ds\|_{\infty,(0,\infty)}$$
$$\leq \|f\|_{X'} \|t^{-1}\ell^{-\mathbb{A}}(t)\ell\ell^{-\mathbb{B}}(t)\varphi_X(t)\|_{\infty,(0,\infty)} = \|f\|_{X'},$$

which proves the converse embedding $X' \hookrightarrow \mathfrak{L}$. The proof is complete. $\square$

Next, we are going to describe associated spaces of $L_{(p,q;\mathbb{A},\mathbb{B})}$ with $0 < p,q \leq \infty$ and $\mathbb{A}, \mathbb{B} \in \mathbb{R}^2$. However, according to (9.5.13) and our previous results on associated spaces to $L_{p,q;\mathbb{A},\mathbb{B}}$, it is enough to consider the case when $0 < p \leq 1, 0 < q \leq \infty$ and $\mathbb{A}, \mathbb{B} \in \mathbb{R}^2$. If $p = 1, 1 \leq q \leq \infty$ and $\mathbb{A}, \mathbb{B} \in \mathbb{R}^2$, the desired description of $(L_{(p,q;\mathbb{A},\mathbb{B})})'$ is given in the following theorem.

Theorem 9.6.9. *Let $1 < q \leq \infty$, $\mathbb{A} = (\alpha_0, \alpha_\infty)$, $\mathbb{B} = (\beta_0, \beta_\infty) \in \mathbb{R}^2$, and assume that the space $X = L_{(1,q;\mathbb{A},\mathbb{B})}$ is nontrivial. Then $X' = \mathfrak{L}$, a space described below:*

(i) *Let $1 < q \leq \infty$, let either $\alpha_0 + \frac{1}{q} > 0$ or $\alpha_0 + \frac{1}{q} = 0$ and $\beta_0 + \frac{1}{q} \geq 0$, and let either $\alpha_\infty + \frac{1}{q} < 0$ or $\alpha_\infty + \frac{1}{q} = 0$ and $\beta_\infty + \frac{1}{q} < 0$. Then $\mathfrak{L}$ reads as*

$$L_{\infty,q';-\mathbb{A}-1,-\mathbb{B}} \qquad \text{if } \alpha_0 + \frac{1}{q} > 0, \alpha_\infty + \frac{1}{q} < 0,$$

Section 9.6 The associate space

$L_{\infty,q';(-\alpha_0-1,-\frac{1}{q'}),(-\beta_0,-\beta_\infty-1)}$ if $\alpha_0 + \frac{1}{q} > 0,\ \alpha_\infty + \frac{1}{q} = 0,$

$\beta_\infty + \frac{1}{q} < 0,$

$L_{\infty,q';(-\frac{1}{q'},-\alpha_\infty-1),(-\beta_0-1,-\beta_\infty)}$ if $\alpha_0 + \frac{1}{q} = 0,\ \beta_0 + \frac{1}{q} > 0,$

$\alpha_\infty + \frac{1}{q} < 0,$

$L_{\infty,q';(-\frac{1}{q'},-\frac{1}{q'}),-\mathbb{B}-1}$ if $\alpha_0 + \frac{1}{q} = 0,\ \beta_0 + \frac{1}{q} > 0,$

$\alpha_\infty + \frac{1}{q} = 0,\ \beta_\infty + \frac{1}{q} < 0,$

$L_{\infty,q';(-\frac{1}{q'},-\alpha_\infty-1),(-\frac{1}{q'},-\beta_\infty),(-1,0)}$ if $\alpha_0 + \frac{1}{q} = 0,\ \beta_0 + \frac{1}{q} = 0,$

$\alpha_\infty + \frac{1}{q} < 0,$

$L_{\infty,q';(-\frac{1}{q'},-\frac{1}{q'}),(-\frac{1}{q'},-\beta_\infty-1),(-1,0)}$ if $\alpha_0 + \frac{1}{q} = 0,\ \beta_0 + \frac{1}{q} = 0,$

$\alpha_\infty + \frac{1}{q} = 0,\ \beta_\infty + \frac{1}{q} < 0.$

(ii) Let $1 < q \leq \infty$, let either $\alpha_0 + \frac{1}{q} < 0$ or $\alpha_0 + \frac{1}{q} = 0$ and $\beta_0 + \frac{1}{q} < 0$, and let either $\alpha_\infty + \frac{1}{q} < 0$ or $\alpha_\infty + \frac{1}{q} = 0$ and $\beta_\infty + \frac{1}{q} < 0$. Then

$$\mathfrak{L} = \{f \in \mathcal{M}(\mathcal{R},\mu);\ \|f\|_{\mathfrak{L}} = \|f\|_\infty + N(f) < \infty\}, \qquad (9.6.27)$$

where

$$N(f) = \begin{cases} \|t^{-\frac{1}{q'}} \ell^{-\alpha_\infty-1}(t)\ell\ell^{-\beta_\infty}(t) f^*(t)\|_{q',(1,\infty)} & \text{if } \alpha_\infty + \frac{1}{q} < 0 \\ \|t^{-\frac{1}{q'}} \ell^{-\frac{1}{q'}}(t)\ell\ell^{-\beta_\infty-1}(t) f^*(t)\|_{q',(1,\infty)} & \text{if } \alpha_\infty + \frac{1}{q} = 0, \\ & \quad \beta_\infty + \frac{1}{q} < 0. \end{cases}$$

(9.6.28)

(iii) Let $q = \infty$, $\alpha_\infty = 0$, and $\beta_\infty = 0$. Then

$$\mathfrak{L} = \{f \in \mathcal{M}(\mathcal{R},\mu);\ \|f\|_{\mathfrak{L}} < \infty\},$$

where

$$\|f\|_{\mathfrak{L}} = \begin{cases} \int_0^1 t^{-1}\ell^{-\alpha_0-1}(t)\ell\ell^{-\beta_0}(t) f^*(t)\,dt & \text{if } \alpha_0 > 0 \\ \int_0^1 t^{-1}\ell^{-1}(t)\ell\ell^{-\beta_0-1}(t) f^*(t)\,dt & \text{if } \alpha_0 = 0,\ \beta_0 > 0 \\ \|f\|_\infty & \text{if either } \alpha_0 < 0 \\ & \text{or } \alpha_0 = 0,\ \beta_0 \leq 0. \end{cases}$$

(9.6.29)

Remark 9.6.10. We see from (9.6.29) that the space $\mathfrak{L}$ of Theorem 9.6.9 (iii) is given by

$$\mathfrak{L} = \begin{cases} L_{\infty,1;\alpha_0-1,-\beta_0}(0,1) & \text{if } \alpha_0 > 0 \\ L_{\infty,1;-1,-\beta_0-1}(0,1) & \text{if } \alpha_0 = 0, \ \beta_0 > 0. \end{cases}$$

Proof of Theorem 9.6.9. (i) By Theorem 9.5.1 (i), any space in part (i) coincides with its analogue in the norm of which f^* is replaced by f^{**}. Thus, by Remark 9.3.2, any such space is a Banach function space (or BFS for short), i.e. the space $\mathfrak{L}$ is a BFS. Consequently, $\mathfrak{L} = \mathfrak{L}''$. Further, by Theorem 9.6.2 (ii), (iii), and (v), and by Theorem 9.6.8 (ii)–(iv) (supplemented by their analogues for spaces with three tiers of logarithms), $\mathfrak{L}' = L_{(1,q;\mathbb{A},\mathbb{B})}$. Hence, $\mathfrak{L} = \mathfrak{L}'' = (L_{(1,q;\mathbb{A},\mathbb{B})})' = X'$.

(ii) In this case we have

$$\|t^{-\frac{1}{q}}\ell^{\alpha_0}(t)\ell\ell^{\beta_0}(t)\|_{q,(\frac{1}{2},1)} \approx 1 \approx \|t^{-\frac{1}{q}}\ell^{\alpha_0}(t)\ell\ell^{\beta_0}(t)\|_{q,(0,1)}.$$

Consequently, we get for every $g \in X$,

$$\int_0^1 g^*(t)\,dt \lesssim \int_0^{\frac{1}{2}} g^*(t)\,dt \approx \|t^{-\frac{1}{q}}\ell^{\alpha_0}(t)\ell\ell^{\beta_0}(t)\|_{q,(\frac{1}{2},1)} \int_0^{\frac{1}{2}} g^*(t)\,dt$$

$$\lesssim \|t^{-\frac{1}{q}}\ell^{\alpha_0}(t)\ell\ell^{\beta_0}(t)\int_0^t g^*(s)\,ds\|_{q,(\frac{1}{2},1)} \leq \|t^{\frac{1}{q'}}\ell^{\alpha_0}(t)\ell\ell^{\beta_0}(t)g^{**}(t)\|_{q,(0,1)}$$

$$\leq \|t^{-\frac{1}{q}}\ell^{\alpha_0}(t)\ell\ell^{\beta_0}(t)\|_{q,(0,1)} \int_0^1 g^*(s)\,ds \approx \int_0^1 g^*(t)\,dt,$$

which implies

$$\|g\|_X \approx \int_0^1 g^*(t)\,dt + \|t^{\frac{1}{q'}}\ell^{\alpha_\infty}(t)\ell\ell^{\beta_\infty}(t)g^{**}(t)\|_{q,(1,\infty)}. \qquad (9.6.30)$$

Now we shall prove the embedding

$$X' \hookrightarrow \mathfrak{L}. \qquad (9.6.31)$$

For this purpose it is enough to verify for all step functions $f \in \mathcal{M}(\mathcal{R}, \mu)$ the inequality

$$\|f\|_\mathfrak{L} \lesssim \|f\|_{X'}. \qquad (9.6.32)$$

For a step function $f \in \mathcal{M}(\mathcal{R}, \mu)$, we define a function ϱ by

$$\varrho(t) = [f^*(1)]^{q'-1}, \qquad 0 < t \leq 1, \qquad (9.6.33)$$

Section 9.6 The associate space

and, for $1 < t \le \infty$, by

$$\varrho(t) = \begin{cases} [f^*(t)]^{q'-1} t^{-1} \ell^{(-\alpha_\infty-1)q'}(t) \ell\ell^{-\beta_\infty q'}(t) & \text{if } \alpha_\infty + \frac{1}{q} < 0 \\ [f^*(t)]^{q'-1} t^{-1} \ell^{-1}(t) \ell\ell^{(-\beta_\infty-1)q'}(t) & \text{if } \alpha_\infty + \frac{1}{q} = 0, \\ & \quad \beta_\infty + \frac{1}{q} < 0. \end{cases}$$
(9.6.34)

Then ϱ is equivalent to a nonincreasing function on $(0, \infty)$ and, by Theorem 9.1.4, there is a $\overline{\varrho} \in \mathcal{M}(\mathcal{R}, \mu)$ such that $\overline{\varrho}^* \approx \varrho$. Moreover, we have from (9.6.27) and (9.6.28) that

$$\|f\|_{\varrho}^{q'} \approx \|f\|_{\infty}^{q'} + \int_1^\infty \varrho(t) f^*(t) \, dt.$$
(9.6.35)

Our assumptions on $\mathbb{A}$ and $\mathbb{B}$ and Lemma 9.4.1 (ii) imply

$$\varphi_X(t) \approx t \qquad \text{for all} \quad t \in (0, 1).$$

Together with (7.9.13), this yields

$$\|f\|_\infty = \lim_{t \to 0+} \frac{1}{t} \int_0^t f^*(s) \, ds \le \|f\|_{X'} \lim_{t \to 0+} \frac{1}{t} \varphi_X(t) \approx \|f\|_{X'}.$$
(9.6.36)

Moreover, by the Hölder inequality,

$$\int_1^\infty \varrho(t) f^*(t) \, dt \le \|\overline{\varrho}\|_X \|f\|_{X'}.$$
(9.6.37)

If we prove that

$$\|\overline{\varrho}\|_X \lesssim \|f\|_{\varrho}^{q'-1},$$
(9.6.38)

we would obtain from (9.6.35), (9.6.27), (9.6.37), (9.6.38) and (9.6.36) that

$$\|f\|_{\varrho}^{q'} \lesssim \|f\|_{\varrho}^{q'-1} \|f\|_\infty + \|\overline{\varrho}\|_X \|f\|_{X'} \lesssim \|f\|_{\varrho}^{q'-1} \|f\|_{X'},$$

and (9.6.32) would follow.

Since (cf. (9.6.33))

$$\int_0^1 \varrho(s) \, ds = [f^*(1)]^{q'-1} \le \|f\|_\infty^{q'-1} \le \|f\|_{\varrho}^{q'-1},$$

and

$$\left\| t^{-\frac{1}{q}} \ell^{\alpha_\infty}(t) \ell\ell^{\beta_\infty}(t) \int_0^1 \varrho(s) \, ds \right\|_{q,(1,\infty)} \approx \int_0^1 \varrho(s) \, ds \le \|f\|_{\varrho}^{q'-1},$$

we have from (9.6.30) that

$$\|\bar{\varrho}\|_X \approx \int_0^1 \varrho(s)\,ds + \|t^{\frac{1}{q'}}\ell^{\alpha_\infty}(t)\ell\ell^{\beta_\infty}(t)\varrho^{**}(t)\|_{q,(1,\infty)}$$

$$\lesssim \|f\|_{\mathfrak{L}}^{q'-1} + \|t^{-\frac{1}{q}}\ell^{\alpha_\infty}(t)\ell\ell^{\beta_\infty}(t)\int_1^t \varrho(s)\,ds\|_{q,(1,\infty)}. \qquad (9.6.39)$$

Thus, (9.6.38) will follow from (9.6.39) and (9.6.27) once we show that

$$\|t^{-\frac{1}{q}}\ell^{\alpha_\infty}(t)\ell\ell^{\beta_\infty}(t)\int_1^t \varrho(s)\,ds\|_{q,(1,\infty)} \lesssim N(f)^{q'-1}.$$

Using (9.6.28) and (9.6.34), this inequality can be rewritten as

$$\|t^{-\frac{1}{q}}\ell^{\alpha_\infty}(t)\ell\ell^{\beta_\infty}(t)\int_1^t \varrho(s)\,ds\|_{q,(1,\infty)} \lesssim \|t^{\frac{1}{q'}}\ell^{\alpha_\infty+1}(t)\ell\ell^{\beta_\infty}(t)\varrho(t)\|_{q,(1,\infty)} \qquad (9.6.40)$$

if $\alpha_\infty + \frac{1}{q} < 0$, and

$$\|t^{-\frac{1}{q}}\ell^{\alpha_\infty}(t)\ell\ell^{\beta_\infty}(t)\int_1^t \varrho(s)\,ds\|_{q,(1,\infty)} \lesssim \|t^{\frac{1}{q'}}\ell^{\frac{1}{q'}}(t)\ell\ell^{\beta_\infty+1}(t)\varrho(t)\|_{q,(1,\infty)} \qquad (9.6.41)$$

if $\alpha_\infty + \frac{1}{q} = 0$ and $\beta_\infty + \frac{1}{q} < 0$. To verify (9.6.40) and (9.6.41) is a standard matter using the well-known criterion for the Hardy inequality (cf. Theorem 3.11.8). This proves (9.6.38), and in turn also (9.6.32).

We shall now prove the converse inclusion to (9.6.31), i.e. $\mathfrak{L} \hookrightarrow X'$. Thus, we need to verify that

$$\|f\|_{X'} \lesssim \|f\|_{\mathfrak{L}} \quad \text{for all} \quad f \in \mathfrak{L}. \qquad (9.6.42)$$

Since for every $f \in \mathcal{M}(\mathcal{R}, \mu)$,

$$\|f\|_{X'} = \sup_{\|g\|_X \le 1} \int_\mathcal{R} f(x)g(x)\,d\mu$$

we see that, with the help of the Hardy–Littlewood inequality (7.2.2), in order to prove (9.6.42) it suffices to show that

$$\int_0^\infty f^*(t)g^*(t)\,dt \lesssim \|f\|_{\mathfrak{L}} \|g\|_X \qquad (9.6.43)$$

for every $f \in \mathfrak{L}$ and $g \in X$. First, we have (cf. (9.6.27) and (9.6.30))

$$\int_0^1 f^*(t)g^*(t)\,dt \le \|f\|_\infty \int_0^1 g^*(t)\,dt \lesssim \|f\|_{\mathfrak{L}} \|g\|_X.$$

Section 9.6 The associate space

Therefore, it remains to prove that

$$\int_1^\infty f^*(t)g^*(t)\,dt \lesssim \|f\|_{\mathcal{Y}} \|g\|_X. \tag{9.6.44}$$

Assume that (9.6.44) is not satisfied. Then there are two sequences of functions $\{f_n\}, \{g_n\}$ such that $\|f_n\|_{\mathcal{Y}} \le 1, \|g_n\|_X \le 1, n \in \mathbb{N}$, and

$$\int_1^\infty f_n^*(t)g_n^*(t)\,dt \to \infty \quad \text{as} \quad n \to \infty. \tag{9.6.45}$$

Setting

$$h_n(x) = \min\{|g_n(x)|, g_n^*(1)\}\operatorname{sign} g_n(x), \quad x \in \mathcal{R}, \ n \in \mathbb{N}, \tag{9.6.46}$$

we have $|h_n| \le |g_n|$ and since X is a BFS (cf. Remark 9.3.2),

$$\|h_n\|_X \le \|g_n\|_X \le 1, \quad n \in \mathbb{N}.$$

Moreover, by (9.6.30),

$$g_n^*(1) \le \int_0^1 g_n^*(t)\,dt \lesssim \|g_n\|_X \le 1, \quad n \in \mathbb{N}. \tag{9.6.47}$$

We now claim that $\{h_n\}$ is uniformly bounded in the space Y, where

$$Y = L_{(1,q;(1,\alpha_\infty),\mathbb{B})}. \tag{9.6.48}$$

To prove (9.6.48), note that (9.6.46) implies

$$h_n^*(s) = \min[g_n^*(s), g_n^*(1)], \quad s \ge 0.$$

Hence

$$h_n^*(s) = g_n^*(1), \ s \in (0,1] \quad \text{and} \quad h_n^*(s) = g_n^*(s), \ s \in (1,\infty).$$

This yields $h_n^{**}(s) = g_n^*(1)$ for $s \in (0,1]$ and $h_n^{**} \le g_n^{**}$. Thus, using (9.6.47),

$$\|h_n\|_Y \lesssim \|t^{1-\frac{1}{q}}\ell(t)\ell\ell^{\beta_0}(t)g_n^*(1)\|_{q,(0,1)} + \|t^{1-\frac{1}{q}}\ell^{\alpha_\infty}(t)\ell\ell^{\beta_\infty}(t)g_n^{**}\|_{q,(1,\infty)}$$
$$\lesssim g_n^*(1)\|t^{1-\frac{1}{q}}\ell(t)\ell\ell^{\beta_0}(t)\|_{q,(0,1)} + \|g_n\|_X \lesssim 1, \tag{9.6.49}$$

which proves the uniform boundedness of $\{h_n\}$ in Y.

Now, by part (i), we can determine the space Y', namely

$$Y' = \begin{cases} L_{\infty,q';(-2,-\alpha_\infty-1),-\mathbb{B}} & \text{if } \alpha_\infty + \frac{1}{q} < 0 \\ L_{\infty,q';(-2,-\frac{1}{q'}),(-\beta_0,-\beta_\infty-1)} & \text{if } \alpha_\infty + \frac{1}{q} = 0, \ \beta_\infty + \frac{1}{q} < 0. \end{cases}$$

Hence, applying (9.6.27), (9.6.28) and the estimate

$$f_n^*(t) \le \|f_n\|_\infty \le \|f_n\|_{\mathfrak{L}} \le 1, \quad n \in \mathbb{N},$$

we obtain

$$\|f_n\|_{Y'} \lesssim \|t^{-\frac{1}{q'}} \ell^{-2}(t) \ell\ell^{-\beta_0}(t) f_n^*(t)\|_{q',(0,1)} + N(f_n) \lesssim \|f_n\|_{\mathfrak{L}} \le 1, \tag{9.6.50}$$

which means that f_n are uniformly bounded in Y'. Now, (9.6.49) and (9.6.50) contradict (9.6.45) since $h_n^*(s) = g_n^*(s)$ for $s \in (1, \infty)$.

(iii) In the case when $q = \infty$, $\alpha_\infty = 0$, and $\beta_\infty = 0$, we have for all $g \in X$,

$$N_1 := \|t\ell^{\alpha_0}(t)\ell\ell^{\beta_0}(t)g^{**}(t)\|_{q,(0,1)} = \|\ell^{\alpha_0}(t)\ell\ell^{\beta_0}(t) \int_0^t g^*(s)\,ds\|_{\infty,(0,1)},$$

$$N_2 := \|t\ell^{\alpha_\infty}(t)\ell\ell^{\beta_\infty}(t)g^{**}(t)\|_{q,(1,\infty)} = \|\int_0^t g^*(s)\,ds\|_{\infty,(1,\infty)}$$

$$= \int_0^\infty g^*(s)\,ds,$$

and therefore

$$\|g\|_{(1,q;\mathbb{A},\mathbb{B})} = \max\{N_1, N_2\}$$

$$= \max\left\{\|\ell^{\alpha_0}(t)\ell\ell^{\beta_0}(t) \int_0^t g^*(s)\,ds\|_{\infty,(0,1)}, \int_0^\infty g^*(s)\,ds\right\}. \tag{9.6.51}$$

If either $\alpha_0 < 0$ or $\alpha_0 = 0$ and $\beta_0 \le 0$, then $N_1 \approx \int_0^1 g^*(s)\,ds$. This implies that $X = L^1$ and thus $X' = L^\infty = \mathfrak{L}$.

Now let either $\alpha_0 > 0$ or $\alpha_0 = 0$ and $\beta_0 > 0$. We shall first show that (9.6.32) holds. Set (cf. (9.6.33) and (9.6.34)) for $t \in (0, \infty)$,

$$\varrho(t) = \begin{cases} t^{-1}\ell^{-\alpha_0-1}(t)\ell\ell^{-\beta_0}(t)\chi_{(0,1)}(t) & \text{if } \alpha_0 > 0 \\ t^{-1}\ell^{-1}(t)\ell\ell^{-\beta_0-1}(t)\chi_{(0,1)}(t) & \text{if } \alpha_0 = 0, \beta_0 > 0. \end{cases}$$

The function ϱ is equivalent to a nonincreasing function on $(0, \infty)$ and, by Theorem 9.1.4, there is a $\overline{\varrho} \in \mathcal{M}(\mathcal{R}, \mu)$ such that $\overline{\varrho}^* \approx \varrho$. This and our assumptions on q, $\mathbb{A}$, and $\mathbb{B}$ yield $\|\overline{\varrho}\|_X \approx 1$. Moreover, (9.6.29) and the Hölder inequality imply that for all step functions $f \in \mathcal{M}(\mathcal{R}, \mu)$,

$$\|f\|_{\mathfrak{L}} = \int_0^\infty \varrho(t) f^*(t)\,dt \le \|\overline{\varrho}\|_X \|f\|_{X'}$$

and (9.6.32) follows.

Section 9.6 The associate space

We have to prove the converse, that is, (9.6.42). To this end it suffices to verify (9.6.43). Since $f^*(1) \lesssim \|f\|_{\mathfrak{L}}$, we have from (9.6.51) that for all $f \in \mathfrak{L}$ and $g \in X$,

$$\int_1^\infty f^*(t)g^*(t)\,dt \lesssim f^*(1)\int_1^\infty g^*(t)\,dt \lesssim \|f\|_{\mathfrak{L}}\,\|g\|_X.$$

Hence, it remains to prove that for all $f \in \mathfrak{L}$ and $g \in X$,

$$\int_0^1 f^*(t)g^*(t)\,dt \lesssim \|f\|_{\mathfrak{L}}\,\|g\|_X. \tag{9.6.52}$$

Assume that (9.6.52) is not true. Then there are two sequences of functions $\{f_n\}, \{g_n\}$ such that $\|f_n\|_{\mathfrak{L}} \leq 1$, $\|g_n\|_X \leq 1$, $n \in \mathbb{N}$, and

$$\int_0^1 f_n^*(t)g_n^*(t)\,dt \to \infty \quad \text{as} \quad n \to \infty. \tag{9.6.53}$$

Define the functions h_n, $n \in \mathbb{N}$, by

$$h_n(x) = [|g_n(x)| - g_n^*(1)]^+ \operatorname{sign} g_n(x), \quad x \in \mathcal{R},$$

and the space Y by

$$Y = L_{(1,\infty;(\alpha_0,-1),\mathbb{B})}.$$

Then, by part (i),

$$Y' = \begin{cases} L_{\infty,1;(-\alpha_0-1,-2),-\mathbb{B}} & \text{if } \alpha_0 > 0 \\ L_{\infty,1;(-1,-2),(-\beta_0-1,-\beta_\infty)} & \text{if } \alpha_0 = 0, \beta_0 > 0. \end{cases} \tag{9.6.54}$$

Since $h_n^*(s) = [g_n^*(s) - g_n^*(1)]^+$, $s \geq 0$, we have $\operatorname{supp} h_n^* \subset [0,1]$ and $h_n^*(s) = g_n^*(s) - g_n^*(1)$ for $s \in (0,1)$. Hence,

$$h_n^{**}(t) = g_n^{**}(t) - g_n^*(1), \quad t \in (0,1), \quad \text{and}$$
$$h_n^{**}(t) = t^{-1}[g_n^{**}(1) - g_n^*(1)], \quad t \in (1,\infty).$$

This yields

$$\|h_n\|_Y \lesssim \|t\ell^{\alpha_0}(t)\ell\ell^{\beta_0}(t)g_n^{**}(t)\|_{\infty,(0,1)} + \|t\ell^{-1}(t)\ell\ell^{\beta_\infty}(t)t^{-1}g_n^{**}(1)\|_{\infty,(1,\infty)}$$

$$\leq \|g_n\|_X + \int_0^1 g_n^*(s)\,ds\,\|\ell^{-1}(t)\ell\ell^{\beta_\infty}(t)\|_{\infty,(1,\infty)}$$

$$= \|g_n\|_X + \int_0^1 g_n^*(s)\,ds,$$

and, on using (9.6.51) with g_n instead of g,

$$\|h_n\|_Y \lesssim \|g_n\|_X \leq 1 \quad \text{for all} \quad n \in \mathbb{N}. \tag{9.6.55}$$

Defining the functions $\psi_n, n \in \mathbb{N}$, by

$$\psi_n(x) = [|f_n(x)| - f^*(1)]^+ \operatorname{sign} f_n(x), \quad x \in \mathcal{R},$$

we have $\psi_n^*(s) = [f_n^*(s) - f_n^*(1)]^+$, $s \geq 0$. Consequently, supp $\psi_n^* \subset [0, 1]$ and $\psi_n^*(s) = f_n^*(s) - f_n^*(1)$ for $s \in (0, 1)$. Together with (9.6.54) and (9.6.29), this yields

$$\|\psi_n\|_{Y'} \leq \|f_n\|_{\mathcal{L}} \leq 1 \quad \text{for all} \quad n \in \mathbb{N}. \tag{9.6.56}$$

By (9.6.51),

$$\int_0^1 g_n^*(s)\,ds \lesssim \|g_n\|_X \lesssim 1,$$

which shows that, for all $n \in \mathbb{N}$,

$$g_n^*(1) \lesssim 1 \quad \text{and} \quad \int_0^1 h_n^*(s)\,ds \lesssim 1. \tag{9.6.57}$$

Since, for every $t \in (0, 1)$,

$$1 \lesssim \min\{t^{-1}\ell^{-\alpha_0-1}(t)\ell\ell^{-\beta_0}(t), t^{-1}\ell^{-1}(t)\ell\ell^{-\beta_0-1}(t)\},$$

we have from (9.6.29) for all $n \in \mathbb{N}$,

$$\int_0^1 f_n^*(s)\,ds \lesssim \|f_n\|_{\mathcal{L}} \lesssim 1,$$

which in turn implies that

$$f^*(1) \lesssim 1 \quad \text{and} \quad \int_0^1 \psi_n^*(t)\,dt \lesssim 1. \tag{9.6.58}$$

Finally, (9.6.55), (9.6.56), (9.6.57) and (9.6.58) contradict (9.6.53) as

$$\int_0^1 f_n^*(t)g_n^*(t)\,dt = \int_0^1 (\psi_n^*(t) + f_n^*(1))(h_n^*(t) + g_n^*(1))\,dt$$

$$= \int_0^1 \psi_n^*(t)h_n^*(t)\,dt + f_n^*(1)\left(\int_0^1 h_n^*(t)\,dt + g_n^*(1)\right) + g_n^*(1)\int_0^1 \psi_n^*(t)\,dt.$$

The proof is complete. □

Section 9.6 The associate space

The following theorem is a complement of Theorem 9.6.9.

Theorem 9.6.11. *Let* $0 < q \le 1$, $\mathbb{A} = (\alpha_0, \alpha_\infty)$, $\mathbb{B} = (\beta_0, \beta_\infty) \in \mathbb{R}^2$, *and assume that the space* $L_{(1,q;\mathbb{A},\mathbb{B})}$ *is nontrivial. Then* $(L_{(1,q;\mathbb{A},\mathbb{B})})' = \mathfrak{L}$, *a space described below:*

(i) *Let either* $\alpha_0 + \frac{1}{q} > 0$ *or* $\alpha_0 + \frac{1}{q} = 0$ *and* $\beta_0 + \frac{1}{q} \ge 0$. *Then* $\mathfrak{L}$ *reads as*

$$L_{\infty,\infty;-\mathbb{A}-\frac{1}{q},-\mathbb{B}} \quad \text{if } \alpha_0 + \frac{1}{q} > 0, \ \alpha_\infty + \frac{1}{q} < 0,$$

$$L_{\infty,\infty;(-\alpha_0-\frac{1}{q},0),(-\beta_0,-\beta_\infty-\frac{1}{q})} \quad \text{if } \alpha_0 + \frac{1}{q} > 0, \ \alpha_\infty + \frac{1}{q} = 0,$$
$$\beta_\infty + \frac{1}{q} < 0,$$

$$L_{\infty,\infty;(0,-\alpha_\infty-\frac{1}{q}),(-\beta_0-\frac{1}{q},-\beta_\infty)} \quad \text{if } \alpha_0 + \frac{1}{q} = 0, \ \beta_0 + \frac{1}{q} > 0,$$
$$\alpha_\infty + \frac{1}{q} < 0,$$

$$L_{\infty,\infty;(0,0),-\mathbb{B}-\frac{1}{q}} \quad \text{if } \alpha_0 + \frac{1}{q} = 0, \ \beta_0 + \frac{1}{q} > 0,$$
$$\alpha_\infty + \frac{1}{q} = 0, \ \beta_\infty + \frac{1}{q} < 0,$$

$$L_{\infty,\infty;(0,-\alpha_\infty-\frac{1}{q}),(0,-\beta_\infty),(-\frac{1}{q},0)} \quad \text{if } \alpha_0 + \frac{1}{q} = 0, \ \beta_0 + \frac{1}{q} = 0,$$
$$\alpha_\infty + \frac{1}{q} < 0,$$

$$L_{\infty,\infty;(0,0),(0,-\beta_\infty-\frac{1}{q}),(-\frac{1}{q},0)} \quad \text{if } \alpha_0 + \frac{1}{q} = 0, \ \beta_0 + \frac{1}{q} = 0,$$
$$\alpha_\infty + \frac{1}{q} = 0, \ \beta_\infty + \frac{1}{q} < 0.$$

(ii) *Let either* $\alpha_0 + \frac{1}{q} < 0$ *or* $\alpha_0 + \frac{1}{q} = 0$ *and* $\beta_0 + \frac{1}{q} < 0$. *Then,*

$$\mathfrak{L} = \begin{cases} L_{\infty,\infty;(0,-\alpha_\infty-\frac{1}{q}),(0,-\beta_\infty)} & \text{if } \alpha_\infty + \frac{1}{q} < 0, \\ L_{\infty,\infty;(0,0),(0,-\beta_\infty-\frac{1}{q})} & \text{if } \alpha_\infty + \frac{1}{q} = 0, \ \beta_\infty + \frac{1}{q} < 0. \end{cases}$$

Proof. The assertion can be proved analogously to Theorem 9.6.8. The details are omitted. □

Remark 9.6.12. If $q = 1$, then there is a simple proof of Theorem 9.6.11. Indeed, by Theorem 9.5.1 (i), any space in Theorem 9.6.11 coincides with its analogue in the norm of which f^* is replaced by f^{**}. Thus, by Remark 9.3.2, any such space is a BFS, which implies that the space $\mathfrak{L}$ is a BFS. Consequently, $\mathfrak{L}'' = \mathfrak{L}$. By Theorem 9.6.2 (iv) (and its analogue for spaces with three tiers of logarithms) and Theorem 9.5.1 (ii), $\mathfrak{L}' = L_{(1,1;\mathbb{A},\mathbb{B})}$. Hence,

$$\mathfrak{L} = \mathfrak{L}'' = (L_{(1,1;\mathbb{A},\mathbb{B})})'.$$

We shall finally list the associate spaces of GLZ spaces of functions defined on a space of finite measure. With no loss of generality, we will restrict ourselves to the case when $\mu(\mathcal{R}) = 1$. We omit the proofs because they are completely analogous to those in the case of an infinite measure space.

Theorem 9.6.13. Let $\mu(\mathcal{R}) = 1$. Let $0 < p, q \leq \infty$, $\alpha, \beta \in \mathbb{R}$, and assume that the space $L_{p,q;\alpha,\beta}$ is nontrivial (cf. (9.5.7)). Then $(L_{p,q;\alpha,\beta})'$ coincides with

$\{0\}$ if either $0 < p < 1$
                        or $p = 1$, $0 < q \leq 1$, $\alpha < 0$
                        or $p = 1$, $0 < q \leq 1$, $\alpha = 0$, $\beta < 0$
                        or $p = 1$, $1 < q \leq \infty$, $\alpha - \frac{1}{q'} < 0$
                        or $p = 1$, $1 < q \leq \infty$, $\alpha - \frac{1}{q'} = 0$, $\beta - \frac{1}{q'} \leq 0$,

$L_{(p',q';-\alpha,-\beta)}$ if $1 < q \leq \infty$ and either $p = 1$, $\alpha > \frac{1}{q'}$
                                     or $p = 1$, $\alpha = \frac{1}{q'}$, $\beta \geq \frac{1}{q'}$
                                     or $1 < p < \infty$,

$L_{(1,q';-\alpha-1,-\beta)}$ if $p = \infty$, $1 < q < \infty$, $\alpha + \frac{1}{q} < 0$,

$L_{(1,q';-\frac{1}{q'},-\beta-1)}$ if $p = \infty$, $1 < q < \infty$, $\alpha + \frac{1}{q} = 0$, $\beta + \frac{1}{q} < 0$,

$L_{1,1;-\alpha,-\beta}$ if $p = \infty$, $q = \infty$, and either $\alpha < 0$
                                or $\alpha = 0$, $\beta \leq 0$,

$L_{(p',\infty;-\alpha,-\beta)}$ if $0 < q \leq 1$, and either $1 < p < \infty$
                                   or $p = 1$, $\alpha > 0$
                                   or $p = 1$, $\alpha = 0$, $\beta \geq 0$,

$L_{(1,\infty;-\alpha-\frac{1}{q},-\beta)}$ if $p = \infty,\ 0 < q \leq 1,\ \alpha + \dfrac{1}{q} < 0,$

$L_{(1,\infty;0,-\beta-\frac{1}{q})}$ if $p = \infty,\ 0 < q \leq 1,\ \alpha + \dfrac{1}{q} = 0,\ \beta + \dfrac{1}{q} < 0.$

Let $0 < p, q \leq \infty$, and $\alpha, \beta \in \mathbb{R}$. Since $L_{(p,q;\alpha,\beta)} = L_{p,q;\alpha,\beta}$ if $1 < p \leq \infty$, $L_{(p,q;\alpha,\beta)} = L^1$ if $0 < p < 1$, and the associated spaces of $L_{p,q;\alpha,\beta}$ have already been described in the previous theorem, it remains to characterize the associated spaces of $L_{(1,q;\alpha,\beta)}$. This is done in the following theorem.

Theorem 9.6.14. *Let $\mu(\mathcal{R}) = 1$. Let $0 < q \leq \infty$, $\alpha, \beta \in \mathbb{R}$, and assume that the space $L_{(1,q;\alpha,\beta)}$ is nontrivial. Then its associate space $(L_{(1,q;\alpha,\beta)})'$ coincides with*

$L_{\infty,q';-\alpha-1,-\beta}$ if $1 < q \leq \infty,\ \alpha + \dfrac{1}{q} > 0,$

$L_{\infty,q';-\frac{1}{q'},-\beta-1}$ if $1 < q \leq \infty,\ \alpha + \dfrac{1}{q} = 0,\ \beta + \dfrac{1}{q} > 0,$

$L_{\infty,q';-\frac{1}{q'},-\frac{1}{q'},-1}$ if $1 < q < \infty,\ \alpha + \dfrac{1}{q} = 0,\ \beta + \dfrac{1}{q} = 0,$

L^∞ if $0 < q \leq \infty,$ *and either* $\alpha + \dfrac{1}{q} < 0$

         or $\alpha + \dfrac{1}{q} = 0,\ \beta + \dfrac{1}{q} < 0$

         or $q = \infty,\ \alpha = 0,\ \beta = 0,$

$L_{\infty,\infty;-\alpha-\frac{1}{q},-\beta}$ if $0 < q \leq 1,\ \alpha + \dfrac{1}{q} > 0,$

$L_{\infty,\infty;0,-\beta-\frac{1}{q}}$ if $0 < q \leq 1,\ \alpha + \dfrac{1}{q} = 0,\ \beta + \dfrac{1}{q} > 0,$

$L_{\infty,\infty;0,0,-\frac{1}{q}}$ if $0 < q \leq 1,\ \alpha + \dfrac{1}{q} = 0,\ \beta + \dfrac{1}{q} = 0.$

9.7 When GLZ is BFS

It is obvious that the functional that governs a GLZ space is not necessarily a Banach function norm. We shall now single out those GLZ spaces which satisfy all the axioms of the Banach function space (BFS). We begin with the spaces $L_{p,q;\mathbb{A},\mathbb{B}}$.

Theorem 9.7.1. *Let $0 < p, q \leq \infty$ and $\mathbb{A}, \mathbb{B} \in \mathbb{R}^2$. Then the space $X = L_{p,q;\mathbb{A},\mathbb{B}}$ is a BFS if and only if one of the following conditions holds:*

$$\begin{cases} 1 < p < \infty, \ 1 \leq q \leq \infty, \\ p = 1, \ q = 1, \ \alpha_0 > 0, \ \alpha_\infty < 0, \\ p = 1, \ q = 1, \ \alpha_0 > 0, \ \alpha_\infty = 0, \ \beta_\infty \leq 0, \\ p = 1, \ q = 1, \ \alpha_0 = 0, \ \beta_0 \geq 0, \ \alpha_\infty < 0, \\ p = 1, \ q = 1, \ \alpha_0 = 0, \ \beta_0 \geq 0, \ \alpha_\infty = 0, \ \beta_\infty \leq 0, \\ p = \infty, \ 1 \leq q \leq \infty, \ \alpha_0 + \frac{1}{q} < 0, \\ p = \infty, \ 1 \leq q \leq \infty, \ \alpha_0 + \frac{1}{q} = 0, \ \beta_0 + \frac{1}{q} < 0, \\ p = \infty, \ q = \infty, \ \alpha_0 = 0, \ \beta_0 = 0. \end{cases} \quad (9.7.1)$$

Remarks 9.7.2. Some particular cases of Theorem 9.7.1 are worth noticing:

(i) A space $L_{1,q;\mathbb{A},\mathbb{B}}$, where $q \neq 1$, is not a BFS, for any $\mathbb{A}$ and $\mathbb{B}$.

(ii) A space $X = L_{1,1;\mathbb{A},\mathbb{B}}$ is a BFS if and only if $\varphi_X \in \mathcal{F}$.

(iii) A space $X = L_{\infty,q;\mathbb{A},\mathbb{B}}$ is a BFS if and only if $1 \leq q \leq \infty$ and X is nontrivial.

Proof of Theorem 9.7.1. (i) Let $1 < p \leq \infty$, $1 \leq q \leq \infty$, and $\mathbb{A}, \mathbb{B} \in \mathbb{R}^2$. Then, by Theorem 9.5.1 (i), $L_{p,q;\mathbb{A},\mathbb{B}} = L_{(p,q;\mathbb{A},\mathbb{B})}$ and by Remark 9.3.2 and Lemma 9.3.1 (ii), the latter space is a BFS if and only if either $1 < p < \infty$, or $p = \infty$ and either $\alpha_0 + \frac{1}{q} < 0$, or $\alpha_0 + \frac{1}{q} = 0$, $\beta_0 + \frac{1}{q} < 0$ or $q = \infty$, $\alpha_0 = 0$, $\beta_0 = 0$.

(ii) Let $p = 1$ and $1 < q \leq \infty$.

Assume first that

$$\text{either} \quad \alpha_0 + \frac{1}{q} < 1 \quad \text{or} \quad \alpha_0 + \frac{1}{q} = 1 \quad \text{and} \quad \beta_0 + \frac{1}{q} \leq 1.$$

Then, by Theorem 9.5.14 (vi), (vii), $L_{1,q;\mathbb{A},\mathbb{B}}$ is not locally embedded into L^1. Therefore, (P5) is not satisfied, and hence $L_{1,q;\mathbb{A},\mathbb{B}}$ is not a BFS.

Assume now that

$$\text{either} \quad \alpha_0 + \frac{1}{q} > 1 \quad \text{or} \quad \alpha_0 + \frac{1}{q} = 1 \quad \text{and} \quad \beta_0 + \frac{1}{q} > 1. \quad (9.7.2)$$

Then, by Theorem 9.6.2 (i),

$$(L_{1,q;\mathbb{A},\mathbb{B}})' = L_{\infty,q';-\mathbb{A},-\mathbb{B}} \quad (9.7.3)$$

and the conditions in (9.7.2) guarantee that the latter space is nontrivial (cf. (9.3.1)). Recalling that $X = L_{1,q;\mathbb{A},\mathbb{B}}$, we get from Lemma 9.4.1 (i)

$$\varphi_X(t) \approx t\ell^{\alpha_0}(t)\ell\ell^{\beta_0}(t), \quad t \in (0, 1).$$

By (9.7.3) and Lemma 9.4.1 (i), we have for $t \in (0, 1)$,

$$\varphi_{X'}(t) \approx \begin{cases} \ell^{-\alpha_0 + \frac{1}{q'}}(t)\ell\ell^{-\beta_0}(t) & \text{if } \alpha_0 + \frac{1}{q} > 1 \\ \ell\ell^{-\beta_0 + \frac{1}{q'}}(t) & \text{if } \alpha_0 + \frac{1}{q} = 1, \beta_0 + \frac{1}{q} > 1. \end{cases}$$

Since $q > 1$, and thus $q' \neq +\infty$, the relation (7.9.7) is not satisfied, which implies that X is not a BFS.

(iii) Let $p = 1, q = 1$ and let one of the conditions in (9.7.1) hold. We have

$$\|f\|_{1,1;\mathbb{A},\mathbb{B}} = \int_0^\infty f^*(t)\ell^{\mathbb{A}}(t)\ell\ell^{\mathbb{B}}(t)\,dt.$$

Each of the conditions in (9.7.1) with $p = 1$ guarantees that $\varphi \in \mathcal{F}$, where

$$\varphi(t) = \int_0^t \ell^{\mathbb{A}}(s)\ell\ell^{\mathbb{B}}(s)\,ds, \qquad t \in [0, \infty)$$

(since $\ell^{\mathbb{A}}(s)\ell\ell^{\mathbb{B}}(s)$, $s \in (0, \infty)$, is equivalent to a nonincreasing function on $(0, \infty)$). Hence $L_{1,1;\mathbb{A},\mathbb{B}} = \Lambda_\varphi$ and therefore $L_{1,1;\mathbb{A},\mathbb{B}}$ is a BFS.

(iv) Let $p = 1, q = 1$, and assume that none of the conditions in (9.7.1) is satisfied. Then, by Lemma 6.5, $\varphi_X \notin \mathcal{F}$, and consequently X is not a BFS.

(v) Let $0 < p < 1$. Then, by Lemma 9.4.1 (i), the fundamental function of $L_{p,q;\mathbb{A},\mathbb{B}}$ is not in $\mathcal{F}$. Consequently, $L_{p,q;\mathbb{A},\mathbb{B}}$ is not a BFS.

(vi) Finally, let $1 \leq p \leq \infty$ and $0 < q < 1$. There are three possibilities: either $X = \{0\}$, or $X \neq \{0\}$ and $\varphi_X \notin \mathcal{F}$, or $X \neq \{0\}$ and $\varphi_X \in \mathcal{F}$. It is clear (cf. (P1) and (P4)) that X is not a BFS when $X = \{0\}$, and also, as mentioned in Section 2 above, that X is not a BFS when $\varphi_X \notin \mathcal{F}$. Using Lemma 9.4.1 and embedding results above, we obtain that Λ_{φ_X} is not embedded into X when $X \neq \{0\}$ and $\varphi_X \in \mathcal{F}$. Thus, again, X is not a BFS.
The proof is complete. $\square$

We now turn our attention to the spaces $L_{(p,q;\mathbb{A},\mathbb{B})}$.

Theorem 9.7.3. *Let $0 < p, q \leq \infty$ and $\mathbb{A}, \mathbb{B} \in \mathbb{R}^2$. Then the space $X = L_{(p,q;\mathbb{A},\mathbb{B})}$ is a BFS if and only if $1 \leq q \leq \infty$ and $X \neq \{0\}$.*

Proof. If $1 \leq q \leq \infty$, then, by Remark 9.3.2, X is a BFS if and only if $X \neq \{0\}$.

Conversely, let either $X = \{0\}$ or $0 < q < 1$. If $X = \{0\}$, then it is evident from (P1) and (P4) that X is not a BFS. If $0 < q < 1$, then Λ_{φ_X} is not embedded into X. To see this, we use Theorem 9.5.1 (ii) to rewrite Λ_{φ_X} as $L_{(1,1;\mathbb{L},\mathbb{E})}$ with appropriate $\mathbb{L}, \mathbb{E} \in \mathbb{R}^2$, and then apply the above embedding results. Thus, X is not a BFS. $\square$

Finally, we present analogues of Theorems 9.7.1 and 9.7.3 for the case when $\mu(\mathcal{R}) < \infty$.

Theorem 9.7.4. *Assume that $\mu(\mathcal{R}) < \infty$. Let $0 < p,q \leq \infty$ and $\alpha, \beta \in \mathbb{R}^2$. Then the space $L_{p,q;\alpha,\beta}$ is a BFS if and only if one of the following conditions holds:*

$$\begin{cases} 1 < p < \infty, \ 1 \leq q \leq \infty, \\ p = 1, \ q = 1, \ \alpha > 0, \\ p = 1, \ q = 1, \ \alpha = 0, \ \beta \geq 0, \\ p = \infty, \ 1 \leq q \leq \infty, \ \alpha + \frac{1}{q} < 0, \\ p = \infty, \ 1 \leq q \leq \infty, \ \alpha + \frac{1}{q} = 0, \ \beta + \frac{1}{q} < 0, \\ p = \infty, \ q = \infty, \ \alpha = 0, \ \beta = 0. \end{cases}$$

Theorem 9.7.5. *Assume that $\mu(\mathcal{R}) < \infty$. Let $0 < p,q \leq \infty$ and $\alpha, \beta \in \mathbb{R}^2$. Then the space $L_{(p,q;\alpha,\beta)}$ is a BFS if and only if $1 \leq q \leq \infty$ and one of the following conditions holds:*

$$\begin{cases} 0 < p < \infty, \\ p = \infty, \ \alpha + \frac{1}{q} < 0, \\ p = \infty, \ \alpha + \frac{1}{q} = 0, \ \beta + \frac{1}{q} < 0, \\ p = \infty, \ q = \infty, \ \alpha = 0, \ \beta = 0. \end{cases}$$

9.8 Comparison of GLZ spaces to Orlicz spaces

We shall now turn our attention to the comparison of Orlicz spaces and GLZ spaces. In particular, we shall give a complete characterization of those GLZ spaces $L_{p,q;\mathbb{A},\mathbb{B}}$ and $L_{(p,q;\mathbb{A},\mathbb{B})}$ which coincide with Orlicz spaces. All such spaces are described by Theorems 9.8.9 and 9.8.12. To avoid trivial cases, we throughout this section use the following.

Convention 9.8.1. Since Orlicz spaces are BFS, we restrict ourselves to the values of $p, q; \mathbb{A}, \mathbb{B}$ that satisfy (9.7.1) in the case of spaces $L_{p,q;\mathbb{A},\mathbb{B}}$, or (9.3.2) and $1 \leq q \leq \infty$ in the case of spaces $L_{(p,q;\mathbb{A},\mathbb{B})}$ (cf. Theorems 9.7.1 and 9.7.3).

Let us first consider the case when $p = q$. We shall first recall a result which follows from [73, Lemmas 2.1 and 2.2], concerning spaces of functions defined on a finite measure space.

Lemma 9.8.2. *Let $\mu(\mathcal{R}) < \infty$. Let one of the following conditions hold:*

$$1 < p < \infty, \tag{9.8.1}$$
$$p = 1, \ \alpha > 0, \tag{9.8.2}$$
$$p = 1, \ \alpha = 0, \ \beta > 0, \tag{9.8.3}$$
$$p = \infty, \ \alpha < 0, \tag{9.8.4}$$
$$p = \infty, \ \alpha = 0, \ \beta < 0. \tag{9.8.5}$$

Then
$$L_{p,p;\alpha,\beta} = L^{\Phi},$$
where the Young function Φ satisfies for large t,

$$\Phi(t) \approx \begin{cases} t^p \ell^{\alpha p}(t) \ell\ell^{\beta p}(t) & \text{if one of (9.8.1)–(9.8.5) holds,} \\ \exp\left(t^{-\frac{1}{\alpha}} \ell^{-\frac{\beta}{\alpha}}(t)\right) & \text{if (9.8.4) holds,} \\ \exp\exp(t^{-\frac{1}{\beta}}) & \text{if (9.8.5) holds.} \end{cases} \tag{9.8.6}$$

Corollary 9.8.3. *Let one of the conditions (9.8.1)–(9.8.5) hold, and let the Young function Φ be given for large t by (9.8.6). Let $T \in (0, \infty)$ and $f \in \mathcal{M}(\mathcal{R}, \mu)$. Then*

$$\int_0^T [f^*(t)]^p \ell^{\alpha}(t) \ell\ell^{\beta}(t) \, dt < \infty$$

if and only if there exists a $\lambda > 0$ such that

$$\int_0^T \Phi(\lambda f^*(t)) \, dt < \infty.$$

When $\mu(\mathcal{R}) = \infty$, the values of $\Phi(t)$ for t small become important. To handle them properly, we shall need some auxiliary results. First, we formulate a modified version of the Young inequality.

Lemma 9.8.4. (i) *Let $\lambda > 0$ and $\varepsilon \in \mathbb{R}$. Then for every $a, b > 1$,*

$$ab \lesssim \exp\left(a^{\frac{1}{\lambda}} \ell^{-\frac{\varepsilon}{\lambda}}(a)\right) + b \ell^{\lambda}(b) \ell\ell^{\varepsilon}(b). \tag{9.8.7}$$

(ii) *Let $\varepsilon > 0$. Then for every $a, b > 1$,*

$$ab \lesssim \exp\exp(a^{\frac{1}{\varepsilon}}) + b \ell\ell^{\varepsilon}(b).$$

Proof. This follows from the usual Young inequality by a straightforward calculation of complementary functions. □

Lemma 9.8.5. *Let $\mu(\mathcal{R}) = \infty$, $1 \le p < \infty$, and let either $\lambda > 0$ or $\lambda = 0$ and $\varepsilon > 0$. Suppose that $f \in \mathcal{M}(\mathcal{R}, \mu)$ is such that $f^*(1) < \infty$, and set $R = \mu(\operatorname{supp} f)$. Then*

$$\int_1^R [f^*(t)]^p \ell^\lambda(t) \ell\ell^\varepsilon(t)\, dt < \infty \tag{9.8.8}$$

if and only if

$$\int_1^R [f^*(t)]^p \ell^\lambda(f^*(t)) \ell\ell^\varepsilon(f^*(t))\, dt < \infty. \tag{9.8.9}$$

Proof. The assertion is trivial when $R < \infty$. Assume that $R = \infty$.

First, let us note that both (9.8.8) and (9.8.9) imply that

$$f^*(t) \to 0 \quad \text{as} \quad t \to \infty. \tag{9.8.10}$$

For $T \in [1, \infty)$, we denote

$$I_1(T) = \int_T^\infty [f^*(t)]^p \ell^\lambda(t) \ell\ell^\varepsilon(t)\, dt,$$

$$I_2(T) = \int_T^\infty [f^*(t)]^p \ell^\lambda(f^*(t)) \ell\ell^\varepsilon(f^*(t))\, dt.$$

We see that for any $T \in [1, \infty)$, $I_i(T) < \infty$ if and only if $I_i(1) < \infty$, $i = 1, 2$. Hence, using also (9.8.10), we may assume with no loss of generality that $f^*(t) \le 1$ for every $t \in [1, \infty)$.

Suppose that (9.8.9) holds. Then we have

$$K_1 := \sup_{1 < t < \infty} t^{\frac{1}{p}} f^*(t) < \infty. \tag{9.8.11}$$

Indeed, assuming that

$$t_k^{\frac{1}{p}} f^*(t_k) \to \infty \quad \text{for some} \quad t_k \to \infty, \tag{9.8.12}$$

and using also the fact that the function $\ell^\lambda(x) \ell\ell^\varepsilon(x)$ is decreasing for $x \in (0, x_0)$ with x_0 small enough, we get

$$\int_1^\infty [f^*(t)]^p \ell^\lambda(f^*(t)) \ell\ell^\varepsilon(f^*(t))\, dt \ge \int_{\frac{t_k}{2}}^{t_k} [f^*(t)]^p \ell^\lambda(f^*(t)) \ell\ell^\varepsilon(f^*(t))\, dt$$

$$\ge \frac{t_k}{2} [f^*(t_k)]^p \ell^\lambda\left(f^*\left(\frac{t_{k_0}}{2}\right)\right) \ell\ell^\varepsilon\left(f^*\left(\frac{t_{k_0}}{2}\right)\right)$$

for all $k \ge k_0$, where $k_0 \in \mathbb{N}$ is chosen so that $f^*(\frac{t_{k_0}}{2}) < x_0$. Combined with (9.8.12), this estimate contradicts (9.8.9), and thus (9.8.11) holds.

By (9.8.11),
$$1 \le \frac{t^{\frac{1}{p}}}{K_1} \le \frac{1}{f^*(t)}$$
if $t > \max\{1, K_1^p\}$. The function $\ell^\lambda(x)\ell\ell^\varepsilon(x)$ is increasing on (x_1, ∞) for some x_1 large enough. Thus, taking $t > T := \max\{1, K_1^p, (x_1 K_1)^p\}$, we get
$$\ell^\lambda\left(\frac{t^{\frac{1}{p}}}{K_1}\right)\ell\ell^\varepsilon\left(\frac{t^{\frac{1}{p}}}{K_1}\right) \le \ell^\lambda\left(\frac{1}{f^*(t)}\right)\ell\ell^\varepsilon\left(\frac{1}{f^*(t)}\right).$$

This yields immediately
$$\ell^\lambda(t)\ell\ell^\varepsilon(t) \lesssim \ell^\lambda(f^*(t))\ell\ell^\varepsilon(f^*(t)), \qquad t \in (T, \infty),$$
and (9.8.8) follows from (9.8.9).

Conversely, suppose that (9.8.8) holds. Let $n_0 \in \mathbb{N} \cup \{0\}$ be such that
$$e^{-\frac{n_0+1}{p}} < f^*(1) \le e^{-\frac{n_0}{p}}.$$

We define
$$t_n = \inf\{t \ge 1; f^*(t) \le e^{-\frac{n}{p}}\}, \quad n \ge n_0.$$

Then $\{t_n\}$ is a nondecreasing sequence such that $t_n \to \infty$ as $n \to \infty$. Since f^* is right-continuous (Proposition 7.1.12), we have
$$e^{-\frac{n+1}{p}} < f^*(t) \le e^{-\frac{n}{p}}, \quad t \in [t_n, t_{n+1}).$$

This yields for all $t \in [t_n, t_{n+1})$,
$$[f^*(t)]^p \approx e^{-n}, \quad \ell^\lambda(f^*(t)) \approx n^\lambda, \quad \ell\ell^\varepsilon(f^*(t)) \approx \log^\varepsilon(n+2), \qquad (9.8.13)$$

and hence
$$I_2(1) = \sum_{n \ge n_0} \int_{t_n}^{t_{n+1}} [f^*(t)]^p \ell^\lambda(f^*(t)) \ell\ell^\varepsilon(f^*(t))\, dt \qquad (9.8.14)$$
$$\approx \sum_{n \ge n_0} e^{-n} n^\lambda (\log^\varepsilon(n+2))(t_{n+1} - t_n) = \sum_{n \ge n_0} e^{-n} a_n b_n,$$

where
$$a_n = \left(\frac{n}{2}\right)^\lambda \log^\varepsilon(n+2), \quad b_n = 2^\lambda(t_{n+1} - t_n).$$

We shall use (9.8.7) to estimate the right-hand side of (9.8.14). Since
$$a_n^{\frac{1}{\lambda}} \ell^{-\frac{\varepsilon}{\lambda}}(a_n) \approx \frac{n}{2}$$

and
$$b_n\ell^\lambda(b_n)\ell\ell^\varepsilon(b_n) \approx (t_{n+1}-t_n)\ell^\lambda(t_{n+1}-t_n)\ell\ell^\varepsilon(t_{n+1}-t_n),$$
we get from (9.8.14) and (9.8.7),
$$I_2(1) \lesssim \sum_{n\geq n_0} e^{-n}e^{\frac{n}{2}} + \sum_{n\geq n_0} e^{-n}(t_{n+1}-t_n)\ell^\lambda(t_{n+1}-t_n)\ell\ell^\varepsilon(t_{n+1}-t_n) = S_1 + S_2,$$
say. Obviously, $S_1 < \infty$. Now, we claim that, for large n, say $n \geq n_1$, we have
$$(t_{n+1}-t_n)\ell^\lambda(t_{n+1}-t_n)\ell\ell^\varepsilon(t_{n+1}-t_n) \tag{9.8.15}$$
$$\lesssim t_{n+1}\ell^\lambda(t_{n+1})\ell\ell^\varepsilon(t_{n+1}) - t_n\ell^\lambda(t_n)\ell\ell^\varepsilon(t_n).$$

Once this is shown, the assertion will follow easily as, by (9.8.15),
$$S_2 \lesssim \sum_{n=n_0}^{n_1-1} e^{-n}(t_{n+1}-t_n)\ell^\lambda(t_{n+1}-t_n)\ell\ell^\varepsilon(t_{n+1}-t_n) + S_3,$$
where
$$S_3 = \sum_{n\geq n_1} e^{-n}[t_{n+1}\ell^\lambda(t_{n+1})\ell\ell^\varepsilon(t_{n+1}) - t_n\ell^\lambda(t_n)\ell\ell^\varepsilon(t_n)].$$
But, by (9.8.13) and (9.8.8),
$$S_3 \lesssim \sum_{n\geq n_1} e^{-n} \int_{t_n}^{t_{n+1}} \ell^\lambda(t)\ell\ell^\varepsilon(t)\,dt \lesssim I_1(1) < \infty.$$

It remains to prove (9.8.15). Assume first that $t_{n+1} \geq 2t_n$, i.e. $t_{n+1} \leq 2(t_{n+1}-t_n)$. Then
$$t_{n+1}\ell^\lambda(t_{n+1})\ell\ell^\varepsilon(t_{n+1}) \leq 2(t_{n+1}-t_n)\ell^\lambda(t_{n+1})\ell\ell^\varepsilon(t_{n+1}). \tag{9.8.16}$$
Since the function $\ell^\lambda(t)\ell\ell^\varepsilon(t)$ is increasing near ∞, we obtain from (9.8.16) that, for large n,
$$t_{n+1}\ell^\lambda(t_{n+1})\ell\ell^\varepsilon(t_{n+1}) \leq 2[t_{n+1}\ell^\lambda(t_{n+1})\ell\ell^\varepsilon(t_{n+1}) - t_n\ell^\lambda(t_n)\ell\ell^\varepsilon(t_n)],$$
and (9.8.15) follows.

Now assume that $t_{n+1} \leq 2t_n$, i.e. $t_{n+1} - t_n \leq t_n$. Since the function $F(t) = t\ell^\lambda(t)\ell\ell^\varepsilon(t)$ is convex near ∞, it is equivalent to an increasing convex function on entire $[1,\infty)$. With no loss of generality, we shall assume that F itself is increasing and convex on $[1,\infty)$. Thus, for every n,
$$F'(t_n)(t_{n+1}-t_n) \lesssim F(t_{n+1}) - F(t_n), \tag{9.8.17}$$
and
$$\ell^\lambda(t_n)\ell\ell^\varepsilon(t_n) \approx F'(t_n). \tag{9.8.18}$$

Section 9.8 Generalized Lorentz–Zygmund spaces and Orlicz spaces

Similarly, the function $\ell^\lambda(t)\ell\ell^\varepsilon(t)$ is equivalent to an increasing function on $[1, \infty)$. Therefore, using also $t_{n+1} - t_n \leq t_n$, (9.8.18) and (9.8.17), we obtain for every n,

$$(t_{n+1} - t_n)\ell^\lambda(t_{n+1} - t_n)\ell\ell^\varepsilon(t_{n+1} - t_n)$$
$$\leq (t_{n+1} - t_n)\ell^\lambda(t_n)\ell\ell^\varepsilon(t_n) \approx (t_{n+1} - t_n)F'(t_n)$$
$$\lesssim t_{n+1}\ell^\lambda(t_{n+1})\ell\ell^\varepsilon(t_{n+1}) - t_n\ell^\lambda(t_n)\ell\ell^\varepsilon(t_n)$$

and (9.8.15) follows. The proof is complete. □

The following simple lemma is of independent interest as it indicates a relation between Orlicz spaces and Marcinkiewicz spaces (cf. [136, Theorem 2]).

Lemma 9.8.6. *Let $f \in \mathcal{M}(\mathcal{R}, \mu)$ and let Φ be a Young function on $(0, \infty)$ such that*

$$\int_0^\infty \Phi\left(\gamma\Phi^{-1}\left(\frac{1}{t}\right)\right) dt < \infty \qquad \text{for some} \quad \gamma > 0. \tag{9.8.19}$$

Then $f \in L^\Phi$ if and only if there exists a constant $K = K(f)$ such that

$$\sup_{0 < t < \infty} \frac{f^*(t)}{\Phi^{-1}\left(\frac{1}{t}\right)} = K < \infty. \tag{9.8.20}$$

Remark 9.8.7. If Φ is a Young function, then we have, by (7.9.4), $\varphi_{L^\Phi}(t) = \frac{1}{\Phi^{-1}\left(\frac{1}{t}\right)}$ for every $t \in (0, \infty)$. Therefore, the condition (9.8.20) can be rewritten as

$$\sup_{0 < t < \infty} f^*(t)\varphi_{L^\Phi}(t) = K < \infty.$$

Proof of Lemma 9.8.6. If (9.8.20) holds, then we have for $\lambda < \gamma K^{-1}$

$$\int_0^\infty \Phi\left(\lambda f^*(t)\right) dt \leq \int_0^\infty \Phi\left(\gamma\Phi^{-1}\left(\frac{1}{t}\right)\right) dt < \infty,$$

in other words, $f \in L^\Phi$.

Conversely, assume that $f \in L^\Phi$. Then there is a $\lambda_0 > 0$ such that

$$\int_0^\infty \Phi\left(\lambda_0 f^*(t)\right) dt \leq \frac{1}{2}. \tag{9.8.21}$$

Let $a > 0$. Then, by (9.8.21),

$$\frac{1}{2} \geq \int_{\frac{a}{2}}^a \Phi\left(\lambda_0 f^*(t)\right) dt \geq \Phi\left(\lambda_0 f^*(a)\right)\frac{a}{2},$$

which is equivalent to

$$f^*(a) \leq \lambda_0^{-1}\Phi^{-1}\left(\frac{1}{a}\right), \qquad a > 0,$$

and (9.8.20) follows with $K = \lambda_0^{-1}$. (Let us note that $f \in L^\Phi$ implies (9.8.20) for *any* Young function Φ, regardless of the validity of (9.8.19).) □

The following example follows by an easy calculation.

Example 9.8.8. Let Φ be a Young function such that, for $0 < t \leq 1$, either

$$\Phi(t) \approx \exp\left(-t^\alpha \log^\beta\left(\frac{1}{t}\right)\right), \qquad \alpha < 0, \ \beta \in \mathbb{R},$$

or

$$\Phi(t) \approx \exp(-\exp t^\beta), \qquad \beta < 0,$$

and, for $1 < t < \infty$, either

$$\Phi(t) \approx \exp(t^\gamma \log^\delta t), \qquad \gamma > 0, \ \delta \in \mathbb{R},$$

or

$$\Phi(t) \approx \exp \exp t^\delta, \qquad \delta > 0.$$

Then Φ satisfies (9.8.19).

Now we are in a position to prove the main result of this section.

Theorem 9.8.9. *Assume that $\mu(\mathcal{R}) = \infty$ and $\mathbb{A} = (\alpha_0, \alpha_\infty)$, $\mathbb{B} = (\beta_0, \beta_\infty) \in \mathbb{R}^2$.*

(i) *Let $1 < p < \infty$. Then $L_{p,p;\mathbb{A},\mathbb{B}} = L^\Phi$, where*

$$\Phi(t) \approx t^p \ell^{\widetilde{\mathbb{A}}p}(t) \ell\ell^{\widetilde{\mathbb{B}}p}(t), \qquad t \in (0, \infty) \tag{9.8.22}$$

(recall that $\widetilde{\mathbb{A}} = (\alpha_\infty, \alpha_0)$ and $\widetilde{\mathbb{B}} = (\beta_\infty, \beta_0)$).

(ii) *Let $\alpha_\infty > 0$ and either $\alpha_0 < 0$ or $\alpha_0 = 0$ and $\beta_0 < 0$. Then $L_{\infty,\infty;\mathbb{A},\mathbb{B}} = L^\Phi$, where*

$$\Phi(t) \approx \exp\left(-t^{-\frac{1}{\alpha_\infty}} \ell^{-\frac{\beta_\infty}{\alpha_\infty}}(t)\right) \qquad \text{for} \quad t \in (0, 1), \tag{9.8.23}$$

and, for $t \in (1, \infty)$,

$$\Phi(t) \approx \begin{cases} \exp\left(t^{-\frac{1}{\alpha_0}} \ell^{-\frac{\beta_0}{\alpha_0}}(t)\right) & \text{if} \ \alpha_0 < 0, \\ \exp\exp(t^{-\frac{1}{\beta_0}}) & \text{if} \ \alpha_0 = 0, \ \beta_0 < 0. \end{cases} \tag{9.8.24}$$

(iii) *Let $\alpha_\infty = 0$, $\beta_\infty > 0$, and either $\alpha_0 < 0$ or $\alpha_0 = 0$ and $\beta_0 < 0$. Then $L_{\infty,\infty;\mathbb{A},\mathbb{B}} = L^\Phi$, where*

$$\Phi(t) \approx \exp(-\exp t^{-\frac{1}{\beta_\infty}}) \qquad \text{for} \quad t \in (0, 1),$$

and Φ satisfies (9.8.24) for $t \in (1, \infty)$.

Section 9.8 Generalized Lorentz–Zygmund spaces and Orlicz spaces

(iv) Let either $\alpha_0 > 0$ or $\alpha_0 = 0$ and $\beta_0 > 0$, and let either $\alpha_\infty < 0$ or $\alpha_\infty = 0$ and $\beta_\infty < 0$. Then $L_{1,1;\mathbb{A},\mathbb{B}} = L^\Phi$, where

$$\Phi(t) \approx t\ell^{\widetilde{\mathbb{A}}}(t)\ell\ell^{\widetilde{\mathbb{B}}}(t), \qquad t \in (0,\infty). \tag{9.8.25}$$

Proof. (i) Let Φ satisfy (9.8.22). Then $\Phi \in \Delta_2$. Hence, $f \in L^\Phi$ if and only if

$$\int_0^\infty \Phi\left(f^*(t)\right) dt < \infty. \tag{9.8.26}$$

Set $R = \mu(\mathrm{supp}\, f)$. Define $T = \inf\{t > 0;\ f^*(t) \le 1\}$. Using the fact that $1 < f^*(t) < \infty$ for $t \in (0,T)$ and $f^*(t) \le 1$ for $t \in [T,\infty)$, and (9.8.22), we get from (9.8.26),

$$\int_0^T (f^*(t))^p \ell^{\alpha_0 p}(f^*(t)) \ell\ell^{\beta_0 p}(f^*(t))\, dt \tag{9.8.27}$$

$$+ \int_T^R (f^*(t))^p \ell^{\alpha_\infty p}(f^*(t)) \ell\ell^{\beta_\infty p}(f^*(t))\, dt < \infty.$$

By Corollary 9.8.3 and Lemma 9.8.5, (9.8.27) holds if and only if

$$\int_0^R (f^*(t))^p \ell^{\mathbb{A}p}(t) \ell\ell^{\mathbb{B}p}(t)\, dt < \infty$$

in the case $\alpha_\infty > 0$ or $\alpha_\infty = 0$, $\beta_\infty > 0$. This proves (i) in such case. If $\alpha_\infty < 0$ or $\alpha_\infty = 0$, $\beta_\infty < 0$, the assertion follows via duality (Theorem 9.6.2). In the remaining case $\alpha_\infty = 0$, $\beta_\infty = 0$, the assertion is obvious.

(ii) Let Φ satisfy (9.8.23) and (9.8.24). Then, by Example 8.7, Φ satisfies (9.8.19). By Lemma 9.8.6, $f \in L^\Phi$ if and only if (9.8.20) holds. However, it is easy to see that (9.8.20) with our Φ is equivalent to $f \in L_{\infty,\infty;\mathbb{A},\mathbb{B}}$.

(iii) The proof is analogous to that of (ii).

(iv) By Theorem 9.6.8 (i) we have

$$\left(L_{1,1;\mathbb{A},\mathbb{B}}\right)' = L_{\infty,\infty;-\mathbb{A},-\mathbb{B}}. \tag{9.8.28}$$

If $\alpha_\infty < 0$, then, by (ii) with $\mathbb{A}, \mathbb{B}$ replaced by $-\mathbb{A}, -\mathbb{B}$, we get

$$L_{\infty,\infty;-\mathbb{A},-\mathbb{B}} = L^\Psi, \tag{9.8.29}$$

where

$$\Psi(t) \approx \begin{cases} \exp\left(-t^{-\frac{1}{\alpha_\infty}} \ell^{-\frac{\beta_\infty}{\alpha_\infty}}(t)\right), & t \in (0,1), \\ \exp\left(t^{\frac{1}{\alpha_0}} \ell^{-\frac{\beta_0}{\alpha_0}}(t)\right), & t \in (1,\infty),\ \alpha_0 > 0, \\ \exp\exp(t^{\frac{1}{\beta_0}}), & t \in (1,\infty),\ \alpha_0 = 0,\ \beta_0 > 0. \end{cases}$$

A direct calculation shows that the complementary function of Ψ is given by (9.8.25). Together with (9.8.28) and (9.8.29) this yields

$$L_{1,1;\mathbb{A},\mathbb{B}} = \left(L_{1,1;\mathbb{A},\mathbb{B}}\right)'' = \left(L_{\infty,\infty;-\mathbb{A},-\mathbb{B}}\right)' = \left(L^{\Psi}\right)' = L^{\Phi},$$

which is the desired result.

If $\alpha_\infty = 0$ and $\beta_\infty < 0$, we adopt an analogous argument using (iii) rather than (ii). □

Lemma 9.8.11 below completes the results of Theorem 9.8.9. To prove it, we shall need the following assertion, which follows easily from (4.2.3), (4.2.4) and (7.9.4).

Lemma 9.8.10. *Let X be an r.i. space. If its fundamental function φ_X is equivalent near 0 or near ∞ either to t or to 1, then X is not an Orlicz space.*

Lemma 9.8.11. *Suppose that $\mu(\mathcal{R}) = \infty$.*

(i) *Let either $\alpha_0 < 0$ or $\alpha_0 = 0$ and $\beta_0 \leq 0$, and let either $\alpha_\infty < 0$ or $\alpha_\infty = 0$ and $\beta_\infty < 0$. Then $L_{\infty,\infty;\mathbb{A},\mathbb{B}}$ is not an Orlicz space.*

(ii) *Let one of the following conditions be satisfied:*

$$\alpha_0 = 0, \quad \beta_0 = 0, \quad \alpha_\infty < 0,$$
$$\alpha_0 = 0, \quad \beta_0 = 0, \quad \alpha_\infty = 0, \quad \beta_\infty \leq 0,$$
$$\alpha_0 > 0, \quad \alpha_\infty = 0, \quad \beta_\infty = 0,$$
$$\alpha_0 = 0, \quad \beta_0 \geq 0, \quad \alpha_\infty = 0, \quad \beta_\infty = 0.$$

Then $L_{1,1;\mathbb{A},\mathbb{B}}$ is not an Orlicz space.

Proof. (i) Set $X = L_{\infty,\infty;\mathbb{A},\mathbb{B}}$. By Corollary 9.5.3, $X = L_{\infty,\infty;(\alpha_0,0),(\beta_0,0)}$. Thus, by Lemma 9.4.1 (i), $\varphi_X(t) \approx 1$ for all $t \in (1,\infty)$, and the result follows from Lemma 9.8.10.

(ii) The proof is analogous. □

Now we turn our attention to spaces $L_{(p,q;\mathbb{A},\mathbb{B})}$. In view of Lemma 9.3.1 (ii), Theorem 9.5.1 (i) and Theorem 9.8.9 (i)–(ii), it suffices to consider the case $p = 1$.

Theorem 9.8.12. *Assume that $\mu(\mathcal{R}) = \infty$ and $\mathbb{A} = (\alpha_0, \alpha_\infty)$, $\mathbb{B} = (\beta_0, \beta_\infty) \in \mathbb{R}^2$. Let*

$$\text{either} \quad \alpha_0 + 1 > 0 \quad \text{or} \quad \alpha_0 + 1 = 0, \; \beta_0 + 1 \geq 0,$$

and

$$\text{either} \quad \alpha_\infty + 1 < 0 \quad \text{or} \quad \alpha_\infty + 1 = 0, \; \beta_\infty + 1 < 0.$$

Section 9.8 Generalized Lorentz–Zygmund spaces and Orlicz spaces

Then $L_{(1,1;\mathbb{A},\mathbb{B})} = L^\Phi$, where, for $t \in (0, 1)$,

$$\Phi(t) \approx \begin{cases} t\ell^{\alpha_\infty+1}(t)\ell\ell^{\beta_\infty}(t) & \text{if } \alpha_\infty + 1 < 0, \\ t\ell\ell^{\beta_\infty+1}(t) & \text{if } \alpha_\infty + 1 = 0, \beta_\infty + 1 < 0, \end{cases}$$

and, for $t \in [1, \infty)$,

$$\Phi(t) \approx \begin{cases} t\ell^{\alpha_0+1}(t)\ell\ell^{\beta_0}(t) & \text{if } \alpha_0 + 1 > 0, \\ t\ell\ell^{\beta_0+1}(t) & \text{if } \alpha_0 + 1 = 0, \beta_0 + 1 > 0, \\ t\ell\ell\ell(t) & \text{if } \alpha_0 + 1 = 0, \beta_0 + 1 = 0. \end{cases}$$

Proof. For example, assume that either $\alpha_0 + 1 > 0$ and $\alpha_\infty + 1 < 0$. Then, by Theorem 9.5.1 (ii) and Theorem 9.8.9 (iv), $L_{(1,1;\mathbb{A},\mathbb{B})} = L_{1,1;\mathbb{A}+1,\mathbb{B}} = L^\Phi$, where $\Phi(t) \approx t\ell^{\widetilde{\mathbb{A}}+1}(t)\ell\ell^{\widetilde{\mathbb{B}}}(t)$, $t \in (0, \infty)$, and the result follows. The proof in other cases is similar. □

To complete our analysis, we shall show that if $p \neq q$, then neither the space $L_{p,q;\mathbb{A},\mathbb{B}}$ nor the space $L_{(p,q;\mathbb{A},\mathbb{B})}$ coincides with an Orlicz space. (Consequently, Theorems 9.8.9, and 9.8.12 cover all possible cases when a GLZ space is an Orlicz space.) It is thus enough to consider those spaces which are BFS. Further reduction is enabled by Lemma 9.8.10.

Let us first deal with the spaces $L_{p,q;\mathbb{A},\mathbb{B}}$. It follows from Theorem 9.7.1, Lemma 9.8.10 and Lemma 9.4.1 (i) that we have only to consider one of the cases

$$\begin{cases} 1 < p < \infty, \ 1 \leq q \leq \infty, \ p \neq q, \\ p = \infty, \ 1 \leq q < \infty, \ \alpha_0 + \frac{1}{q} < 0, \ \alpha_\infty + \frac{1}{q} > 0, \\ p = \infty, \ 1 \leq q < \infty, \ \alpha_0 + \frac{1}{q} < 0, \ \alpha_\infty + \frac{1}{q} = 0, \ \beta_\infty + \frac{1}{q} \geq 0, \\ p = \infty, \ 1 \leq q < \infty, \ \alpha_0 + \frac{1}{q} = 0, \ \beta_0 + \frac{1}{q} < 0, \ \alpha_\infty + \frac{1}{q} > 0, \\ p = \infty, \ 1 \leq q < \infty, \ \alpha_0 + \frac{1}{q} = 0, \ \beta_0 + \frac{1}{q} < 0, \\ \qquad \alpha_\infty + \frac{1}{q} = 0, \ \beta_\infty + \frac{1}{q} \geq 0. \end{cases} \quad (9.8.30)$$

We shall use the following auxiliary result.

Lemma 9.8.13. *Let X and Y be two r.i. spaces such that $X \neq Y$ and $\varphi_X \approx \varphi_Y$. Suppose that there is a Young function Φ such that $Y = L^\Phi$. Then $X \neq L^\Psi$ for any Young function Ψ.*

Proof. Assume the contrary, that is, $X = L^\Psi$ for some Young function Ψ. Then, by (7.9.4),

$$\frac{1}{\Phi^{-1}(\frac{1}{t})} \approx \varphi_X(t) \approx \varphi_Y(t) \approx \frac{1}{\Psi^{-1}(\frac{1}{t})}, \qquad t \in (0, \infty).$$

Thus $\Phi \approx \Psi$ on $(0, \infty)$, and therefore $X = L^\Phi = L^\Psi = Y$, which is a contradiction. □

Corollary 9.8.14. *Let $0 < p < \infty$, $0 < q \le \infty$, $p \ne q$, and $\mathbb{A} = (\alpha_0, \alpha_\infty)$, $\mathbb{B} = (\beta_0, \beta_\infty) \in \mathbb{R}^2$. Set $X = L_{p,p;\mathbb{A},\mathbb{B}}$ and $Y = L_{p,q;\mathbb{A},\mathbb{B}}$. Suppose that there is a Young function Φ such that $X = L^\Phi$. Then Y is not an Orlicz space.*

Proof. Embedding results above show that $X \ne Y$. Moreover, by Lemma 9.4.1 (i), $\varphi_X \approx \varphi_Y$ on $(0, \infty)$. The result now follows from Lemma 9.8.13. □

Now we are in a position to finish the analysis concerning the spaces $L_{p,q;\mathbb{A},\mathbb{B}}$. The following theorem completes the picture.

Theorem 9.8.15. *Let $0 < p, q \le \infty$ and $\mathbb{A} = (\alpha_0, \alpha_\infty)$, $\mathbb{B} = (\beta_0, \beta_\infty) \in \mathbb{R}^2$. Set $X = L_{p,q;\mathbb{A},\mathbb{B}}$. Let one of the conditions in (9.8.30) be satisfied. Then X is not an Orlicz space.*

Proof. If the first condition in (9.8.30) holds, then the result follows immediately from Corollary 9.8.14. In all other cases we use a method which will be illustrated on the case

$$p = \infty, \quad 1 \le q < \infty, \quad \alpha_0 + \frac{1}{q} < 0, \quad \alpha_\infty + \frac{1}{q} > 0. \tag{9.8.31}$$

Other cases are left to the reader.

Assume (9.8.31). Then, by Lemma 9.4.1 (i), $\varphi_X(t) \approx \ell^{\mathbb{A}+\frac{1}{q}}(t)\ell\ell^{\mathbb{B}}(t)$ for $t \in (0, \infty)$. Consequently, φ_X is on $(0, \infty)$ equivalent to an increasing concave function φ such that $\varphi(0_+) = 0$ and $\varphi(\infty_-) = \infty$. Therefore, setting $\Phi(t) = 1/[\varphi^{-1}(\frac{1}{t})]$, $t \in (0, \infty)$, we obtain (cf. (7.9.4)) that Φ is equivalent on $(0, \infty)$ to a Young function. Observe that

$$\Phi(t) \approx \begin{cases} \exp\left(-t^{-\frac{1}{\alpha_\infty+1/q}} \ell^{-\frac{\beta_\infty}{\alpha_\infty+1/q}}(t)\right) & \text{if } t \in (0, 1) \\ \exp\left(t^{-\frac{1}{\alpha_0+1/q}} \ell^{-\frac{\beta_0}{\alpha_0+1/q}}(t)\right) & \text{if } t \in [1, \infty). \end{cases}$$

By Example 9.8.8, Φ satisfies (9.8.19). Hence, $f \in L^\Phi$ if and only if (9.8.20) holds.

We have to show that $X \ne L^\Phi$ (then the result will follow from Lemma 9.8.13). Note that the function $f(t) = \frac{1}{\varphi(t)} = \Phi^{-1}(\frac{1}{t})$, $t \in (0, \infty)$, satisfies (9.8.20), but

$$\|f\|_X^q \ge \int_0^1 t^{-1} \ell^{-1}(t)\, dt = \infty.$$

The proof (in the case (9.8.31)) is thus complete. □

Now let us deal with the spaces $L_{(p,q;\mathbb{A},\mathbb{B})}$. Since $L_{(p,q;\mathbb{A},\mathbb{B})} = \{0\}$ for $0 < p < 1$, and $L_{(p,q;\mathbb{A},\mathbb{B})} = L_{p,q;\mathbb{A},\mathbb{B}}$ when $1 < p < \infty$, it is enough to consider the case $p = 1$. Moreover, using Theorem 9.7.3, Lemma 9.8.10 and Lemma 9.4.1 (ii), we see that we may restrict ourselves to the situation when $1 < q \leq \infty$ and one of the following conditions is satisfied:

$$\begin{cases} \alpha_0 + \frac{1}{q} > 0, \ \alpha_\infty + \frac{1}{q} < 0, \\ \alpha_0 + \frac{1}{q} > 0, \ \alpha_\infty + \frac{1}{q} = 0, \ \beta_\infty + \frac{1}{q} < 0, \\ \alpha_0 + \frac{1}{q} = 0, \ \beta_0 + \frac{1}{q} > 0, \ \beta_\infty + \frac{1}{q} < 0, \\ \alpha_0 + \frac{1}{q} = 0, \ \beta_0 + \frac{1}{q} > 0, \ \alpha_\infty + \frac{1}{q} = 0, \ \beta_\infty + \frac{1}{q} < 0, \\ 1 < q < \infty, \ \alpha_0 + \frac{1}{q} = 0, \ \beta_0 + \frac{1}{q} = 0, \ \alpha_\infty + \frac{1}{q} < 0, \\ 1 < q < \infty, \ \alpha_0 + \frac{1}{q} = 0, \ \beta_0 + \frac{1}{q} = 0, \ \alpha_\infty + \frac{1}{q} = 0, \ \beta_\infty + \frac{1}{q} < 0. \end{cases}$$
(9.8.32)

Theorem 9.8.16. *Let $1 < q \leq \infty$ and $\mathbb{A} = (\alpha_0, \alpha_\infty)$, $\mathbb{B} = (\beta_0, \beta_\infty) \in \mathbb{R}^2$. Set $X = L_{(1,q;\mathbb{A},\mathbb{B})}$. Let one of the conditions in (9.8.32) be satisfied. Then X is not an Orlicz space.*

Proof. It follows from Theorem 9.6.9 (i), complemented with its analogue involving three tiers of logarithms, and Theorem 9.8.15, that X' is not an Orlicz space. Therefore, neither is X. □

9.9 Absolute continuity of norm

The aim of this section is to characterize those GLZ spaces $L_{p,q;\mathbb{A},\mathbb{B}}$ and $L_{(p,q;\mathbb{A},\mathbb{B})}$ that have absolutely continuous (quasi) norm.

If $\mu(\mathcal{R}) < \infty$, then it is easy to see that the space $L_{p,q;\mathbb{A},\mathbb{B}}$ (or $L_{(p,q;\mathbb{A},\mathbb{B})}$) with $0 < q < \infty$ has absolutely continuous (quasi) norm. Indeed, taking $f \in X = L_{p,q;\mathbb{A},\mathbb{B}}$ and $\{E_n\}_{n=1}^\infty \subset \mathcal{R}$ satisfying $E_n \searrow \emptyset$ μ-a.e., we have $\mu(E_n) \leq \mu(\mathcal{R}) < \infty$ for all $n \in \mathbb{N}$, which implies

$$\lim_{n \to \infty} \mu(E_n) = \mu\left(\bigcap_{n=1}^\infty E_n\right) = \mu(\emptyset) = 0.$$

Thus, as $n \to \infty$,

$$(f\chi_{E_n})^* \leq f^*\chi_{(0,\mu(E_n))} \to 0,$$

and, by the Lebesgue dominated convergence theorem,

$$\|f\chi_{E_n}\|_X \leq \left(\int_0^\infty \left[t^{\frac{1}{p}-\frac{1}{q}}\ell^{\mathbb{A}}(t)\ell\ell^{\mathbb{B}}(t)f^*(t)\chi_{(0,\mu(E_n))}(t)\right]^q dt\right)^{\frac{1}{q}} \to 0, \quad n \to \infty,$$

and the result follows.

However, when $\mu(\mathcal{R}) = \infty$, it may happen that $E_n \searrow \emptyset$ as $n \to \infty$ but $\mu(E_n) = \infty$ for every $n \in \mathbb{N}$. Then $\chi_{(0,\mu(E_n))} = \chi_{(0,\infty)}$, and we do not obtain $f^*\chi_{(0,\mu(E_n))} \to 0$ from the trivial estimate $(f\chi_{E_n})^* \leq f^*\chi_{(0,\mu(E_n))}$ as above.

In the case of infinite measure a deeper analysis is needed and we find the following lemmas useful in this.

Lemma 9.9.1. *Let $f_n, f \in \mathcal{M}_+(\mathcal{R}, \mu)$, $n \in \mathbb{N}$, be such that $\limsup_{n\to\infty} f_n \le f$ μ-a.e. Assume that there are $g \in \mathcal{M}_+(\mathcal{R}, \mu)$ and $n_0 \in \mathbb{N}$ satisfying*

$$g \ge f_n \quad \mu\text{-a.e.} \quad \text{for all} \quad n \ge n_0, \tag{9.9.1}$$

$$\mu_g(\lambda) < \infty \quad \text{for all} \quad \lambda \in [0, \infty). \tag{9.9.2}$$

Then for all $\lambda \in [0, \infty)$,

$$\limsup_{n\to\infty} \mu_{f_n}(\lambda) \le \mu_f(\lambda). \tag{9.9.3}$$

Proof. For $\lambda \in [0, \infty)$ and $F \in \mathcal{M}(\mathcal{R}, \mu)$ set

$$E_\lambda(F) = \{x \in \mathcal{R}; |F(x)| > \lambda\}.$$

Since (μ-a.e.)

$$f(x) \ge \limsup_{n\to\infty} f_n(x) = \lim_{n\to\infty} h_n(x),$$

where $h_n(x) = \sup_{m \ge n} f_m(x)$, the function $h(x) := \lim_{n\to\infty} h_n(x)$ satisfies $h_n \searrow h$ and $h \le f$. Consequently,

$$\mu_h(\lambda) \le \mu_f(\lambda), \quad \lambda \in [0, \infty). \tag{9.9.4}$$

Moreover, for all $\lambda \in [0, \infty)$,

$$E_\lambda(h) = \bigcap_{n=1}^{\infty} E_\lambda(h_n) = \bigcap_{n=1}^{\infty} \bigcup_{m \ge n} E_\lambda(f_m). \tag{9.9.5}$$

The assumptions (9.9.1) and (9.9.2) imply

$$\mu\left(\bigcup_{m \ge n_0} E_\lambda(f_m)\right) \le \mu(E_\lambda(g)) = \mu_g(\lambda) < \infty.$$

Together with (9.9.5), this yields

$$\mu_h(\lambda) = \lim_{n\to\infty} \mu\left(\bigcup_{m \ge n} E_\lambda(f_m)\right).$$

Further, observe that

$$\mu\left(\bigcup_{m \ge n} E_\lambda(f_m)\right) \ge \sup_{k \ge n} \mu(E_\lambda(f_k)) = \sup_{k \ge n} \mu_{f_k}(\lambda).$$

Thus,

$$\mu_h(\lambda) \ge \lim_{n\to\infty} \sup_{k \ge n} \mu_{f_k}(\lambda) = \limsup_{n\to\infty} \mu_{f_n}(\lambda),$$

which, together with (9.9.4), implies (9.9.3). □

Section 9.9 Absolute continuity of norm

Combining Lemma 9.9.1 with a symmetric assertion, cf. (7.1.5), one can prove the following result.

Lemma 9.9.2. *Let $f_n, f \in \mathcal{M}_+(\mathcal{R}, \mu)$, $n \in \mathbb{N}$, be such that $\lim_{n\to\infty} f_n = f$ μ-a.e. Assume that there are $g \in \mathcal{M}_+(\mathcal{R}, \mu)$ and $n_0 \in \mathbb{N}$ such that (9.9.1) and (9.9.2) hold. Then for all $\lambda \in [0, \infty)$,*
$$\mu_f(\lambda) = \lim_{n\to\infty} \mu_{f_n}(\lambda).$$

Lemma 9.9.3. *Let $f_n, f \in \mathcal{M}_+(\mathcal{R}, \mu)$, $n \in \mathbb{N}$, be such that $\limsup_{n\to\infty} f_n \leq f$ μ-a.e. Assume that there are $g \in \mathcal{M}_+(\mathcal{R}, \mu)$ and $n_0 \in \mathbb{N}$ satisfying*

$$g \geq f_n \quad \mu\text{-a.e.} \quad \text{for all} \quad n \geq n_0, \tag{9.9.6}$$

$$g^*(t) < \infty \quad \text{for all} \quad t \in (0, \infty), \tag{9.9.7}$$

$$\lim_{t\to\infty} g^*(t) = 0. \tag{9.9.8}$$

Then

$$\limsup_{n\to\infty} f_n^* \leq f^*. \tag{9.9.9}$$

Proof. Assuming that $\mu_g(\lambda_0) = \infty$ for some $\lambda_0 \in [0, \infty)$, we get for any $t > 0$,
$$g^*(t) = \inf\{\lambda; \mu_g(\lambda) \leq t\} \geq \lambda_0,$$
which contradicts (9.9.8). Consequently, $\mu_g(\lambda) < \infty$ for all $\lambda \in [0, \infty)$. Thus, by Lemma 9.9.1,
$$\limsup_{n\to\infty} \mu_{f_n} \leq \mu_f.$$

Let m denote the Lebesgue measure on $(0, \infty)$. Applying Lemma 9.9.1 again, this time to $\{\mu_{f_n}\}$, μ_g, μ_f, and m instead of $\{f_n\}$, g, f, and μ, we get
$$\limsup_{n\to\infty} m_{\mu_{f_n}} \leq m_{\mu_f},$$
which is (9.9.9), since for any $h \in \mathcal{M}(\mathcal{R}, \mu)$ we have $m_{\mu_h} = h^*$ (cf. (7.1.7)). □

Combining Lemma 9.9.3 with a symmetric assertion (cf. (7.1.11)), one can prove the following result.

Lemma 9.9.4. *Let $f_n, f \in \mathcal{M}_+(\mathcal{R}, \mu)$, $n \in \mathbb{N}$, be such that $\lim_{n\to\infty} f_n = f$ μ-a.e. Assume that there are $g \in \mathcal{M}_+(\mathcal{R}, \mu)$ and $n_0 \in \mathbb{N}$ such that (9.9.6)–(9.9.8) hold. Then $\lim_{n\to\infty} f_n^* = f^*$.*

Now we are in a position to prove the main result of this section.

Theorem 9.9.5. *Let $0 < p, q \le \infty$, $\mathbb{A}, \mathbb{B} \in \mathbb{R}^2$ and let X be one of the spaces $L_{p,q;\mathbb{A},\mathbb{B}}$ or $L_{(p,q;\mathbb{A},\mathbb{B})}$. Assume that $X \neq \{0\}$. Then X has absolutely continuous (quasi) norm if and only if $0 < q < \infty$.*

Proof. (i) Assume first that $X = L_{p,q;\mathbb{A},\mathbb{B}}$. Let $0 < q < \infty$. It is a consequence of (P3) (cf. Section 2) that X has absolutely continuous (quasi) norm if (ACN) holds for every $f \in X \cap \mathcal{M}_+(\mathcal{R}, \mu)$. For such f we clearly have $f^*(t) < \infty$, $t \in (0, \infty)$, and $\lim_{t \to \infty} f^*(t) = 0$. Let $\{E_n\} \subset \mathcal{R}$ satisfy $E_n \searrow \emptyset$ μ-a.e. Setting $f_n = f \chi_{E_n}$, we have $0 \le f_n \le f$, $n \in \mathbb{N}$, and, by Lemma 9.9.4, $f_n^*(t) \to 0$ as $n \to \infty$ for all $t \in (0, \infty)$. Since $f_n^* \le f^*$, $n \in \mathbb{N}$, the Lebesgue dominated convergence theorem shows that

$$\|f\chi_{E_n}\|_{p,q;\mathbb{A},\mathbb{B}} = \left(\int_0^\infty [t^{\frac{1}{p}-\frac{1}{q}} \ell^{\mathbb{A}}(t) \ell\ell^{\mathbb{B}}(t) f_n^*(t)]^q \, dt \right)^{\frac{1}{q}} \to 0.$$

Hence X has absolutely continuous (quasi) norm.

Let $q = \infty$. Since $X \neq \{0\}$, either $p < \infty$, or $p = \infty$, $\alpha_0 < 0$, or $p = \infty$, $\alpha_0 = 0$, $\beta_0 \le 0$. Thus, there exists $n_0 \in \mathbb{N}$ such that the function $h(t) = t^{-\frac{1}{p}} \ell^{-\alpha_0}(t) \ell\ell^{-\beta_0}(t) \chi_{(0,\frac{1}{n_0})}(t)$, $t \in (0, \infty)$, is nonincreasing on $(0, \infty)$. By Theorem 9.1.4, there is a $g \in \mathcal{M}(\mathcal{R}, \mu)$ satisfying $g^* = h$. Furthermore, by Theorem 9.1.3, there is a measure-preserving transformation σ from the set $G = \operatorname{supp} g$ onto $[0, \frac{1}{n_0}] = \operatorname{supp} h$ such that $g = h \circ \sigma$ μ–a.e. on G. Set $h_n = h \chi_{(0,\frac{1}{n})}$ and $g_n = h_n \circ \sigma$ for $n \in \mathbb{N}$, $n \ge n_0$. Then (cf. Proposition 9.1.2) $g_n^* = h_n$. It is clear from the definition of g_n that the sequence $\{E_n\}$ of μ-measurable subsets of $\mathcal{R}$ given by $E_n := \operatorname{supp} g_n$, $n \ge n_0$, satisfies $E_{n+1} \subset E_n \subset G$ and $g_n = g \chi_{E_n}$ μ–a.e. on G. Since $\mu(E_n) = |(0, \frac{1}{n})| = \frac{1}{n}$ (where $|E|$ denotes the usual Lebesgue measure of a set E), we see that $E_n \searrow \emptyset$ as $n \to \infty$. Moreover,

$$\|g\chi_{E_n}\|_X = \|g_n\|_{p,\infty;\mathbb{A},\mathbb{B}} = \sup_{0<t<\frac{1}{n}} t^{\frac{1}{p}} \ell^{\mathbb{A}}(t) \ell\ell^{\mathbb{B}}(t) h(t) = \sup_{0<t<\frac{1}{n}} 1 = 1$$

for every $n \ge n_0$. Since $g \in X$, the space X does not have absolutely continuous (quasi) norm.

(ii) Now let $X = L_{(p,q;\mathbb{A},\mathbb{B})}$. Since $X \neq \{0\}$, one of the conditions in (9.3.2) is satisfied. In particular, $1 \le p \le \infty$. By Theorem 9.5.1 (i), $X = L_{p,q;\mathbb{A},\mathbb{B}}$ if $1 < p \le \infty$. This case is covered by part (i). It remains to consider the case when $p = 1$.

Let $0 < q < \infty$ and $f \in X = L_{(1,q;\mathbb{A},\mathbb{B})} \cap \mathcal{M}_+(\mathcal{R}, \mu)$. For such f we clearly have $f^{**}(t) < \infty$, $t \in (0, \infty)$, and $\lim_{t \to \infty} f^{**}(t) = 0$, which in turn implies $f^*(t) < \infty$ for $t \in (0, \infty)$, and $\lim_{t \to \infty} f^*(t) = 0$. Let $\{E_n\} \subset \mathcal{R}$ satisfy $E_n \searrow \emptyset$ μ-a.e. Define f_n as in part (i). Since $f_n^* \le f^*$, $n \in \mathbb{N}$, and

Section 9.9 Absolute continuity of norm

$f_n^*(t) \to 0$ as $n \to \infty$ for all $t \in (0, \infty)$, the Lebesgue dominated convergence theorem implies that for every $s \in (0, \infty)$,

$$f_n^{**}(s) = s^{-1} \int_0^s f_n^*(t)\, dt \to 0.$$

Since moreover $f_n^{**} \leq f^{**}$, $n \in \mathbb{N}$, one more application of the Lebesgue dominated convergence theorem shows that

$$\|f\chi_{E_n}\|_{(1,q;\mathbb{A},\mathbb{B})} = \left(\int_0^\infty [s^{1-\frac{1}{q}} \ell^{\mathbb{A}}(s) \ell\ell^{\mathbb{B}}(s) f_n^{**}(s)]^q\, ds \right)^{\frac{1}{q}} \to 0 \text{ as } n \to \infty.$$

Finally, let $p = 1$ and $q = \infty$. Assume first that

$$\text{either} \qquad \alpha_0 > 0 \qquad \qquad (9.9.10)$$
$$\text{or} \qquad \alpha_0 = 0 \text{ and } \beta_0 > 0. \qquad (9.9.11)$$

If (9.9.10) holds, then there is a $n_0 \in \mathbb{N}$ such that the function

$$h(t) = t^{-1}\ell^{-\alpha_0 - 1}(t)\ell\ell^{-\beta_0}(t)\chi_{(0,\frac{1}{n_0})}(t), \qquad t \in (0, \infty), \qquad (9.9.12)$$

is nonincreasing on $(0, \infty)$. The same argument as the one used in part (i) above implies that there is a $g \in \mathcal{M}(\mathcal{R}, \mu)$ and a sequence $\{E_n\} \subset \mathcal{R}$ with $E_n \searrow \emptyset$ as $n \to \infty$ such that the functions $g_n := g\chi_{E_n}$ satisfy $g_n^* = h_n := h\chi_{(0,\frac{1}{n})}$, $n \geq n_0$. Hence, for all $n \geq n_0$ and every $t \in (0, \frac{1}{n})$,

$$g_n^{**}(t) = \frac{1}{t}\int_0^t h(s)\, ds \approx \frac{1}{t}\ell^{-\alpha_0}(t)\ell\ell^{-\beta_0}(t).$$

Consequently,

$$\|g\chi_{E_n}\|_X = \|g_n\|_{(1,\infty;\mathbb{A},\mathbb{B})} \geq \sup_{0<t<\frac{1}{n}} t\ell^{\mathbb{A}}(t)\ell\ell^{\mathbb{B}}(t)g_n^{**}(t) \approx 1$$

for every $n \geq n_0$. Since $g \in X$, the space X does not have absolutely continuous (quasi) norm.

When (9.9.11) is satisfied, we use the same argument as above with (9.9.12) replaced by

$$h(t) = t^{-1}\ell^{-1}(t)\ell\ell^{-\beta_0 - 1}(t)\chi_{(0,\frac{1}{n_0})}(t), \qquad t \in (0, \infty),$$

and the result follows again.

Assume now that either $\alpha_0 < 0$ or $\alpha_0 = 0$ and $\beta_0 \leq 0$. Then, by Corollary 9.5.5, $X = L_{(1,\infty;\mathbb{A},\mathbb{B})} = L_{(1,\infty;(0,\alpha_\infty),(0,\beta_\infty))}$. The same argument as above with (9.9.12) replaced by $h = \chi_{(0,1)}$ shows that $L_{(1,\infty;(0,\alpha_\infty),(0,\beta_\infty))}$ ($= X$) does not have absolutely continuous (quasi) norm. □

9.10 Lorentz–Zygmund spaces

The GLZ spaces studied in this chapter are a direct generalization of the Lorentz–Zygmund spaces. For the sake of completeness, we shall briefly recall their definition and basic properties.

In this section we will assume throughout that (R, μ) is a resonant measure space and that $\mu(\mathcal{R}) = 1$. All the results can be easily generalized to the case $\mu(\mathcal{R}) < \infty$.

Definition 9.10.1. Let $0 < p, q \le \infty$ and $\alpha \in \mathbb{R}$. The *Lorentz–Zygmund space* $L_{p,q;\alpha} = L_{p,q;\alpha}(\mathcal{R}, \mu)$ is the collection of all $f \in \mathcal{M}(\mathcal{R}, \mu)$ such that $\|f\|_{L_{p,q;\alpha}} < \infty$, where

$$\|f\|_{L_{p,q;\alpha}} = \|t^{\frac{1}{p}-\frac{1}{q}} \log\left(\frac{e}{t}\right)^{\alpha} f^*(t)\|_{L^q(0,1)}. \tag{9.10.1}$$

Remarks 9.10.2. (i) The space $L_{p,q,\alpha}$ is nontrivial (that is, not equal to $\{0\}$) if and only if one of the following conditions holds:

$$\begin{cases} 0 < p < \infty, \\ p = \infty, \ \alpha + \frac{1}{q} < 0, \\ p = q = \infty, \ \alpha = 0. \end{cases} \tag{9.10.2}$$

(ii) The following relations between function spaces hold:

$$L_{p,q;0} = L^{p,q}, \tag{9.10.3}$$
$$L_{p,p;\alpha} = L^p (\log L)^{\alpha p}, \quad 0 < p < \infty, \quad \alpha \in \mathbb{R}, \tag{9.10.4}$$
$$L_{\infty,\infty;-a} = \exp L^{\frac{1}{a}}, \quad a > 0. \tag{9.10.5}$$

(iii) The functional $\|\cdot\|_{L_{p,q;\alpha}}$ is not always a norm. For example, when $p < 1$ and/or $q < 1$, then the triangle inequality is not satisfied.

(iv) Let $0 < p_i, q_i \le \infty$, $i = 0, 1$, and let $\alpha, \beta \in \mathbb{R}$. Assume that $p_0 < p_1$. Then the embedding $L_{p_1,q_1;\alpha} \hookrightarrow L_{p_0,q_0;\beta}$ holds. When $p_0 = p_1$, a finer interplay between the remaining parameters comes into picture. This is a subject of the next theorem [199].

Theorem 9.10.3 (Sharpley). Let $0 < p, q, r \le \infty$, and let $\alpha, \beta \in \mathbb{R}$. Assume that $L_{p,q;\alpha} \ne \{0\}$. Then the embedding $L_{p,q;\alpha} \hookrightarrow L_{p,r;\beta}$ holds if and only if one of the following conditions holds:

$$0 < q \le r \le \infty, \quad p = \infty, \quad \alpha + \frac{1}{q} \ge \beta + \frac{1}{r}, \tag{9.10.6}$$
$$0 < q \le r \le \infty, \quad p < \infty, \quad \alpha \ge \beta, \tag{9.10.7}$$
$$0 < r < q \le \infty, \quad \alpha + \frac{1}{q} > \beta + \frac{1}{r}. \tag{9.10.8}$$

The following result is a special case of a more general theorem from [172].

Theorem 9.10.4. *Let $0 < p, q \leq \infty$ and $\alpha \in \mathbb{R}$. Then the space $X = L_{p,q;\alpha}$ is equivalent to a Banach function space if and only if one of the following conditions holds:*

$$\begin{cases} 1 < p < \infty, \ 1 \leq q \leq \infty, \\ p = 1, \ q = 1, \ \alpha \geq 0, \\ p = \infty, \ 1 \leq q \leq \infty, \ \alpha + \frac{1}{q} < 0, \\ p = \infty, \ q = \infty, \ \alpha = 0. \end{cases} \quad (9.10.9)$$

In such cases, the fundamental function of X satisfies

$$\varphi_X(t) \approx \begin{cases} t^{\frac{1}{p}} \log\left(\frac{e}{t}\right)^{\alpha} & \text{when} \quad p < \infty, \\ \log\left(\frac{e}{t}\right)^{\alpha + \frac{1}{q}} & \text{when} \quad p = \infty, \ \alpha + \frac{1}{q} < 0, \\ 1 \text{ for } t \in (0,1), \varphi_X(0) = 0 & \text{when} \quad p = \infty, \ q = \infty, \ \alpha = 0. \end{cases}$$

Furthermore, we have

$$X' = L_{p',q';-\alpha}.$$

9.11 Lorentz–Karamata spaces

Let us at the end of the chapter just briefly mention the so-called Lorentz–Karamata spaces, introduced in [69] in connection with some specific tasks concerning Sobolev inequalities. These spaces are not covered by the GLZ spaces but they constitute their further natural generalization.

We assume here that $\mu(\mathcal{R}) = 1$.

Definition 9.11.1. A positive function b is said to be *slowly varying* (s.v.) on $(1, \infty)$, in the sense of Karamata, if for each $\varepsilon > 0$, $t^{\varepsilon} b(t)$ is equivalent to an increasing function and $t^{-\varepsilon} b(t)$ is equivalent to a decreasing function.

Examples 9.11.2. The following functions are slowly varying on $(1, \infty)$:
$b(t) = (e + \log t)^{\alpha} (\log(e + \log t))^{\beta}$, $\alpha, \beta \in \mathbb{R}$,
$b(t) = \exp(\sqrt{\log t})$.

Remarks 9.11.3. Suppose that b is slowly varying on $(1, \infty)$. Then

(i) b^r is slowly varying on $(1, \infty)$ for all $r \in \mathbb{R}$;

(ii) $\int_{t^{-1}}^{1} s^{-1} b(s^{-1}) \, ds$ is slowly varying on $(1, \infty)$ and (see [247, Chapter 2, p. 186])

$$\lim_{t \to \infty} \frac{b(t)}{\int_{t^{-1}}^{1} s^{-1} b(s^{-1}) \, ds} = 0; \quad (9.11.1)$$

(iii) $\lim_{t\to\infty} \frac{b(ct)}{b(t)} = 1$ for all $c > 0$.

Definition 9.11.4. Let $p, q \in (0, \infty]$ and let b be a slowly varying function on $(1, \infty)$. Assume that $\|t^{-\frac{1}{q}} b(t^{-1})\|_{L^q(0,1)} < \infty$ when $p = \infty$. Then the *Lorentz–Karamata* (L–K) *space* $L_{p,q;b} = L_{p,q;b}(\mathcal{R}, \mu)$ is the collection of all $f \in \mathcal{M}(\mathcal{R}, \mu)$ such that $\|f\|_{L_{p,q;b}} < \infty$, where

$$\|f\|_{L_{p,q;b}} = \|t^{\frac{1}{p}-\frac{1}{q}} b(\tfrac{1}{t}) f^*(t)\|_{L^q(0,1)}. \tag{9.11.2}$$

Remark 9.11.5. (i) The functional $\|f\|_{L_{p,q;b}}$ is a norm if $p > q$ or if $p = q$ and b is nondecreasing on $(1, \infty)$. Moreover, it is equivalent to the r.i. norm

$$\|f\|_{L_{(p,q;b)}} = \|t^{\frac{1}{p}-\frac{1}{q}} b(\tfrac{1}{t}) f^{**}(t)\|_{L^q(0,1)}$$

if and only if $p > 1$.

Chapter 10
Classical Lorentz spaces

The Lorentz spaces studied in Chapter 8 can be viewed also as special instances of the so-called *classical Lorentz spaces*, of types Λ and Γ. In fact, the Lorentz spaces were first introduced by Lorentz in [135] as "spaces Λ." The spaces of type Γ were introduced later by Sawyer in [194], in connection with certain duality questions concerning the spaces of type Λ. Both types of spaces have been intensively studied ever since. Recently, a third type, called *type S*, was added, in connection with the study of spaces of functions whose mean oscillation is controlled in a certain way. We shall systematically study all three types of classical Lorentz spaces, their basic properties and their mutual relations.

Since 1990, an enormous effort has been spent by many authors in the hunt of a characterization of embeddings between classical Lorentz spaces of both types Λ and Γ. Thanks to this extensive research powerful new techniques and methods were developed, and all of the desired characterizations have been obtained. (This was subsequently done in [5, 32, 35, 36, 37, 38, 89, 93, 95, 194, 200, 201, 202, 203, 216] and many more. The last missing case was characterized in [34].)

There is a vast literature on the subject available; the results presented here are taken mostly from [33, 34, 35, 56]. In particular, in Section 10.2 we study the question when a classical Lorentz space can justifiably be called "a space" – here we follow the explanation of [56].

10.1 Definition and basic properties

Convention 10.1.1. In this chapter, we shall use the following notation. If v and w are weights on $(0, \mu(\mathcal{R}))$, then we shall denote $V(t) = \int_0^t v(s)\,ds$ and $W(t) = \int_0^t w(s)\,ds$ for $t \in (0, \mu(\mathcal{R}))$. By $A \lesssim B$ we mean that $A \leq CB$ with some positive C independent of appropriate quantities. If $A \lesssim B$ and $B \lesssim A$, we write $A \approx B$. We shall use the symbol $f \downarrow$ to denote that f is nonincreasing on $(0, \infty)$. Embedding constants and other important quantities are denoted as $A_{(10.3.1)}$, $A_{(10.3.2)}, \ldots$, where the subscript indicates the label of the formula where they are introduced.

We also consider only the case when $(\mathcal{R}, \mu)$ is nonatomic unless stated otherwise. Also, Convention 7.1.8 is still in power.

Definition 10.1.2. Let $0 < p \leq \infty$ and let w be a weight on $(\mathcal{R}, \mu)$. The *classical Lorentz space* $\Lambda^p(w) = \Lambda^p(w)(\mathcal{R}, \mu)$ is the collection of all functions $f \in$

$\mathcal{M}(\mathcal{R}, \mu)$ such that $\|f\|_{\Lambda^p(w)} < \infty$, where

$$\|f\|_{\Lambda^p(w)} := \begin{cases} \left(\int_0^{\mu(\mathcal{R})} (f^*(t))^p w(t)\, dt\right)^{\frac{1}{p}} & \text{when } p < \infty, \\ \sup_{0 < t < \mu(\mathcal{R})} f^*(t) w(t) & \text{when } p = \infty. \end{cases} \quad (10.1.1)$$

The space $\Gamma^p(w) = \Gamma^p(w)(\mathcal{R}, \mu)$ is the collection of all functions $f \in \mathcal{M}(\mathcal{R}, \mu)$ such that $\|f\|_{\Gamma^p(w)} < \infty$, where

$$\|f\|_{\Gamma^p(w)} := \begin{cases} \left(\int_0^{\mu(\mathcal{R})} (f^{**}(t))^p w(t)\, dt\right)^{\frac{1}{p}} & \text{when } p < \infty, \\ \sup_{0 < t < \mu(\mathcal{R})} f^{**}(t) w(t) & \text{when } p = \infty. \end{cases} \quad (10.1.2)$$

Notation 10.1.3. It turns out that for some specific tasks such as characterization of embeddings between function spaces, it is useful to introduce the following notation. Let $p \in (0, \infty)$ and let w be a weight on $[0, \mu(\mathcal{R}))$. Suppose that $W(t) < \infty$ for all $t \in (0, \mu(\mathcal{R}))$. We then denote by $\Lambda^{p,\infty}(w) = \Lambda^{p,\infty}(w)(\mathcal{R}, \mu)$ the collection of all functions $f \in \mathcal{M}(\mathcal{R}, \mu)$ such that $\|f\|_{\Lambda^{p,\infty}(w)} < \infty$, where

$$\|f\|_{\Lambda^{p,\infty}(w)} := \sup_{0 < t < \mu(\mathcal{R})} f^*(t) W^{\frac{1}{p}}(t).$$

Similarly, we denote by $\Gamma^{p,\infty}(w) = \Gamma^{p,\infty}(w)(\mathcal{R}, \mu)$ the collection of all functions $f \in \mathcal{M}(\mathcal{R}, \mu)$ such that $\|f\|_{\Gamma^{p,\infty}(w)} < \infty$, where

$$\|f\|_{\Gamma^{p,\infty}(w)} := \sup_{0 < t < \mu(\mathcal{R})} f^{**}(t) W^{\frac{1}{p}}(t).$$

Remark 10.1.4. The main reason for introduction of the spaces $\Lambda^{p,\infty}(w)$ and $\Gamma^{p,\infty}(w)$ is that they play a role of a certain natural "weak" modification of the corresponding spaces $\Lambda^p(w)$ and $\Gamma^p(w)$, respectively. In particular, we immediately see that

$$\Lambda^p(w) \hookrightarrow \Lambda^{p,\infty}(w) \quad \text{and} \quad \Gamma^p(w) \hookrightarrow \Gamma^{p,\infty}(w). \quad (10.1.3)$$

Remarks 10.1.5. (i) Since $f^* \leq f^{**}$, we immediately have, for every $p \in (0, \infty)$ and every weight w on $[0, \mu(\mathcal{R}))$,

$$\Gamma^p(w) \hookrightarrow \Lambda^p(w) \quad \text{and} \quad \Gamma^{p,\infty}(w) \hookrightarrow \Lambda^{p,\infty}(w). \quad (10.1.4)$$

(ii) When w is quasi-concave on $[0, \mu(\mathcal{R}))$, then we have $\Gamma^\infty(w) = M_w$, where M_w is the Marcinkiewicz space introduced in Definition 7.10.1.

We will need some special classes of weights.

Section 10.1 Definition and basic properties

Definition 10.1.6. We say that a weight v on $(0, \mu(\mathcal{R}))$ belongs to the *class B_p* if

$$\int_t^{\mu(\mathcal{R})} \frac{v(s)}{s^p}\, ds \lesssim \frac{1}{t^p} \int_0^t v(s)\, ds, \quad t \in (0, \mu(\mathcal{R})). \tag{10.1.5}$$

We say that a weight v on $(0, \mu(\mathcal{R}))$ belongs to the *reverse class B_p*, written $w \in RB_p$, if

$$\frac{1}{t^p} \int_0^t v(s)\, ds \lesssim \int_t^{\mu(\mathcal{R})} \frac{v(s)}{s^p}\, ds, \quad t \in (0, \mu(\mathcal{R})).$$

We say that a weight v on $(0, \mu(\mathcal{R}))$ belongs to the *class $B_{1,\infty}$* if

$$\frac{1}{t} \int_0^t w(y)\, dy \le \frac{C}{s} \int_0^s w(y)\, dy, \quad 0 < s \le t < \mu(\mathcal{R}). \tag{10.1.6}$$

Remarks 10.1.7. The weak Lorentz spaces were introduced in [37] and further investigated in [36, 38] and [32].

Lorentz [135] proved that, for $p \ge 1$, $\|f\|_{\Lambda^p(w)}$ is a norm if and only if v is nonincreasing. The class of weights for which $\|f\|_{\Lambda^p(w)}$ is merely *equivalent* to a norm, that is, $\Lambda^p(w)$ is equivalent to a Banach space, is however considerably wider. For $p \in (1, \infty)$, $\Lambda^p(w)$ is equivalent to a Banach space if and only if $w \in B_p$. On the other hand, $\Lambda^1(w)$ is equivalent to a Banach space if and only if $w \in B_{1,\infty}$ [32, Theorem 2.3]. It is worth noticing that it is an essentially weaker requirement on w than $w \in B_1$. In this sense, (10.1.6) is not the limiting case of (10.1.5) when $p \to 1+$. On the other hand, Ariño and Muckenhoupt [5] established that, for $p \in (0, \infty)$, $w \in B_p$ if and only if $\Lambda^p(w) = \Gamma^p(w)$. Sawyer [194, Theorem 4] then showed that $w \in B_p$ if and only if

$$\left(\int_0^t w(s)\, ds \right)^{\frac{1}{p}} \left(\int_0^t \left(\frac{1}{s} \int_0^s w(y)\, dy \right)^{1-p'} ds \right)^{\frac{1}{p'}} \le Ct$$

for some $C > 0$ and all $t \in (0, \mu(\mathcal{R}))$. On integration by parts one can prove that this is equivalent to

$$\left(\int_0^t \left(\frac{1}{s} \int_0^s w(y)\, dy \right)^{-p'} w(s)\, ds \right)^{\frac{1}{p'}} \left(\int_0^t w(s)\, ds \right)^{\frac{1}{p}} \le Ct \tag{10.1.7}$$

for some $C > 0$ and all $t \in (0, \mu(\mathcal{R}))$. But now, observe that (10.1.6) *is* the limiting case of (10.1.7) when $p \to 1+$ (even though not of (10.1.5)).

Definition 10.1.8. Let X be a linear set of functions from $\mathcal{M}(\mathcal{R}, \mu)$ containing characteristic functions of sets of finite measure, endowed with a positively homogeneous functional $\|\cdot\|_X$, defined for every $f \in \mathcal{M}(\mathcal{R}, \mu)$ and such that $f \in X$ if and only if $\|f\|_X < \infty$. If $\|f\|_X = \|g\|_X$ whenever $f^* = g^*$, and moreover $0 \le f \le g$ implies $\|f\|_X \le \|g\|_X$, then we say that X is a *rearrangement-invariant* (r.i.) *lattice*. We further define

$$X' := \left\{ f \in \mathcal{M}(\mathcal{R}, \mu); \int_{\mathcal{R}} |f(x)g(x)| \, d\mu(x) < \infty \text{ for every } g \in X \right\},$$

and

$$\|f\|_{X'} := \sup \left\{ \int_{\mathcal{R}} |f(x)g(x)| \, d\mu(x); \|g\|_X \le 1 \right\}.$$

If, in particular, X is a Banach function space, then X' is its associate space.

Definition 10.1.9. Let X be an r.i. lattice.

(i) For each finite $t \in [0, \mu(\mathcal{R})]$ let E be any subset of $\mathcal{R}$ such that $\mu(E) = t$, and $\varphi_X(t) = \|\chi_E\|_X$. Similarly as in the case of Banach function spaces, the function φ_X so defined is called the *fundamental function* of X. Since μ is nonatomic, for every finite $t \in (0, \mu(\mathcal{R})]$ there is a μ–measurable subset E of $\mathcal{R}$ such that $\mu(E) = t$, and therefore $(\chi_E)^* = \chi_{(0,t)}$.

(ii) Set

$$\psi_X(t) := \sup \left\{ \int_0^t f^*(s) \, ds; \|f\|_X \le 1 \right\}, \quad t \in (0, \mu(\mathcal{R})), \tag{10.1.8}$$

and

$$\varrho_X(t) := \sup_{f \ne 0} \frac{f^*(t)}{\|f\|_X}, \quad t \in (0, \mu(\mathcal{R})). \tag{10.1.9}$$

Remarks 10.1.10. (i) If X is a Banach function space, then $\psi_X = \varphi_{X'}$ and X' is its associate space.

(ii) For any r.i. lattice X and every $t \in (0, \mu(\mathcal{R}))$,

$$\varrho_X(t) = \frac{1}{\varphi_X(t)}.$$

Indeed, let $t > 0$ and let $E \subset \mathcal{R}$ be such that $\mu(E) = t$. Testing the supremum in the definition of ϱ_X on $f = \chi_E$, we get $\varrho_X(t) \ge \frac{1}{\varphi_X(t)}$. Conversely, let $E_t := \{x \in \mathcal{R}; |f(x)| \ge f^*(t)\}$. Then $\mu(E_t) \ge t$, and therefore

$$\|f\|_X \ge \|f\chi_{E_t}\|_X \ge f^*(t)\|\chi_{E_t}\|_X \ge f^*(t)\varphi_X(t).$$

Hence, $\varrho_X(t) \le \frac{1}{\varphi_X(t)}$.

For "bad enough" weight v, the classical or weak Lorentz spaces may be trivial, that is, equal to $\{0\}$. If such a space is nontrivial, then it is an r.i. lattice. Our next lemma gives necessary and sufficient conditions for classical and weak Lorentz spaces to be nontrivial and formulas for fundamental functions in these cases.

Lemma 10.1.11. *Let $p \in (0, \infty)$ and let w be a weight on $(0, \mu(\mathcal{R}))$.*

(i) *The spaces $\Lambda^p(w)$ and $\Lambda^{p,\infty}(w)$ are nontrivial if and only if w is integrable near zero. In such cases,*

$$\varphi_{\Lambda^p(w)}(t) = \varphi_{\Lambda^{p,\infty}(w)}(t) = W^{\frac{1}{p}}(t), \quad t \in (0, \mu(\mathcal{R})). \tag{10.1.10}$$

(ii) *The space $\Gamma^p(w)$ is nontrivial if and only if w is integrable near zero and $\int_t^\infty s^{-p} w(s)\, ds < \infty$ for every $t > 0$. In such cases,*

$$\varphi_{\Gamma^p(w)}(t) = \left(W(t) + t^p \int_t^\infty \frac{w(s)}{s^p}\, ds \right)^{\frac{1}{p}}, \quad t \in (0, \mu(\mathcal{R})). \tag{10.1.11}$$

(iii) *The space $\Gamma^{p,\infty}(w)$ is nontrivial if and only if w is integrable on every interval $(0, a)$, $a \in (0, \mu(\mathcal{R}))$, and $\limsup_{t \to \mu(\mathcal{R})} t^{-1} W^{\frac{1}{p}}(t) < \infty$ (of course, this last condition is relevant only if $\mu(\mathcal{R}) = \infty$). In such cases,*

$$\varphi_{\Gamma^{p,\infty}(w)}(t) = t \sup_{s \geq t} \frac{W^{\frac{1}{p}}(s)}{s}, \quad t \in (0, \mu(\mathcal{R})). \tag{10.1.12}$$

Proof. All the assertions follow easily from definitions and the relations in (10.1.4) and (10.1.3). □

Example 10.1.12. If $p \in (0, \infty)$ and $w(t) = t^{p-1}$, then $\Gamma^p(w)$ is trivial but $\Gamma^{p,\infty}(w)$, $\Lambda^p(w)$, and $\Lambda^{p,\infty}(w)$ are nontrivial. More precisely,

$$\Gamma^p(w) = \{0\}, \quad \Gamma^{p,\infty}(w) = L^1, \quad \Lambda^p(w) = L^{1,p} \quad \text{and} \quad \Lambda^{p,\infty}(w) = L^{1,\infty}.$$

This example also shows that there is no general inclusion between the spaces $\Lambda^p(w)$ and $\Gamma^{p,\infty}(w)$ (here $\Lambda^p(w) \hookrightarrow \Gamma^{p,\infty}(w)$ when $0 < p \leq 1$, but $\Gamma^{p,\infty}(w) \hookrightarrow \Lambda^p(w)$ when $1 < p < \infty$, and both these embeddings are proper in general).

Remarks 10.1.13. Let w be a weight on $(0, \mu(\mathcal{R}))$. Then, by the Fubini theorem,

$$\Gamma^1(w) = \Lambda^1(\tilde{w}), \quad \text{where } \tilde{w}(t) := \int_t^{\mu(\mathcal{R})} s^{-1} w(s)\, ds, \quad t \in (0, \mu(\mathcal{R})). \tag{10.1.13}$$

Similarly, for $p \in (0, \infty)$,

$$\Lambda^\infty(w) = \Lambda^\infty(\tilde{w}), \quad \text{where } \tilde{w}(t) := \sup_{0 < s < t} w(s), \quad t \in (0, \mu(\mathcal{R})). \tag{10.1.14}$$

Indeed, we have

$$\sup_{0 < t < \mu(\mathcal{R})} \tilde{w}(t) f^*(t) = \sup_{0 < t < \mu(\mathcal{R})} \sup_{0 < s < t} w(s) f^*(s)$$

$$= \sup_{0 < s < \mu(\mathcal{R})} w(s) \sup_{s < t < \mu(\mathcal{R})} f^*(t) = \sup_{0 < t < \mu(\mathcal{R})} w(t) f^*(t).$$

Therefore, we can always assume with no loss of generality that in the space $\Lambda^\infty(w)$, the weight w is nondecreasing.

Now, (10.1.14) can be for $p \in (0, \infty)$ rewritten as

$$\Lambda^{p,\infty}(w) = \Lambda^{1,\infty}(\tilde{w}).$$

For the spaces of the type Γ, we similarly have

$$\Gamma^{p,\infty}(w) = \Gamma^{1,\infty}(\tilde{w}), \quad \text{where } \tilde{w}(t) = \tfrac{1}{p} W(t)^{\frac{1}{p}-1} w(t), \quad t \in (0, \mu(\mathcal{R})).$$

10.2 Functional properties of classical Lorentz spaces

We already know that the functionals $\|f\|_{\Lambda^p(w)}, \|f\|_{\Gamma^p(w)}$ do not necessarily satisfy all the requirements of a norm for every p and w, not even when $p \geq 1$. In particular, they may be trivial (that is, finite only when f is the function which is identically equal to zero) or they may not satisfy the triangle inequality. In fact, it turns out that the functional $\|f\|_{\Lambda^p(w)}$ may not even be a quasinorm [37], and the "space" $\|f\|_{\Lambda^p(w)}$ does not have to be necessarily a linear set [56]. We shall now pursue some of these facts in detail. We follow closely the expository of [56].

Theorem 10.2.1. *The functional $\|\cdot\|_{\Lambda^p(w)}$ satisfies the triangle inequality if and only if w is equivalent to a nonincreasing function.*

Proof. Assume that $\|\cdot\|_{\Lambda^p(w)}$ satisfies the triangle inequality. Let a and h be positive real numbers satisfying $a + 2h \leq \mu(\mathcal{R})$ and let δ be a positive real number. We then define the functions

$$f(x) = \begin{cases} 1 + \delta & \text{if } x \in (0, a + h) \\ 1 & \text{if } x \in [a + h, a + 2h) \\ 0 & \text{if } x \in [a + 2h, \mu(\mathcal{R})) \end{cases}$$

and

$$g(x) = \begin{cases} 1 & \text{if } x \in (0, h) \\ 1 + \delta & \text{if } x \in [h, a + 2h) \\ 0 & \text{if } x \in [a + 2h, \mu(\mathcal{R})). \end{cases}$$

Then

$$(f + g)^*(t) = \begin{cases} 2 + 2\delta & \text{if } x \in (0, a) \\ 2 + \delta & \text{if } x \in [a, a + 2h) \\ 0 & \text{if } x \in [a + 2h, \mu(\mathcal{R})). \end{cases}$$

Moreover, $\|f\|_{\Lambda^p(w)} = \|g\|_{\Lambda^p(w)}$, hence the triangle inequality

$$\|f + g\|_{\Lambda^p(w)} \leq \|f\|_{\Lambda^p(w)} + \|g\|_{\Lambda^p(w)}$$

Section 10.2 Functional properties

implies

$$\left((2+2\delta)^p \int_0^a w(t)\,dt + (2+\delta)^p \int_a^{a+2h} w(t)\,dt\right)^{\frac{1}{p}}$$
$$\leq 2\left((1+\delta)^p \int_0^{a+h} w(t)\,dt + \int_{a+h}^{a+2h} w(t)\,dt\right)^{\frac{1}{p}},$$

that is,

$$(2+\delta)^p \int_a^{a+2h} w(t)\,dt \leq (2+2\delta)^p \int_a^{a+h} w(t)\,dt + 2^p \int_{a+h}^{a+2h} w(t)\,dt.$$

This can be rewritten as

$$\frac{(1+\delta)^p - (1+\frac{\delta}{2})^p}{(1+\frac{\delta}{2})^p - 1} \int_a^{a+h} w(t)\,dt \geq \int_{a+h}^{a+2h} w(t)\,dt.$$

We now define $W(t) := \int_0^t w(s)\,ds$, $t \in (0, \mu(\mathcal{R}))$, and let $\delta \to 0_+$. We obtain

$$W(a+h) \geq \frac{1}{2}(W(a) + W(a+2h)).$$

This shows that W is concave on $(0, \mu(\mathcal{R}))$, which entails that w is equivalent to a nonincreasing function.

Conversely, assume now that w is nonincreasing on $(0, \mu(\mathcal{R}))$. Then

$$\|f\|_{\Lambda^p(w)} = \sup_{\tilde{w} \sim w} \left(\int_0^{\mu(\mathcal{R})} f^*(t)^p \tilde{w}(t)\,dt\right)^{\frac{1}{p}},$$

where the supremum is taken over all functions $\tilde{w}$ equimeasurable with w. Thus, $f, g \in \Lambda^p(w)$ implies $f + g \in \Lambda^p(w)$ and the desired triangle inequality. $\square$

We shall now turn to the question when a classical Lorentz space is a linear set. We first formulate an auxiliary assertion.

Lemma 10.2.2. *Suppose that w and v are nonnegative measurable functions on $(0, \mu(\mathcal{R}))$ and a is a positive number such that $w(t) = 0$ for a.e. $t \in (0, a)$ and $w(t) = v(t)$ for all $t \geq a$, and the functions $V(t) = \int_0^t v(x)\,dx$ and $W(t) = \int_0^t w(x)\,dx$ are finite for all $t > 0$.*

Let $f : \mathcal{R} \to \mathcal{R}$ be a μ-measurable function. Then the following are equivalent:

(i) $f \in \Lambda^p(w)$;

(ii) $\min\{\lambda, |f|\} \in \Lambda^p(v)$ *for some positive number* λ;

(iii) $\min\{\lambda, |f|\} \in \Lambda^p(v)$ *for every positive number* λ.

Proof. Clearly we can assume without loss of generality that f is nonnegative.

It follows easily from the definition of nonincreasing rearrangements and distribution functions that, for each constant $\lambda > 0$, the nonincreasing rearrangement of the function $\min\{f, \lambda\}$ satisfies

$$(\min\{f, \lambda\})^*(t) = \min\{f^*(t), \lambda\} \quad \text{for each } t \in (0, \mu(\mathcal{R})). \tag{10.2.1}$$

Now suppose that $f \in \Lambda^p(w)$. If $f^*(a) > 0$ we choose $\lambda = f^*(a)$. If $f^*(a) = 0$ then we can choose λ however we wish, for example $\lambda = 1$. Note that in this latter case we have $f^*(t) = 0$ for all $t \in [a, \infty)$. In both cases we have

$$\int_0^\infty \left(\min\{f^*(t), \lambda\}\right)^p v(t)\,dt \le \int_0^a \lambda^p v(t)\,dt + \int_a^\infty (f^*(t))^p w(t)\,dt$$
$$\le \lambda^p V(a) + \|f\|_{\Lambda^p(w)}^p < \infty,$$

where $V(t) := \int_0^t v(s)\,ds$, $t \in (0, \mu(\mathcal{R}))$. This shows that (i) implies (ii).

Now suppose that $\min\{f, \lambda_0\} \in \Lambda^p(v)$ for some $\lambda_0 > 0$. Then obviously $\min\{f, \lambda\} \in \Lambda^p(v)$ for all $\lambda \in (0, \lambda_0]$. For each $\lambda > \lambda_0$, let $A = \{t > 0;\ f^*(t) > \lambda_0\}$ and $B = (0, \infty) \setminus A$. Then we have

$$\|\min\{f, \lambda\}\|_{\Lambda^p(v)}^p = \int_0^\infty \min\{f^*(t), \lambda\}^p v(t)\,dt$$
$$\le \int_A \lambda^p v(t)\,dt + \int_B \min\{f^*(t), \lambda_0\}^p v(t)\,dt.$$

The integral over B is of course finite. The set A must be of the form $(0, \alpha)$ for some $\alpha \in (0, \infty]$. If $\alpha < \infty$ then the integral over A equals $\lambda^p V(\alpha)$ and is finite. If $\alpha = \infty$ then $\min\{f^*(t), \lambda_0\} = \lambda_0$ for all $t > 0$ and the assumption that $\min\{f, \lambda_0\} \in \Lambda^p(v)$ implies that $\int_0^\infty v(t)\,dt < \infty$. Thus in both cases we deduce that $\min\{f, \lambda\} \in \Lambda^p(v)$, which shows that (ii) implies (iii).

Finally, suppose that f satisfies condition (iii). If $f^*(a) = 0$ then obviously $\|f\|_{\Lambda^p(w)} = 0$. If $f^*(a) > 0$ then we have that $\min\{f, f^*(a)\} \in \Lambda^p(v)$. By (10.2.1) we have $f^*(t) = (\min\{f, f^*(a)\})^*(t)$ for all $t > a$, and so

$$\|f\|_{\Lambda^p(w)}^p = \int_a^\infty \left((\min\{f, f^*(a)\})^*(t)\right)^p w(t)\,dt \le \|\min\{f, f^*(a)\}\|_{\Lambda^p(v)}^p < \infty.$$

Thus in both cases we deduce that $f \in \Lambda^p(w)$, which shows that (iii)$\Longrightarrow$(i) and completes the proof of the lemma. $\square$

Remark 10.2.3. Let us briefly consider what would happen had we not imposed the condition that W is finite. The four spaces introduced in Definition 10.1.2 could of course still be defined without this condition. If $W(t) = \infty$ for all $t > 0$, then each

Section 10.2 Functional properties

of these spaces is trivial, i.e. it contains only the zero element. In the remaining case, where

$$0 < t_0 = \sup\{t > 0; \ W(t) < \infty\} < \infty,$$

the space $\Gamma^{p,\infty}(w)$ is again trivial, and neither $\Lambda^p(w)$ nor $\Lambda^{p,\infty}(w)$ is a linear space. To see this, take $f = \chi_{(0,t_1)}$ and $g(s) = f(-s)$ with $\frac{t_0}{2} < t_1 < t_0$. Then f and g are both in $\Lambda^p(w)$ and $\Lambda^{p,\infty}(w)$ but $(f+g)^*(s) = f^*(\frac{s}{2}) = \chi_{(0,2t_1)}(s)$ is not in either of these spaces. Note that here W does not satisfy the Δ_2-condition.

We stress that (except during the brief discussion in the preceding remark) we always assume that $W(t)$ is finite for every $t > 0$.

Theorem 10.2.4. *The following are equivalent:*

(i) $\Lambda^p(w)$ *is not a linear space.*

(ii) *There exists a sequence of positive numbers t_n tending either to 0 or to ∞, such that $W(2t_n) > 2^n W(t_n)$ for all $n \in \mathbb{N}$.*

(iii) *There exists a sequence of positive numbers t_n such that $W(2t_n) > 2^n W(t_n) > 0$ for all $n \in \mathbb{N}$.*

Corollary 10.2.5. *The following are equivalent:*

(i)' $\Lambda^p(w)$ *is a linear space.*

(ii)' *There exist positive constants α, β and C such that $W(2t) \leq CW(t)$ for all $t \leq \alpha$ and all $t \geq \beta$.*

(iii)' *There exists a constant C' such that one of the following two conditions hold: Either*

(iii-A)'
$$W(2t) \leq C'W(t) \tag{10.2.2}$$

for all $t > 0$, or

(iii-B)' *$W(t) = 0$ on some interval $(0, a)$ and (10.2.2) holds for all t in some interval (b, ∞).*

Joint proof of Theorem 10.2.4 and Corollary 10.2.5. We begin with the easy proof that (ii) implies (iii). If (ii) holds and the sequence $\{t_n\}$ tends to ∞, then $W(2t_1) > 0$ and so $W(t) > 0$ for all $t \geq 2t_1$. If the original sequence $\{t_n\}$ does not satisfy $W(t_n) > 0$ for all n then we simply replace it by a subsequence of numbers in $[2t_1, \infty)$ and we are done. Alternatively if (ii) holds and the sequence $\{t_n\}$ tends to 0, then, since $W(2t_n) > 0$ for all $n \in \mathbb{N}$, we deduce that $W(t) > 0$ for all positive t and we obtain condition (iii).

Let us next prove that (iii) implies (i). Given a sequence $\{t_n\}$ satisfying (iii) we set $f := \sum_{n=1}^{\infty} \lambda_n \chi_{[0,t_n)}$ where

$$\lambda_n := \frac{1}{n^{p+\frac{1}{p}} (W(t_n))^{\frac{1}{p}}}.$$

Clearly $f^* = f$. Consequently, in the case where $p \leq 1$, we have

$$\|f\|_{\Lambda^p(w)}^p = \int_0^{\infty} \left(\sum_{n=1}^{\infty} \lambda_n \chi_{[0,t_n)}(t) \right)^p w(t)\, dt$$

$$\leq \int_0^{\infty} \left(\sum_{n=1}^{\infty} \lambda_n^p \chi_{[0,t_n)}(t) \right) w(t)\, dt$$

$$= \sum_{n=1}^{\infty} \lambda_n^p W(t_n) = \sum_{n=1}^{\infty} \frac{1}{n^{p^2+1}} < \infty.$$

On the other hand, if $p > 1$ we have, by the Minkowski inequality,

$$\|f\|_{\Lambda^p(w)} = \left(\int_0^{\infty} \left(\sum_{n=1}^{\infty} \lambda_n \chi_{[0,t_n)}(t) \right)^p w(t)\, dt \right)^{\frac{1}{p}}$$

$$\leq \sum_{n=1}^{\infty} \left(\int_0^{\infty} (\lambda_n \chi_{[0,t_n)}(t))^p w(t)\, dt \right)^{\frac{1}{p}}$$

$$= \sum_{n=1}^{\infty} \lambda_n (W(t_n))^{\frac{1}{p}} = \sum_{n=1}^{\infty} \frac{1}{n^{p+\frac{1}{p}}} < \infty.$$

Thus, in both cases, $f \in \Lambda^p(w)$. The function g defined by $g(x) = f(-x)$ satisfies $g^* = f^*$ and so it too is in $\Lambda^p(w)$.

Now consider the function $h = f + g$. Its distribution function is twice the distribution function of f and so $h^*(t) = f^*(\frac{t}{2}) = \sum_{n=1}^{\infty} \lambda_n \chi_{[0,2t_n)}(t)$. So, for each $m \in \mathbb{N}$, we have that

$$\|h\|_{\Lambda^p(w)}^p = \int_0^{\infty} \left(\sum_{n=1}^{\infty} \lambda_n \chi_{[0,2t_n)}(t) \right)^p w(t)\, dt \geq \int_0^{\infty} \lambda_m^p \chi_{[0,2t_m)}(t) w(t)\, dt$$

$$= \lambda_m^p W(2t_m) = \frac{1}{m^{p^2+1}} \frac{W(2t_m)}{W(t_m)} > \frac{2^m}{m^{p^2+1}}.$$

Taking the limit as m tends to ∞, we see that $h \notin \Lambda^p(w)$. This shows that $\Lambda^p(w)$ is not a linear space and completes the proof that (iii) implies (i).

It remains to show that (i) implies (ii). We will do this indirectly, i.e. by showing that if (ii) does not hold then (i) does not hold. In the course of doing this we will also prove some of the implications of the corollary.

Section 10.2 Functional properties

Suppose then that (ii) does not hold. Consider the sets E_n defined by $E_n = \{t > 0;\ W(2t) > 2^n W(t)\}$. They of course satisfy $E_{n+1} \subset E_n$. The fact that (ii) does not hold means that there exist a pair of positive numbers α and β with $\alpha < \beta$ and some integer $N = N(\alpha, \beta)$ such that $E_n \subset (\alpha, \beta)$ for all $n \geq N(\alpha, \beta)$. This establishes that condition (ii)$'$ of the corollary holds for any choice of α and β as above, provided we choose $C = 2^{N(\alpha,\beta)}$.

Thus it is clear that, if we can show that conditions (i)$'$, (ii)$'$ and (iii)$'$ of the corollary satisfy the two implications (ii)$' \Longrightarrow$ (iii)$'$ and (iii)$' \Longrightarrow$ (i)$'$, then this will certainly complete the proof that (i) implies (ii) and so will complete the proof of the theorem. At the same time we will have completed a considerable part of the proof of the corollary.

Accordingly, we shall now prove that (ii)$' \Longrightarrow$ (iii)$'$.

Let α, β and C be positive constants for which (ii)$'$ holds. If $\beta < \alpha$ then $W(2t) \leq CW(t)$ for all $t > 0$, i.e. W satisfies condition (iii-A)$'$ with $C' = C$. Thus we can suppose that $\alpha \leq \beta$. Let us first consider the case where $W(t) > 0$ for all $t > 0$. Then $\frac{W(2t)}{W(t)}$ is a continuous function for all $t > 0$ and is therefore bounded by some constant C_1 on the interval $[\alpha, \beta]$. It follows that $W(2t) \leq \max\{C, C_1\} W(t)$ for all $t > 0$ so again we have obtained condition (iii-A)$'$, this time with $C' = \max\{C, C_1\}$.

It remains to consider the case where $W(t) = 0$ for some positive t. In this case there exists $t_0 > 0$ such that $W(t) = 0$ if and only if $t \in (0, t_0]$. Thus we obtain condition (iii-B)$'$ for $a = t_0$, $b = \beta$ and $C' = C$.

Our final step will be to show that (iii)$' \Longrightarrow$ (i)$'$, i.e. that either of the conditions (iii-A)$'$ or (iii-B)$'$ is sufficient to imply that $\Lambda^p(w)$ is a linear space. In the case of condition (iii-A)$'$, which is exactly the Δ_2-condition used in [99, Theorem 1] (see also [37, 113]), we can apply [99, Theorem 1] (see also [37, Corollary 2.2, p. 482], [113, p. 6]) to obtain that $\|\cdot\|_{\Lambda^p(w)}$ is a quasinorm, which in turn immediately implies that $\Lambda^p(w)$ is a linear space.

If condition (iii-B)$'$ holds, then we need a somewhat longer argument.

We define an auxiliary weight function $v : (0, \infty) \to [0, \infty)$ by $v(t) = \frac{1}{a}$ for $t \in (0, a]$ and $v(t) = w(t)$ for $t \in (a, \infty)$. Let $V(t) = \int_0^t v(x)\,dx$. We claim that V satisfies the Δ_2-condition. To prove this claim we first note that $V(t) > 0$ for all positive t. Furthermore, for all $t \in (0, \frac{a}{2}]$ we have $\frac{V(2t)}{V(t)} = 2$. So, by obvious continuity considerations, it will suffice to show that $V(2t)V(t)$ is bounded on the interval $[\gamma, \infty)$ for some $\gamma > 0$. There are two cases to be considered. Suppose first that W is bounded, i.e. $W(t) \leq M$ for all $t > 0$. Then, for all $t \geq \gamma$, we have

$$\frac{V(2t)}{V(t)} = \frac{1 + W(2t)}{1 + W(t)} \leq \frac{1 + M}{1}.$$

Alternatively, if $\lim_{t \to \infty} W(t) = \infty$, we can choose γ sufficiently large so that $W(\gamma) \geq 1$ and $\gamma \geq b$. Then, for all $t \geq \gamma$, we have

$$\frac{V(2t)}{V(t)} = \frac{1 + W(2t)}{1 + W(t)} \leq \frac{W(\gamma) + W(2t)}{W(t)} \leq \frac{2W(2t)}{W(t)} \leq 2C'.$$

This proves our claim. Consequently it also shows, again by [99, Theorem 1] (see also [37, Corollary 2.2], [113, p. 6]) that $\Lambda^p(v)$ is a linear space. We are now ready to use Lemma 10.2.2 to deduce that $\Lambda^p(w)$ is also a linear space:

Let f and g be arbitrary functions in $\Lambda^p(w)$. We need to show that $f + g \in \Lambda^p(w)$. Clearly $|f|$ and $|g|$ are also in $\Lambda^p(w)$ and it will suffice to show that $|f| + |g| \in \Lambda^p(w)$, i.e. we can assume that f and g are nonnegative functions. By Lemma 10.2.2, the functions $\min\{f, 1\}$ and $\min\{g, 1\}$ are both in $\Lambda^p(v)$. Furthermore, it is easy to check that

$$\min\{f + g, 1\} \leq 2\left(\min\{f, 1\} + \min\{g, 1\}\right). \tag{10.2.3}$$

(Obviously, (10.2.3) holds at the points where $f + g \leq 1$, and if $f + g > 1$ then at least one of f and g must be greater than $\frac{1}{2}$ which again ensures that (10.2.3) holds.) We deduce that $\min\{f + g, 1\} \in \Lambda^p(v)$ and another application of Lemma 10.2.2 shows that $f + g \in \Lambda^p(w)$.

This completes the proof of the theorem. But we still need to obtain one last implication in the corollary, namely that (i)' $\Longrightarrow$ (ii)'.

Again we shall use an indirect approach, showing that if (ii)' does not hold then neither does (i)':

Indeed if (ii)' does not hold, then, for each positive α, β and C, there must exist $t = t(\alpha, \beta, C) \in (0, \alpha] \cup [\beta, \infty)$ such that $W(2t) > CW(t)$. In particular the sequence $\{t(\frac{1}{m}, m, 2^m)\}_{m \in \mathbb{N}}$, must have a subsequence with the properties stated in condition (ii) of the theorem. We have already seen that this implies (iii) which in turn implies (i), i.e. that $\Lambda^p(w)$ is not a linear space, exactly as is required to complete our proof. $\square$

The conditions for linearity of spaces $\Lambda^{p,\infty}(w)$ are the same as for linearity of $\Lambda^p(w)$. Below we formulate only one condition. Notice that (iii) of Theorem 10.2.4 is the same as (ii) below.

Theorem 10.2.6. *The following are equivalent:*

(i) $\Lambda^{p,\infty}(w)$ *is not a linear space.*

(ii) *There exists a sequence of positive numbers t_n such that*

$$\frac{W(2t_n)}{W(t_n)} \to \infty, \quad n \to \infty.$$

Proof. Clearly, the parameter p is immaterial here so we can with no loss of generality assume that $p = 1$. Suppose that (ii) is satisfied. Then, as shown in the proof of Theorem 10.2.4, there exists a monotone (either increasing or decreasing) sequence $\{t_n\}$ such that $W(t_n) > 0$ and

$$\frac{W(2t_n)}{W(t_n)} \to \infty, \quad n \to \infty.$$

Section 10.2 Functional properties

We set
$$f(t) = \sum_{n=1}^{\infty} \frac{1}{W(t_n)} \chi_{[s_n, t_n)},$$

where
$$s_n = \begin{cases} t_{n-1} & \text{when } \{t_n\} \text{ is increasing,} \\ t_{n+1} & \text{when } \{t_n\} \text{ is decreasing,} \end{cases}$$

and $t_0 = 0$. In both cases,
$$\|f\|_{\Lambda^{1,\infty}(w)} = \sup_{0<t<\infty} f^*(t)W(t) = \sup_{n\in\mathbb{N}} \sup_{s_n<t<t_n} f^*(t)W(t)$$
$$= \sup_{n\in\mathbb{N}} f^*(s_n)W(t_n) = 1,$$

while, for h as in the proof of Theorem 10.2.4,
$$\|h\|_{\Lambda^{1,\infty}(w)} = \sup_{0<t<\infty} f^*(\frac{t}{2})W(t) = \sup_{0<t<\infty} f^*(t)W(2t)$$
$$= \sup_{n\in\mathbb{N}} \sup_{s_n<t<t_n} f^*(t)W(2t) = \sup_{n\in\mathbb{N}} f^*(s_n)W(2t_n)$$
$$= \frac{W(2t_n)}{W(t_n)} \to \infty.$$

The proof of the converse implication is analogous to that in Theorem 10.2.4. □

We shall now present some examples. We assume with no loss of generality that $\mu(\mathcal{R}) = \infty$.

Example 10.2.7. We will find a measurable function $w : (0, \infty) \to [0, \infty)$ such that the two functions $W(x) = \int_0^x w(t)\,dt$ and $\Phi(x) = \int_x^\infty t^{-P} w(t)\,dt$ are both finite for all $x > 0$ but the set $\Lambda^P(w) = \{f; \int_0^\infty f^*(t)^P w(t)\,dt < \infty\}$ is not a linear space.

These conditions on W and Φ are apparently necessary and sufficient to ensure that the space $\Gamma^P(w)$ is nontrivial so it seems relevant to impose them here.

Initially the w which we construct can assume the value 0. But, as we shall see, it is easy to modify this to an example where w is strictly positive.

Here is the construction:

Let us first define a sequence of positive numbers w_n recursively by setting $w_1 = 1$ and
$$w_n = (2^n - 1)(w_1 + w_2 + \ldots + w_{n-1}) \tag{10.2.4}$$

for all $n > 1$. Then we define a second sequence of positive numbers s_n recursively by setting $s_1 = 2$ and, for each $n > 1$,
$$s_n = \max\left\{2s_{n-1}, w_n^{\frac{1}{p}} 2^{\frac{n}{p}} + 1\right\}. \tag{10.2.5}$$

The function $w : (0, \infty) \to [0, \infty)$ is defined by $w = \sum_{n=1}^{\infty} w_n \chi_{(s_{n-1}, s_n]}$. It follows from (10.2.5) that $\frac{w_n}{(s_n-1)^p} \leq 2^{-n}$ and so

$$\Phi(0) = \int_0^\infty t^{-p} w(t)\, dt = \sum_{n=1}^\infty w_n \int_{s_{n-1}}^{s_n} t^{-p}\, dt \leq \sum_{n=1}^\infty w_n (s_n - 1)^{-p} \leq 1.$$

Obviously, we also have $\int_0^x w(t)\, dt < \infty$ for each $x > 0$. Thus w satisfies the conditions mentioned above which ensure that $\Gamma^p(w)$ is nontrivial.

We claim that for this particular choice of w, the set $\Lambda^p(w)$ is not a linear space.

To show this let us first observe that, by (10.2.5) we have $s_{n-1} \leq \frac{s_n}{2} \leq s_n - 1$ and so

$$\int_0^{s_{n-1}} w(t)\, dt = \int_0^{\frac{s_n}{2}} w(t)\, dt = w_1 + w_2 + \ldots + w_{n-1}$$

It follows, using (10.2.4), that

$$W(s_n) = \int_0^{s_n} w(t)\, dt = w_1 + w_2 + \ldots + w_{n-1} + w_n$$
$$= 2^n (w_1 + w_2 + \ldots + w_{n-1}) = 2^n \int_0^{\frac{s_n}{2}} w(t)\, dt = 2^n W(\frac{s_n}{2}).$$

Thus the sequence $\{t_n\}$ given by $t_n = \frac{s_n}{2}$ tends to ∞ and satisfies

$$W(2t_n) = 2^n W(t_n).$$

In other words w satisfies condition (iii) of Theorem 10.2.4. Consequently $\Lambda^p(w)$ is not a linear space.

To get another less exotic example, i.e. where the weight function is strictly positive, we simply replace w by $w + u$ where u is any strictly positive function for which the above function f satisfies $\int_0^\infty f(t)^p u(t)\, dt < \infty$ and $\int_t^\infty s^{-p}(w(s) + u(s))\, ds < \infty$.

Example 10.2.8. Let $w = \chi_{[1,\infty)}$. Then $\Lambda^p(w)$ is not quasi-normable but is a linear space.

10.3 Embeddings between classical Lorentz spaces

In this section we shall study various inclusion and embedding relations between classical Lorentz spaces. So far we know only certain trivial relations such as those mentioned in (10.1.4) and (10.1.3). Furthermore, as we know already from Example 10.1.12, there exist nontrivial relations for example between spaces of type $\Lambda^p(w)$

and $\Gamma^{q,\infty}(w)$. We shall now present characterizations of all possible embeddings between classical Lorentz spaces.

Some embeddings can be characterized in quite a general setting by means of the functions ψ_X and ϱ_X, defined in (10.1.8) and (10.1.9), respectively.

Most of the material in this section has a form of a survey. In particular, we will not give all the details of the proofs of most of the known results and we will instead refer the reader to the corresponding sources.

We recall that by $\overline{X}$ we denote the representation space of X.

Theorem 10.3.1. *Let X be a rearrangement-invariant lattice and $\|f\|_X = \|f^*\|_{\overline{X}(0,\mu(\mathcal{R}))}$ where $\|\cdot\|_{\overline{X}(0,\mu(\mathcal{R}))}$ is some functional on $\mathcal{M}_+(0,\mu(\mathcal{R}))$, endowed with the one-dimensional Lebesgue measure, let $p, q \in (0, \infty)$ and let v, w be weights on $(0, \mu(\mathcal{R}))$.*

(i) *The embedding $\Lambda^{p,\infty}(v) \hookrightarrow X$ holds if and only if*

$$A_{(10.3.1)} := \|V^{-\frac{1}{p}}\|_{\overline{X}(0,\mu(\mathcal{R}))} < \infty, \qquad (10.3.1)$$

and moreover the optimal constant C of the embedding equals $A_{(10.3.1)}$.

(ii) *The embedding $X \hookrightarrow \Gamma^{q,\infty}(w)$ holds if and only if*

$$A_{(10.3.2)} := \sup_{t \in (0,\mu(\mathcal{R}))} \frac{W^{\frac{1}{q}}(t)\psi_X(t)}{t} < \infty, \qquad (10.3.2)$$

and moreover the optimal constant C of the embedding equals $A_{(10.3.2)}$.

(iii) *The embedding $X \hookrightarrow \Lambda^{q,\infty}(w)$ holds if and only if*

$$A_{(10.3.3)} := \sup_{t \in (0,\mu(\mathcal{R}))} W^{\frac{1}{q}}(t)\varrho_X(t) = \sup_{t \in (0,\mu(\mathcal{R}))} \frac{W^{\frac{1}{q}}(t)}{\varphi_X(t)} < \infty, \qquad (10.3.3)$$

and moreover the optimal constant C of the embedding equals $A_{(10.3.3)}$.

Proof. (i) This is a particular case of a more general result in [208, Proposition 2.7].

(ii) Changing the suprema, we get

$$C = \sup_{\|f\|_X \leq 1} \sup_{t > 0} f^{**}(t) W^{\frac{1}{q}}(t) = \sup_{t > 0} \frac{W^{\frac{1}{q}}(t)}{t} \sup_{\|f\|_X \leq 1} \int_0^t f^*(s)\,ds = A_{(10.3.2)}.$$

(iii) Again, changing the suprema and using Remark 10.1.10 (ii), we obtain

$$C = \sup_{f \neq 0} \sup_{t > 0} \frac{f^*(t) W^{\frac{1}{q}}(t)}{\|f\|_X} = \sup_{t > 0} W^{\frac{1}{q}}(t)\varrho_X(t) = A_{(10.3.3)}.$$

The proof is complete. □

Corollary 10.3.2. *If X is a rearrangement-invariant Banach function space, $q \in (0, \infty)$ and w is a weight, then*

$$X \hookrightarrow \Gamma^{q,\infty}(w) \quad \text{if and only if} \quad X \hookrightarrow \Lambda^{q,\infty}(w). \tag{10.3.4}$$

Proof. The "only if" part is a direct consequence of (10.1.4). The "if" part follows from Theorem 10.3.1 (ii) and (iii) combined with the relation

$$\frac{\psi_X(t)}{t} = \varrho_X(t), \quad t \in (0, \mu(\mathcal{R})), \tag{10.3.5}$$

which follows from (7.9.7). □

Remarks 10.3.3. (i) The statements (10.3.5) and (10.3.4) do not hold in general (that is, without the assumption that X is a rearrangement-invariant Banach function space). For example, take $p \in (0,1)$, $q \in (0,\infty)$, $w(t) = t^{\frac{q}{p}-1}$, and $X = L^p$. Then $\psi_X(t) = \infty$ for every $t > 0$, thus $A_{(10.3.2)} = \infty$, but $A_{(10.3.3)} < \infty$.

(ii) If $\Lambda^{q,\infty}(w)$ is a Banach function space, then [208, Theorem 3.1] $\Gamma^{q,\infty}(w) = \Lambda^{q,\infty}(w)$. In particular, in such case (10.3.4) is true.

(iii) Let X be an r.i. lattice. There is a close connection between $\|g\|_{X'}$ and the optimal constant C of the embedding $X \hookrightarrow \Lambda^1(g^*)$. Indeed, we have

$$\|g\|_{X'} = \sup_{f \neq 0} \frac{\int_{\mathcal{R}} |f(x)g(x)| \, d\mu(x)}{\|f\|_X} = \sup_{f \neq 0} \frac{\int_0^\infty f^*(t)g^*(t) \, dt}{\|f\|_X} = C. \tag{10.3.6}$$

Notation 10.3.4. We shall denote by $\mathcal{G}$ the cone of all quasi-concave functions on $(0, \infty)$, and $\mathcal{G}_0 = \{\varphi \in \mathcal{G}; \lim_{x \to 0_+} \varphi(x) = 0\}$.

Definition 10.3.5. Let $p \in (0, \infty)$ and let v be a weight. We denote

$$\mathcal{V}_p(t) = \sup_{s \geq t} \frac{V^{\frac{1}{p}}(s)}{s}, \quad t > 0. \tag{10.3.7}$$

Remarks 10.3.6. (i) Let us recall [15, Lemma 5.4.3] that $\varphi \in \mathcal{G}$ if and only if there is a constant $\lambda \geq 0$ and a nonincreasing function ψ such that $\varphi(t) \leq \lambda + \int_0^t \psi(s) \, ds \leq 2\varphi(t)$ for $t > 0$. Similarly, $\varphi \in \mathcal{G}_0$ if and only if there is a nonincreasing function ψ such that $\varphi(t) \leq \int_0^t \psi(s) \, ds \leq 2\varphi(t)$.

(ii) If $\mathcal{V}_p(t) < \infty$ for every $t > 0$, then $\frac{1}{\mathcal{V}_p} \in \mathcal{G}$. Moreover,

$$t\mathcal{V}_p(t) = \sup_{s > 0} \min\{1, \tfrac{t}{s}\} V^{\frac{1}{p}}(s),$$

hence $tV_p(t)$ is the least quasi-concave majorant of $V^{\frac{1}{p}}(t)$. Therefore, $tV_p(t) \in \mathcal{G}_0$.

(iii) It follows readily from (10.3.7) that there exists a decomposition $(0, \infty) = E \cup \bigcup_{j=1}^{\infty}(a_j, b_j)$, such that

$$\mathcal{V}_p(t) = \frac{V^{\frac{1}{p}}(t)}{t}\chi_E(t) + \sum_{j=1}^{\infty} c_j \chi_{(a_j,b_j)}(t), \qquad c_j = \frac{V^{\frac{1}{p}}(b_j)}{b_j}. \qquad (10.3.8)$$

As long as $a_j \neq 0$, we also have

$$c_j = \frac{V^{\frac{1}{p}}(a_j)}{a_j}.$$

(iv) The introduction of $\mathcal{V}_p(t)$ allows an alternative version of $\|f\|_{\Gamma^{p,\infty}(v)}$, namely,

$$\|f\|_{\Gamma^{p,\infty}(v)} = \sup_{t>0}\left(\int_0^t f^*(s)\,ds\right)\mathcal{V}_p(t) = \sup_{t>0} t\mathcal{V}_p(t) f^{**}(t). \qquad (10.3.9)$$

Indeed, we have

$$\|f\|_{\Gamma^{p,\infty}(v)} \leq \sup_t\left(\int_0^t f^*(s)\,ds\right)\sup_{s\geq t}\frac{V^{\frac{1}{p}}(s)}{s}$$

$$= \sup_s \frac{V^{\frac{1}{p}}(s)}{s}\sup_{t\leq s}\int_0^t f^*(s)\,ds = \|f\|_{\Gamma^{p,\infty}(v)}.$$

(v) There exist a $\lambda \geq 0$ and a nonincreasing function v_p such that

$$\frac{1}{2}\left(\lambda + \int_0^t v_p(s)\,ds\right) \leq \frac{1}{\sup_{s\geq t} s^{-1}V^{\frac{1}{p}}(s)} \leq \lambda + \int_0^t v_p(s)\,ds.$$

Consequently,

$$\sup_{t>0}\frac{\int_0^t f^*(s)\,ds}{\lambda + \int_0^t v_p(s)\,ds} \leq \|f\|_{\Gamma^{p,\infty}(v)} \leq 2\sup_{t>0}\frac{\int_0^t f^*(s)\,ds}{\lambda + \int_0^t v_p(s)\,ds}. \qquad (10.3.10)$$

Convention 10.3.7. In the rest of this section we shall assume that $\mu(\mathcal{R}) = \infty$. If $\mu(\mathcal{R}) < \infty$, then the relevant embedding constants just have to be modified in an obvious way, replacing the interval $(0, \infty)$ by $(0, \mu(\mathcal{R}))$. In particular, by "weights" we mean weights on $(0, \infty)$.

10.3.1 Embeddings of type $\Lambda \hookrightarrow \Lambda$

Theorem 10.3.8 (the case $\Lambda^p(v) \hookrightarrow \Lambda^q(w)$). *Let $p, q \in (0, \infty)$ and let v, w be weights.*

(i) *Let $0 < p \leq q < \infty$. Then the inequality*

$$\left(\int_0^\infty (f^*(t))^q w(t) \, dt \right)^{\frac{1}{q}} \leq C \left(\int_0^\infty (f^*(t))^p v(t) \, dt \right)^{\frac{1}{p}} \quad (10.3.11)$$

holds if and only if

$$A_{(10.3.12)} := \sup_{t>0} W^{\frac{1}{q}}(t) V^{-\frac{1}{p}}(t) < \infty, \quad (10.3.12)$$

and moreover the optimal constant C in (10.3.11) equals $A_{(10.3.12)}$.

(ii) *Let $0 < q < p < \infty$ and let r be given by $\frac{1}{r} = \frac{1}{q} - \frac{1}{p}$. Then the inequality (10.3.11) holds if and only if*

$$A_{(10.3.13)} := \left(\int_0^\infty \left(\frac{W(t)}{V(t)} \right)^{\frac{r}{p}} w(t) \, dt \right)^{\frac{1}{r}} \quad (10.3.13)$$

$$= \left(\frac{q}{r} \frac{W^{\frac{r}{q}}(\infty)}{V^{\frac{r}{p}}(\infty)} + \frac{q}{p} \int_0^\infty \left(\frac{W(t)}{V(t)} \right)^{\frac{r}{q}} v(t) \, dt \right)^{\frac{1}{r}} < \infty,$$

and $C \approx A_{(10.3.13)}$.

These results can be found in [194, Remark (i), p. 148] for $1 < p, q < \infty$, and in [216, Proposition 1] for all values $0 < p, q < \infty$. Part (i) also follows from a more general result in [37, Corollary 2.7].

Theorem 10.3.9 (the case $\Lambda^p(v) \hookrightarrow \Lambda^{q,\infty}(w)$). *Let $p, q \in (0, \infty)$ and let v, w be weights. The inequality*

$$\sup_{t>0} f^*(t) W^{\frac{1}{q}}(t) \leq C \left(\int_0^\infty (f^*(t))^p v(t) \, dt \right)^{\frac{1}{p}}, \quad (10.3.14)$$

holds if and only if $A_{(10.3.12)} < \infty$, and moreover the optimal constant C in (10.3.14) equals $A_{(10.3.12)}$.

Proof. This follows from Theorem 10.3.1 (iii) with $X = \Lambda^p(v)$ combined with (10.1.10). □

Theorem 10.3.10 (the case $\Lambda^{p,\infty}(v) \hookrightarrow \Lambda^q(w)$). *Let $p,q \in (0,\infty)$ and let v, w be weights. The inequality*

$$\left(\int_0^\infty (f^*(t))^q w(t)\, dt\right)^{\frac{1}{q}} \leq C \sup_{t>0} f^*(t) V^{\frac{1}{p}}(t) \tag{10.3.15}$$

holds if and only if

$$A_{(10.3.16)} := \left(\int_0^\infty V^{-\frac{q}{p}}(t) w(t)\, dt\right)^{\frac{1}{q}} < \infty, \tag{10.3.16}$$

and the optimal constant C in (10.3.15) equals $A_{(10.3.16)}$.

Proof. This follows from Theorem 10.3.1 (i) with $X = \Lambda^q(w)$. A direct proof is an easy exercise. □

Theorem 10.3.11 (the case $\Lambda^{p,\infty}(v) \hookrightarrow \Lambda^{q,\infty}(w)$). *Let $p,q \in (0,\infty)$ and let v, w be weights. Then the inequality*

$$\sup_{t>0} f^*(t) W^{\frac{1}{q}}(t) \leq C \sup_{t>0} f^*(t) V^{\frac{1}{p}}(t) \tag{10.3.17}$$

holds if and only if $A_{(10.3.12)} < \infty$, and moreover the optimal constant C in (10.3.17) equals $A_{(10.3.12)}$.

Proof. The assertion immediately follows from Theorem 10.3.1 (i) or (iii) and (10.1.10). □

10.3.2 Embeddings of type $\Lambda \hookrightarrow \Gamma$

Theorem 10.3.12 (the case $\Lambda^p(v) \hookrightarrow \Gamma^q(w)$). *Let $p,q \in (0,\infty)$ and let v, w be weights.*

(i) *Let $1 < p \leq q < \infty$ Then the inequality*

$$\left(\int_0^\infty (f^{**}(t))^q w(t)\, dt\right)^{\frac{1}{q}} \leq C \left(\int_0^\infty (f^*(t))^p v(t)\, dt\right)^{\frac{1}{p}} \tag{10.3.18}$$

holds if and only if $A_{(10.3.12)} < \infty$ and

$$A_{(10.3.19)} := \sup_{t>0} \left(\int_t^\infty \frac{w(s)}{s^q}\, ds\right)^{\frac{1}{q}} \left(\int_0^t \frac{v(s) s^{p'}}{V^{p'}(s)}\, ds\right)^{\frac{1}{p'}} < \infty, \tag{10.3.19}$$

and moreover the optimal constant C in (10.3.18) satisfies $C \approx A_{(10.3.12)} + A_{(10.3.19)}$.

(ii) Let $0 < p \leq 1$, $0 < p \leq q < \infty$. Then (10.3.18) holds if and only if $A_{(10.3.12)} < \infty$ and

$$A_{(10.3.20)} := \sup_{t>0} t \left(\int_t^\infty \frac{w(s)}{s^q} \, ds \right)^{\frac{1}{q}} V^{-\frac{1}{p}}(t) < \infty, \qquad (10.3.20)$$

and $C \approx A_{(10.3.12)} + A_{(10.3.20)}$.

(iii) Let $1 < p < \infty$, $0 < q < p < \infty$, $q \neq 1$. Then (10.3.18) holds if and only if $A_{(10.3.13)} < \infty$ and, for r given by $\frac{1}{r} = \frac{1}{q} - \frac{1}{p}$,

$$A_{(10.3.21)} := \left(\int_0^\infty \left[\left(\int_t^\infty \frac{w(s)}{s^q} \, ds \right)^{\frac{1}{q}} \left(\int_0^t \frac{v(s) s^{p'}}{V^{p'}(s)} \, ds \right)^{\frac{q-1}{q}} \right]^r \frac{v(t) t^{p'}}{V^{p'}(t)} \, dt \right)^{\frac{1}{r}}$$

$$\approx \left(\int_0^\infty \left[\left(\int_t^\infty \frac{w(s)}{s^q} \, ds \right)^{\frac{1}{p}} \left(\int_0^t \frac{v(s) s^{p'}}{V^{p'}(s)} \, ds \right)^{\frac{1}{p'}} \right]^r \frac{w(t)}{t^q} \, dt \right)^{\frac{1}{r}} < \infty,$$

(10.3.21)

and $C \approx A_{(10.3.13)} + A_{(10.3.21)}$.

(iv) Let $1 = q < p < \infty$. Then (10.3.18) holds if and only if $A_{(10.3.13)} < \infty$ and

$$A_{(10.3.22)} := \left(\int_0^\infty \left(\frac{W(t) + t \int_t^\infty \frac{w(s)}{s} \, ds}{V(t)} \right)^{p'-1} \int_t^\infty \frac{w(s)}{s} \, ds \, dt \right)^{\frac{1}{p'}}$$

$$\approx \frac{W(\infty)}{V^{\frac{1}{p}}(\infty)} + \left(\int_0^\infty \left(\frac{W(t) + t \int_t^\infty \frac{w(s)}{s} \, ds}{V(t)} \right)^{p'} v(t) \, dt \right)^{\frac{1}{p'}} < \infty,$$

(10.3.22)

and $C \approx A_{(10.3.13)} + A_{(10.3.22)}$.

(v) Let $0 < q < p = 1$. Then (10.3.18) holds if and only if $A_{(10.3.13)} < \infty$ and $A_{(10.3.23)} < \infty$, where

$$A_{(10.3.23)} := \left(\int_0^\infty \left(\int_t^\infty \frac{w(s)}{s^q} \, ds \right)^{\frac{q}{1-q}} \left(\operatorname*{ess\,inf}_{0 < s < t} \frac{V(s)}{s} \right)^{\frac{q}{q-1}} \frac{w(t)}{t^q} \, dt \right)^{\frac{1-q}{q}},$$

(10.3.23)

and $C \approx A_{(10.3.13)} + A_{(10.3.23)}$.

(vi) *Let $0 < q < p < 1$. If $A_{(10.3.13)} < \infty$ and*

$$A_{(10.3.24)} := \left(\int_0^\infty \left(\int_t^\infty \frac{w(s)}{s^q} ds \right)^{\frac{r}{q}} V^{-\frac{r}{p}}(t) t^{r-1} dt \right)^{\frac{1}{r}} < \infty, \quad (10.3.24)$$

then (10.3.18) holds. Moreover, $A_{(10.3.13)} \lesssim C \lesssim A_{(10.3.13)} + A_{(10.3.24)}$.

Case (i) is shown in [194, Theorem 2], case (ii) independently in [36, Proposition 2.6b] and [216, Theorem 3b] (for a particular case see also [131, Theorem 2.2]), case (iii) in [194, Theorem 2] for $1 < q < p < \infty$ and in [216, Theorem 3a] for $0 < q < 1 < p < \infty$, case (iv) follows from (10.1.13) and Theorem 10.3.8, case (v) can be found in [203, Theorem 4.1], and finally case (vi) is treated in [216, Proposition 2], see also [92, Theorem 6].

Theorem 10.3.13 (the case $\Lambda^p(v) \hookrightarrow \Gamma^{q,\infty}(w)$). *Let $q \in (0, \infty)$ and let v, w be weights.*

(i) *Let $0 < p \leq 1$. The inequality*

$$\sup_{t>0} f^{**}(t) W^{\frac{1}{q}}(t) \leq C \left(\int_0^\infty (f^*(t))^p v(t) dt \right)^{\frac{1}{p}} \quad (10.3.25)$$

holds if and only if

$$A_{(10.3.26)} := \sup_{0 < s \leq t < \infty} \frac{W^{\frac{1}{q}}(t) s}{t V^{\frac{1}{p}}(s)} < \infty, \quad (10.3.26)$$

and moreover the optimal constant C in (10.3.25) equals $A_{(10.3.26)}$.

(ii) *Let $1 < p < \infty$. Then (10.3.25) holds if and only if*

$$A_{(10.3.27)} := \sup_{t>0} \frac{W^{\frac{1}{q}}(t)}{t} \left(\int_0^t \left(\frac{s}{V(s)} \right)^{p'-1} ds \right)^{\frac{1}{p'}} < \infty, \quad (10.3.27)$$

and $C \approx A_{(10.3.27)}$.

Proof. Both assertions follow from [36, Theorem 3.3]. □

Remark 10.3.14. By Theorem 10.3.8,

$$\psi_{\Lambda^p(v)}(t) = \sup_{f \neq 0} \frac{\int_0^\infty f^*(s) \chi_{(0,t)}(s) ds}{\|f\|_{\Lambda^p(v)}}$$

$$\begin{cases} = \sup_{0 < s \leq t} \frac{s}{V^{\frac{1}{p}}(s)} & \text{if } 0 < p \leq 1, \\ \approx \left(\int_0^t \left(\frac{s}{V(s)} \right)^{p'-1} ds \right)^{\frac{1}{p'}} & \text{if } 1 < p < \infty. \end{cases}$$

Applying Theorem 10.3.1 (ii), we get a new proof of Theorem 10.3.13.

Theorem 10.3.15 (the case $\Lambda^{p,\infty}(v) \hookrightarrow \Gamma^q(w)$). *Let $p, q \in (0, \infty)$ and let v, w be weights. The inequality*

$$\left(\int_0^\infty (f^{**}(t))^q w(t) \, dt \right)^{\frac{1}{q}} \leq C \sup_{t>0} f^*(t) V^{\frac{1}{p}}(t) \tag{10.3.28}$$

holds if and only if

$$A_{(10.3.29)} := \left(\int_0^\infty \left(\frac{1}{t} \int_0^t V^{-\frac{1}{p}}(s) \, ds \right)^q w(t) \, dt \right)^{\frac{1}{q}} < \infty, \tag{10.3.29}$$

and moreover the optimal constant C in (10.3.28) equals $A_{(10.3.29)}$.

This result can be found in [208, Theorem 4.1 (i)] (cf. also Theorem 10.3.1 (i) with $X = \Gamma^q(w)$).

Theorem 10.3.16 (the case $\Lambda^{p,\infty}(v) \hookrightarrow \Gamma^{q,\infty}(w)$). *Let $p, q \in (0, \infty)$ and let v, w be weights. The inequality*

$$\sup_{t>0} f^{**}(t) W^{\frac{1}{q}}(t) \leq c \sup_{t>0} f^*(t) V^{\frac{1}{p}}(t) \tag{10.3.30}$$

holds if and only if

$$A_{(10.3.31)} := \sup_{t>0} \frac{W^{\frac{1}{q}}(t)}{t} \int_0^t V^{-\frac{1}{p}}(s) \, ds < \infty, \tag{10.3.31}$$

and moreover the optimal constant C in (10.3.30) equals $A_{(10.3.31)}$.

This result can be found in [208, Theorem 4.1 (ii)]. It also follows from Theorem 10.3.1 (i) with $X = \Gamma^{q,\infty}(w)$ or Theorem 10.3.1 (ii) with $X = \Lambda^{p,\infty}(v)$ combined with an easy observation that $\psi_X(t)$ from (10.1.8) equals $\int_0^t V^{-\frac{1}{p}}(s) \, ds$.

10.3.3 Embeddings of type $\Gamma \hookrightarrow \Lambda$

Theorem 10.3.17 (the case $\Gamma^p(v) \hookrightarrow \Lambda^q(w)$). *Let $p, q \in (0, \infty)$ and let v, w be weights. When $q < p$, we write $r := \frac{pq}{p-q}$. Let v satisfy the nondegeneracy conditions (cf. [95])*

$$\int_0^\infty \frac{v(s)}{(s+1)^p} \, ds < \infty, \quad \int_0^1 \frac{v(s)}{s^p} \, ds = \int_1^\infty v(s) \, ds = \infty. \tag{10.3.32}$$

(i) *If $0 < p \leq q < \infty$, $1 \leq q < \infty$, then the inequality*

$$\left(\int_0^\infty (f^*(t))^q w(t) \, dt \right)^{\frac{1}{q}} \leq C \left(\int_0^\infty (f^{**}(t))^p v(t) \, dt \right)^{\frac{1}{p}} \tag{10.3.33}$$

Section 10.3 Embeddings

holds if and only if

$$A_{(10.3.34)} := \sup_{t>0} \frac{W^{\frac{1}{q}}(t)}{\left(V(t) + t^p \int_t^\infty s^{-p} v(s)\, ds\right)^{\frac{1}{p}}} < \infty, \qquad (10.3.34)$$

and moreover the optimal constant C in (10.3.33) is equal to $A_{(10.3.34)}$.

(ii) *If $1 \le q < p < \infty$, then (10.3.33) holds for some $C > 0$ and all f if and only if*

$$A_{(10.3.35)} := \left(\int_0^\infty \frac{t^{r+p-1} \left[\sup_{y \in [t,\infty)} y^{-r} W(y)^{\frac{r}{q}}\right] \times V(t) \int_t^\infty s^{-p} v(s)\, ds}{\left(V(t) + t^p \int_t^\infty s^{-p} v(s)\, ds\right)^{\frac{r}{p}+2}}\, dt\right)^{\frac{1}{r}} < \infty$$

(10.3.35)

and moreover the optimal constant C in (10.3.33) satisfies $C \approx A_{(10.3.35)}$.

(iii) *If $0 < p \le q < 1$, then (10.3.33) holds for some $C > 0$ and all f if and only if*

$$A_{(10.3.36)} := \sup_{t \in (0,\infty)} \frac{W(t)^{\frac{1}{q}} + t \left(\int_t^\infty W(s)^{\frac{q}{1-q}} w(s) s^{-\frac{q}{1-q}}\, ds\right)^{\frac{1-q}{q}}}{\left(V(t) + t^p \int_t^\infty s^{-p} v(s)\, ds\right)^{\frac{1}{p}}} < \infty$$

(10.3.36)

and moreover the optimal constant C in (10.3.33) satisfies $C \approx A_{(10.3.36)}$.

(iv) *If $0 < q < 1$ and $0 < q < p$, then (10.3.33) holds for some $C > 0$ and all f if and only if $A_{(10.3.37)} < \infty$, where*

$$A_{(10.3.37)} := \left(\int_0^\infty \frac{\left[W(t)^{\frac{1}{1-q}} + t^{\frac{q}{1-q}} \int_t^\infty W(s)^{\frac{q}{1-q}} w(s)\right.}{\left. \times s^{-\frac{q}{1-q}}\, ds\right]^{\frac{r(1-q)}{q}-1} W(t)^{\frac{q}{1-q}} w(t)}{\left(V(t) + t^p \int_t^\infty s^{-p} v(s)\, ds\right)^{\frac{r}{p}}}\, dt\right)^{\frac{1}{r}}$$

(10.3.37)

and moreover the optimal constant C in (10.3.33) satisfies $C \approx A_{(10.3.37)}$. Moreover, $A_{(10.3.37)} \approx A_{(10.3.38)}$, where

$$A_{(10.3.38)} = \left(\int_0^\infty \frac{\left[W(t)^{\frac{1}{1-q}} + t^{\frac{q}{1-q}} \int_t^\infty W(s)^{\frac{q}{1-q}} w(s) s^{-\frac{q}{1-q}} \, ds \right]^{\frac{r(1-q)}{q}}}{\left(V(t) + t^p \int_t^\infty s^{-p} v(s) \, ds \right)^{\frac{r}{p}+2}} \right.$$

$$\left. \times V(t) \int_t^\infty s^{-p} v(s) \, ds \, t^{p-1} \, dt \right)^{\frac{1}{r}}.$$

(10.3.38)

Proof. The assertion (i) is proved in [163, Theorem 3.2] (for $1 \leq p = q < \infty$), [131, Theorem 2.1] (for $1 \leq p \leq q < \infty$), and [215, p. 473] (for $0 < p \leq q < \infty$, $1 \leq q < \infty$). The proof with the best constant can be found in [102, Theorem 3.2 (a) and (c)] or [9], where a multidimensional case is treated.

The assertions (ii), (iii) and (iv) follow from a more general result in [89, Theorem 4.2]. □

Theorem 10.3.18 (the case $\Gamma^p(v) \hookrightarrow \Lambda^{q,\infty}(w)$). *Let $p, q \in (0, \infty)$ and let v, w be weights. The inequality*

$$\sup_{t>0} f^*(t) W^{\frac{1}{q}}(t) \leq C \left(\int_0^\infty (f^{**}(t))^p v(t) \, dt \right)^{\frac{1}{p}} \qquad (10.3.39)$$

holds if and only if $A_{(10.3.34)} < \infty$ *(cf. (10.3.34)). Moreover, the optimal constant C in (10.3.39) equals $A_{(10.3.34)}$.*

Proof. This follows from Theorem 10.3.1 (iii) with $X = \Gamma^p(v)$, and (10.1.11). □

Theorem 10.3.19 (the case $\Gamma^{p,\infty}(v) \hookrightarrow \Lambda^{q,\infty}(w)$). *Let $p, q \in (0, \infty)$ and let v, w be weights. The inequality*

$$\sup_{t>0} f^*(t) W^{\frac{1}{q}}(t) \leq C \sup_{t>0} f^{**}(t) V^{\frac{1}{p}}(t) \qquad (10.3.40)$$

holds if and only if

$$A_{(10.3.41)} := \sup_{t>0} \frac{W^{\frac{1}{q}}(t)}{t \sup_{s \geq t} s^{-1} V^{\frac{1}{p}}(s)} < \infty. \qquad (10.3.41)$$

Moreover, the optimal constant C in (10.3.40) equals $A_{(10.3.41)}$.

Proof. This follows from Theorem 10.3.1 (iii) with $X = \Gamma^{p,\infty}(v)$, and (10.1.12). □

10.3.4 Embeddings of type $\Gamma \hookrightarrow \Gamma$

Theorem 10.3.20 (the case $\Gamma^p(v) \hookrightarrow \Gamma^q(w)$). *Let $p, q \in (0, \infty)$ and let v, w be weights. Let v satisfy the nondegeneracy conditions (10.3.32). When $q < p$, we write $r := \frac{pq}{p-q}$.*

(i) *Let $0 < p \leq q < \infty$. Then the inequality*

$$\left(\int_0^\infty (f^{**}(t))^q w(t) \, dt \right)^{\frac{1}{q}} \leq C \left(\int_0^\infty (f^{**}(t))^p v(t) \, dt \right)^{\frac{1}{p}} \tag{10.3.42}$$

holds if and only if

$$A_{(10.3.43)} := \sup_{t \in (0, \infty)} \frac{\left(W(t) + t^q \int_t^\infty s^{-q} w(s) \, ds \right)^{\frac{1}{q}}}{\left(V(t) + t^p \int_t^\infty s^{-p} v(s) \, ds \right)^{\frac{1}{p}}} < \infty. \tag{10.3.43}$$

Moreover, the optimal constant C in (10.3.42) satisfies $C \approx A_{(10.3.43)}$.

(ii) *Let $0 < q < p < \infty$. Then (10.3.42) holds if and only if*

$$A_{(10.3.44)} := \left(\int_0^\infty \frac{\left(W(t) + t^q \int_t^\infty y^{-q} w(y) \, dy \right)^{\frac{r}{q}}}{\left(V(t) + t^p \int_t^\infty s^{-p} v(s) \, ds \right)^{\frac{r}{p}+2}} \right.$$

$$\left. \times V(t) \int_t^\infty s^{-p} v(s) \, ds \, t^{p-1} \, dt \right)^{\frac{1}{r}} < \infty,$$

$$\tag{10.3.44}$$

and moreover the optimal constant C in (10.3.42) satisfies $C \approx A_{(10.3.44)}$. Moreover, $A_{(10.3.44)} \approx A_{(10.3.45)}$, where

$$A_{(10.3.45)} := \left(\int_0^\infty \frac{\left(W(t) + t^q \int_t^\infty s^{-q} w(s) \, ds \right)^{\frac{r}{q}-1} w(t)}{\left(V(t) + t^p \int_t^\infty s^{-p} v(s) \, ds \right)^{\frac{r}{p}}} \, dt \right)^{\frac{1}{r}} < \infty. \tag{10.3.45}$$

Proof. The assertion follows from a more general result in [89, Theorem 5.1]. Particular cases can also be found in [95, Theorem 3.2]. The case $q \geq 1$ of (i) was also treated in [215, Theorem 3.3]. A proof with a sharp constant is given in [102, Theorem 3.7]. □

Theorem 10.3.21 (the case $\Gamma^p(v) \hookrightarrow \Gamma^{q,\infty}(w)$). *Let $p, q \in (0, \infty)$ and let v, w be weights. The inequality*

$$\sup_{t>0} f^{**}(t) W^{\frac{1}{q}}(t) \leq C \left(\int_0^\infty (f^{**}(t))^p v(t) \, dt \right)^{\frac{1}{p}} \tag{10.3.46}$$

holds if and only if $A_{(10.3.34)} < \infty$ (cf. (10.3.34)). Moreover, the optimal constant C in (10.3.46) satisfies $C \approx A_{(10.3.34)}$.

Proof. The lower bound, $C \geq A_{(10.3.34)}$, follows from Theorem 10.3.18 and formula (10.1.4). As for the upper bound, for every $t > 0$ we have

$$f^{**}(t) W^{\frac{1}{q}}(t) \lesssim A_{(10.3.34)} \left(f^{**}(t) V^{\frac{1}{p}}(t) + \int_0^t f^*(y) \, dy \left(\int_t^\infty \frac{v(s)}{s^p} \, ds \right)^{\frac{1}{p}} \right)$$

$$\lesssim A_{(10.3.34)} \left(f^{**}(t) V^{\frac{1}{p}}(t) + \left(\int_t^\infty (f^{**}(s))^p \, v(s) \, ds \right)^{\frac{1}{p}} \right).$$

Taking the supremum over $t > 0$ and using (10.1.4) we get $C \lesssim A_{(10.3.34)}$. □

Remark 10.3.22. For $0 < p \leq 1$, the upper bound in Theorem 10.3.21 follows also from Theorem 10.3.17 (i) and Remark 10.3.14. Yet another proof can be obtained from Theorem 10.3.1 (ii) and Theorem 10.3.17 (i).

Theorem 10.3.23 (the case $\Gamma^{p,\infty}(v) \hookrightarrow \Gamma^q(w)$). Let $p, q \in (0, \infty)$ and let v, w be weights. The inequality

$$\left(\int_0^\infty (f^{**}(t))^q \, w(t) \, dt \right)^{\frac{1}{q}} \leq C \sup_{t > 0} f^{**}(t) V^{\frac{1}{p}}(t) \tag{10.3.47}$$

holds if and only if

$$A_{(10.3.48)} := \left(\int_0^\infty \frac{w(t)}{\left(t \sup_{s \geq t} s^{-1} V^{\frac{1}{p}}(s) \right)^q} \, dt \right)^{\frac{1}{q}} < \infty. \tag{10.3.48}$$

Moreover, the optimal constant C in (10.3.47) satisfies $C \approx A_{(10.3.48)}$.

Proof. The upper bound $C \leq A_{(10.3.48)}$ follows from (10.3.9).

Conversely, let f be any function such that $f^* = v_p$. Then, by (10.3.10), $\|f\|_{\Gamma^{p,\infty}(v)} \leq 2$, and, by (10.3.47),

$$A_{(10.3.48)} \leq \max\{1, 2^{\frac{q-1}{q}}\} \left[2C + \lambda \left(\int_0^\infty \frac{w(t)}{t^q} \, dt \right)^{\frac{1}{q}} \right].$$

If $\lambda = 0$, we are done. Assume that $\lambda > 0$. For arbitrary fixed $a > 0$, take any f such that $f^* = \frac{\lambda}{a} \chi_{(0,a)}$. Now, observe that $\sup_{s \geq 0} s^{-1} V^{\frac{1}{p}}(s) \leq \frac{2}{\lambda}$, whence

$$\|f\|_{\Gamma^{p,\infty}(v)} \leq \int_0^\infty f^*(y) \, dy \, \sup_{s \geq 0} s^{-1} V^{\frac{1}{p}}(s) \leq 2.$$

Thus, using the estimate $f^{**}(t) \geq \frac{\lambda}{t}\chi_{(a,\infty)}(t)$ and (10.3.47), we finally obtain

$$\lambda \left(\int_a^\infty \frac{w(t)}{t^q} \, dt \right)^{\frac{1}{q}} \leq C,$$

and the result follows on letting $a \to 0_+$. □

Theorem 10.3.24 (the case $\Gamma^{p,\infty}(v) \hookrightarrow \Gamma^{q,\infty}(w)$). *Let $p, q \in (0, \infty)$ and let v, w be weights. The inequality*

$$\sup_{t>0} f^{**}(t) W^{\frac{1}{q}}(t) \leq C \sup_{t>0} f^{**}(t) V^{\frac{1}{p}}(t) \tag{10.3.49}$$

holds if and only if

$$A_{(10.3.50)} := \sup_{t>0} \frac{W^{\frac{1}{q}}(t)}{t \sup_{s \geq t} s^{-1} V^{\frac{1}{p}}(s)} < \infty. \tag{10.3.50}$$

Moreover, the optimal constant C in (10.3.49) satisfies $C \approx A_{(10.3.50)}$.

Proof. The upper bound follows directly from the definition of $A_{(10.3.50)}$ similarly as in (10.3.9). To get the lower bound we test (10.3.49) with $\chi_{(0,a)}$, $a > 0$. For $0 < p \leq 1$, the assertion follows also from Theorem 10.3.1 (ii). □

10.3.5 The Halperin level function

Let us now recall the notion of the Halperin level function, introduced in [97, Definition 3.2]. We shall follow a slightly modified approach of [200, Section 4]. Let w be a weight on $(0, \infty)$, satisfying $\limsup_{t \to \infty} \frac{W(t)}{t} < \infty$. Then there exists a sequence $\{I_j\}_{j=1}^\infty$ of pairwise disjoint intervals in $[0, \infty)$ of finite measure and a set $E \subset [0, \infty)$ satisfying

$$E \cap \bigcup_{j=1}^\infty I_j = \emptyset$$

and such that, with a defined by $a := \sup \left\{ s; s \in E \cup \left(\bigcup_{j=1}^\infty I_j \right) \right\}$ (with $a = \infty$ as a possibility), the function w°, defined as

$$w^\circ(x) = \begin{cases} w(x), & x \in E, \\ \frac{1}{|I_j|} \int_{I_j} w(x) \, dx, & x \in I_j, \ j \in \mathbb{N}, \\ \limsup_{t \to \infty} \frac{W(t)}{t}, & x \in (a, \infty), \end{cases}$$

is nonincreasing on $[0, \infty)$. The function $w°$ so defined is called the *Halperin level function* of w. We note that, in particular, w is nonincreasing on E and, moreover, for every $x \in (0, \infty)$, one has

$$\int_0^x w(t)\, dt \leq \int_0^x w°(t)\, dt \quad \text{(with equality for } x \in E\text{)} \tag{10.3.51}$$

and

$$\lim_{x \to \infty} \frac{\int_0^x w(t)\, dt}{\int_0^x w°(t)\, dt} = 1.$$

It was shown in [200, Section 4] that $w°$ coincides almost everywhere with the derivative of the least concave majorant of W.

The following observation will be useful in this text. We shall use the symbol by $f \downarrow$ to indicate that f is nonincreasing.

Proposition 10.3.25. *Let $q \in (0, 1]$. Let $\mathcal{N}$ be a functional defined on locally integrable functions on $(0, \infty)$, with the property*

$$\int_0^t f(s)\, ds \leq \int_0^t g(s)\, ds, \quad f, g \downarrow, \ t > 0 \quad \Rightarrow \quad \mathcal{N}(f) \leq \mathcal{N}(g).$$

For a weight w on $(0, \infty)$, satisfying $\limsup_{t \to \infty} \frac{W(t)}{t} < \infty$, define

$$\beta(w) := \sup_{f \downarrow} \frac{\|f\|_{\Lambda^q(w)}}{\mathcal{N}(f)}.$$

Then

$$\beta(w) = \beta(w°),$$

where $w°$ is the level function of w.

Proof. First, $\beta(w) \leq \beta(w°)$ follows from (10.3.51) and the Hardy lemma. As for the converse inequality, let $[0, \infty) = E \cup \bigcup_j I_j \cup [a, \infty)$ be the partition of $(0, \infty)$ from the definition of $w°$. We shall assume that $a < \infty$. The proof needs only trivial modifications when $a = \infty$. Let f be a nonincreasing function on $(0, \infty)$ and let $t > a$. Then we define

$$f_t(x) = \begin{cases} f(x), & x \in E, \\ \frac{1}{|I_j|} \int_{I_j} f(t)\, dt, & x \in I_j, \ j \in \mathbb{N}, \\ \frac{1}{t-a} \int_a^t f(s)\, ds, & x \in [a, t), \\ 0, & x \in [t, \infty). \end{cases} \tag{10.3.52}$$

It is easy to see that f_t is nonincreasing for any $t > a$, and that

$$\int_0^s f_t(y)\, dy \leq \int_0^s f(y)\, dy, \quad s > 0. \tag{10.3.53}$$

Section 10.3 Embeddings

It follows immediately from (10.3.53) that for every $t > 0$

$$\mathcal{N}(f_t) \leq \mathcal{N}(f). \tag{10.3.54}$$

We now claim that

$$\int_0^\infty f^q(y) w^\circ(y) \, dy \leq \sup_{t>a} \int_0^\infty f_t^q(y) w(y) \, dy. \tag{10.3.55}$$

To prove (10.3.55), note first that

$$\limsup_{s \to \infty} \frac{W(s)}{s} = \limsup_{s \to \infty} \frac{1}{s-a} \int_a^s w(y) \, dy,$$

whence, by the Hölder inequality (recall $q \leq 1$),

$$\int_a^\infty f^q(y) w^\circ(y) \, dy = \limsup_{t \to \infty} \frac{W(t)}{t} \left(\int_a^\infty f^q(t) \, dt \right) \tag{10.3.56}$$

$$= \limsup_{t \to \infty} \left(\frac{1}{t-a} \int_a^t f^q(s) \, ds \right) \left(\int_a^t w(s) \, ds \right)$$

$$\leq \sup_{t>a} \left(\frac{1}{t-a} \int_a^t f(s) \, ds \right)^q \left(\int_a^t w(s) \, ds \right)$$

$$= \sup_{t>a} \int_a^\infty f_t^q(s) w(s) \, ds.$$

Now, by the definition of w°, (10.3.52), and (10.3.56),

$$\int_0^\infty f^q(s) w^\circ(s) \, ds = \int_E f^q(s) w(s) \, ds + \sum_j \frac{1}{|I_j|} \int_{I_j} f^q(s) \, ds \int_{I_j} w(s) \, ds$$

$$+ \int_a^\infty f^q(s) w^\circ(s) \, ds \leq \sup_{t>a} \int_0^\infty f_t^q(s) w(s) \, ds,$$

and (10.3.55) follows. Finally, combining (10.3.55), (10.3.54), and the fact that f_t is nonincreasing, we get

$$\beta(w^\circ) \leq \sup_{f \downarrow} \sup_{t>a} \frac{\left(\int_0^\infty (f_t)^q(s) w(s) \, ds \right)^{\frac{1}{q}}}{\mathcal{N}(f)} \leq \sup_{f \downarrow} \sup_{t>a} \frac{\left(\int_0^\infty (f_t)^q(s) w(s) \, ds \right)^{\frac{1}{q}}}{\mathcal{N}(f_t)}$$

$$\leq \sup_{g \downarrow} \frac{\left(\int_0^\infty g^q(s) w(s) \, ds \right)^{\frac{1}{q}}}{\mathcal{N}(g)} = \beta(w),$$

as desired. $\square$

Example 10.3.26. Setting, for $p \in (0, \infty)$, either

$$\mathcal{N}(f) = \left(\int_0^\infty \left(\frac{1}{t} \int_0^t f(s) \, ds \right)^p v(t) \, dt \right)^{\frac{1}{p}}$$

or

$$\mathcal{N}(f) = \sup_{0 < t < \infty} \left(\frac{1}{t} \int_0^t f(s) \, ds \right) V^{\frac{1}{p}}(t),$$

we obtain from Proposition 10.3.25 that, for $q \in (0, 1)$, $\Gamma^p(v) \hookrightarrow \Lambda^q(w)$ is equivalent to $\Gamma^p(v) \hookrightarrow \Lambda^q(w^\circ)$, and the same for $\Gamma^p(v)$ replaced by $\Gamma^{p,\infty}(v)$. Moreover, the corresponding optimal embedding constants coincide.

10.3.6 Embeddings of type $\Gamma^{p,\infty}(v) \hookrightarrow \Lambda^q(w)$

This section is devoted to the inequality

$$\left(\int_0^\infty (f^*(t))^q w(t) \, dt \right)^{\frac{1}{q}} \leq C \sup_{t>0} f^{**}(t) V^{\frac{1}{p}}(t). \tag{10.3.57}$$

By (10.3.9) and (10.3.10), (10.3.57) is equivalent to

$$\left(\int_0^\infty (f^*(t))^q w(t) \, dt \right)^{\frac{1}{q}} \leq C \sup_{t>0} \left(\int_0^t f^*(s) \, ds \right) \mathcal{V}_p(t), \tag{10.3.58}$$

where $\mathcal{V}_p(t)$ is from (10.3.7), and also to

$$\left(\int_0^\infty (f^*(t))^q w(t) \, dt \right)^{\frac{1}{q}} \leq \widetilde{C} \sup_{t>0} \frac{\int_0^t f^*(s) \, ds}{\lambda + \int_0^t v_p(s) \, ds}, \tag{10.3.59}$$

where λ and v_p are from (10.3.10). This could be rewritten as

$$C \approx \sup \left\{ \left(\int_0^\infty (f^*(t))^q w(t) \, dt \right)^{\frac{1}{q}} ; \int_0^t f^*(s) \, ds \leq \lambda + \int_0^t v_p(s) \, ds \right\}. \tag{10.3.60}$$

We shall focus on the case $q = 1$. First observe that the quantity

$$A_{(10.3.61)} := \sup_{t>0} \frac{W(t)}{t \mathcal{V}_p(t)} \approx \sup_{t>0} \frac{W(t)(\lambda + \int_0^t v_p(s) \, ds)}{t}, \tag{10.3.61}$$

is a lower bound for the optimal constant C in (10.3.58) with $q = 1$. Indeed, $A_{(10.3.61)} \lesssim C$ follows on testing (10.3.58) with $\chi_{(0,a)}$, $a > 0$. In particular, (10.3.58) implies $A_{(10.3.61)} < \infty$. Therefore,

$$\frac{W(t)}{t} \leq A_{(10.3.61)} \mathcal{V}_p(t), \qquad t > 0,$$

Section 10.3 Embeddings

and, as $\mathcal{V}_p$ is nonincreasing, also

$$\limsup_{t\to\infty} \frac{W(t)}{t} < \infty.$$

This guarantees the existence of w° (cf. Section 10.3.5).

Theorem 10.3.27 (the case $\Gamma^{p,\infty}(v) \hookrightarrow \Lambda^1(w)$). *Let $p \in (0,\infty)$ and let v, w be weights. The inequality (10.3.57) holds with $q = 1$ for every $f \in \Gamma^{p,\infty}(v)$ if and only if $A_{(10.3.61)} < \infty$, and*

$$A_{(10.3.62)} := \int_0^\infty w^\circ(t) v_p(t)\, dt < \infty. \tag{10.3.62}$$

Moreover, the optimal constant C in (10.3.57) satisfies

$$C \approx A_{(10.3.61)} + A_{(10.3.62)}.$$

Proof. By Proposition 10.3.25 and Example 10.3.9,

$$C = \sup \frac{\int_0^\infty f^*(t) w^\circ(t)\, dt}{\|f\|_{\Gamma^{p,\infty}(v)}}.$$

We claim that

$$C \approx \int_0^{\|w^\circ\|_\infty} \frac{1}{\mathcal{V}_p(w_*^\circ(y))}\, dy. \tag{10.3.63}$$

Indeed, using the formula for the distribution function, and [36, Theorem 2.1], we get

$$C = \sup\left\{\int_0^\infty f^*(s) w^\circ(s)\, ds;\ \int_0^t f^*(s)\, ds \le \frac{1}{\mathcal{V}_p(t)}\ \text{for all}\ t \in (0,\infty)\right\}$$

$$= \sup\left\{\int_0^\infty \left(\int_0^{w_*^\circ(s)} f^*(y)\, dy\right) ds;\ \int_0^t f^*(s)\, ds \le \frac{1}{\mathcal{V}_p(t)}\ \text{for all}\ t \in (0,\infty)\right\}$$

$$\le \int_0^{\|w^\circ\|_\infty} \frac{1}{\mathcal{V}_p(w_*^\circ(y))}\, dy.$$

Conversely, let $r > 0$ and set

$$g_r(t) = \begin{cases} \dfrac{t}{r\mathcal{V}_p(r)}, & t \le r \\[2pt] \dfrac{1}{\mathcal{V}_p(t)}, & t > r. \end{cases}$$

Then, since g_r is a quasi-concave function, there exists a nonincreasing function f_r so that $g_r(t) \approx \int_0^t f_r(s)\,ds$. Since, obviously, $\int_0^t f_r(s)\,ds \lesssim \frac{1}{V_p(t)}$, we get, for every $r > 0$,

$$C \geq \int_0^\infty f_r(t) w^\circ(t)\,dt = \int_0^\infty \int_0^{w_*^\circ(t)} f_r(s)\,ds\,dt$$
$$\approx \int_0^\infty g_r(w_*^\circ(s))\,ds \geq \int_0^{w^\circ(r)} \frac{1}{V_p(w_*^\circ(y))}\,dy,$$

and (10.3.63) follows on letting $r \to 0_+$.

Finally, observe that

$$C \approx \int_0^{\|w^\circ\|_\infty} \frac{1}{V_p(w_*^\circ(y))}\,dy = \lambda \|w^\circ\|_\infty + \int_0^{\|w^\circ\|_\infty} \left(\int_0^{w_*^\circ(t)} v_p(s)\,ds \right) dt$$
$$\approx \lambda \|w^\circ\|_\infty + A_{(10.3.62)}.$$

It remains to show $\lambda \|w^\circ\|_\infty \leq A_{(10.3.61)}$, or (cf. (10.3.61)), in particular, $\|w^\circ\|_\infty \leq \sup_t \frac{W(t)}{t}$. This is obvious when, for some $\delta > 0$, $(0, \delta)$ is one of the intervals I_j, since then $\|w^\circ\|_\infty = \delta^{-1} \int_0^\delta w(s)\,ds$. If $t_k \to 0_+$ for some sequence t_k in E, then the monotonicity of w° and the equality (10.3.51) for $x = t_k$ yields

$$\|w^\circ\|_\infty = \lim_{k \to \infty} w^\circ(t_k) \leq \lim_{k \to \infty} \frac{\int_0^{t_k} w^\circ(s)\,ds}{t_k} = \lim_{k \to \infty} \frac{W(t_k)}{t_k} \leq \sup_t \frac{W(t)}{t}.$$

The proof is complete. □

Remarks 10.3.28. (i) It follows from the proof of Theorem 10.3.27 that, if w is nonincreasing, then

$$\sup \left\{ \int_0^\infty f(t) w(t)\,dt;\ \int_0^t f(s)\,ds \leq \frac{1}{V_p(t)} \text{ for all } t \in (0, \infty) \right\} = C,$$

that is, we get the same constant without the restriction to nonincreasing f.

(ii) By a simple calculation it can be also shown that the optimal constant C in (10.3.57) with $q = 1$ satisfies

$$C \approx A_{(10.3.61)} + \int_0^\infty \frac{1}{V_p(t)}\,d(-w^\circ(t)) \approx A_{(10.3.61)} + \int_0^\infty w^\circ(t)\,d\left(\frac{1}{V_p(t)}\right).$$

10.3.7 The single-weight case $\Gamma^{1,\infty}(v) \hookrightarrow \Lambda^1(v)$

Now we return to the problem, when the norm in $\Lambda^1(v)$ can be expressed in terms of f^{**}. By Theorem 10.3.27 we can characterize the embedding $\Gamma^{1,\infty}(v) \hookrightarrow \Lambda^1(v)$.

By Proposition 10.3.25, we may restrict ourselves to the case when v is nonincreasing. By [135], then $\Lambda^1(v)$ is a Banach space, and by Theorem 10.3.13 (i) it is embedded into $\Gamma^{1,\infty}(v)$. Thus the embedding $\Gamma^{1,\infty}(v) \hookrightarrow \Lambda^1(v)$ is actually equivalent to $\Gamma^{1,\infty}(v) = \Lambda^1(v)$, and consequently, there is an expression of the $\Lambda^1(v)$ norm in terms of f^{**}.

Theorem 10.3.29. *Let v be a nonincreasing weight. Then the following statements are equivalent:*

$$\int_0^\infty f^*(s) v(s)\, ds \le C \sup_{t>0} f^{**}(t) V(t), \tag{10.3.64}$$

$$A_{(10.3.65)} := \int_0^\infty \frac{t}{V(t)}\, d(-v(t)) < \infty, \tag{10.3.65}$$

$$A_{(10.3.66)} := \int_0^\infty \frac{V(t) - tv(t)}{V^2(t)} v(t)\, dt < \infty, \tag{10.3.66}$$

$$\lim_{t \to 0+} v(t) < \infty \quad \text{and} \quad \begin{cases} \text{either} & \lim_{t\to\infty} V(t) < \infty \text{ and } \lim_{t\to\infty} v(t) = 0, \\ \text{or} & \lim_{t\to\infty} V(t) < \infty = \infty \text{ and } \lim_{t\to\infty} v(t) > 0. \end{cases} \tag{10.3.67}$$

Moreover, the optimal constant C in (10.3.64) satisfies $C \approx A_{(10.3.65)} + 1$.

Proof. We first claim that

$$A_{(10.3.65)} = \begin{cases} A_{(10.3.66)} & \text{if } V(\infty) = \infty, \\ A_{(10.3.66)} + 1 & \text{if } V(\infty) < \infty. \end{cases} \tag{10.3.68}$$

Let $A_{(10.3.65)} < \infty$. Then, for $t > 0$,

$$\infty > A_{(10.3.65)} = \int_0^\infty \frac{d(V(s) - sv(s))}{V(s)} \ge \int_0^t \frac{d(V(s) - sv(s))}{V(s)}$$

$$\ge \frac{1}{V(t)} \int_0^t d(V(s) - sv(s)) = \frac{V(t) - tv(t)}{V(t)},$$

since $V(t) - tv(t) = \int_0^t (v(s) - v(t))\, ds \to 0$ as $t \to 0+$, and therefore, by the absolute continuity of integral,

$$\lim_{t \to 0+} \frac{V(t) - tv(t)}{V(t)} = 0, \quad \lim_{t \to 0+} \frac{tv(t)}{V(t)} = 1. \tag{10.3.69}$$

Let $V(\infty) < \infty$. Then

$$V(t) - tv(t) = \int_0^t (v(s) - v(t))\, ds = \int_0^t \int_s^t d(-v(y))\, ds = \int_0^t s\, d(-v(s)).$$

Hence $\lim_{t\to\infty}(V(t) - tv(t)) = V(\infty) < \infty$, and consequently

$$\lim_{t\to\infty} \frac{V(t) - tv(t)}{V(t)} = 1, \qquad \lim_{t\to\infty} \frac{tv(t)}{V(t)} = 0. \qquad (10.3.70)$$

Integrating by parts we get

$$A_{(10.3.65)} = A_{(10.3.66)} + \left[\frac{-tv(t)}{V(t)}\right]_{t=0}^{t=\infty}. \qquad (10.3.71)$$

By (10.3.69), (10.3.70) and (10.3.71) we have $A_{(10.3.65)} = A_{(10.3.66)} + 1$.

Now let $V(\infty) = \infty$. Then, assuming that $A_{(10.3.66)} < \infty$, we get for $t > 0$,

$$\infty > A_{(10.3.66)} = \int_0^\infty (V(s) - sv(s))\,d\left(-\frac{1}{V(s)}\right) \geq \int_t^\infty (V(s) - sv(s))\,d\left(-\frac{1}{V(s)}\right)$$

$$\geq (V(t) - tv(t))\left(\frac{1}{V(t)} - \frac{1}{V(\infty)}\right) = \frac{V(t) - tv(t)}{V(t)},$$

since $(V(t) - tv(t))$ is nondecreasing in t, and by the absolute continuity of integral,

$$\lim_{t\to\infty} \frac{V(t) - tv(t)}{V(t)} = 0, \qquad \lim_{t\to\infty} \frac{tv(t)}{V(t)} = 1. \qquad (10.3.72)$$

Thus, by (10.3.69), $A_{(10.3.65)} = A_{(10.3.66)}$, proving (10.3.68) and in turn the equivalence of (10.3.65) and (10.3.66).

Since v is nonincreasing, we have $\mathcal{V}_p(t) = \mathcal{V}_1(t) = \frac{V(t)}{t}$, whence $A_{(10.3.61)} = 1$ and $A_{(10.3.62)} = A_{(10.3.66)}$. Thus, Theorem 10.3.27 yields the equivalence of (10.3.64) and (10.3.66) as well as the best constants relation.

It remains to show that (10.3.64) holds if and only if one of the options in (10.3.67) takes place. The "if" part is easy to verify. To prove the "only if" part, let $A_{(10.3.65)} < \infty$. Then, by (10.3.69), $2tv(t) \geq V(t)$ for $t \in (0, t_0)$ with some $t_0 > 0$. Hence

$$\infty > A_{(10.3.65)} \geq \int_0^{t_0} \frac{tv(t)}{V(t)} \frac{d(-v(t))}{v(t)} \geq \frac{1}{2} \log \frac{v(0+)}{v(t_0)},$$

and consequently $0 < v(0+) < \infty$. If $V(\infty) < \infty$, then $v(t) \leq \frac{V(t)}{t} \to 0$ as $t \to \infty$, and therefore $v(\infty) = 0$. If $V(\infty) = \infty$ and $A_{(10.3.66)} < \infty$, then by (10.3.72) $2tv(t) \geq V(t)$ on (t_1, ∞) for some $t_1 > 0$. Thus,

$$\infty > A_{(10.3.65)} \geq \frac{1}{2} \int_{t_1}^\infty \frac{d(-v(t))}{v(t)} \geq \frac{1}{2} \log \frac{v(t_1)}{v(\infty-)},$$

and therefore $0 < v(\infty-) < \infty$. This shows that (10.3.64) implies (10.3.67), and the proof is complete. $\square$

We finish this section with a characterization of all possible fundamental functions for which the corresponding Marcinkiewicz and Lorentz endpoint spaces coincide. This result is of independent interest, and it is a consequence of Theorem 10.3.29, more precisely of the equivalence of (10.3.64) and (10.3.67). It was also mentioned in Remark 7.12.15. Let us recall that the endpoint spaces Λ_φ and M_φ for a given suitable concave function φ were defined in Section 7.10, more precisely, see Definitions 7.10.1 and 7.10.13, cf. also Definition 7.10.17.

Corollary 10.3.30. *Let φ be a concave nondecreasing function on $[0, \infty)$ such that $\varphi(t) = 0$ if and only if $t = 0$, and $\frac{\varphi(t)}{t}$ is nonincreasing on $(0, \infty)$. Then the equality $\Lambda_\varphi = M_\varphi$ is true if and only if φ is on $(0, \infty)$ equivalent to one of the following four functions:*

$$t, \ 1, \ \max\{1; t\} \text{ and } 1 + t.$$

The latter function can be equivalently rewritten in the form $\min\{1; t\}$.

10.4 Associate spaces of classical and weak Lorentz spaces

In this section we apply our embedding results to the characterization of the associate spaces of classical and weak Lorentz spaces.

Theorem 10.4.1. *Let $p \in (0, \infty)$ and let v be a weight.*

(i) *Let $X = \Lambda^p(v)$. Then, for $0 < p \leq 1$,*

$$\|g\|_{X'} \approx \sup_{t>0} g^{**}(t) \frac{t}{V^{\frac{1}{p}}(t)},$$

and, for $1 < p < \infty$,

$$\|g\|_{X'} \approx \left(\int_0^\infty (g^{**}(t))^{p'} \frac{t^{p'} v(t)}{V^{p'}(t)} \, dt \right)^{\frac{1}{p'}} + V^{-\frac{1}{p}}(\infty) \int_0^\infty g^*(t) \, dt.$$

(ii) *Let $X = \Lambda^{p,\infty}(v)$. Then*

$$\|g\|_{X'} \approx \int_0^\infty g^*(t) V^{-\frac{1}{p}}(t) \, dt.$$

(iii) *Let $X = \Gamma^p(v)$. Let v satisfy (10.3.32). Then, for $0 < p \leq 1$,*

$$\|g\|_{X'} \approx \sup_{t>0} g^{**}(t) \frac{t}{\left(V(t) + t^p \int_t^\infty s^{-p} v(s) \, ds \right)^{\frac{1}{p}}},$$

and, for $1 < p < \infty$,

$$\left(\int_0^\infty (g^{**}(t))^{p'} t^{p'} \sum_k t \left(V(t) + t^p \int_t^\infty s^{-p} v(s)\, ds\right)^{-p'} \delta_{\mu_k}(t)\, dt\right)^{\frac{1}{p'}}$$

(cf. Theorem 10.3.17 above for the definition of μ_k and δ_{μ_k}).

(iv) Let $X = \Gamma^{p,\infty}(v)$. Then

$$\|g\|_{X'} \approx \sup_{t>0} g^{**}(t) \frac{1}{\mathcal{V}_p(t)} + \int_0^\infty g^*(t) v_p(t)\, dt,$$

where $\mathcal{V}_p$ is from (10.3.7) and v_p is from (10.3.10).

Proof. All the results follow from the embedding theorems above (namely Theorems 3.1, 3.3, 5.1 and 7.1), combined with (10.3.6). □

Corollary 10.4.2. *Assume that X is a rearrangement-invariant Banach function space and that φ_X is its fundamental function. Assume that φ_X is absolutely continuous and $\varphi_X \in \mathcal{F}$. Then*

$$(\Lambda_X)' = M_{X'}. \qquad (10.4.1)$$

Proof. It follows from $\varphi_X \in \mathcal{F}$ that φ_X is concave and that for every $f \in \mathcal{M}(\mathcal{R}, \mu)$ one has

$$\|f\|_{\Lambda_X} = \int_0^\infty f^*(t)\, d\varphi_X(t),$$

hence $\Lambda_X = \Lambda^1(d\varphi_X)$. Consequently, applying Theorem 10.4.1 (i) to $p = 1$ and $v(t) = d\varphi_X(t)$, $t \in (0, \infty)$, we get

$$\|f\|_{(\Lambda_X)'} = \sup_{t \in (0,\infty)} g^{**}(t) \frac{t}{\int_0^t d\varphi_X(s)\, ds}.$$

Using further the absolute continuity of φ_X and the relation (7.9.7), we obtain

$$\|f\|_{(\Lambda_X)'} = \sup_{t \in (0,\infty)} g^{**}(t) \frac{t}{\varphi_X(t)} = \sup_{t \in (0,\infty)} g^{**}(t) \varphi_{X'}(t) = \|f\|_{M_{X'}},$$

proving the claim. The proof is complete. □

Remarks 10.4.3. Some particular cases of the results of Theorem 10.4.1 can be reformulated in terms of equivalence of function spaces (cf. also [194] and [95]):

(i) If $0 < p \leq 1$ and $tV^{-\frac{1}{p}}(t)$ is nondecreasing, then

$$(\Lambda^p(v))' = \Gamma^{1,\infty}\left(\frac{d}{dt}\left(\frac{t}{V^{\frac{1}{p}}(t)}\right)\right).$$

(ii) If $1 < p < \infty$ and $V(\infty) = \infty$, then [194, Remark, p. 147]

$$(\Lambda^p(v))' = \Gamma^{p'}\left(\frac{t^{p'}v(t)}{V^{p'}(t)}\right).$$

(iii) For $0 < p < \infty$, $(\Lambda^{p,\infty}(v))' = \Lambda^1(V^{-\frac{1}{p}})$.

(iv) If $0 < p \leq 1$ and $t/[\varrho_p(t)]$ is nondecreasing, then

$$(\Gamma^p(v))' = \Gamma^{1,\infty}\left(\frac{d}{dt}\frac{t}{\varrho_p(t)}\right),$$

where

$$\varrho_p(t) = \left(\int_0^\infty \frac{v(s)}{(s+t)^p}\,ds\right)^{\frac{1}{p}}.$$

(v) If $1 < p < \infty$, then $(\Gamma^p(v))' = \Gamma^{p'}(\sigma)$, where

$$\sigma = t^{p'}\sum_k t\left(V(t) + t^p\int_t^\infty s^{-p}v(s)\,ds\right)^{-p'}\delta_{\mu_k}(t).$$

(vi) If $0 < p < \infty$, v is nonincreasing and $tV^{-\frac{1}{p}}(t)$ is nondecreasing, then

$$(\Gamma^{p,\infty}(v))' = \Gamma^{1,\infty}(u) \cap \Lambda^1(u),$$

where

$$u(t) = \frac{d}{dt}\left(\frac{t}{V^{\frac{1}{p}}(t)}\right).$$

10.5 Comparison of classical Lorentz spaces to Orlicz spaces

It is of interest to compare Lorentz spaces to Orlicz spaces. In this direction, there is the following result.

Proposition 10.5.1. *Let Φ be a Young function and let L^Φ be the corresponding Orlicz space endowed with the Luxemburg norm. Let $\omega_\Phi : [0,1) \to [0,\infty)$ be the function given by*

$$\omega_\Phi(s) = \begin{cases} \dfrac{1}{\Phi^{-1}\left(\frac{1}{s}\right)} & \text{when } s \in (0,1), \\ 0 & \text{when } s = 0. \end{cases}$$

Suppose that there exists a $\delta \in (0, 1)$ such that

$$\int_0^{\mu(\mathcal{R})} \Phi\left(\delta\Phi^{-1}\left(\frac{1}{s}\right)\right) ds < \infty. \tag{10.5.1}$$

Then $L^\Phi(\mathcal{R}, \mu) = \Gamma^\infty(\omega_\Phi)(\mathcal{R}, \mu)$.

Proof. By Example 7.9.4 (iii), ω_Φ is the fundamental function of the space $L^\Phi(\mathcal{R}, \mu)$. Moreover, by Remark 10.1.5, the space $\Gamma^\infty(\omega_\Phi)(\mathcal{R}, \mu)$ coincides with the Marcinkiewicz space $M_{\omega_\Phi}(\mathcal{R}, \mu)$. Therefore, the embedding

$$L^\Phi(\mathcal{R}, \mu) \hookrightarrow \Gamma^\infty(\omega_\Phi)(\mathcal{R}, \mu)$$

is a straightforward consequence of Proposition 7.10.6. On the other hand, (10.5.1) implies that the function $g_0(t) := \Phi^{-1}\left(\frac{1}{t}\right)$, $t \in (0, \mu(\mathcal{R}))$ belongs to $L^\Phi(0, \mu(\mathcal{R}))$. Next, any function g such that $\|g\|_{\Gamma^\infty(\omega_\Phi)(\mathcal{R},\mu)} \leq 1$ fulfills the inequality

$$g^*(t) \leq g^{**}(t) \leq \Phi^{-1}\left(\frac{1}{t}\right) \quad \text{for } t \in (0, 1). \tag{10.5.2}$$

Therefore, every such g satisfies

$$\|g\|_{L^\Phi(\mathcal{R},\mu)} = \|g^*\|_{L^\Phi(0,\mu(\mathcal{R}))} \leq \|g_0\|_{L^\Phi(0,\mu(\mathcal{R}))} < \infty,$$

and the reverse embedding

$$\Gamma^\infty(\omega_\Phi)(\mathcal{R}, \mu) \hookrightarrow L^\Phi(\mathcal{R}, \mu)$$

follows from the lattice property of Banach function spaces. □

10.6 Function spaces measuring symmetrized mean oscillation

In this section we will study function spaces whose norms are defined in terms of the functional $f^{**} - f^*$.

The functional $f^{**} - f^*$ has been shown to be useful in various parts of analysis including the interpolation theory (see [33] for some history and references). In [13], the functional

$$\left(\int_0^\infty t^{\frac{1}{p}}\left[f^{**}(t) - f^*(t)\right]^q \frac{dt}{t}\right)^{\frac{1}{q}}$$

was introduced (for $1 < p < \infty$ and $0 < q \leq \infty$) and interesting applications were found. In particular, the *weak L^∞* space, determined by the seminorm $\sup_{t \in (0,\infty)} (f^{**}(t) - f^*(t))$ (corresponding formally to the case $q = \infty$) was created and proved useful in situations in which the classical L^∞ fails.

Section 10.6 Spaces measuring oscillation

It should be noticed that in the study of function spaces defined in terms of the functional $f^{**} - f^*$ certain care must be exercised. In particular, this functional vanishes on constant functions and, moreover, the operation $f \mapsto (f^{**} - f^*)$ is not subadditive. Therefore, quantities involving $f^{**} - f^*$ do not have norm properties, which makes the study of the corresponding function spaces difficult.

Recently, various structures involving the quantity $f^{**} - f^*$ appear quite regularly as natural function spaces in various situations. For example, they play an important role in certain "optimal partner target space problems" (see, e.g., [117, 154, 10]), in the duality problem for classical Lorentz spaces of type Γ [201], in the investigation of the boundedness of maximal Calderón–Zygmund singular integral operators on classical Lorentz spaces [18], and so on. For more detailed history and more references we refer the reader to [33].

The main object of study in this section will be the fairly general class of weighted function spaces denoted by $S^p(v)$, which was introduced in [33].

Definition 10.6.1. Let $0 < p < \infty$ and let v be a weight on $(0, \infty)$, that is, a measurable nonnegative function. Then, the space $S^p(v)$ is the collection of all measurable functions on $(0, \infty)$ such that $\|f\|_{S^p(v)} < \infty$, where

$$\|f\|_{S^p(v)} := \left(\int_0^\infty [f^{**}(t) - f^*(t)]^p v(t)\, dt \right)^{\frac{1}{p}}.$$

Remark 10.6.2. As already noted above, the functional $\|f\|_{S^p(v)}$ is not a norm because it vanishes on constant functions. To overcome this problem, one can either factor out constants or assume that $f^*(\infty) = 0$. Even then, however, it is not necessarily a norm.

In view of Remark 10.6.2 it is desirable to investigate when $\|f\|_{S^p(v)}$ is at least equivalent to a norm, and to carry out a thorough research of relations of the spaces $S^p(v)$ to other, more familiar function spaces. Our principal objective will be to study embedding relations between the spaces $S^p(v)$ and the spaces of type Λ and Γ.

One of the main sources of motivation for this research is the well-known inequality

$$t^{\frac{1}{n}} \left(f^{**}(t) - f^*(t) \right) \leq C (\nabla f)^{**}(t),$$

involving the *gradient* ∇f of a differentiable function f of several variables defined on a suitable domain in $\mathbb{R}^n$ (we recall that by *gradient* we mean the vector formed of all first-order partial derivatives of f). This estimate folds for every appropriate function f and every $t > 0$. Hence, taking into account Proposition 8.1.8, we get, for $p > 1$,

$$\left\| t^{\frac{1}{n}} \left(f^{**}(t) - f^*(t) \right) \right\|_p \leq C \, \|\nabla f\|_p,$$

which gives a lower bound for an L^p-norm of the gradient in terms of the norm in the space $S^p(t^{\frac{p}{n}})$. Since estimates of similar kinds are of crucial importance in the theory of partial differential equations, it clearly follows that some knowledge about relations between the spaces $S^p(v)$ and other function spaces might come handy.

We shall need certain mean integral operators.

Notation 10.6.3. Let h be a locally integrable function on $(0, \infty)$. Then we denote by P the *average Hardy operator*

$$(Ph)(t) := \frac{1}{t} \int_0^t h(s)\, ds, \quad h \geq 0, \tag{10.6.1}$$

and by Q its associate operator (under the pairing $\int_0^\infty fg$), that is,

$$(Qh)(t) := \int_t^\infty \frac{h(s)}{s}\, ds, \quad h \geq 0. \tag{10.6.2}$$

Remark 10.6.4. The characterization of one of the embeddings which we have in mind will require a necessary and sufficient condition for the two-operator weighted norm inequality

$$\|Ph\|_{L^q(w)} \lesssim \|Qh\|_{L^p(v)} \tag{10.6.3}$$

for every positive measurable h on $(0, \infty)$. Moreover, the inequality (10.6.3) is clearly of independent interest (see, e.g., its particular instances when $1 < p = q < \infty$ and $v = w$ in [163]).

Our aim will be to give necessary and sufficient conditions for (10.6.3) to hold. We restrict ourselves to the case $1 \leq q \leq \infty$ (the technical reason for this restriction is the use of duality). In the proofs of theorems that follow we shall often need some results from the theory of weighted inequalities that go beyond the scope of this book; we will thus only give appropriate references.

Remark 10.6.5. The task ahead of us can be reformulated also as follows: we need to characterize the quantity

$$A := \sup_{h \geq 0} \frac{\|Ph\|_{L^q(w)}}{\|Qh\|_{L^p(v)}}, \tag{10.6.4}$$

where $0 < p \leq \infty$, $1 \leq q \leq \infty$, v, w are weights on $(0, \infty)$ and

$$\|f\|_{L^p(v)} := \begin{cases} \left(\int_0^\infty |f(x)|^p v(x)\, dx\right)^{\frac{1}{p}} & \text{if } 0 < p < \infty \\ \operatorname*{ess\,sup}_{x \in (0,\infty)} |f(x)| v(x) & \text{if } p = \infty. \end{cases}$$

We start with recalling a useful inequality, which is just a particular case of the general result in [36, Theorem 3.2] (cf. also [203]). Let $0 < p \leq 1$ and let v be a weight on $(0, \infty)$. Then,

$$\left(\int_0^t f^*(s)^{\frac{1}{p}} v(s)\, ds\right)^p \leq p \int_0^t f^*(s) V(s)^{p-1} v(s)\, ds, \quad t \in (0, \infty). \tag{10.6.5}$$

Section 10.6 Spaces measuring oscillation

Notation 10.6.6. Given a function $f \in \mathcal{M}_+(0, \infty)$, a weight v and $p \in (0, \infty]$, we define
$$B(f) := \sup_{g \geq 0} \frac{\int_0^\infty f(t)g(t)\, dt}{\|Qg\|_{L^p(v)}}. \tag{10.6.6}$$

We will now find a necessary and sufficient condition for the reverse Hardy inequality involving the operator Q.

Theorem 10.6.7. *Let f be a nonnegative measurable function on $(0, \infty)$. Let v be a weight on $(0, \infty)$ and let $B(f)$ be given by (10.6.6).*

(i) *Assume that $1 < p < \infty$. Then*
$$B(f) \approx \left(\int_0^\infty \left[\operatorname*{ess\,sup}_{0 < s \leq t} s f(s) \right]^{p'} \frac{v(t)}{V^{p'}(t)} \, dt \right)^{\frac{1}{p'}} + \frac{\operatorname*{ess\,sup}_{0 < s < \infty} s f(s)}{V(\infty)^{\frac{1}{p}}}. \tag{10.6.7}$$

(ii) *Assume that $0 < p \leq 1$. Then*
$$B(f) \approx \operatorname*{ess\,sup}_{0 < s < \infty} \frac{s f(s)}{V(s)^{\frac{1}{p}}}.$$

(iii) *Assume that $p = \infty$. Then*
$$B(f) \approx \int_0^\infty \operatorname*{ess\,sup}_{0 < s \leq t} s f(s) \, d\left(\frac{-1}{\operatorname*{ess\,sup}_{0 < s \leq t} v(s)} \right) + \frac{\operatorname*{ess\,sup}_{0 < s \leq \infty} s f(s)}{\operatorname*{ess\,sup}_{0 < s \leq \infty} v(s)}.$$

Proof. (i) The assertion is a simple modification of a recent result of [202, Corollary 3.8] (for various related results see also [96]).

(ii) We recall the inequality (cf. [36, 203, 216])
$$\int_0^\infty h^*(t) V(t)^{\frac{1}{p}-1} v(t) \, dt \lesssim \left(\int_0^\infty h^*(t)^p v(t) \, dt \right)^{\frac{1}{p}}, \qquad h \in \mathcal{M}_+(0, \infty). \tag{10.6.8}$$

Using (10.6.8) and the Fubini theorem,
$$B(f) \leq \operatorname*{ess\,sup}_{0 < t < \infty} \frac{t f(t)}{V(t)^{\frac{1}{p}}} \sup_{g \geq 0} \frac{\int_0^\infty \frac{g(s)}{s} V(s)^{\frac{1}{p}} \, ds}{\|Qg\|_{L^p(v)}}$$
$$\approx \operatorname*{ess\,sup}_{0 < t < \infty} \frac{t f(t)}{V(t)^{\frac{1}{p}}} \sup_{g \geq 0} \frac{\int_0^\infty \frac{g(s)}{s} \int_0^s V(y)^{\frac{1}{p}-1} v(y) \, dy \, ds}{\|Qg\|_{L^p(v)}}$$

$$= \operatorname*{ess\,sup}_{0<t<\infty} \frac{tf(t)}{V(t)^{\frac{1}{p}}} \sup_{g\geq 0} \frac{\int_0^\infty (Qg)(s) V(s)^{\frac{1}{p}-1} v(s)\,ds}{\|Qg\|_{L^p(v)}}$$

$$\lesssim \operatorname*{ess\,sup}_{0<t<\infty} \frac{tf(t)}{V(t)^{\frac{1}{p}}}.$$

(The last step is (10.6.8) applied to $h^* = Qg$.)

Conversely, given $\varepsilon > 0$ and $x \in (0, \infty)$, set

$$g_{\varepsilon,x}(t) := \frac{t}{\varepsilon} \chi_{(x-\varepsilon, x)}(t), \qquad t \in (0, \infty).$$

Then,

$$(Qg_{\varepsilon,x})(t) \leq \chi_{(0,x)}(t),$$

whence

$$B(f) \geq \sup_{x,\varepsilon} \frac{\int_0^\infty f(t) g_{\varepsilon,x}(t)\,dt}{\|Qg_{\varepsilon,x}\|_{L^p(v)}} \geq \sup_{x,\varepsilon} \frac{\frac{1}{\varepsilon}\int_{x-\varepsilon}^x tf(t)\,dt}{\|\chi_{(0,x)}\|_{L^p(v)}} \geq \operatorname*{ess\,sup}_{0<x<\infty} \frac{xf(x)}{V(x)^{\frac{1}{p}}},$$

proving the claim.

(iii) Let $p = \infty$. We claim that then

$$B(f) = \sup_{g \geq 0} \frac{\int_0^\infty g(t) \operatorname{ess\,sup}_{0<s\leq t} sf(s)\,dt}{\operatorname{ess\,sup}_{0<t<\infty} v(t) \int_t^\infty g(s)\,ds}. \tag{10.6.9}$$

By [202, Theorem 2.1, (2.3)], we have

$$\sup_{g\geq 0} \frac{\int_0^\infty g(t) \operatorname{ess\,sup}_{0<s\leq t} sf(s)\,dt}{\operatorname{ess\,sup}_{0<t<\infty} v(t) \int_t^\infty g(s)\,ds}$$

$$= \sup_{g\geq 0} \sup_{\int_t^\infty h \leq \int_t^\infty g} \frac{\int_0^\infty h(t) tf(t)\,dt}{\operatorname{ess\,sup}_{0<t<\infty} v(t) \int_t^\infty g(s)\,ds}$$

$$\leq \sup_{g\geq 0} \sup_{\int_t^\infty h \leq \int_t^\infty g} \frac{\int_0^\infty h(t) tf(t)\,dt}{\operatorname{ess\,sup}_{0<t<\infty} v(t) \int_t^\infty h(s)\,ds}$$

$$\leq \sup_{h\geq 0} \frac{\int_0^\infty h(t) tf(t)\,dt}{\operatorname{ess\,sup}_{0<t<\infty} v(t) \int_t^\infty h(s)\,ds}.$$

Since the converse inequality is trivial, this proves (10.6.9).

Section 10.6 Spaces measuring oscillation

We now define the set

$$\mathcal{H} := \left\{ h \in \mathcal{M}_+(0, \infty); \int_0^t h(s)\, ds \leq \operatorname*{ess\,sup}_{0 < s \leq t} sf(s)\, ds,\ t \in (0, \infty) \right\}$$

and claim that

$$\int_0^\infty g(t) \operatorname*{ess\,sup}_{0 < s \leq t} sf(s)\, dt = \sup_{h \in \mathcal{H}} \int_0^\infty g(t) \int_0^t h(s)\, ds\, dt.$$

Indeed, we first observe that the inequality "$\geq$" is obvious. To get the converse one, note that, for every nondecreasing function Φ on $(0, \infty)$, there is a sequence $\{H_n\}_{n=1}^\infty$ of smooth increasing functions such that $H_n \nearrow \Phi$ as $n \to \infty$. By the Fatou Lemma,

$$\int_0^\infty g(t)\Phi(t)\, dt \leq \limsup_{n \to \infty} \int_0^\infty g(t) H_n(t)\, dt.$$

Applying this to the (nondecreasing) function $\Phi(t) = \operatorname*{ess\,sup}_{0 < s \leq t} sf(s)$ and noting that the functions H_n, being smooth, can be represented as $H_n(t) = \int_0^t h_n(s)\, ds$ for some positive measurable functions h_n on $(0, \infty)$, we obtain

$$\int_0^\infty g(t) \operatorname*{ess\,sup}_{0 < s \leq t} sf(s)\, dt \leq \sup_{h \in \mathcal{H}} \int_0^\infty g(t) \int_0^t h(s)\, ds\, dt,$$

proving our claim. Thus, we have

$$\int_0^\infty g(t) \operatorname*{ess\,sup}_{0 < s \leq t} sf(s)\, dt = \sup_{h \in \mathcal{H}} \int_0^\infty g(t) \int_0^t h(s)\, ds\, dt$$

$$= \sup_{h \in \mathcal{H}} \int_0^\infty h(t) \int_t^\infty g(s)\, ds\, dt.$$

Inserting this to (10.6.9), we have

$$B(f) = \sup_{g \geq 0} \sup_{h \in \mathcal{H}} \frac{\int_0^\infty h(t) \int_t^\infty g(s)\, ds\, dt}{\operatorname*{ess\,sup}_{0 < t < \infty} v(t) \int_t^\infty g(s)\, ds}.$$

Thus, by the monotonicity of $\int_t^\infty g(s)\, ds$, the Fubini theorem, and Theorems 10.3.10 and 10.4.1 we have

$$B(f) = \sup_{h \in \mathcal{H}} \sup_{k \in \mathcal{M}_+(0,\infty)} \frac{\int_0^\infty h(t) k^*(t)\, dt}{\operatorname*{ess\,sup}_{0 < t < \infty} v(t) k^*(t)}$$

$$= \sup_{h \in \mathcal{H}} \sup_{k \in \mathcal{M}_+(0,\infty)} \frac{\int_0^\infty h(t) k^*(t)\, dt}{\operatorname*{ess\,sup}_{0 < t < \infty} \left(\operatorname*{ess\,sup}_{0 < s \leq t} v(s) \right) k^*(t)}$$

$$= \sup_{h \in \mathcal{H}} \int_0^\infty \frac{h(t)}{\operatorname*{ess\,sup}_{0 < s \leq t} v(s)}\, dt.$$

Finally, the integration by parts for the Stieltjes integral and the Fatou lemma give

$$B(f) = \sup_{h \in \mathcal{H}} \left[\int_0^\infty \left(\int_0^t h(s)\,ds \right) d\left(\frac{-1}{\operatorname{ess\,sup}_{0<s\leq t} v(s)} \right) + \frac{\int_0^\infty h(t)\,dt}{\operatorname{ess\,sup}_{0<s<\infty} v(s)} \right]$$

$$= \left[\int_0^\infty \operatorname*{ess\,sup}_{0<s\leq t} sf(s)\, d\left(\frac{-1}{\operatorname{ess\,sup}_{0<s\leq t} v(s)} \right) + \frac{\operatorname{ess\,sup}_{0<s<\infty} sf(s)}{\operatorname{ess\,sup}_{0<s<\infty} v(s)} \right].$$

Of course, the last summand disappears if $\operatorname{ess\,sup}_{0<s<\infty} v(s) = \infty$. □

Given weights u, v and $t \in (0, \infty)$, we define

$$\bar{u}(t) := \operatorname*{ess\,sup}_{0<s<t} u(s)$$

and

$$\sigma_p(t) := \begin{cases} \left(\int_t^\infty \frac{v(s)^{1-p'}}{s^{p'}}\,ds \right)^{\frac{1}{p'}} & \text{if } p > 1 \\ \operatorname*{ess\,sup}_{t\leq s<\infty} \frac{1}{sv(s)} & \text{if } p = 1. \end{cases}$$

Proposition 10.6.8. *Let $0 < p < \infty$, $1 \leq q < \infty$, $0 < \alpha < \infty$ and let v, w be weights on $(0, \infty)$. If $q < p$, set $r = \frac{pq}{p-q}$. Then the inequality*

$$\left(\int_0^\infty \left[\sup_{0<s\leq t} s^\alpha (Qg)(s) \right]^q w(t)\,dt \right)^{\frac{1}{q}} \lesssim \left(\int_0^\infty g(t)^p v(t)\,dt \right)^{\frac{1}{p}} \tag{10.6.10}$$

holds for all $g \geq 0$ if and only if one of the following conditions holds:

(i) $1 \leq p \leq q < \infty$ and

$$A_{(10.6.11)} := \sup_{0<t<\infty} \left(t^{\alpha q} \int_t^\infty w(s)\,ds + \int_0^t s^{\alpha q} w(s)\,ds \right)^{\frac{1}{q}} \sigma_p(t) < \infty; \tag{10.6.11}$$

(ii) $0 < q < p$, $1 \leq p < \infty$,

$$A_{(10.6.12)} := \left(\int_0^\infty \left(\int_0^t s^{\alpha q} w(s)\,ds \right)^{\frac{r}{q}} w(t) t^{\alpha q} \left[\sigma_p(t) \right]^r dt \right)^{\frac{1}{r}} < \infty \tag{10.6.12}$$

Section 10.6 Spaces measuring oscillation

and

$$A_{(10.6.13)} := \left(\int_0^\infty \left(\int_t^\infty w(s)\,ds \right)^{\frac{r}{p}} w(t) \sup_{0<s\leq t} \left(s^{\alpha r} [\sigma_p(s)]^r \right) dt \right)^{\frac{1}{r}} < \infty. \tag{10.6.13}$$

Moreover, the best constant in (10.6.10) is comparable to $A_{(10.6.11)}$ in the case (i) and to $A_{(10.6.12)} + A_{(10.6.13)}$ in the case (ii).

Proof. We will prove both assertions for the case when $\alpha = 1$ (the general case can be then obtained by a simple change of variables).

(i) Assume that $1 \leq p \leq q < \infty$. Then, by a consecutive change of variables,

$$\int_0^\infty \left[\sup_{0<s\leq t} s(Qg)(s) \right]^q w(t)\,dt = \int_0^\infty \left[\sup_{0<s\leq t} s \int_0^{\frac{1}{s}} \frac{g(\frac{1}{y})}{y}\,dy \right]^q w(t)\,dt$$

$$= \int_0^\infty \left[\sup_{\frac{1}{t} \leq s<\infty} \frac{1}{s} \int_0^s \frac{g(\frac{1}{y})}{y}\,dy \right]^q w(t)\,dt$$

$$= \int_0^\infty \left[\sup_{t\leq s<\infty} \frac{1}{s} \int_0^s \frac{g\left(\frac{1}{y}\right)}{y}\,dy \right]^q \frac{w\left(\frac{1}{t}\right)}{t^2}\,dt,$$

while

$$\int_0^\infty g(t)^p v(t)\,dt = \int_0^\infty t^{-p} g\left(\frac{1}{t}\right)^p \tilde{v}_p(t)\,dt.$$

Thus, setting $h(y) := g(\frac{1}{y})\frac{1}{y}$, we obtain that (10.6.10) is equivalent to

$$\left(\int_0^\infty \left[\sup_{t\leq s<\infty} \frac{1}{s} \int_0^s h(y)\,dy \right]^q \frac{w(\frac{1}{t})}{t^2}\,dt \right)^{\frac{1}{q}} \lesssim \left(\int_0^\infty h(t)^p \tilde{v}_p(t)\,dt \right)^{\frac{1}{p}}. \tag{10.6.14}$$

Then, by [90, Theorem 4.1], (10.6.14) holds if and only if

$$\sup_{0<t<\infty} \left(\frac{1}{t^q} \int_0^t \frac{w(\frac{1}{s})}{s^2}\,ds + \int_t^\infty \frac{w(\frac{1}{s})}{s^{q+2}}\,ds \right)^{\frac{1}{q}} \sigma_p\left(\frac{1}{t}\right) < \infty. \tag{10.6.15}$$

By calculation, we have, for every $t \in (0, \infty)$,

$$\int_0^t \frac{w(\frac{1}{s})}{s^2}\,ds = \int_{\frac{1}{t}}^\infty w(s)\,ds, \tag{10.6.16}$$

and

$$\int_t^\infty \frac{w(\frac{1}{s})}{s^{q+2}}\,ds = \int_0^{\frac{1}{t}} s^q w(s)\,ds, \tag{10.6.17}$$

Altogether, (10.6.10) holds if and only if

$$\sup_{0<t<\infty} \left(\frac{1}{t^q}\int_{\frac{1}{t}}^\infty w(s)\,ds + \int_0^{\frac{1}{t}} s^q w(s)\,ds\right)^{\frac{1}{q}} \sigma_p\left(\frac{1}{t}\right) < \infty,$$

which is clearly equivalent to (10.6.11). The best-constant relation follows from the argument.

(ii) Again, (10.6.10) is equivalent to (10.6.14). By [90, Theorem 4.4], (10.6.14) holds if and only if

$$\left(\int_0^\infty \left(\int_t^\infty \frac{w(\frac{1}{s})}{s^{q+2}}\,ds\right)^{\frac{r}{p}} \sigma_p\left(\frac{1}{t}\right)^r \frac{w(\frac{1}{t})}{t^{q+2}}\,dt\right)^{\frac{1}{r}} < \infty \tag{10.6.18}$$

and

$$\left(\int_0^\infty \left(\int_0^t \frac{w(\frac{1}{s})}{s^2}\,ds\right)^{\frac{r}{p}} \operatorname*{ess\,sup}_{t\le s<\infty} s^{-q}\sigma_p\left(\frac{1}{s}\right)^r \frac{w(\frac{1}{t})}{t^2}\,dt\right)^{\frac{1}{r}} < \infty. \tag{10.6.19}$$

Using (10.6.16)–(10.6.17) and changing variables, we get (10.6.18) and (10.6.19) equivalent to (10.6.12) and (10.6.13), respectively. $\square$

We are now in a position to characterize (10.6.3).

Theorem 10.6.9. *Let $0 < p \le \infty$, $1 \le q \le \infty$. When $0 < p < q < \infty$, we set $r = \frac{pq}{p-q}$. Let v, w be weights on $(0, \infty)$. Then (10.6.3) holds if and only if one of the following conditions holds:*

(i) $1 < p \le q < \infty$ and

$$\sup_{0<t<\infty} \left(\int_t^\infty \frac{w(s)}{s^q}\,ds\right)^{\frac{1}{q}} \left(\int_0^t \frac{s^{p'}v(s)}{V(s)^{p'}}\,ds + \frac{t^{p'}}{V(t)^{p'-1}}\right)^{\frac{1}{p'}} < \infty;$$

(ii) $1 < p < \infty$, $q = \infty$ and

$$\sup_{0<t<\infty} \left(\int_0^t \frac{s^{p'}v(s)}{V(s)^{p'}}\,ds + \frac{t^{p'}}{V(t)^{p'-1}}\right)^{\frac{1}{p'}} \operatorname*{ess\,sup}_{t\le s<\infty} \frac{w(s)}{s} < \infty;$$

Section 10.6 Spaces measuring oscillation

(iii) $p = q = \infty$ and

$$\sup_{0<t<\infty} \left(\frac{t}{\operatorname{ess\,sup}_{0<s\leq t} v(s)} + \int_0^t s\,d\left(\frac{-1}{\operatorname{ess\,sup}_{0<y\leq s} v(y)} \right) \right) \operatorname{ess\,sup}_{t\leq s<\infty} \frac{w(s)}{s} < \infty;$$

(iv) $1 < q < p < \infty$,

$$\left(\int_0^\infty \left(\sup_{0<s\leq t} s^q \int_s^\infty \frac{w(y)}{y^q}\,dy \right)^{\frac{r}{q}} \frac{v(t)}{V(t)^{\frac{r}{q}}}\,dt \right)^{\frac{1}{r}} < \infty,$$

$$\frac{\sup_{0<t<\infty} t \left(\int_t^\infty \frac{w(s)}{s^q}\,ds \right)^{\frac{1}{q}}}{V(\infty)^{\frac{1}{p}}} < \infty$$

and

$$\left(\int_0^\infty \left(\int_0^t \frac{s^{p'} v(s)}{V(s)^{p'}}\,ds \right)^{\frac{r}{p'}} \frac{t^{p'} v(t)}{V(t)^{p'}} \left(\int_t^\infty \frac{w(s)}{s^q}\,ds \right)^{\frac{r}{q}} dt \right)^{\frac{1}{r}} < \infty;$$

(v) $p = \infty$, $1 < q < \infty$,

$$\int_0^\infty \left(\int_0^t s\,d\left(\frac{-1}{\operatorname{ess\,sup}_{0<y\leq s} v(y)} \right) \right)^q t \left(\int_t^\infty \frac{w(s)}{s^q}\,ds \right)^{\frac{1}{q}}$$

$$\times d\left(\frac{-1}{\operatorname{ess\,sup}_{0<s\leq t} v(s)} \right) < \infty,$$

$$\frac{\operatorname{ess\,sup}_{0<s<\infty} w(s)}{\operatorname{ess\,sup}_{0<s<\infty} v(s)} < \infty$$

and

$$\int_0^\infty \left(\int_t^\infty d\left(\frac{-1}{\operatorname{ess\,sup}_{0<y\leq s} v(y)} \right) \right)^{q-1} \sup_{0<s\leq t} s \left(\int_s^\infty \frac{w(y)}{y^q}\,dy \right)^{\frac{1}{q}}$$

$$\times d\left(\frac{-1}{\operatorname{ess\,sup}_{0<s\leq t} v(s)} \right) < \infty;$$

(vi) $0 < p \leq 1 < q < \infty$ and

$$\sup_{0<t<\infty} \frac{t}{V(t)^{\frac{1}{p}}} \left(\int_t^\infty \frac{w(s)}{s^q}\,ds \right)^{\frac{1}{q}} < \infty;$$

(vii) $0 < p \leq 1$, $q = \infty$ and

$$\sup_{0<t<\infty} \left(\sup_{0<s\leq t} \frac{s}{V(s)^{\frac{1}{p}}} \right) \operatorname*{ess\,sup}_{t\leq s<\infty} \frac{w(s)}{s} < \infty;$$

(viii) $q = 1 < p < \infty$,

$$\left(\int_0^\infty \left(\sup_{0<s\leq t} s \int_s^\infty \frac{w(y)}{y} \, dy \right)^{p'} \frac{v(t)}{V(t)^{p'}} \, dt \right)^{\frac{1}{p'}} < \infty$$

and

$$\frac{\sup_{0<t<\infty} t \int_t^\infty \frac{w(s)}{s} \, ds}{V(\infty)^{\frac{1}{p}}} < \infty;$$

(ix) $0 < p \leq 1 = q$ and

$$\sup_{0<t<\infty} \frac{t}{V(t)^{\frac{1}{p}}} \int_t^\infty \frac{w(s)}{s} \, ds < \infty;$$

(x) $p = \infty$, $q = 1$,

$$\int_0^\infty \sup_{0<s\leq t} s \int_s^\infty \frac{w(s)}{s} \, d\left(\frac{-1}{\operatorname*{ess\,sup}_{0<s\leq t} v(s)} \right) < \infty$$

and

$$\frac{\sup_{0<t<\infty} t \int_t^\infty \frac{w(s)}{s} \, ds}{\operatorname*{ess\,sup}_{0<s<\infty} v(s)} < \infty;$$

(xi) $p = 1$, $0 < q < 1$ and

$$\left(\int_0^\infty \left(\int_t^\infty \frac{w(s)}{s^q} \right)^{\frac{q}{1-q}} \left[\operatorname*{ess\,sup}_{0<s\leq t} \frac{1}{(Pv)(s)} \right]^{\frac{q}{1-q}} \frac{w(t)}{t^q} \, dt \right)^{\frac{q}{1-q}} < \infty.$$

Proof. By the standard duality argument, when $1 < q < \infty$, we have for A from (10.6.4)

$$A = \sup_{h \geq 0} \sup_{g \geq 0} \frac{\int_0^\infty (Ph)(t) g(t) \, dt}{\|g\|_{L^{q'}(w^{1-q'})} \|Qh\|_{L^p(v)}}$$

$$= \sup_{g \geq 0} \frac{1}{\|g\|_{L^{q'}(w^{1-q'})}} \sup_{h \geq 0} \frac{\int_0^\infty h(t)(Qg)(t) \, dt}{\|Qh\|_{L^p(v)}}.$$

Similarly, when $q = 1$,

$$A = \sup_{h \geq 0} \frac{\int_0^\infty (Ph)(t)w(t)\,dt}{\|Qh\|_{L^p(v)}} = \sup_{h \geq 0} \frac{\int_0^\infty h(t)(Qw)(t)\,dt}{\|Qh\|_{L^p(v)}}$$

and when $q = \infty$,

$$A = \sup_{h \geq 0}\sup_{g \geq 0} \frac{\int_0^\infty (Ph)(t)g(t)\,dt}{\left\|\frac{g}{w}\right\|_{L^1}\|Qh\|_{L^p(v)}} = \sup_{g \geq 0} \frac{1}{\left\|\frac{g}{w}\right\|_{L^1}}\sup_{h \geq 0}\frac{\int_0^\infty h(t)(Qg)(t)\,dt}{\|Qh\|_{L^p(v)}}.$$

Using Theorem 10.6.7 and (10.6.5), we get the following characterization of the quantity A: if $1 < p, q < \infty$, then

$$A \approx \sup_{g \geq 0} \frac{1}{\|g\|_{L^{q'}(w^{1-q'})}} \left(\int_0^\infty \left[\sup_{0 < s < t} s(Qg)(s)\right]^{p'} \frac{v(t)}{V(t)^{p'}}\,dt\right)^{\frac{1}{p'}}$$

$$+ \sup_{g \geq 0} \frac{\sup_{0 < t < \infty} t(Qg)(t)}{\|g\|_{L^{q'}(w^{1-q'})} V(\infty)^{\frac{1}{p}}}, \qquad (10.6.20)$$

if $1 < p < q = \infty$, then

$$A \approx \sup_{g \geq 0} \frac{1}{\left\|\frac{g}{w}\right\|_{L^1}} \left(\int_0^\infty \left[\sup_{0 < s < t} s(Qg)(s)\right]^{p'} \frac{v(t)}{V(t)^{p'}}\,dt\right)^{\frac{1}{p'}}$$

$$+ \sup_{g \geq 0} \frac{\sup_{0 < t < \infty} t(Qg)(t)}{\left\|\frac{g}{w}\right\|_{L^1} V(\infty)^{\frac{1}{p}}}, \qquad (10.6.21)$$

if $p = q = \infty$, then

$$A \approx \sup_{g \geq 0} \frac{1}{\left\|\frac{g}{w}\right\|_{L^1}} \int_0^\infty \left[\sup_{0 < s < t} s(Qg)(s)\right] d\left(\frac{-1}{\operatorname{ess\,sup}_{0 < s \leq t} v(s)}\right)$$

$$+ \sup_{g \geq 0} \frac{\sup_{0 < t < \infty} t(Qg)(t)}{\left\|\frac{g}{w}\right\|_{L^1} \operatorname{ess\,sup}_{0 < s < \infty} v(s)}, \qquad (10.6.22)$$

if $p = \infty$ and $1 < q < \infty$, then

$$A \approx \sup_{g \geq 0} \frac{1}{\|g\|_{L^{q'}(w^{1-q'})}} \int_0^\infty \left[\sup_{0 < s < t} s(Qg)(s)\right] d\left(\frac{-1}{\operatorname{ess\,sup}_{0 < s \leq t} v(s)}\right)$$

$$+ \sup_{g \geq 0} \frac{\sup_{0 < t < \infty} t(Qg)(t)}{\|g\|_{L^{q'}(w^{1-q'})} \operatorname{ess\,sup}_{0 < s < \infty} v(s)}, \qquad (10.6.23)$$

if $0 < p \leq 1 < q < \infty$, then

$$A \approx \sup_{g \geq 0} \frac{1}{\|g\|_{L^{q'}(w^{1-q'})}} \sup_{0<s<\infty} \frac{s(Qg)(s)}{V(s)^{\frac{1}{p}}}, \qquad (10.6.24)$$

if $0 < p \leq 1$ and $q = \infty$, then

$$A \approx \sup_{g \geq 0} \frac{1}{\|\frac{g}{w}\|_{L^1}} \sup_{0<s<\infty} \frac{s(Qg)(s)}{V(s)^{\frac{1}{p}}}, \qquad (10.6.25)$$

if $q = 1 < p < \infty$, then

$$A \approx \left(\int_0^\infty \left[\sup_{0<s<t} s(Qw)(s) \right]^{p'} \frac{v(t)}{V(t)^{p'}} \, dt \right)^{\frac{1}{p'}} + \frac{\sup_{0<t<\infty} t(Qw)(t)}{V(\infty)^{\frac{1}{p}}}, \qquad (10.6.26)$$

if $0 < p \leq q = 1$, then

$$A \approx \sup_{0<s<\infty} \frac{s(Qw)(s)}{V(s)^{\frac{1}{p}}}, \qquad (10.6.27)$$

if $p = \infty$ and $q = 1$, then

$$A \approx \int_0^\infty \sup_{0<s\leq t} s(Qw)(s) \, d\left(\frac{-1}{\text{ess sup}_{0<s\leq t} v(s)} \right) + \frac{\sup_{0<t<\infty} t(Qw)(t)}{\text{ess sup}_{0<s<\infty} v(s)} \qquad (10.6.28)$$

and if $p = 1$ and $0 < q \leq 1$, then

$$A = \sup_{h \geq 0} \frac{\|Ph\|_{L^q(w)}}{\|h\|_{L^1(Pv)}}. \qquad (10.6.29)$$

Let us note that by a standard duality argument, we have for $1 < q < \infty$

$$\sup_{g \geq 0} \frac{\sup_{0<t<\infty} t(Qg)(t)}{\|g\|_{L^{q'}(w^{1-q'})}} = \sup_{0<t<\infty} t \left(\int_t^\infty \frac{w(s)}{s^q} \, ds \right)^{\frac{1}{q}} \qquad (10.6.30)$$

and, by an inequality from [203], we have

$$\sup_{g \geq 0} \frac{\sup_{0<t<\infty} t(Qg)(t)}{\|\frac{g}{w}\|_{L^1}} = \text{ess sup}_{0<s<\infty} w(s). \qquad (10.6.31)$$

Now, taking into account (10.6.30) and (10.6.31), we can establish all the statements of the theorem. Precisely, the assertions (i), (ii), and (iii) follow from Proposition 10.6.8 (i) combined with (10.6.20), (10.6.21) and (10.6.22), respectively. The assertions (iv) and (v) follow from Proposition 10.6.8 (ii) combined with (10.6.20) and (10.6.23), respectively. The assertions (vi) and (vii) follow from the weighted Hardy inequality (cf. Theorem 3.11.8) applied to (10.6.24) and (10.6.25), respectively. The assertions (viii), (ix), and (x) are immediate consequences of (10.6.26), (10.6.27) and (10.6.28), respectively. Finally, (xi) follows from (10.6.29) and the characterization of the appropriate weighted Hardy inequality from [203, Theorem 3.3]. □

Section 10.7 The missing case

Remark 10.6.10. The case (xi) is added to the theorem despite its different nature. Indeed, it is the only case in which we allow $q < 1$. In this particular case (i.e. when $p = 1$), the characterization is very easy. All the other cases are omitted.

10.7 The missing case of an embedding

We shall now concentrate on establishing necessary and sufficient conditions on weights v, w such that the inequality

$$\left(\int_0^\infty f^{**}(t)^q w(t)\, dt\right)^{\frac{1}{q}} \lesssim \left(\int_0^\infty f^*(t)^p v(t)\, dt\right)^{\frac{1}{p}}$$

holds when $0 < q < p \le 1$. We note that so far this case has not been covered by preceding results in this chapter.

Theorem 10.7.1. *Let $0 < q < p \le 1$ and let $r = \frac{pq}{p-q}$. Let v, w be weights on $(0, \infty)$. Then the inequality*

$$\left(\int_0^\infty f^{**}(t)^q w(t)\, dt\right)^{\frac{1}{q}} \lesssim \left(\int_0^\infty f^*(t)^p v(t)\, dt\right)^{\frac{1}{p}} \tag{10.7.1}$$

holds if and only if

$$A_{(10.7.2)} := \left(\int_0^\infty \left[\frac{W(t)}{V(t)}\right]^{\frac{r}{p}} w(t)\, dt\right)^{\frac{1}{r}} < \infty \tag{10.7.2}$$

and

$$A_{(10.7.3)} := \left(\int_0^\infty \sup_{0 < s \le t} \frac{s^r}{V(s)^{\frac{r}{p}}} \left(\int_t^\infty \frac{w(s)}{s^q}\, ds\right)^{\frac{r}{p}} \frac{w(t)}{t^q}\, dt\right)^{\frac{1}{r}} < \infty, \tag{10.7.3}$$

and the best constant in (10.7.1) is comparable to $A_{(10.7.2)} + A_{(10.7.3)}$.

Proof. First assume that (10.7.1) is true. Then, since

$$\sup_{0 < s \le t} s f^*(s) \le \int_0^t f^*(s)\, ds, \qquad t \in (0, \infty),$$

we obtain

$$\left(\int_0^\infty \left[\sup_{0 < s \le t} s f^*(s)\right]^q \frac{w(t)}{t^q}\, dt\right)^{\frac{1}{q}} \lesssim \left(\int_0^\infty f^*(t)^p v(t)\, dt\right)^{\frac{1}{p}}. \tag{10.7.4}$$

Assume that $f^*(t)^p = \int_t^\infty h(s)\,ds$ for some $h \in \mathcal{M}_+(0,\infty)$. Then,

$$\sup_{0<s\le t} sf^*(s) = \left[\sup_{0<s\le t} s^p \int_t^\infty h(y)\,dy\right]^{\frac{1}{p}},$$

hence, by the Fubini theorem, (10.7.4) yields

$$\left(\int_0^\infty \left[\sup_{0<s\le t} s^p \int_s^\infty h(y)\,dy\right]^{\frac{q}{p}} \frac{w(t)}{t^q}\,dt\right)^{\frac{1}{q}} \lesssim \left(\int_0^\infty h(t)V(t)\,dt\right)^{\frac{1}{p}} \quad (10.7.5)$$

for all $h \in \mathcal{M}_+(0,\infty)$. Substituting $h(t) \to \frac{g(t)}{t}$, we get

$$\left(\int_0^\infty \left[\sup_{0<s\le t} s^p (Qg)(s)\right]^{\frac{q}{p}} \frac{w(t)}{t^q}\,dt\right)^{\frac{p}{q}} \lesssim \int_0^\infty g(t)\frac{V(t)}{t}\,dt \quad (10.7.6)$$

for all $g \in \mathcal{M}_+(0,\infty)$. By Proposition 10.6.8 (ii), applied to the choice of parameters $q = \frac{q}{p}$, $p = 1$, $\alpha = p$, $w(t) = w(t)t^{-q}$ and $v(t) = \frac{V(t)}{t}$, (10.7.6) holds if and only if both (10.7.2) and (10.7.3) are satisfied.

Conversely, assume that (10.7.2) and (10.7.3) hold. We will be done if we can show

$$\left(\int_0^\infty \left(\frac{1}{t}\int_0^t f^*(s)^{\frac{1}{p}}\,ds\right)^q w(t)\,dt\right)^{\frac{p}{q}} \lesssim \int_0^\infty f^*(t)v(t)\,dt, \quad f \in \mathcal{M}(0,\infty).$$

By (10.6.5), this will follow if we prove

$$\left(\int_0^\infty \left(\frac{1}{t^p}\int_0^t f^*(s)s^{p-1}\,ds\right)^{\frac{q}{p}} w(t)\,dt\right)^{\frac{p}{q}} \lesssim \int_0^\infty f^*(t)v(t)\,dt \quad (10.7.7)$$

for all $f \in \mathcal{M}(0,\infty)$.

Given $f \in \mathbb{A}$, there is a sequence $\{h_n\}$ of positive functions such that

$$F_n(t) := \int_t^\infty h_n(s)\,ds \nearrow f, \quad t \in (0,\infty).$$

Moreover, we have

$$\frac{1}{t^p}\int_0^t F_n^*(s)s^{p-1}\,ds \approx \int_t^\infty h_n(s)\,ds + \frac{1}{t^p}\int_0^t h_n(s)s^p\,ds$$

and, by the Fubini theorem,

$$\int_0^\infty F_n^*(t)v(t)\,dt = \int_0^\infty h_n(t)V(t)\,dt.$$

Summarizing, by the Fatou lemma, we see that (10.7.7) holds if and only if both the inequalities

$$\left(\int_0^\infty \left(\int_t^\infty h(s)\,ds\right)^{\frac{q}{p}} w(t)\,dt\right)^{\frac{p}{q}} \lesssim \int_0^\infty h(t)V(t)\,dt \qquad (10.7.8)$$

and

$$\left(\int_0^\infty \left(\frac{1}{t^p}\int_0^t h(s)s^p\,ds\right)^{\frac{q}{p}} w(t)\,dt\right)^{\frac{p}{q}} \lesssim \int_0^\infty h(t)V(t)\,dt \qquad (10.7.9)$$

are satisfied for all $h \in \mathcal{M}(0, \infty)$. Now, by [203, Theorem 3.3] and by its analogue for integral $\int_t^\infty$ in place of $\int_0^t$, necessary and sufficient conditions for (10.7.8) and (10.7.9) are (10.7.2) and (10.7.3), respectively. □

10.8 General relations between the spaces of type S, Λ, Γ

Having the theory of embeddings of the spaces of type Λ and Γ complete, it is time for the spaces of type S to come into the play. The rest of this chapter, that is, this final section, will be devoted to the study of all possible embeddings between the spaces of type S, Λ and Γ.

Remark 10.8.1. As mentioned above, the functional $f^{**}(t) - f^*(t)$ is zero when f is constant on $(0, \infty)$. Hence, constant functions always belong to any $S^p(v)$ whereas they do not necessarily belong to analogous spaces of type Λ and Γ. For this reason, we will study the appropriate embeddings restricted to the set

$$\mathbb{A} := \left\{ f \in \mathcal{M}_+(0, \infty); \lim_{t \to \infty} f^*(t) = 0 \right\}.$$

We will denote this restriction by writing, for example,

$$S^p(v) \hookrightarrow \Lambda^q(w), \qquad f \in \mathbb{A},$$

meaning

$$\left(\int_0^\infty f^*(t)^q w(t)\,dt\right)^{\frac{1}{q}} \leq C \left(\int_0^\infty (f^{**}(t) - f^*(t))^p v(t)\,dt\right)^{\frac{1}{p}} \qquad \text{for all } f \in \mathbb{A},$$

and so on.

We shall now define a useful operator (cf. [33]).

Notation 10.8.2. For a given positive nonincreasing function f on $(0, \infty)$ such that $\lim_{t \to \infty} f(t) = 0$ and for every $t \in (0, \infty)$, we define the operator T by

$$(Tf)(t) := \frac{1}{t}\left((Pf)\left(\frac{1}{t}\right) - f\left(\frac{1}{t}\right)\right). \tag{10.8.1}$$

Remark 10.8.3. It is not difficult to verify that

$$T \circ T = \mathrm{id} \tag{10.8.2}$$

and, for $f \in \mathbb{A}$,

$$f^{**}(t) - f^*(t) = \frac{1}{t}(Tf^*)\left(\frac{1}{t}\right). \tag{10.8.3}$$

Note that, for $f \in \mathbb{A}$, Tf^* is nonincreasing, and also that given $p \in (0, \infty)$, we have

$$\|f\|_{S^p(v)} = \|Tf^*\|_{\Lambda^p(\tilde{v}_p)}. \tag{10.8.4}$$

Notation 10.8.4. Let v be a weight on $(0, \infty)$ and $t \in (0, \infty)$. We then denote

$$\tilde{v}_p(t) := v\left(\frac{1}{t}\right) t^{p-2} \quad \text{and} \quad \tilde{w}_q(t) := w\left(\frac{1}{t}\right) t^{q-2}. \tag{10.8.5}$$

We start with a simple but important relation.

Proposition 10.8.5. *Let $0 < p < \infty$ and let v be a weight on $(0, \infty)$. Then*

$$\Gamma^p(v) = \Lambda^p(v) \cap S^p(v). \tag{10.8.6}$$

Proof. Since both the quantities f^* and $f^{**} - f^*$ are majorized by f^{**}, we clearly have

$$\Gamma^p(v) \subset \Lambda^p(v) \cap S^p(v).$$

The converse inclusion follows at once from

$$\|f\|_{\Gamma^p(v)} = \|f^{**} - f^* + f^*\|_{L^p(v)} \leq \|f^{**} - f^*\|_{L^p(v)} + \|f^*\|_{L^p(v)}$$
$$= \|f\|_{S^p(v)} + \|f\|_{\Lambda^p(v)}. \qquad \square$$

From Proposition 10.8.5 we immediately have the following.

Corollary 10.8.6. *Let $0 < p < \infty$ and let v be a weight on $(0, \infty)$. Then, the following statements are equivalent:*

(i) $$\Lambda^p(v) \hookrightarrow S^p(v);$$

(ii) $$\Lambda^p(v) = \Gamma^p(v);$$

(iii) $$v \in B_p.$$

Section 10.8 Embeddings

Proof. The equivalence (i)⇔(ii) follows at once from (10.8.6). The equivalence (ii) ⇔ (iii) was proved in [5] for $1 \leq p < \infty$ and in [216] for $0 < p < 1$. □

Corollary 10.8.7. *Let $1 \leq p < \infty$ and let v be a weight on $(0, \infty)$. Then, the following statements are equivalent:*

(i)
$$S^p(v) \hookrightarrow \Lambda^p(v);$$

(ii)
$$S^p(v) = \Gamma^p(v);$$

(iii)
$$v \in RB_p.$$

Proof. Again, the equivalence (i) ⇔ (ii) follows from (10.8.6). A similar argument to that in the proof of (10.7.7) now yields (ii) equivalent to

$$\int_0^\infty \left(\int_t^\infty h(s)\,ds \right)^p v(t)\,dt \lesssim \int_0^\infty \left(\frac{1}{t} \int_0^t sh(s)\,ds \right)^p v(t)\,dt \qquad (10.8.7)$$

for all $h \in \mathcal{M}_+(0, \infty)$. However, (10.8.7) was shown in [163, Theorem 4.1] to be equivalent to $v \in RB_p$. □

Remark 10.8.8. The equivalence in the preceding corollary holds in fact when $p \in (0, \infty)$.

10.8.1 Embeddings of type $S \hookrightarrow S$

Our aim in this subsection is to characterize pairs of weights v, w such that the embedding

$$S^p(v) \hookrightarrow S^q(w) \qquad (10.8.8)$$

holds.

Remark 10.8.9. We note that every $f \in \mathcal{M}_+(0, \infty)$ can be written in the form $f = g + c$, where $g \in \mathbb{A}$ and $c \geq 0$.

Theorem 10.8.10. *Let $0 < p, q < \infty$ and let v, w be weights on $(0, \infty)$. Then, (10.8.8) is equivalent to*

$$\Lambda^p(\tilde{v}_p) \hookrightarrow \Lambda^q(\tilde{w}_q), \qquad (10.8.9)$$

with $\tilde{v}_p$ and $\tilde{w}_q$ from (10.8.5).

Proof. Assume first that (10.8.9) holds, and let $f \in \mathbb{A}$. Then, by (10.8.9) and a double use of (10.8.4),

$$\|f\|_{S^q(w)} = \|Tf^*\|_{\Lambda^q(\tilde{w}_q)} \leq C\|Tf^*\|_{\Lambda^p(\tilde{v}_p)} = \|f\|_{S^p(v)},$$

proving (10.8.8) for $f \in \mathbb{A}$. Since (10.8.8) is trivial for constant functions, it follows from Remark 10.8.9 that (10.8.8) holds for every $f \in \mathcal{M}_+(0, \infty)$.

Conversely, assume that (10.8.8) is satisfied. For every $f \in \mathbb{A}$, there exists a $g \in \mathbb{A}$ such that $f^* = Tg^*$. Thus, using (10.8.4) again,

$$\|f\|_{\Lambda^q(\tilde{w}_q)} = \|Tg^*\|_{\Lambda^q(\tilde{w}_q)} = \|g^*\|_{S^q(w)}$$
$$\leq C\|g^*\|_{S^p(v)} = C\|Tg^*\|_{\Lambda^p(\tilde{v}_p)} = C\|f\|_{\Lambda^p(\tilde{v}_p)},$$

proving (10.8.9) for $f \in \mathbb{A}$.

In view of Remark 10.8.9, it only remains to show that (10.8.9) holds for constant functions, that is, we have to prove the inequality

$$\left(\int_0^\infty \tilde{w}_q(t)\,dt\right)^{\frac{1}{q}} \leq C\left(\int_0^\infty \tilde{v}_p(t)\,dt\right)^{\frac{1}{p}}.$$

Assuming that the right-hand side is finite (otherwise there is nothing to prove) and given $T \in (0, \infty)$, we apply the result just established to the function $f = \chi_{(0,T)}$, which obviously belongs to $\mathbb{A}$. We get

$$\left(\int_0^T \tilde{w}_q(t)\,dt\right)^{\frac{1}{q}} \leq C\left(\int_0^T \tilde{v}_p(t)\,dt\right)^{\frac{1}{p}}$$

with C independent of T. The result now follows on letting $T \to \infty$. □

Corollary 10.8.11. (i) *If $0 < p \leq q < \infty$, then the embedding (10.8.8) holds if and only if*

$$A_{(10.8.10)} := \sup_{t \in (0,\infty)} \frac{\left(\int_t^\infty s^{-q}w(s)\,ds\right)^{\frac{1}{q}}}{\left(\int_t^\infty s^{-p}v(s)\,ds\right)^{\frac{1}{p}}} < \infty, \tag{10.8.10}$$

and moreover the optimal constant C of the embedding (10.8.8) satisfies $C \approx A_{(10.8.10)}$.

(ii) *If $0 < q < p < \infty$, then (10.8.8) holds if and only if*

$$A_{(10.8.11)} := \left(\int_0^\infty \left[\frac{\int_t^\infty s^{-q}w(s)\,ds}{\int_t^\infty s^{-p}v(s)\,ds}\right]^{\frac{r}{p}} \frac{w(t)}{t^q}\,dt\right)^{\frac{1}{r}} < \infty, \tag{10.8.11}$$

and moreover the optimal constant C of the embedding (10.8.8) satisfies $C \approx A_{(10.8.11)}$.

Proof. The claim follows from Theorem 10.8.10, Theorem 10.3.8 and, again, a change of variables in the integrals. □

Section 10.8 Embeddings 431

10.8.2 Embeddings of type $\Gamma \hookrightarrow S$ and $S \hookrightarrow \Gamma$

In this subsection we will characterize embeddings between spaces of type Γ and S (in both directions). We will once again employ the operator T defined in (10.8.1).

Theorem 10.8.12. *Let $0 < p, q < \infty$ and let v, w be weights on $(0, \infty)$. Then the embedding*

$$S^p(v) \hookrightarrow \Gamma^q(w), \quad f \in \mathbb{A}, \qquad (10.8.12)$$

is equivalent to

$$\Lambda^p(\tilde{v}_p) \hookrightarrow \Gamma^q(\tilde{w}_q). \qquad (10.8.13)$$

Similarly, the embedding

$$\Gamma^p(v) \hookrightarrow S^q(w), \quad f \in \mathbb{A}, \qquad (10.8.14)$$

is equivalent to

$$\Gamma^p(\tilde{v}_p) \hookrightarrow \Lambda^q(\tilde{w}_q). \qquad (10.8.15)$$

Proof. The proof is analogous to that of Theorem 10.8.10. We use (10.8.4) and

$$\|f\|_{\Gamma^p(w)} = \|Tf\|_{\Gamma^p(\tilde{w}_p)}, \qquad (10.8.16)$$

the latter being easily verified for every $f \in \mathbb{A}$ by a change of variables. $\square$

Theorem 10.8.13. *Let $0 < p, q < \infty$ and let v, w be weights on $(0, \infty)$.*

(i) *If $1 < p \leq q < \infty$, then the embedding (10.8.12) holds if and only if*

$$A_{(10.8.10)} < \infty \qquad (10.8.17)$$

and

$$A_{(10.8.18)} := \sup_{t \in (0,\infty)} W(t)^{\frac{1}{q}} \left(\int_t^\infty \frac{v(s)}{\left(s^p \int_{\frac{1}{s}}^\infty v(y) y^{-p}\, dy\right)^{p'}}\, ds \right)^{\frac{1}{p'}} < \infty, \qquad (10.8.18)$$

and moreover the optimal constant C of the embedding (10.8.12) satisfies $C \approx A_{(10.8.10)} + A_{(10.8.18)}$.

(ii) *If $0 < p < 1$ and $0 < p \leq q < \infty$, then (10.8.12) holds if and only if $A_{(10.8.10)} < \infty$ and*

$$A_{(10.8.19)} := \sup_{t \in (0,\infty)} \frac{W(t)^{\frac{1}{q}}}{t \left(\int_t^\infty v(s) s^{-p}\, ds \right)^{\frac{1}{p}}}, \qquad (10.8.19)$$

and moreover the optimal constant C of the embedding (10.8.12) satisfies $C \approx A_{(10.8.10)} + A_{(10.8.19)}$.

(iii) If $1 < p < \infty$, $0 < q < p < \infty$ and $q \neq 1$, then (10.8.12) holds if and only if

$$A_{(10.8.11)} < \infty \qquad (10.8.20)$$

and

$$A_{(10.8.21)} := \left(\int_0^\infty \frac{W(t)^{\frac{r}{q}} \left(\int_t^\infty \frac{v(s)}{\left(s^p \int_s^\infty \frac{v(y)}{y^p} dy\right)^{p'}} ds \right)^{\frac{r(q-1)}{q}}}{\left(t^p \int_t^\infty \frac{v(y)}{y^p} dy\right)^{p'}} v(t) \, dt \right)^{\frac{1}{r}} < \infty, \qquad (10.8.21)$$

and moreover the optimal constant C of the embedding (10.8.12) satisfies $C \approx A_{(10.8.11)} + A_{(10.8.21)}$.

(iv) If $1 = q < p < \infty$, then (10.8.12) holds if and only if $A_{(10.8.11)} < \infty$ and

$$A_{(10.8.22)} := \frac{\int_0^\infty \frac{w(s)}{s} ds}{\left(\int_0^\infty \frac{v(s)}{s^p} ds\right)^{\frac{1}{p}}} + \left(\int_0^\infty \left[\frac{\frac{W(t)}{t} + \int_t^\infty \frac{w(s)}{s} ds}{\int_t^\infty \frac{v(s)}{s^p} ds} \right]^{p'} \frac{v(t)}{t^p} dt \right)^{\frac{1}{p'}} < \infty, \qquad (10.8.22)$$

and moreover the optimal constant C of the embedding (10.8.12) satisfies $C \approx A_{(10.8.11)} + A_{(10.8.22)}$.

(v) If $0 < q < p \leq 1$, then (10.8.12) holds if and only if

$$A_{(10.8.23)} := \left(\int_0^\infty \frac{\left(\int_t^\infty \frac{w(s)}{s^q} ds \right)^{\frac{r}{p}}}{\left(\int_t^\infty \frac{v(s)}{s^p} ds \right)^{\frac{r}{p}}} \frac{w(t)}{t^q} dt \right)^{\frac{1}{r}} < \infty \qquad (10.8.23)$$

and

$$A_{(10.8.24)} := \left(\int_0^\infty W(t)^{\frac{r}{p}} \operatorname*{ess\,sup}_{t \leq s < \infty} \frac{1}{s^r \left(\int_s^\infty \frac{v(y)}{y^p} dy \right)^{\frac{r}{p}}} \frac{w(t)}{t^q} dt \right)^{\frac{1}{r}} < \infty \qquad (10.8.24)$$

and moreover the optimal constant C of the embedding (10.8.12) satisfies $C \approx A_{(10.8.23)} + A_{(10.8.24)}$.

Proof. By Theorem 10.8.12, (10.8.12) is equivalent to (10.8.13). The result thus follows from Theorem 10.3.12 and a change of variables. □

Theorem 10.8.14. *Let $0 < p, q < \infty$ and let v, w be weights on $(0, \infty)$.*

(i) *If $0 < p \leq q < \infty$ and $1 \leq q < \infty$, then the embedding (10.8.14) holds if and only if*

$$A_{(10.8.25)} := \sup_{t \in (0,\infty)} \frac{\left(\int_t^\infty \frac{w(s)}{s^q} ds\right)^{\frac{1}{q}}}{\left(\frac{V(t)}{t^p} + \int_t^\infty \frac{v(s)}{s^p} ds\right)^{\frac{1}{p}}} < \infty, \qquad (10.8.25)$$

and moreover the optimal constant C of the embedding (10.8.14) satisfies $C \approx A_{(10.8.25)}$.

(ii) *If $1 \leq q < p < \infty$, then (10.8.14) holds if and only if*

$$A_{(10.8.26)} := \left(\int_0^\infty \frac{\sup_{s \in (0,t)} s^r \left(\int_s^\infty \frac{w(y)}{y^q} dy \right)^{\frac{r}{q}}}{t^{r+p+1} \left(\frac{V(t)}{t^p} + \int_t^\infty \frac{v(s)}{s^p} ds \right)^{\frac{r}{p}+2}} \right.$$

$$\left. \times V(t) \int_t^\infty \frac{v(s)}{s^p} ds \, dt \right)^{\frac{1}{r}} < \infty, \quad (10.8.26)$$

and moreover the optimal constant C of the embedding (10.8.14) satisfies $C \approx A_{(10.8.26)}$.

(iii) *If $0 < p \leq q < 1$, then (10.8.14) holds if and only if*

$$A_{(10.8.27)} :=$$

$$\sup_{t \in (0,\infty)} \frac{\left(\int_t^\infty \frac{w(s)}{s^q} ds \right)^{\frac{1}{q}} + \frac{1}{t} \left(\int_0^t \left[s \int_s^\infty \frac{w(y)}{y^q} dy \right]^{\frac{q}{1-q}} \frac{w(s)}{s^q} ds \right)^{\frac{1-q}{q}}}{\left(\frac{V(t)}{t^p} + \int_t^\infty \frac{v(s)}{s^p} ds \right)^{\frac{1}{p}}} < \infty,$$

$$(10.8.27)$$

and moreover the optimal constant C of the embedding (10.8.14) satisfies $C \approx A_{(10.8.27)}$.

(iv) If $0 < q < 1$ and $0 < q < p$, then (10.8.14) holds if and only if

$$A_{(10.8.28)} := \left(\frac{\left[\left(\int_t^\infty \frac{w(s)}{s^q} ds \right)^{\frac{1}{1-q}} + \frac{1}{t^{\frac{q}{1-q}}} \times \int_0^t \left(s \int_s^\infty \frac{w(y)}{y^q} dy \right)^{\frac{q}{1-q}} \frac{w(s)}{s^q} ds \right]^{\frac{r(1-q)}{q} - 1}}{\left(\frac{V(t)}{t^p} + \int_t^\infty \frac{v(s)}{s^p} ds \right)^{\frac{1}{p}}} \right.$$

$$\left. \times \left(\int_t^\infty \frac{w(s)}{s^q} ds \right)^{\frac{q}{1-q}} \frac{w(t)}{t^q} dt \right)^{\frac{1}{r}},$$

(10.8.28)

and moreover the optimal constant C of the embedding (10.8.14) *satisfies* $C \approx A_{(10.8.28)}$.

Proof. The proof follows the pattern of that of Theorem 10.8.13. We omit the details. □

10.8.3 Embeddings of type $\Lambda \hookrightarrow S$ and $S \hookrightarrow \Lambda$

We start with a simple observation that the embeddings $S^p(v) \hookrightarrow \Lambda^q(w)$ and $\Lambda^p(v) \hookrightarrow S^q(w)$ (restricted to $f \in \mathbb{A}$) are interchangeable (with an appropriate change of weights) and therefore it is enough to investigate only one of them.

Proposition 10.8.15. *Let $0 < p, q < \infty$ and let v, w be weights on $(0, \infty)$. Then the embedding*

$$S^p(v) \hookrightarrow \Lambda^q(w), \quad f \in \mathbb{A}, \qquad (10.8.29)$$

is equivalent to the embedding

$$\Lambda^p(\tilde{v}_p) \hookrightarrow S^q(\tilde{w}_q), \quad f \in \mathbb{A}, \qquad (10.8.30)$$

where $\tilde{v}_p$ and $\tilde{w}_q$ are from (10.8.5).

Proof. Using the operator T defined in (10.8.1), we see that (10.8.29) reads

$$\left(\int_0^\infty f^*(t)^q w(t) dt \right)^{\frac{1}{q}} \lesssim \left(\int_0^\infty \left[\frac{1}{t}(Tf^*)\left(\frac{1}{t}\right) \right]^p v(t) dt \right)^{\frac{1}{p}}.$$

As already noted above, for every $f \in \mathbb{A}$, there exists a $g \in \mathbb{A}$ such that $f^* = Tg^*$. Therefore, (10.8.29) is equivalent to

$$\left(\int_0^\infty (Tg^*)(t)^q w(t)\,dt\right)^{\frac{1}{q}} \lesssim \left(\int_0^\infty \left[\frac{1}{t}g^*\left(\frac{1}{t}\right)\right]^p v(t)\,dt\right)^{\frac{1}{p}}.$$

Changing the variables, we get

$$\left(\int_0^\infty [g^{**}(t) - g^*(t)]^q \tilde{w}_q(t)\,dt\right)^{\frac{1}{q}} \lesssim \left(\int_0^\infty g^*(t)^p \tilde{v}_p(t)\,dt\right)^{\frac{1}{p}},$$

which is (10.8.30), as desired. □

We next reduce (10.8.29) to (10.6.3).

Proposition 10.8.16. *Let $0 < p, q < \infty$ and let v, w be weights on $(0, \infty)$. Then the embedding*

$$\Lambda^p(v) \hookrightarrow S^q(w), \quad f \in \mathbb{A}, \tag{10.8.31}$$

holds if and only if the inequality (10.6.3) is satisfied for every nonnegative function h.

Proof. We shall proceed as in the proof of (10.7.7). Given $f \in \mathbb{A}$, there is a sequence $\{h_n\}$ of positive functions such that

$$F_n(t) := \int_t^\infty h_n(s)\,ds \nearrow f, \quad t \in (0, \infty).$$

Moreover,

$$f^{**}(t) - f^*(t) = \lim_{n \to \infty} \left(F_n^{**}(t) - F_n^*(t)\right), \quad t \in (0, \infty).$$

Thus, by the Fatou lemma,

$$\|f\|_{S^q(w)} \le \liminf_{n \to \infty} \|F_n\|_{S^q(w)} \lesssim \liminf_{n \to \infty} \|F_n\|_{\Lambda^p(v)} = \|f\|_{\Lambda^p(v)}.$$

On the other hand, we have

$$F_n^{**}(t) - F_n^*(t) = \frac{1}{t}\int_0^t \int_s^\infty h_n(y)\,dy\,ds - \int_t^\infty h_n(s)\,ds = \frac{1}{t}\int_0^t sh_n(s)\,ds.$$

Altogether, (10.8.31) is equivalent to

$$\left(\int_0^\infty \left[\frac{1}{t}\int_0^t sh(s)\,ds\right]^q w(t)\,dt\right)^{\frac{1}{q}} \lesssim \left(\int_0^\infty \left[\int_t^\infty h(s)\,ds\right]^p v(t)\,dt\right)^{\frac{1}{p}},$$

which is, after the substitution $sh(s) \to h(s)$, exactly the inequality (10.6.3). □

Corollary 10.8.17. *Let $0 < p < \infty$, $1 \le q < \infty$, and let v, w be weights on $(0, \infty)$. Then, the embedding*
$$\Lambda^p(v) \hookrightarrow S^q(w), \quad f \in \mathbb{A},$$
holds if and only if one of the conditions (i)–(xi) *of Theorem 10.6.9 holds.*

Remark 10.8.18. Necessary and sufficient conditions for the embedding
$$S^p(v) \hookrightarrow \Lambda^q(w), \quad f \in \mathbb{A},$$
where $0 < p \le \infty$, $1 \le q \le \infty$, and v, w be weights on $(0, \infty)$, can be obtained from the combination of Corollary 10.8.17 and Proposition 10.8.15.

Chapter 11

Variable-exponent Lebesgue spaces

11.1 Introduction

The variable-exponent Lebesgue spaces were among others studied in connection with boundary value Dirichlet problems whose weak solution required *a priori* estimates by two (or more) power functions t_i^p, $i = 1, \ldots, k$ on the corresponding number of components Ω_i, $i = 1, \ldots, k$, of the given domain $\Omega \in \mathbb{R}^n$. Their first appearance in the literature can be traced as early as 1931 in the article [173] by Orlicz. Their first systematic investigation was carried out in the 1950s by Nakano in [163], and continued later by Musielak [160]. The theory of variable exponent spaces of functions defined on the real line was also developed by Tsenov [230], Sharapudinov [198] and Zhikov [243, 244].

Since the mid-1970s, several Polish mathematicians including Hudzik, Kamińska and Musielak investigated such spaces in a more general framework of the so-called *modular spaces*, see, e.g., the monograph [160]. The spaces introduced there are known today as Musielak–Orlicz spaces and are still intensively studied.

In the late 1980s these spaces were investigated in connection with certain specific applications in the continuum mechanics (see, for example, Zhikov [243]). Some of their fundamental properties were established in [122] by Kováčik and Rákosník using elementary methods without emphasizing the immediate relation to Musielak–Orlicz spaces. In 2001 Fan and Zhao [77] independently reproved the basic results by recourse to the general theory of Musielak–Orlicz spaces. A deeper study of the norm in $L^{p(x)}$ can be found in [70] by Edmunds, Lang and Nekvinda. Yet other authors considered inequalities of Sobolev type and related questions.

During the 1990s several papers on variable exponent spaces appeared, but the development of the theory was not systematic. The key motivation these days seemed to stem from the fact that many classical results from Lebesgue space theory could be generalized to this setting, but not to general Musielak–Orlicz spaces. Although this proved to be the case, it was often caused by enormous technical complications that the authors had to overcome.

Recently, these spaces have seen a true renaissance, caused by the discovery of Růžička [190] that they constitute a natural functional setting for the mathematical model of electrorheological fluids which involves a nonlinear system of partial differential equations with coefficients of variable rate of growth. As a natural consequence, attention of various authors has been attracted to this area of functional analysis, and many new deep results were obtained. A systematic treatment of these spaces can be found in [63].

11.2 Basic properties

Notation 11.2.1. If X is a Banach space, we denote by B_X its unit ball and by S_X its unit sphere, that is,
$$B_X := \{f \in X;\ \|f\|_X \leq 1\},$$
and
$$S_X := \{f \in X;\ \|f\|_X = 1\}.$$

Throughout this chapter, we work in the following set-up.

Let $(\Omega, \mathcal{S}, \mu)$ be a measure space with a σ–finite complete measure μ. Let $\mathcal{M}(\Omega)$ be the set of all μ–measurable functions on Ω. Let $\mathcal{P}(\Omega)$ denote the family of all $p \in \mathcal{M}(\Omega)$ for which
$$1 \leq p(x) \leq \infty, \quad x \in \Omega.$$

Definition 11.2.2. For a function $p \in \mathcal{P}(\Omega)$, we denote
$$p_{\max} := \operatorname*{ess\,sup}_{x \in \Omega} p(x) \quad \text{and} \quad p_{\min} := \operatorname*{ess\,inf}_{x \in \Omega} p(x).$$

The *generalized Lebesgue space* $L^{p(x)}(\Omega, \mathcal{S}, \mu)$ is the collection of all $\mathcal{S}$-measurable functions f on Ω for which there exists a $\lambda > 0$ such that
$$\int_{\Omega_0 \cup \Omega_1} \left(\frac{|f(x)|}{\lambda}\right)^{p(x)} d\mu(x) + \operatorname*{ess\,sup}_{x \in \Omega_\infty} |f(x)| < \infty,$$
where
$$\Omega_1 := \{x \in \Omega;\ p(x) = 1\},\ \Omega_\infty := \{x \in \Omega;\ p(x) = \infty\} \text{ and } \Omega_0 = \Omega \setminus (\Omega_1 \cup \Omega_\infty).$$

The natural question is whether $L^{p(x)}$ is a normed linear space. Of course, similarly as in the case of Orlicz spaces, although this space obviously has many properties analogous to those of the (constant exponent) Lebesgue space L^p, it cannot be normed in any analogous way, since the expression
$$\left(\int_\Omega |f(x)|^{p(x)} dx\right)^{\frac{1}{p(x)}}$$
does not make sense. The situation is solved in the same manner as in the case of Orlicz spaces, that is, with the help of the Luxemburg norm.

Definition 11.2.3. For a function $p \in \mathcal{P}(\Omega)$, we denote
$$\varrho_p(f) := \int_{\Omega_0} (|f(x)|)^{p(x)} d\mu(x) + \operatorname*{ess\,sup}_{x \in \Omega_\infty} |f(x)| < \infty.$$

Section 11.2 Basic properties

Then, ϱ_p is a convex modular on $\mathcal{M}(\Omega)$ (see Section 1.5). We thus can furnish the set $L^{p(x)}$ with the *Luxemburg norm*, defined as

$$\|f\|_p := \inf\left\{\lambda > 0;\ \varrho_p\left(\frac{f}{\lambda}\right) \leq 1\right\}.$$

In order to be able to introduce an analogue of the Lebesgue conjugate space, we shall need the notion of the conjugate function.

Definition 11.2.4. For $p \in \mathcal{P}(\Omega)$ and $x \in \Omega$, we denote the *conjugate function* p' by

$$p'(x) := \begin{cases} \frac{p(x)}{p(x)-1}, & \text{if } 1 < p(x) < \infty, \\ 1 & \text{if } p(x) = \infty, \\ \infty & \text{if } p(x) = 1. \end{cases}$$

Equipped with the conjugate function, we can formulate the Hölder inequality.

Theorem 11.2.5. *Let $p \in \mathcal{P}(\Omega)$. Then the inequality*

$$\int_\Omega |f(x)g(x)|\,dx \leq r_p \|f\|_p \|g\|_{p'} \tag{11.2.1}$$

holds for all $f \in L^{p(x)}(\Omega)$ and $g \in L^{p'(x)}(\Omega)$, where

$$r_p := \|\chi_{\Omega_1}\|_\infty + \|\chi_{\Omega_0}\|_\infty + \|\chi_{\Omega_\infty}\|_\infty + \frac{1}{p_{\min}} - \frac{1}{p_{\max}}. \tag{11.2.2}$$

Proof. We can assume that $\|f\|_p \neq 0$, $\|g\|_{p'} \neq 0$ and $|\Omega_0| > 0$ since otherwise there is nothing to prove. For almost every $x \in \Omega_0$, we have $1 < p(x) < \infty$, $|f(x)| < \infty$ and $|g(x)| < \infty$. Set

$$a := \frac{f(x)}{\|f\|_p},\ b := \frac{g(x)}{\|g\|_{p'}},\ p = p(x) \text{ and } q = q(x).$$

Then the Young inequality (3.1.5) holds. We integrate it over Ω_0 and use (11.2.3) to obtain

$$\int_{\Omega_0} \frac{|f(x)g(x)|}{\|f\|_p \|g\|_{p'}}\,dx \leq \operatorname*{ess\,sup}_{x\in\Omega_0} \frac{1}{p(x)} \varrho_p\left(\frac{f}{\|f\|_p}\right) + \operatorname*{ess\,sup}_{x\in\Omega_0} \frac{1}{p'(x)} \varrho_{p'}\left(\frac{g}{\|g\|_{p'}}\right)$$

$$\leq 1 + \frac{1}{p_{\min}} - \frac{1}{p_{\max}}.$$

Consequently,

$$\int_{\Omega_0} \frac{|f(x)g(x)|}{\|f\|_p \|g\|_{p'}}\,dx \leq \left(1 + \frac{1}{p_{\min}} - \frac{1}{p_{\max}}\right) \|f\|_p \|g\|_{p'} \|\chi_{\Omega_0}\|_\infty +$$
$$+ \|f\chi_{\Omega_1}\|_1 \|g\chi_{\Omega_\infty}\|_\infty + \|g\chi_{\Omega_1}\|_1 \|f\chi_{\Omega_\infty}\|_\infty \leq r_p \|f\|_p \|g\|_{p'},$$

as desired. The proof is complete. □

It is very useful to note that, similarly as in the case of an Orlicz space, the space $L^{p(x)}$ can be equipped with another reasonable function norm.

Definition 11.2.6. For $p \in \mathcal{P}(\Omega)$ and $f \in \mathcal{M}(\Omega)$, we define the functional

$$|\!|\!|f|\!|\!|_p := \sup_{\varrho_{p'}(g) \leq 1} \int_\Omega f(x)g(x)\,dx.$$

Exercise 11.2.7. (i) If $f \in L^{p(x)}(\Omega)$ satisfies $0 < \|f\|_p < \infty$, then

$$\varrho_p\left(\frac{f}{\|f\|_p}\right) \leq 1. \tag{11.2.3}$$

(ii) Assume that $p_{\max} < \infty$. Then, for every $f \in L^{p(x)}(\Omega)$ such that $0 < \|f\|_p < \infty$, we have

$$\varrho_p\left(\frac{f}{\|f\|_p}\right) = 1. \tag{11.2.4}$$

(iii) If $f \in L^{p(x)}(\Omega)$ satisfies $\|f\|_p \leq 1$, then

$$\varrho_p(f) \leq \|f\|_p. \tag{11.2.5}$$

(iv) If $f \in L^{p(x)}(\Omega)$ is such that $|\!|\!|f|\!|\!|_p \leq 1$, then

$$\varrho_p(f) \leq c_p |\!|\!|f|\!|\!|_p, \tag{11.2.6}$$

where

$$c_p := \|\chi_{\Omega_1}\|_\infty + \|\chi_{\Omega_0}\|_\infty + \|\chi_{\Omega_\infty}\|_\infty.$$

All the assertions can be proved easily along the arguments similar to those in Sections 4.8 or 1.5. Detailed proofs for this specific situation can be found in [122].

It is not hard to verify that $|\!|\!|\cdot|\!|\!|$ has the properties of a norm on $\mathcal{M}(\Omega)$. We shall now prove the equivalence of the norms $\|\cdot\|_p$ and $|\!|\!|\cdot|\!|\!|_p$.

Theorem 11.2.8. *Let* $p \in \mathcal{P}(\Omega)$. *Then, for every function* $f \in \mathcal{M}(\Omega)$,

$$c_p \|f\|_p \leq |\!|\!|f|\!|\!|_p \leq r_p \|f\|_p, \tag{11.2.7}$$

where

$$c_p := \|\chi_{\Omega_1}\|_{L^\infty} + \|\chi_{\Omega_0}\|_{L^\infty} + \|\chi_{\Omega_\infty}\|_{L^\infty}$$

and r_p *is the constant from* (11.2.2).

Proof. Let $f \in L^{p(x)}(\Omega)$. Assume that $\varrho_{p'}(g) \le 1$. Then $\|g\|_{p'} \le 1$ and the Hölder inequality (11.2.1) yields

$$\int_\Omega f(x) g(x) \, dx \le r_p \|f\|_p \|g\|_{p'} \le r_p \|f\|_p.$$

This gives the second inequality in (11.2.7).

To prove the first inequality in (11.2.7), assume that $0 < \|f\|_p < \infty$. Then

$$\|fc_p^{-1}\|f\|^{-1}\|_p = c_p^{-1} \le 1,$$

and, by (11.2.6), we have

$$\varrho_p\left(\frac{f}{c_p \|f\|_p}\right) \le c_p^{-1} c_p = 1,$$

as desired. The proof is complete. □

Exercise 11.2.9. Let $\{f_n\}_{n=1}^\infty$ and f be functions in $L^{p(x)}(\Omega)$. If

$$\lim_{n \to \infty} \|f_n - f\|_p = 0,$$

then

$$\lim_{n \to \infty} \varrho_p(f_n - f) = 0.$$

Moreover, the statements are equivalent when $p_{\max} < \infty$. (See also Section 1.5.)

We continue pursuing important properties of the spaces $L^{p(x)}$. In our next theorem we show that they are Banach spaces.

Theorem 11.2.10. *Let $p \in \mathcal{P}(\Omega)$. Then, the space $L^{p(x)}$ is complete.*

Proof. Let $\{f_n\}_{n=1}^\infty$ be a Cauchy sequence in $L^{p(x)}(\Omega)$ and let $\varepsilon > 0$. Then there exists an $n_0 \in \mathbb{N}$ such that for every $m, n \in \mathbb{N}$, $m, n \ge n_0$, and for every $g \in L^{p'(x)}(\Omega)$ such that $\varrho_{p'}(g) \le 1$, one has

$$\int_\Omega |f_m(x) - f_n(x)| |g(x)| \, dx < \varepsilon. \qquad (11.2.8)$$

Let

$$\Omega = \bigcup_{k=1}^\infty G_k,$$

where G_k are pairwise disjoint subsets of Ω of finite measure. For $k \in \mathbb{N}$, we define the functions

$$g_k(x) := \frac{\chi_{G_k}(x)}{1 + |G_k|}.$$

Then
$$\varrho_{p'}(g_k) \leq \int_{G_k} \frac{dx}{(1+|G_k|)^{p(x)}} + \frac{1}{1+|G_k|} \leq 1.$$

Plugging this into (11.2.8), we obtain

$$\int_{G_k} |f_m(x) - f_n(x)||g(x)|\, dx < \varepsilon(1+|G_k|), \quad m, n \geq n_0,\ k \in \mathbb{N}.$$

Therefore, $\{f_n\}_{n=1}^\infty$ is a Cauchy sequence in every $L^1(G_k)$. That is a complete space, hence $\{f_n\}_{n=1}^\infty$ is convergent. By Theorem 3.13.2 (iii) and the Riesz theorem (Theorem 1.22.8), for every $k \in \mathbb{N}$ there exists a subsequence $\{f_n^{(k)}\}_{n=1}^\infty$ and the limit function $f^{(k)} \in L^1(G_k)$ such that $f_n^{(k)}(x) \to f^{(k)}(x)$ for a.e. $x \in G_k$. Summing up over $k \in \mathbb{N}$, we get

$$\lim_{m \to \infty} f_m^{(m)}(x) = \sum_{k=1}^\infty f^{(k)}(x) \chi_{G_k}(x) = f(x) \quad \text{for a.e. } x \in \Omega.$$

Using (11.2.8) with f_m replaced by $f_m^{(m)}$ and using the Fatou lemma (Theorem 1.21.6), we get

$$\int_\Omega |f(x) - f_n(x)||g(x)|\, dx \leq \sup_{m \in \mathbb{N}} \int_\Omega |f_m^{(m)}(x) - f_n(x)||g(x)|\, dx \leq \varepsilon$$

for every $n \in \mathbb{N}$, $n \geq n_0$ and every g satisfying $\varrho_{p'}(g) \leq 1$. This establishes

$$\|f_n - f\|_p \leq \varepsilon,$$

as desired. The proof is complete. $\square$

Lemma 11.2.11. Let $f \in L^{p(x)}$, $f \not\equiv 0$. Assume that

$$T = \operatorname*{ess\,sup}_{\{x;\, f(x)\neq 0\}} p(x) < \infty.$$

Then

$$\int_{\Omega_0} \left(\frac{|f|}{\|f\|_p}\right)^p d\mu + \operatorname*{ess\,sup}_{\Omega_\infty} \frac{|f|}{\|f\|_p} = 1.$$

Proof. First, suppose that

$$K := \int_{\Omega_0} \left(\frac{|f|}{\|f\|_p}\right)^p d\mu + \operatorname*{ess\,sup}_{\Omega_\infty} \frac{|f|}{\|f\|_p} < 1.$$

Choose $0 < \lambda < \|f\|_p$ such that

$$\left(\frac{\|f\|_p}{\lambda}\right)^T K \leq 1.$$

Section 11.2 Basic properties

Then

$$\int_{\Omega_0} \left(\frac{|f|}{\lambda}\right)^p d\mu + \underset{\Omega_\infty}{\mathrm{ess\,sup}}\, \frac{|f|}{\lambda}$$

$$\leq \left(\frac{\|f\|_p}{\lambda}\right)^T \left(\frac{\lambda}{\|f\|_p}\right)^T \left(\int_{\Omega_0} \left(\frac{|f|}{\lambda}\right)^p d\mu + \underset{\Omega_\infty}{\mathrm{ess\,sup}}\, \frac{|f|}{\lambda}\right)$$

$$\leq \left(\frac{\|f\|_p}{\lambda}\right)^T \left(\int_{\Omega_0} \left(\frac{\lambda}{\|f\|_p}\right)^p \left(\frac{|f|}{\lambda}\right)^p d\mu + \underset{\Omega_\infty}{\mathrm{ess\,sup}}\, \frac{|f|}{\|f\|_p}\right)$$

$$\leq \left(\frac{\|f\|_p}{\lambda}\right)^T \left(\int_{\Omega_0} \left(\frac{|f|}{\|f\|_p}\right)^p d\mu + \underset{\Omega_\infty}{\mathrm{ess\,sup}}\, \frac{|f|}{\|f\|_p}\right) \leq 1.$$

This is a contradiction with the definition of $\|f\|_p$. So we have $K \geq 1$. We will prove that $K \leq 1$. Choose a sequence $\lambda_n \searrow \|f\|_p$. Then

$$\int_{\Omega_0} \left(\frac{|f|}{\lambda_n}\right)^p d\mu + \underset{\Omega_\infty}{\mathrm{ess\,sup}}\, \frac{|f|}{\lambda_n} \leq 1,$$

hence

$$\int_{\Omega_0} \left(\frac{|f|}{\lambda_n}\right)^p d\mu \leq 1 - \underset{\Omega_\infty}{\mathrm{ess\,sup}}\, \frac{|f|}{\|f\|_p}$$

for all n. Using the fact that $\frac{|f|}{\lambda_n} \nearrow \frac{|f|}{\|f\|_p}$ and the Lebesgue monotone convergence theorem, we obtain

$$\int_{\Omega_0} \left(\frac{|f|}{\|f\|_p}\right)^p d\mu + \underset{\Omega_\infty}{\mathrm{ess\,sup}}\, \frac{|f|}{\|f\|_p} \leq 1. \qquad \square$$

Remark 11.2.12. In fact, the space $L^{p(x)}(\Omega)$ is a Banach function space for every $p \in \mathcal{P}(\Omega)$. The verification of the axioms is left to the reader as an exercise.

We shall finish this section with an interesting observation that, in general, it is not possible to recover the space $L^{p(x)}$ by extrapolating from uniform estimates in constant exponent Lebesgue space (not even in a very simple case when Ω is an interval and p is linear). The example is due to Kirchheim who did not publish it; the proof is taken over from [190, Proposition 2.35].

Example 11.2.13. Let $\Omega = [0, 1]$ and let $p(x) := 2 + x$. Then there exist a function $g \notin L^{p(x)}(0, 1)$ and a constant c_0 such that, for every $q \in [2, 3]$, the estimate

$$\int_{q-2}^{1} g(x)^q \, dx \leq c_0$$

holds.

Proof. For every $b \in (0, 1)$, one has
$$\lim_{x \to b-} (b-x)^{-\frac{x}{x+2}} = \infty.$$

Using this fact, we shall construct inductively the sequences $\{a_k\}_{k=1}^{\infty}$ and $\{b_k\}_{k=1}^{\infty}$ according to the following rules: let $b_1 \in (0, 1)$ be chosen arbitrarily and assume that, for some $k \in \mathbb{N}$, $b_1, \ldots, b_k$ and $a_1, \ldots a_{k-1}$ have been fixed. Then there exists an $a_k \in (0, b_k)$ such that
$$(b_k - a_k)^{-\frac{a_k}{a_k+2}} = 2^{k+1}. \tag{11.2.9}$$

We define $b_{k+1} \in (0, a_k)$ by the formula
$$(b_k - a_k)^{\frac{b_{k+1}-a_k}{a_k+2}} = 2^k. \tag{11.2.10}$$

Such b_{k+1} exists since the function
$$F : x \mapsto (b_k - a_k)^{\frac{x-a_k}{a_k+2}}$$
is continuous and nonincreasing on $(0, a_k)$ and satisfies
$$F(0+) = 2^{k+1} \quad \text{and} \quad F(a_k-) = 1.$$

Denote
$$\lambda_k := (b_k - a_k)^{-\frac{1}{a_k+2}}. \tag{11.2.11}$$

Then, by (11.2.10),
$$\lambda_k^{b_{k+1}-a_k} = 2^{-k}. \tag{11.2.12}$$

Let the function g be defined for $x \in (0, 1)$ by
$$g(x) := \sum_{k=1}^{\infty} \lambda_k \chi_{(a_k, b_k)}(x).$$

Then $g \notin L^{p(x)}(0, 1)$, because, using (11.2.11), we get
$$\int_0^1 g(x)^{p(x)} \, dx = \sum_{k=1}^{\infty} \int_{a_k}^{b_k} \lambda_k^{2+x} \, dx \geq \sum_{k=1}^{\infty} (b_k - a_k) \lambda_k^{2+a_k} = \sum_{k=1}^{\infty} 1 = \infty.$$

Assume now that $q > 2$. Let $k_0 \in \mathbb{N}$ be such that
$$b_{k_0+1} \leq q - 2 < b_{k_0}.$$

Then
$$\int_{q-2}^1 g(x)^q \, dx = \sum_{k=1}^{\infty} (b_k - a_k) \lambda_k^q + \int_{q-2}^{b_{k_0}} \lambda_{k_0}^q \chi_{(a_{k_0}, b_{k_0})}(x) \, dx =: I_1 + I_2,$$

say. Since $q \leq b_{k_0} + 2$ and $\lambda_k > 1$, one has, using also the monotonicity of the sequence $\{b_k\}$, (11.2.11) and (11.2.12).

$$I_1 \leq \sum_{k=1}^{k_0-1} \lambda_k^{b_{k_0}+2}(b_k - a_k)$$

$$\leq \sum_{k=1}^{k_0-1} \lambda_k^{b_{k+1}-a_k} \lambda_k^{a_k+2}(b_k - a_k) \leq \sum_{k=1}^{\infty} 2^{-k} = 1.$$

By (11.2.11) once again, we get the estimate

$$I_2 \leq \lambda_{k_0}^{b_{k_0}+2}(b_{k_0} - a_{k_0}) \leq \sup_{k \in \mathbb{N}} \lambda_k^{b_k-a_k} \lambda_k^{a_k+2}(b_k - a_k)$$

$$= \sup_{k \in \mathbb{N}} (b_k - a_k)^{-\frac{b_k-a_k}{a_k+2}} \leq \sup_{k \in \mathbb{N}} (b_k - a_k)^{-\frac{b_k-a_k}{2}} \leq \sup_{x \in (0,1)} x^{-\frac{x}{2}} = e^{\frac{1}{2e}}.$$

Now let $q = 2$. Then

$$\int_0^1 g(x)^2 \, dx = \sum_{k=1}^{\infty} \lambda_k^2(b_k - a_k) = \sum_{k=1}^{\infty} (b_k - a_k)^{\frac{a_k}{a_k+2}} = \sum_{k=1}^{\infty} 2^{-(k+1)} = \frac{1}{2}.$$

In any case, we have

$$\int_{q-2}^1 g(x)^q \, dx \leq 1 + e^{\frac{1}{2e}},$$

establishing the claim. The proof is complete. □

11.3 Embedding relations

As usual, one of the most important and interesting questions concerning function spaces is how they embed into other function spaces and classes. The basic relation of such kind is presented in the following theorem.

Theorem 11.3.1. *Let* $p, q \in \mathcal{P}(\Omega)$ *and assume that* $\mu(\Omega) < \infty$. *Then, the continuous embedding*

$$L^{q(x)}(\Omega) \hookrightarrow L^{p(x)}(\Omega) \tag{11.3.1}$$

holds if and only if

$$p(x) \leq q(x) \quad \mu\text{-a.e. on } \Omega. \tag{11.3.2}$$

Proof. Sufficiency. Let (11.3.2) be satisfied. Then

$$\Omega_\infty^p := \{x \in \Omega;\ p(x) = \infty\} \subset \{x \in \Omega;\ q(x) = \infty\} =: \Omega_\infty^q.$$

By (11.2.5), we have

$$\varrho_q(f) = \int_{\Omega\setminus\Omega_\infty^q} |f(x)|^{q(x)}\, dx + \operatorname*{ess\,sup}_{x\in\Omega_\infty^q} |f(x)| \le 1.$$

This implies, in particular, that $|f(x)| \le 1$ for a.e. $x \in \Omega_\infty^q$. Therefore,

$$\varrho_p(f) \le |\{x \in \Omega \setminus \Omega_\infty^q;\ |f(x)| \le 1\}| + \int_{\Omega\setminus\Omega_\infty^q} |f(x)|^{q(x)}\, dx +$$
$$+ |\Omega_\infty^q \setminus \Omega_\infty^p| + \operatorname*{ess\,sup}_{x\in\Omega_\infty^q} |f(x)| \le |\Omega| + \varrho_q(f) \le |\Omega| + 1.$$

The convexity of the functional ϱ_p now yields

$$\varrho_p\left(\frac{f}{|\Omega|+1}\right) \le \frac{\varrho_p(f)}{|\Omega|+1} \le 1.$$

We have shown that for every $f \in L^{q(x)}(\Omega)$ such that $\|f\|_q \le 1$, one has

$$\|f\|_p \le |\Omega| + 1.$$

This implies (11.3.1) on applying Theorem 1.5.13 with $h(t) := |\Omega| + 1$.

Necessity. Now assume for a contradiction that (11.3.2) does not hold. Then there exists a subset G of Ω of positive measure and such that

$$p(x) > q(x) \quad \text{for every } x \in G.$$

Assume first that

$$|\Omega_\infty^p \setminus G| > 0.$$

Then there exists a set $A \subset \Omega_\infty^p \setminus G$ satisfying $0 < |A| < \infty$ and a number $r \in (1, \infty)$ so that

$$1 \le q(x) \le r < \infty = p(x) \quad \text{for all } x \in A.$$

We find a sequence of pairwise disjoint sets $\{A_k\}_{k=1}^\infty$ such that

$$A = \bigcup_{k=1}^\infty A_k \quad \text{and} \quad |A_k| = 2^{-k}|A| \text{ for every } k \in \mathbb{N}, \tag{11.3.3}$$

and set

$$f(x) := \sum_{k=1}^\infty \left(\frac{3}{2}\right)^{\frac{k}{r}} \chi_{A_k}(x), \quad x \in \Omega.$$

Then,

$$\|f\|_p \ge \|f\chi_A\|_\infty = \infty.$$

On the other hand,

$$\varrho_q(f) = \int_A |f(x)|^{q(x)} \, dx = \sum_{k=1}^{\infty} \int_{A_k} \left(\frac{3}{2}\right)^{\frac{kq(x)}{r}} dx$$

$$\leq \sum_{k=1}^{\infty} \left(\frac{3}{2}\right)^k |A_k| = |A| \sum_{k=1}^{\infty} \left(\frac{3}{4}\right)^k = 2|A| < \infty.$$

We have thus constructed a function f satisfying $f \in L^{q(x)}(\Omega) \setminus L^{p(x)}(\Omega)$, which however contradicts (11.3.1).
  When

$$|\Omega_\infty^p \setminus G| = 0,$$

then $1 \leq q(x) < p(x) < \infty$ for a.e. $x \in G$. Hence there exist a set $A \subset G$ satisfying $0 < |A| < \infty$ and some $a > 0$ and $r \in (1, \infty)$ such that

$$q(x) + a \leq p(x) \leq r \quad \text{for every } x \in A.$$

Once again, we find the sequence $\{A_k\}_{k=1}^{\infty}$ of pairwise disjoint subsets of A satisfying (11.3.3). We then set

$$f(x) := \sum_{k=1}^{\infty} \left(2^k k^{-2}\right)^{\frac{1}{q(x)}} \chi_{A_k}(x), \quad x \in \Omega.$$

Then

$$\varrho_q(f) = \sum_{k=1}^{\infty} 2^k k^{-2} |A_k| = |A| \sum_{k=1}^{\infty} k^{-2} < \infty,$$

hence $f \in L^{q(x)}(\Omega)$. On the other hand, however, for every $\lambda \in (0, 1]$, we have

$$\varrho_p(\lambda f) \geq \lambda^r \sum_{k=1}^{\infty} \int_{A_k} \left(2^k k^{-2}\right)^{\frac{p(x)}{q(x)}} \geq \lambda^r \sum_{k=1}^{\infty} \left(2^k k^{-2}\right)^{1+\frac{a}{r}} |A_k|$$

$$= \lambda^r |A| \sum_{k=1}^{\infty} 2^{\frac{ak}{r}} k^{-2(1+\frac{a}{r})} = \infty,$$

in other words $f \notin L^{p(x)}(\Omega)$, and we get the same contradiction with (11.3.1) again. The proof is complete. □

11.4 Density of smooth functions

In this section we point out that the class of smooth functions is dense in $L^{p(x)}(\Omega)$ provided that p is bounded.

Theorem 11.4.1. Let $p \in \mathcal{P}(\Omega)$ and assume that $p \in L^\infty(\Omega)$. Then, the set $C(\Omega) \cap L^{p(x)}(\Omega)$ is dense in $L^{p(x)}(\Omega)$. If, moreover, Ω is open, then the set $C_0^\infty(\Omega) \cap L^{p(x)}(\Omega)$ is dense in $L^{p(x)}(\Omega)$.

Proof. We first note that when $p_{\max} < \infty$, then the set of all bounded functions on Ω is dense in $L^{p(x)}(\Omega)$. Indeed, for $n \in \mathbb{N}$, denote

$$G_n := \{x \in \Omega \setminus \Omega_\infty; \ |x| < n\},$$

and set

$$f_n(x) := \begin{cases} f(x) & \text{if } |f(x)| \leq n \text{ and } x \in G_n \cup \Omega_\infty, \\ n \operatorname{sign} f(x) & \text{if } |f(x)| > n \text{ and } x \in G_n \cup \Omega_\infty, \\ 0 & \text{for every other } x \in \Omega. \end{cases}$$

Then every function f_n, $n \in \mathbb{N}$, is bounded on Ω. By the Lebesgue dominated convergence theorem (Theorem 1.21.5), we get $\varrho_p(f_n - f) \to 0$. Finally, since $p \in L^\infty$, we get from Exercise 11.2.9 (see also Proposition 1.5.12), we get $\|f_n - f\|_p \to 0$, proving the claim.

Now let $f \in L^{p(x)}(\Omega)$ and $\varepsilon > 0$. By the observation just made, there exists a bounded function $g \in L^{p(x)}(\Omega)$ such that

$$\|f - g\|_p < \varepsilon. \tag{11.4.1}$$

By the Luzin theorem (Theorem 1.21.12), there exists a function $h \in C(\Omega)$ and an open set $U \subset \Omega$ such that

$$|U| < \min\left\{1, \left(\frac{\varepsilon}{2\|g\|_\infty}\right)^{p_{\max}}\right\},$$

$g(x) = h(x)$ for all $x \in \Omega \setminus U$ and

$$\sup |h(x)| = \sup_{\Omega \setminus U} |g(x)| \leq \|g\|_\infty.$$

Consequently,

$$\varrho_p\left(\frac{g-h}{\varepsilon}\right) \leq \max\left\{1, \left(\frac{2\|g\|_\infty}{\varepsilon}\right)^{p_{\max}}\right\} |U| \leq 1,$$

that is, $\|g - h\|_p \leq \varepsilon$. Combined with (11.4.1), this yields

$$\|f - h\|_p < 2\varepsilon. \tag{11.4.2}$$

Assume that Ω is open. Since $p \in L^\infty(\Omega)$, one has $C_0^\infty(\Omega) \subset L^{p(x)}(\Omega)$ and

$$\varrho_p\left(\frac{h}{\varepsilon}\right) < \infty.$$

Thus, there exists an open bounded set $G \subset \Omega$ such that

$$\varrho_p\left(\frac{h\chi_{\Omega\setminus G}}{\varepsilon}\right) \leq 1,$$

that is,

$$\|h - h\chi_G\|_p \leq \varepsilon. \tag{11.4.3}$$

Let P be a polynomial satisfying

$$\sup_{x \in G} |h(x) - P(x)| < \varepsilon \min\{1, |G|^{-1}\}.$$

Then

$$\varrho_p\left(\frac{h\chi_G - P\chi_G}{\varepsilon}\right) \leq \min\{1, |G|^{-1}\}|G| \leq 1.$$

In other words, we obtain

$$\|h\chi_G - P\chi_G\|_p \leq \varepsilon. \tag{11.4.4}$$

Finally, similarly as above, we get a sufficiently small number a such that the set

$$K_a := \{x \in G;\ \operatorname{dist}(x, \partial G) \geq a\},$$

where ∂G is the boundary of G, is compact, and satisfies

$$\|P = \chi_G - P\chi_{K_a}\|_p \leq \varepsilon.$$

For a function $\varphi \in C_0^\infty(G)$ such that $0 \leq \varphi(x) \leq 1$ on G and $\varphi \equiv 1$ on K_a we then have

$$\|P\chi_G - P\varphi\|_p \leq \varepsilon.$$

Altogether, combining all the estimates, we get

$$\|f - P\varphi\|_p \leq 4\varepsilon.$$

Since, clearly, $P\varphi \in C_0^\infty(\Omega)$, the assertion follows. The proof is complete. □

Corollary 11.4.2. *If $p \in \mathcal{P}(\Omega) \cap L^\infty(\Omega)$, then $L^{p(x)}(\Omega)$ is separable.*

Proof. Let $\{G_n\}_{n=1}^\infty$ be a sequence of bounded sets such that $G_n \subset G_{n+1} \subset \Omega$ for all $n \in \mathbb{N}$ and

$$\Omega = \bigcup_{n=1}^\infty G_n.$$

Then it follows from the proof of Theorem 11.4.1 that the set of all functions of the form $P\chi_{G_n}$, $n \in \mathbb{N}$, where P is a polynomial on $\mathbb{R}^N$ with rational coefficients, is dense in $L^{p(x)}(\Omega)$, and the claim follows. □

11.5 Reflexivity and uniform convexity

Recall that a Banach space X is said to be *uniformly convex* if for every $\varepsilon \in (0, 2]$ there exists a $\delta > 0$ such that

$$\left\| \frac{x+y}{2} \right\|_X \leq 1 - \delta \quad \text{whenever} \quad x, y \in S_X, \ \|x - y\|_X \geq \varepsilon.$$

As already observed in Remark 3.19.7, every uniformly convex space is reflexive.

Lemma 11.5.1. *Assume that μ is nonatomic. Then the following statements are equivalent:*

(i) $L^{p(x)}$ *has absolutely continuous norm;*

(ii) $p_{\max} < \infty$.

Proof. Assume that $p_{\max} = \infty$ and $\mu(\Omega_\infty) = 0$. Define

$$\Omega_n := \{x \in \Omega;\ n \leq p(x) < n+1\}, \quad n \in \mathbb{N}.$$

Then there exists a subsequence of natural numbers $\{n_k\}_{k=1}^\infty$ such that $\mu(\Omega_{n_k}) > 0$. Let $c_k > 0$ be such that

$$\int_{\Omega_{n_k}} c_k^{p(x)}\, d\mu(x) = 1, \quad k \in \mathbb{N}.$$

We set

$$f(x) := \sum_{k=1}^\infty c_k \chi_{\Omega_{n_k}}(x), \quad x \in \Omega, \quad \text{and} \quad E_j := \bigcup_{k=j}^\infty \Omega_{n_k}.$$

Then $E_n \to \emptyset$. Now

$$\|f\|_p = \inf\left\{\lambda > 0;\ \sum_{k=1}^\infty \int_{\Omega_{n_k}} \left(\frac{c_k}{\lambda}\right)^{p(x)} d\mu(x) \leq 1\right\}$$

$$\leq \inf\left\{\lambda > 1;\ \sum_{k=1}^\infty \left(\frac{1}{\lambda}\right)^{n_k} \leq 1\right\} \leq 2,$$

and so $f \in L^{p(x)}$.

To finish the proof of the implication (i) $\Rightarrow$ (ii) it suffices to write

$$\|f \chi_{E_j}\|_p \geq \inf\left\{\lambda > 0;\ \int_{\Omega_{n_\ell}} \left(\frac{c_\ell}{\lambda}\right)^{p(x)} d\mu(x) \leq 1\right\} = 1.$$

If $\mu(\Omega_\infty) > 0$, choose $A \subset \Omega_\infty$ with $0 < \mu(A) < \infty$. Then $\|\chi_A\|_p > 0$ and, obviously, χ_A does not have an absolutely continuous norm.

Section 11.5 Reflexivity and uniform convexity   451

To prove the converse implication, we assume that $p_{\max} < \infty$ and choose $f \in L^{p(x)}$ with $\|f\|_p = 1$ and $\varepsilon > 0$. Let $\{E_n\}$ be a sequence of sets such that $\mu(E_n) \searrow 0$. Choose $l \in \mathbb{N}$ with $\|f\chi_{E_l}\|_p \geq 1 - \varepsilon$. Set $\varphi = f\chi_{\Omega \setminus E_l}$ and $\psi = f\chi_{E_l}$. By Lemma 11.2.11,

$$\int_\Omega \left|\frac{\varphi(x)}{\|\varphi\|_p}\right|^{p(x)} d\mu(x) = 1 \quad \text{and} \quad \int_\Omega \left|\frac{\psi(x)}{\|\psi\|_p}\right|^{p(x)} d\mu(x) = 1$$

and so $\|\varphi\|_p^{p_{\max}} \leq \int_\Omega |\varphi|^p \, d\mu$ and $\|\psi\|_p^{p_{\max}} \leq \int_\Omega |\psi|^p \, d\mu$. Moreover, $\int_\Omega |\varphi|^p \, d\mu + \int_\Omega |\psi|^p \, d\mu \leq 1$. So we have

$$\|\psi\|_p^{p_{\max}} \leq \int_\Omega |\psi|^p \, d\mu \leq 1 - \int_\Omega |\varphi|^p \, d\mu \leq 1 - \|\varphi\|_p^{p_{\max}} \leq 1 - (1-\varepsilon)^{p_{\max}},$$

and therefore $\|\psi\|_p \leq (1 - (1-\varepsilon)^{p_{\max}})^{1/p_{\max}}$. □

Remark 11.5.2. Note that the assumption that μ is nonatomic was needed only for the implication (i)⇒(ii).

Theorem 11.5.3. *Assume that μ is nonatomic. Then the following statements are equivalent:*

(i) $L^{p(x)}$ *is reflexive;*

(ii) *the spaces $L^{p(x)}$ and $L^{p'(x)}$ have absolutely continuous norm;*

(iii) $L^{p(x)}$ *is uniformly convex;*

(iv) $1 < p_{\min} \leq p_{\max} < \infty$.

Proof. As mentioned above, the space $L^{p(x)}$ is a Banach function space. Therefore, it is reflexive if and only if it has absolutely continuous norm and its associate space has absolutely continuous norm (see Corollary 6.4.6). This shows (i)⇔(ii).

We shall show (ii)⇒(iv). By symmetry, it suffices to show that if p is essentially unbounded, then the space $L^{p(x)}$ does not have an absolutely continuous norm. But this immediately follows from Lemma 11.5.1.

Since every uniformly convex space is reflexive, we have (iii)⇒(i).

To round the proof off, it just remains to show (iv)⇒(iii). Fix $\varepsilon \in (0, 1)$ and $u, v \in S_{L^{p(x)}}$. We aim to show that there exists a $\delta > 0$ such that

$$\|\tfrac{1}{2}(u+v)\|_p > 1 - \delta \quad \text{implies} \quad \|u - v\|_p \leq \varepsilon.$$

To this end, we set

$$s := \tfrac{1}{2}(u+v) \quad \text{and} \quad t := \tfrac{1}{2}(u-v).$$

We then define
$$\Gamma = \{x \in \Omega;\ p_{\min} \le p(x) \le p_{\max}\}$$
and
$$S := \{x \in \Gamma;\ |t(x)| < \varepsilon |s(x)|\} \quad \text{and} \quad T := \{x \in \Gamma;\ |t(x)| \ge \varepsilon |s(x)|\},$$
and note that $\mu(\Omega \setminus \Gamma) = 0$.

Then
$$\int_S |t(x)|^{p(x)}\,d\mu(x) \le \int_S \varepsilon^{p(x)}|s(x)|^{p(x)}\,d\mu(x) \le \int_\Omega \varepsilon^{p(x)}|s(x)|^{p(x)}\,d\mu(x) \tag{11.5.1}$$
$$\le \varepsilon^{p_{\min}} \int_\Omega |s(x)|^{p(x)}\,d\mu(x) = \varepsilon^{p_{\min}}.$$

Since, for $t \in (1, \infty)$, the function $\lambda \mapsto |\lambda|^t$ is strictly convex on $\mathbb{R}$, we have
$$\tfrac{1}{2}\left(|\lambda+1|^t + |\lambda-1|^t\right) > |\lambda|^t, \quad \lambda \in \mathbb{R}. \tag{11.5.2}$$

Next we observe that the function
$$f : (p, \lambda) \mapsto \tfrac{1}{2}\left(|\lambda+1|^p + |\lambda-1|^p\right) - |\lambda|^p, \quad p \in (1, \infty),\ \lambda \in \mathbb{R},$$
is continuous and strictly positive on $(1, \infty) \times \mathbb{R}$. Hence, there exists an $\alpha > 0$ such that $f(p, \lambda) \ge \alpha$ for every $p \in [p_{\max}, p_{\min}]$ and $\lambda \in [-1/\varepsilon, 1/\varepsilon]$. Therefore
$$\tfrac{1}{2}\left(|\lambda+1|^{p(x)} + |\lambda-1|^{p(x)}\right) - |\lambda|^{p(x)} \ge \alpha \tag{11.5.3}$$
for every $x \in \Gamma$ and $\lambda \in [-1/\varepsilon, 1/\varepsilon]$. An appeal to (11.5.3) reveals that
$$\tfrac{1}{2}\left(|s(x) + t(x)|^{p(x)} + |s(x) - t(x)|^{p(x)}\right) \ge |s(x)|^{p(x)} + \alpha |t(x)|^{p(x)}, \quad x \in T,$$
while by (11.5.2),
$$\tfrac{1}{2}\left(|s(x) + t(x)|^{p(x)} + |s(x) - t(x)|^{p(x)}\right) \ge |s(x)|^{p(x)}, \quad x \in S,$$
(consider $\lambda = s(x)/t(x)$). Consequently,
$$1 = \int_\Omega \tfrac{1}{2}\left(|s(x) + t(x)|^{p(x)} + |s(x) - t(x)|^{p(x)}\right) d\mu(x)$$
$$\ge \int_\Omega |s(x)|^{p(x)}\,d\mu(x) + \int_T \alpha |t(x)|^{p(x)}\,d\mu(x).$$

It follows that
$$\int_T |t(x)|^{p(x)}\,d\mu(x) < \varepsilon^{p_{\min}} \quad \text{provided} \quad \int_\Omega |s(x)|^{p(x)}\,d\mu(x) > 1 - \alpha\varepsilon^{p_{\min}}.$$

Set $\Delta = \alpha \varepsilon^{p_{\min}}$ and assume that

$$\|\tfrac{1}{2}(u+v)\|_p > 1 - \Delta.$$

Then $\int_\Omega |s(x)|^{p(x)} \, d\mu(x) > 1 - \Delta$ and invoking (11.5.1), we obtain

$$\int_\Omega |t(x)|^{p(x)} \, d\mu(x) < 2\varepsilon^{p_{\min}}.$$

Hence,

$$\|u - v\|_p = \|2t\|_p \le 2 \left(2\varepsilon^{p_{\min}}\right)^{\frac{1}{p_{\max}}},$$

and this suffices to complete the proof. □

Remark 11.5.4. We provide relatively simple proofs of the equivalence of assertions (i)–(iv) although some separate implications of Theorem 11.5.3 are not new. The characterization of reflexivity was shown by Ková čik and Rákosník in [122] while uniform convexity of $L^{p(x)}$-spaces was treated by other authors, too, see for example the paper [77] by Fan and Zhao (in particular, Theorem 1.10) and the references therein. Note that the assumption of μ nonatomic was needed only in the proof of the implication (ii)⇒(iv).

The uniform convexity of classical reflexive Lebesgue L^p-spaces (with constant exponent) was shown by Clarkson in [53] who established this result using the Clarkson inequalities (see Remark 3.1.10). The idea of the proof of (iv)⇒(iii) (still for the classical L^p-spaces) goes back to McShane [150], cf. also Fabian et al. [75, Theorem 9.3]. The present proof follows Malý's reasoning in the classical case. The idea of his proof can be found in Lukeš [139, Theorem 21.9].

11.6 Radon–Nikodým property

A Banach space X is said to have the *Radon–Nikodým property*, shortly RNP, if given a finite measure space $(\Omega, \mathcal{S}, \mu)$ and a vector measure $\nu : \mathcal{S} \to X$ of finite variation and absolutely continuous with respect to μ, then there exists an integrable function $g : \Omega \to X$ such that

$$\nu(E) = \int_E g \, d\mu \quad \text{for any } E \in \mathcal{S}.$$

Note that the Radon–Nikodým property is hereditary to closed subspaces (see Diestel and Uhl [64, Theorem III.3.2]).

The next lemma is a variation of the Dunford theorem (see [65] or Diestel and Uhl [64, Theorem III.1.6], and its proof follows the lines of the proof presented therein).

Lemma 11.6.1. *Let X be a Banach function space on (Ω, μ) with absolutely continuous norm and let $\Omega = \bigcup_{n=1}^{\infty} A_n$ be a union of measurable sets. If the subspaces $X_n := \{f \in X;\ f = 0 \text{ on } \Omega \setminus A_n\}$ have the Radon–Nikodým property, then X has the Radon–Nikodým property as well.*

Proof. Let (Γ, Σ, κ) be a finite measure space and let $\nu : \Sigma \to X$ be a vector measure of finite variation which is absolutely continuous with respect to κ. Define projections $P_n : X \to X_n$ by $P_n(f) = f\chi_{A_n}$ and set $\nu_n = P_n(\nu)$. Then each ν_n is an X_n-valued vector measure of finite variation which is absolutely continuous with respect to κ and so (using the fact that each space X_n has the Radon–Nikodým property) there exists an integrable function $g_n : \Gamma \to X_n$ satisfying

$$\nu_n(E) = \int_\Gamma g_n \, d\kappa \quad \text{for each} \quad E \in \Sigma.$$

Now, for $E \in \Sigma$ and $n \in \mathbb{N}$, we have

$$\int_E \left\| \sum_{k=1}^n g_k \right\|_X d\kappa \leq |\nu|(E), \tag{11.6.1}$$

and so, for κ–almost all $\alpha \in \Gamma$, there exists a $g(\alpha) \in X$ such that $g(\alpha) = \sum_{k=1}^{\infty} g_k(\alpha)$. By (11.6.1) and the Fatou lemma (Theorem 1.21.6), g is also integrable. Finally,

$$\nu(E) = \lim_{n\to\infty} \sum_{k=1}^n \nu_k(E) = \lim_{n\to\infty} \sum_{k=1}^n \int_E g_n \, d\kappa = \int_E g \, d\kappa. \qquad \square$$

Proposition 11.6.2. *Let X be a Banach function space with $X_a \neq X$. Then ℓ^∞ isomorphically embeds into X. In particular, X lacks the Radon–Nikodým property.*

Proof. Choose $f \in X \setminus X_a$ with $\|f\|_X = 1$. There exist $\delta > 0$ and a decreasing sequence $\{A_n\}$ of subsets of Ω such that $\mu(A_n) \to 0$ and $1 \geq \|f\chi_{A_n}\|_X \searrow 2\delta$. Denote $E_n := \Omega \setminus A_n$, $n \in \mathbb{N}$. There exists an $i \in \mathbb{N}$ such that $\|f\chi_{E_i}\|_X > \delta$. Set $g_1 = f\chi_{E_i}$ and $f_1 = f\chi_{A_i}$. Note that f_1 and g_1 have disjoint supports and $1 \geq \|f_1 \chi_{A_n}\|_X \searrow 2\delta$. So we can apply the preceding procedure to f_1 in place of f, and we obtain functions g_2 and f_2 with disjoints supports such that $1 \geq \|g_2\|_X \geq \delta$ and $1 \geq \|f_2 \chi_{A_n}\|_X \searrow 2\delta$. Using it inductively we can obtain a sequence $\{g_n\}$ of functions with pairwise disjoint supports such that $1 \geq \|g_n\|_X \searrow \delta$ and $\|\sum g_n\|_X \leq 1$. Set

$$Y := \left\{ g \in X;\ \text{there exists } \{c_n\} \in \ell^\infty \text{ such that } g = \sum_{n=1}^{\infty} c_n g_n \right\}.$$

Choose $c = \{c_n\} \in \ell^\infty$ and $\varepsilon > 0$. There exists an $i \in \mathbb{N}$ such that $\|c\|_\infty \le |c_i| + \varepsilon$. Then

$$\delta\|c\|_\infty \le \delta|c_i| + \delta\varepsilon \le \|c_i g_i\|_X + \delta\varepsilon \le \Big\|\sum_{n=1}^\infty c_n g_n\Big\|_X + \delta\varepsilon$$

$$\le \Big\|\sum_{n=1}^\infty \|c\|_\infty g_n\Big\|_X + \delta\varepsilon \le \|c\|_\infty \Big\|\sum_{n=1}^\infty g_n\Big\|_X + \delta\varepsilon$$

$$\le \|c\|_\infty + \delta\varepsilon.$$

Hence

$$\delta\|c\|_\infty \le \Big\|\sum_{n=1}^\infty c_n g_n\Big\|_X \le \|c\|_\infty.$$

The mapping $\{c_n\} \mapsto \sum_{n=1}^\infty c_n g_n$ is an isomorphic embedding of ℓ^∞ into X. Since the space ℓ^∞ fails to have the Radon–Nikodým property, and the RNP is hereditary to closed subspaces, we conclude that X cannot have the Radon–Nikodým property. □

Remark 11.6.3. By an unpublished result of Lotz, if X is a Banach lattice, then X^* has the RNP if and only if no sublattice of X^* is isomorphic to c_0 or $L^1([0,1])$; cf. Meyer-Nieberg [168, Theorem 5.4.14]. Concerning the $L^{p(x)}$-spaces, we get the following result.

Theorem 11.6.4. *Assume that μ is nonatomic. Then the following statements are equivalent:*

(i) $L^{p(x)}$ *has the Radon–Nikodým property;*

(ii) $\mu(\{x \in \Omega;\ p(x) = 1\}) = 0$ *and* $p_{\max} < \infty$.

Proof. Assume that $\mu(\{x \in \Omega;\ p(x) = 1\}) > 0$. Then $L^{p(x)}$ contains the space $L^1(\Omega_1, \mathcal{S}|_{\Omega_1}, \mu|_{\Omega_1})$ which lacks the RNP. Since the RNP is hereditary to closed subspaces, $L^{p(x)}$ lacks it as well. In the case when $p_{\max} = \infty$, $L^{p(x)}$ does not have an absolutely continuous norm according to Lemma 11.5.1, and therefore by Lemma 11.6.2 lacks the RNP.

Assume now that (ii) holds. Decompose Ω into countably many pairwise disjoint sets $\{\Lambda_n\}_{n=1}^\infty$ with ess $\inf\{p(t);\ t \in \Lambda_n\} > 1$ for each $n \in \mathbb{N}$. To finish the proof, it is sufficient to use Theorem 11.5.3, Lemma 11.6.1 and the fact that reflexive spaces have the RNP. □

11.7 Daugavet property

A Banach space X is said to have the *Daugavet property*, and is called a *Daugavet space*, if

$$\|\operatorname{Id} + T\|_X = 1 + \|T\|_X$$

for every compact operator $T : X \to X$. Daugavet (cf. [59]) has shown that the space $C([0,1])$ possesses this property. Three years later, Lozanovskij proved in [138] that the space $L^1([0,1])$ also has the Daugavet property. The same result is contained in the paper by Babenko and Pichugov [7].

Later on, the Daugavet property of various Banach spaces was investigated. Particular attention was paid to function spaces. Chauveheid has shown in [39] that the space $C(K)$ has the Daugavet property if and only if the compact space K has no isolated points and that the space $L^1(\mu)$ is Daugavet if μ is a nonatomic measure. In this case the space $L^\infty(\mu)$ is a Daugavet space. The same results can be found also in a paper by Kamowitz [114]. Since Daugavet spaces are not reflexive (cf. Corollary 2.5 of [234] by Werner), the spaces $L^p(\mu)$ for $p \in (1, \infty)$ lack the Daugavet property.

Moreover, miscellaneous conditions characterizing Daugavet spaces were established; see, for example, the survey paper of Werner [234] where the following geometric characterization of Daugavet spaces can be traced. For $\varepsilon > 0$ and $f \in S_{X^*}$ define a slice $S(f, \varepsilon) := \{x \in B_X;\ f(x) \geq 1 - \varepsilon\}$.

Theorem 11.7.1. *Let X be a Banach space. Then the following two statements are equivalent:*

(i) *X has the Daugavet property;*

(ii) *for every $\varepsilon > 0$, $f \in S_{X^*}$ and $x \in S_X$, there exists $y \in S(f, \varepsilon)$ such that $\|x + y\|_X \geq 2 - \varepsilon$.*

We will also need the following lemma.

Lemma 11.7.2. *Assume that $p < \infty$ μ-a.e. Let $f, g \in L^{p(x)}$, $\varepsilon > 0$ and $K > 0$. Then there exists $T > 1$ such that*

$$\|f + g\|_p \leq \max(\|f\|_p, \|g\|_p) + \varepsilon$$

whenever $\|f\|_p \leq K$, $\|g\|_p \leq K$, g vanishes outside the set $\{x \in \Omega;\ p(x) \geq T\}$ and

$$\{x \in \Omega;\ g(x) \neq 0\} \cap \{x \in \Omega;\ f(x) \neq 0\} = \emptyset.$$

Proof. Denote $\beta := \max(\|f\|_p, \|g\|_p)$ and pick $T > 1$ for the moment arbitrarily. Then

$$\int_\Omega \frac{|f(x) + g(x)|^{p(x)}}{(\beta + \varepsilon)^{p(x)}} d\mu(x) \leq \int_\Omega \left(\frac{|f(x)|}{(\beta + \varepsilon)}\right)^{p(x)} d\mu(x) + \int_\Omega \left(\frac{|g(x)|}{(\beta + \varepsilon)}\right)^{p(x)} d\mu(x)$$

$$= \int_\Omega \left(\frac{|f(x)|}{\|f\|_p} \cdot \frac{\|f\|_p}{(\beta + \varepsilon)}\right)^{p(x)} d\mu(x) + \int_\Omega \left(\frac{|g(x)|}{\|g\|_p} \cdot \frac{\|g\|_p}{(\beta + \varepsilon)}\right)^{p(x)} d\mu(x)$$

$$\leq \frac{\|f\|_p}{\beta + \varepsilon} \int_\Omega \left(\frac{|f(x)|}{\|f\|_p}\right)^{p(x)} d\mu(x) + \left(\frac{\|g\|_p}{\beta + \varepsilon}\right)^T \int_\Omega \left(\frac{|g(x)|}{\|g\|_p}\right)^{p(x)} d\mu(x)$$

$$\leq \frac{\|f\|_p}{\|f\|_p + \varepsilon} + \left(\frac{\|g\|_p}{\|g\|_p + \varepsilon}\right)^T \leq \frac{1}{1 + \varepsilon K^{-1}} + \frac{1}{(1 + \varepsilon K^{-1})^T}.$$

Choose now $T > 1$ in such a way that

$$\frac{1}{1+\varepsilon K^{-1}} + \frac{1}{(1+\varepsilon K^{-1})^T} \leq 1.$$

Then,

$$\|f+g\|_p \leq \beta + \varepsilon = \max\left(\|f\|_p, \|g\|_p\right) + \varepsilon. \qquad \square$$

Theorem 11.7.3. *Assume that μ is a nonatomic measure on Ω. The following statements are equivalent:*

(i) $L^{p(x)}$ *has the Daugavet property;*

(ii) $\mu\left(\{x \in \Omega;\ 1 < p(x) < \infty\}\right) = 0.$

Proof. For the proof of (i) $\Rightarrow$ (ii), assume that $L^{p(x)}$ is a Daugavet space and that there exists an interval $[a,b] \subset (1,\infty)$ such that $\mu\left(p^{-1}([a,b])\right) > 0$. Denote $A := p^{-1}([a,b])$, $q := p|_A$, and

$$Y := \{g \in L^{p(x)};\ g = 0 \text{ on } \Omega \setminus A\},$$
$$X := L^{q(x)}(A, \mathcal{S}|_A, \mu|_A).$$

Then Y is a closed subspace of $L^{p(x)}$ isomorphic to X. By Theorem 11.5.3, X is reflexive, and consequently, so is Y.

Hence, Y does not possess the Daugavet property. According to Theorem 11.7.1, there exist $0 < \varepsilon < 1$, $F \in S_{Y^*}$ and $f \in S_Y$ such that, for every $g \in B_Y$ satisfying $F(g) \geq 1 - \varepsilon$, we have $\|f + g\|_Y < 2 - \varepsilon$. Let P_A be the projection of X onto Y defined as

$$P_A : \varphi \mapsto \varphi \chi_A, \quad \varphi \in X.$$

If $G := F \circ P_A$ (the composition of operators), then $G \in X^*$ and $\|G\|_{X^*} = \|F\|_{X^*} = 1$. Using Lemma 11.7.2, there exists $T > 1$ such that for every $h \in B_X$

$$\|f + P_A(h) + P_B(h)\|_X \leq \max\left(\|f + P_A(h)\|_X, \|P_B(h)\|_X\right) + \frac{\varepsilon}{3}$$

where

$$B := p^{-1}([T, \infty)) \quad \text{and} \quad P_B : \varphi \mapsto \varphi \chi_B, \quad \varphi \in X.$$

Set further

$$C := \Omega_0 \setminus (A \cup B) \quad \text{and} \quad P_C : \varphi \mapsto \varphi \chi_C, \quad \varphi \in X$$

and

$$P_\infty : \varphi \mapsto \varphi \chi_{\Omega_\infty}, \quad \varphi \in X.$$

We may assume that $T > \beta$. For any $\varphi \in X$,
$$\varphi = P_A(\varphi) + P_B(\varphi) + P_C(\varphi) + P_\infty(\varphi) \quad \text{on } \Omega.$$
Since our aim is to show that X does not possess the Daugavet property, it suffices to find $\gamma > 0$ in such a way that for every $s \in B_X$ satisfying $G(s) \geq 1 - \gamma$ we have $\|f + s\|_X \leq 2 - \frac{\varepsilon}{3}$.

Fix now $\omega > 0$ and $s \in B_X$ such that $G(s) \geq 1 - \omega$. Then
$$1 \geq \|s\|_X \geq \|P_A(s) + P_C(s) + P_\infty(s)\|_X \geq \|P_A(s)\|_X \geq 1 - \omega.$$
Hence
$$\int_\Omega |P_C(s)|^p \, d\mu + \int_\Omega |P_A(s)|^p \, d\mu + \operatorname*{ess\,sup}_{\Omega_\infty} |P_\infty(s)| \leq 1$$
and
$$\int_\Omega \left| \frac{P_A(s)}{1-\omega} \right|^p d\mu \geq 1.$$
It follows that
$$\int_\Omega |P_C(s)|^p \, d\mu + \operatorname*{ess\,sup}_{\Omega_\infty} |P_\infty(s)| \leq 1 - \int_\Omega |P_A(s)|^p \, d\mu \leq 1 - (1-\omega)^\beta.$$
Analogously, using Lemma 11.2.11,
$$1 = \int_\Omega \left| \frac{P_C(s)}{\|P_C(s) + P_\infty(s)\|_X} \right|^p d\mu + \operatorname*{ess\,sup}_{\Omega_\infty} \frac{|P_\infty(s)|}{\|P_C(s) + P_\infty(s)\|_X}$$
$$\leq \frac{1}{\|P_C(s) + P_\infty(s)\|_X^T} \left(\int_\Omega |P_C(s)|^p \, d\mu + \operatorname*{ess\,sup}_{\Omega_\infty} |P_\infty(s)| \right).$$
Summing up, we get
$$\|P_C(s) + P_\infty(s)\|_X^T \leq 1 - (1-\omega)^\beta.$$
Choose now $\gamma > 0$ in such a way that
$$\left(1 - (1-\gamma)^\beta \right)^{1/T} \leq \frac{\varepsilon}{3}.$$
Finally,
$$\|f + s\|_X \leq \|f + P_A(s) + P_B(s)\|_X + \|P_C(s) + P_\infty(s)\|_X$$
$$\leq \max(\|f + P_A(s)\|_X, \|P_B(s)\|_X) + \frac{\varepsilon}{3} + \frac{\varepsilon}{3}$$
$$\leq \max(2 - \varepsilon, 1) + \frac{\varepsilon}{3} + \frac{\varepsilon}{3} \leq 2 - \frac{\varepsilon}{3}.$$

For the implication (ii) $\Rightarrow$(i) it is sufficient to use the fact that in this case the space $L^{p(x)}$ is isometrically isomorphic to the space $L^1(\Omega_0) \oplus_1 L^\infty(\Omega_\infty)$ (the direct sum of $L^1(\Omega_0)$ and $L^\infty(\Omega_\infty)$ equipped with the ℓ^1–norm) and the fact that an ℓ^1-sum of Daugavet spaces is Daugavet space as well (see Wojtaszczyk [238, Theorem 1]). □

Bibliography

[1] D. R. Adams, A sharp inequality of J. Moser for higher order derivatives, *Annals of Math.* **128** (1988), 385–398.

[2] D. R. Adams and L. I. Hedberg, *Function spaces and potential theory*, Springer, Berlin, 1976.

[3] R. A. Adams, *Sobolev spaces*, Academic Press, New York, 1975.

[4] F. Albiac and N. J. Kalton, *Topics in Banach space theory*, Graduate texts in mathematics 233, Springer, Berlin, 2006.

[5] M. Ariño and B. Muckenhoupt, Maximal functions on classical Lorentz spaces and Hardy's inequality with weights for non-increasing functions, *Trans. Amer. Math. Soc.* **320** (1990), 727–735.

[6] A. Avantaggiati, On compact imbedding theorems in weighted Sobolev spaces, *Czechoslovak Math. J.* **29** (1979), 635–648.

[7] V. F. Babenko and S. A. Pichugov, On a property of compact operators in the space of integrable functions, *Ukrain. Mat. Zh.* **33** (1981), 491–492.

[8] S. Bakry, T. Coulhon, M. Ledoux and L. Saloff-Coste, Sobolev inequalities in disguise, *Indiana Univ. Math. J.* **44** (1995), 1033–1074.

[9] S. Barza, L. E. Persson and J. Soria, Sharp weighted multidimensional integral inequalities for monotone functions, *Math. Nachr.* **210** (2000), 43–58.

[10] J. Bastero, M. Milman and F. Ruiz, A note on $L(\infty, q)$ spaces and Sobolev embeddings, *Indiana Univ. Math. J.* **52** (2003), 1215–1230.

[11] A. Benedek and R. Panzone, The space L^p, with mixed norm, *Duke Math. J.* **28** (1961), 301–324.

[12] C. Bennett and K. Rudnick, On Lorentz–Zygmund spaces, *Dissert. Math.* **175** (1980), 1–72.

[13] C. Bennett, R. A. De Vore and R. Sharpley, Weak-L^∞ and BMO, *Ann. of Math.* **113** (1981), 601-611.

[14] C. Bennett and R. Sharpley, *Interpolation of Operators*, Pure and Applied Mathematics Vol. 129, Academic Press, Boston, 1988.

[15] J. Bergh and J. Löfström, *Interpolation Spaces. An introduction*, Grundlehren der Mathematischen Wissenschaften, No. 223. Springer-Verlag, Berlin-New York, 1976.

[16] J.-E. Björk, On extensions of Lipschitz functions, *Ark. Math.* **7** (1969), 513–515.

[17] D. W. Boyd, Indices of function spaces and their relationship to interpolation, *Canad. J Math.* **21** (1969), 1245–1254.

[18] S. Boza and J. Martín, Equivalent expressions for norms in classical Lorentz spaces, *Forum Math.* **17 (3)** (2005), 361–374.

[19] J. S. Bradley, Hardy inequalities with mixed norms, *Canad. Math. Bull.* **21** (1978), 405–408.

[20] M. Bramanti and C. Cerutti, $W_p^{1,2}$ solvability for the Cauchy–Dirichlet problem for parabolic equations with VMO coefficients, *Comm. Part. Diff. Eq.* **18** (1993), 1735–1763.

[21] H. Brézis and S. Wainger, A note on limiting cases of Sobolev embeddings and convolution inequalities, *Comm. Partial Diff. Eq.* **5** (1980), 773–789.

[22] J. E. Brothers and W. P. Ziemer, Minimal rearrangements of Sobolev functions, *J. Reine Angew. Math.* **384** (1988), 153–179.

[23] P. L. Butzer and H. Berens, *Semigroups of operators and approximation*, Springer, New York, 1967.

[24] A. P. Calderón, Spaces between L^1 and L^∞ and the theorem of Marcinkiewicz, *Studia Math.* **26** (1966), 273–299.

[25] S. Campanato, Proprietà di inclusione per spazi di Morrey, *Ricerche Mat.* **12** (1963), 67–86.

[26] S. Campanato, Proprietà di hölderianità di alcune classi di funzioni, *Ann. Scuola Norm. Sup. Pisa* **17** (1963), 175–188.

[27] S. Campanato, Proprietà di una famiglia di spazi funzionali, *Ann. Scuola Norm. Sup. Pisa* **18** (1964), 137–160.

[28] S. Campanato, *Appunti del corso di analisi superiore*, Istituto matematico Università degli studi, Pisa, 1965.

[29] S. Campanato, Equazioni ellittiche del II deg ordine espazi $\mathfrak{L}^{(2,\lambda)}$, *Ann. Mat. Pura Appl.* **69** (1965), 321–381.

[30] S. Campanato, Equazioni paraboliche del secondo ordine e spazi $\mathfrak{L}^{2,\theta}(\Omega,\delta)$, *Ann. Mat. Pura Appl.* **73** (1966), 55–102.

[31] S. Campanato, *Alcune osservazioni relative alle soluzioni di equazioni ellittiche di ordine 2m*, Atti del Convegno sulle Equazioni alle Derivate Parziali (Bologna, 1967) pp. 17–25 Edizioni "Oderisi", Gubbio.

[32] M. Carro, A. García del Amo and J. Soria, Weak-type weights and normable Lorentz spaces, *Proc. Amer. Math. Soc.* **124** (1996), 849–857.

[33] M. Carro, A. Gogatishvili, J. Martín and L. Pick, Functional properties of rearrangement invariant spaces defined in terms of oscillations, *J. Funct. Anal.* **229, 2** (2005), 375–404.

[34] M. Carro, A. Gogatishvili, J. Martín and L. Pick, Weighted inequalities involving two Hardy operators with applications to embeddings of function spaces, *J. Operator Theory* **59,2** (2008), 101–124.

[35] M. Carro, L. Pick, J. Soria and V. Stepanov, On embeddings between classical Lorentz spaces, *Math. Ineq. Appl* **4** (2001), 397–428.

[36] M. Carro and J. Soria, Boundedness of some integral operators, *Canad. J. Math.* **45** (1993), 1155–1166.

[37] M. Carro and J. Soria, Weighted Lorentz spaces and the Hardy operator, *J. Funct. Anal.* **112** (1993), 480–494.

[38] M. Carro and J. Soria, The Hardy-Littlewood maximal function and weighted Lorentz spaces, *J. London Math. Soc.* **55** (1997), 146–158.

[39] P. Chauveheid, On a property of compact operators in Banach spaces, *Bull. Soc. Roy. Sci. Liege* **51** (1982), 371–378.

[40] F. Chiarenza, L^p-regularity for systems of PDE's with coefficients in VMO, Nonlinear Analysis, Function Spaces and Applications Vol. **5**, Prometheus, Prague, 1995, 1–32.

[41] F. Chiarenza, M. Frasca and P. Longo, $W^{2,p}$-solvability of the Dirichlet problem for non divergence elliptic equations with VMO coefficients, *Trans. Amer. Math. Soc.* **330** (1993), 841–853.

[42] A. Cianchi, A sharp embedding theorem for Orlicz–Sobolev spaces, *Indiana Univ. Math. J.* **45** (1996), 39–65.

[43] A. Cianchi, Continuity properties of functions from Orlicz–Sobolev spaces and embedding theorems, *Ann. Sc. Norm. Sup. Pisa* **23** (1996), 575–608.

[44] A. Cianchi, *Rearrangement estimates and applications to Sobolev and related inequalities*, Function Spaces and Interpolation, Proceedings of the Spring School in Analysis, Paseky nad Jizerou 2001, J. Lukeš and L. Pick (Eds.), MATFYZPRESS, Prague 2001, 1–44.

[45] A. Cianchi, Symmetrization and second-order Sobolev inequalities, *Ann. Mat. Pura Appl.* **183** (2004), 45–77.

[46] A. Cianchi, D. E. Edmunds and P. Gurka, On weighted Poincaré inequalities, *Math. Nachr.* **180** (1996), 15–41.

[47] A. Cianchi, R. Kerman, B. Opic and L. Pick, A sharp rearrangement inequality for fractional maximal operator, *Studia Math.* **138** (2000), 277–284.

[48] A. Cianchi and L. Pick, Sobolev embeddings into BMO, VMO, and L_∞, *Ark. Mat.* **36** (1998), 317–340.

[49] A. Cianchi and L. Pick, Sobolev embeddings into spaces of Campanato, Morrey and Hölder type, *J. Math. Anal. Appl.* **282** (2003), 128–150.

[50] Z. Ciesielski, On the isomorphisms of the spaces H_α and m, *Bull. Acad. Polon. Sci. Sér. Sci. Math. Astronom. Phys.* **8** (1960), 217–222.

[51] Z. Ciesielski, A construction of basis in $C^{(1)}(I^2)$, *Studia Math.* **33** (1969), 243–247.

[52] Z. Ciesielski and J. Domsta, Construction of an orthonormal basis in $C^m(I^d)$ and $W_p^m(I^d)$, *Studia Math.* **41** (1972), 211–224.

[53] J. A. Clarkson, Uniformly convex spaces, *Trans. Amer. Math. Soc.* **40** (1936), 396–414.

[54] F. Cobos, A. Gogatishvili, B. Opic and L. Pick, *Interpolation of uniformly absolutely continuous operators*, To appear in Math. Nachr., published online, DOI: 10.1002/mana.201100205, 2012.

[55] F. Conti, Su alcuni spazi funzionali e loro applicazioni ad equazioni differenziali di tipo ellittico, *Boll. Un. Mat. Ital.* **4** (1969), 554–569.

[56] M. Cwikel, A. Kamińska, L. Maligranda and L. Pick, Are generalized Lorentz "spaces" really spaces?, *Proc. Amer. Math. Soc.* **132** (2004), 3615–3625.

[57] M. Cwikel and E. Pustylnik, Sobolev type embeddings in the limiting case, *J. Fourier Anal. Appl.* **4** (1998), 433–446.

[58] G. Da Prato, Spazi $\mathfrak{L}^{(p,\theta)}(\Omega,\delta)$ e loro proprietà, *Ann. Mat. Pura Appl.* **69** (1965), 383–392.

[59] I. K. Daugavet, On a property of compact operators in the space $\mathcal{C}$, *Uspekhi Mat. Nauk* **18** (1963), 157–158.

[60] M. M Day, The spaces L^p with $0 < p < 1$, *Bull. Amer. Math. Soc.* **46** (1940), 816–823.

[61] K. De Leeuw, Banach spaces of Lipschitz functions, *Studia Math.* **21** (1961/1962), 55–66.

[62] R. A. De Vore and K. Scherer, Interpolation of linear operators on Sobolev spaces, *Ann. of Math.* **109** (1979), 583–599.

[63] L. Diening, P. Harjulehto, P. Hästö and M. Ružička, *Lebesgue and Sobolev spaces with variable exponents*, Lecture Notes in Mathematics, 2017. Springer, Heidelberg, 2011.

[64] J. Diestel and J. J. Uhl, *Vector measures*, AMS, 1977.

[65] N. Dunford, Integration and linear operations, *Trans. Amer. Math. Soc.* **40** (1936), 474–494.

[66] N. Dunford and J. T. Schwartz, *Linear Operators I. General theory.*, Interscience Publishers, Inc., New York–London, 1958.

[67] D. E. Edmunds, P. Gurka and B. Opic, Double exponential integrability of convolution operators in generalized Lorentz–Zygmund spaces, *Indiana Univ. Math. J.* **44** (1995), 19–43.

[68] D. E. Edmunds, P. Gurka and B. Opic, On embeddings of logarithmic Bessel potential spaces, *J. Funct. Anal.* **146** (1997), 116–150.

[69] D. E. Edmunds, R. Kerman and L. Pick, Optimal Sobolev embeddings involving rearrangement-invariant quasinorms, *J. Funct. Anal.* **170** (2000), 307–355.

[70] D. E. Edmunds, J. Lang and A. Nekvinda, On $L^{p(x)}$ norms, *Proc. Roy. Soc. London Ser. A.* **445** (1999), 219–225.

[71] R. Engelking, *General topology*, Heldermann Verlag, Berlin, 1989.

[72] W. D. Evans and B. Opic, Real interpolation with logarithmic functors and reiteration, *Canad. J. Math.* **52** (2000), 920–960.

[73] W. D. Evans, B. Opic and L. Pick, Interpolation of operators on scales of generalized Lorentz–Zygmund spaces, *Math. Nachr.* **182** (1996), 127–181.

[74] W. D. Evans, B. Opic and L. Pick, Real interpolation with logarithmic functors, *J. of Inequal. et Appl.* **7** (2002), 187–269.

[75] M. Fabian et al., *Functional analysis and infinite–dimensional geometry*, CMS Books in Mathematics 8, Springer-Verlag, 2001.

[76] K. Fan and I. Glicksberg, Some geometric properties of the spheres in a normed linear space, *Duke Math. J.* **25** (1958), 553–568.

[77] X. Fan and D. Zhao, On the spaces $L^{p(x)}(\Omega)$ and $W^{m,p(x)}(\Omega)$, *J. Math. Anal. Appl.* **263** (2001), 424–446.

[78] H. Federer, *Geometric Measure Theory*, Springer, Berlin, 1969 (Second edition 1996).

[79] C. Fefferman and E. Stein, H^p spaces of several variables, *Acta Math.* **129** (1972), 137–193.

[80] P. Fernández-Martínez, A. Manzano and E. Pustylnik, Absolutely continuous embeddings of rearrangement-invariant spaces, *Mediterr. J. Math.* **7** (2010), 539–552.

[81] A. Fiorenza, A summability condition on the gradient ensuring BMO, *Rev. Mat. Univ. Complut. Madrid* **11** (1998).

[82] L. Frampton and A. J. Tromba, On the classification of spaces of Hölder continuous functions, *J. Funct. Anal.* **10** (1972), 336–345.

[83] N. Fusco, P. L. Lions and C. Sbordone, Sobolev imbedding theorems in borderline cases, *Proc. Amer. Math. Soc.* **124** (1996), 561–565.

[84] G. Gaimnazarov, Imbedding theorems for the $L_p(-\infty, \infty)$ classes of functions (Russian), *Izv. Vysš. Učebn. Zaved. Matematika* **4(119)** (1972), 44–54.

[85] V. F. Gaposhkin, The Haar system as an unconditional basis in $L_p[0, 1]$. (Russian), *Mat. Zametki* **15** (1974), 191–196.

[86] J. B. Garnett and P. W. Jones, The distance in BMO to L_∞, *Ann. of Math.* **108** (1978), 373–393.

[87] G. Glaeser, Étude de quelques algèbres tayloriennes, *J. Analyse Math.* **6** (1958), 1–124.

[88] E. Giusti, Equazioni quasi ellittiche e spazi $\mathcal{L}^{p,\theta}(\Omega, \delta)$, *Ann. Mat. Pura Appl.* **75** (1967), 313–353.

[89] A. Gogatishvili and L. Pick, Discretization and anti-discretization of rearrangement-invariant norms, *Publ. Mat.* **47** (2003), 311–358.

[90] A. Gogatishvili, B. Opic and L. Pick, Weighted inequalities for Hardy-type operators involving suprema, *Collect. Math.* **57, 3** (2006), 227–255.

[91] M. L. Goldman, On integral inequalities on a cone of functions with monotonicity properties, *Soviet Math. Dokl.* **44** (1992), 581–587.

[92] M. L. Goldman, On integral inequalities on the set of functions with some properties of monotonicity, *Function spaces, Differential Operators and Nonlinear Analysis, Teubner Texte Zur Math.* **133** (1993), 274–279.

[93] M. L. Goldman, Sharp estimates for the norms of Hardy-type operators on the cones of quasimonotone functions, *Proc. Steklov Inst. Math.* **232** (2001), 1–29.

[94] M. L. Goldman and R. Kerman, *The dual of the cone of decreasing functions in a weighted Orlicz class and the associate of an Orlicz-Lorentz space*, Manuscript, 2001.

[95] M. L. Goldman, H. P. Heinig and V. D. Stepanov, On the principle of duality in Lorentz spaces, *Canad. J. Math.* **48** (1996), 959–979.

[96] K.-G. Grosse–Erdmann, *The Blocking Technique, Weighted Mean Operators and Hardy's Inequality*, Lect. Notes Math. 1679, Springer, Berlin, 1998.

[97] I. Halperin, Function spaces, *Canad. J. Math.* **5** (1953), 273–288.

[98] I. Halperin and H. W. Ellis, Function spaces determined by a levelling length function, *Canad. J. Math.* **5** (1953), 576–592.

[99] A. Haaker, *On the conjugate space of Lorentz space*, Technical Report, Lund.

[100] K. Hansson, Imbedding theorems of Sobolev type in potential theory, *Math. Scand.* **45** (1979), 77–102.

[101] G. H. Hardy, J. E. Littlewood and G. Pólya, *Inequalities*, Cambridge University Press, Cambridge 1934, 2nd edition 1952.

[102] H. P. Heinig and L. Maligranda, Weighted inequalities for monotone and concave functions, *Studia Math.* **116** (1995), 133–165.

[103] J. A. Hempel, G. R. Morris and N. S. Trudinger, On the sharpness of a limiting case of the Sobolev imbedding theorem, *Bull. Australian Math. Soc.* **3** (1970), 369–373.

[104] L.-I. Hedberg, On certain convolution inequalities, *Proc. Amer. Math. Soc.* **36** (1972), 505–510.

[105] E. Hewitt and K. Stromberg, *Real and abstract analysis*, Springer, New York, 1965.

[106] T. Holmstedt, Interpolation of quasi-normed spaces, *Math. Scand.* **26** (1970), 177–199.

[107] S. Janson, On functions with conditions on the mean oscillation, *Ark. Math.* **14** (1976), 189–196.

[108] M. Jaroszewska, Some properties of the spaces $W_s^{(p,\lambda)}$, *Comment. Math. Prace Mat.* **17 (1973/74)** (1973/74), 359–372.

[109] B. Jawerth and M. Milman, *Extrapolation theory with applications*, Memoirs Amer. Math. Soc. **440** (1991).

[110] F. John and L. Nirenberg, On functions of bounded mean oscillation, *Comm. Pure Appl. Math.* **14** (1961), 415–426.

[111] R. A. Johnson, Atomic and nonatomic measures, *Proc. Amer. Math. Soc.* **25** (1970), 650–655.

[112] V. P. Kabaila, On embeddings of the space $L_p(\mu)$ into $L_r(\nu)$, *Lit. Mat. Sb.* **21** (1981), 143–148.

[113] A. Kamińska and L. Maligranda, Order convexity and concavity in Lorentz spaces with arbitrary weight, *Research Report, Luleå University of Technology* **4** (1999), 1–21.

[114] H. Kamowitz, A property of compact operators, *Proc. Amer. Math. Soc.* **91** (1984), 231–236.

[115] R. A. Kerman, Function spaces continuously paired by operators of convolution-type, *Canad. Math. Bull.* **22** (1979), 499–507.

[116] R. A. Kerman, An integral extrapolation theorem with applications, *Studia Math.* **76** (1983), 183–195.

[117] R. Kerman and L. Pick, Optimal Sobolev imbeddings, *Forum Math* **18, 4** (2006), 535–570.

[118] R. Kerman and L. Pick, Optimal Sobolev imbedding spaces, *Studia Math* **192, 3** (2009), 195–217.

[119] A. N. Kolmogorov, Über Kompaktheit der Funktionenmengen bei der Konvergenz im Mittel, *Nachr. Ges. Göttingen Math.-Phys.* **9** (1931), 60–63.

[120] P. P. Korovkin, *Linear operators and approximation theory*, Gordon and Breach, New York; Hindustan Publishing Corp. (India), Delhi, 1960.

[121] G. Köthe, *Topological vector spaces. Vol. I.*, Translated from the German by D. J. H. Garling. Die Grundlehren der mathematischen Wissenschaften, Band 159 Springer-Verlag New York Inc., New York 1969 xv+456 pp.

[122] O. Kováčik and J. Rákosník, On spaces $L^{p(x)}$ and $W^{k,p(x)}$, *Czechoslovak Math. J.* **41** (1991), 592–618.

[123] M. A. Krasnosel'skii and Ya.B. Rutitskii, *Convex functions and Orlicz spaces*, Noordhoff, Groningen, 1961.

[124] M. A. Krasnosel'skii, P. P. Zabreiko, E. I. Pustylnik and P. E. Sobolevskii, *Integral operators in spaces of summable functions (Russian)*, Izdat. "Nauka", Moscow, 1966.

[125] M. Krbec and L. Pick, On imbeddings between weighted Orlicz spaces, *Z. Anal. Anwend.* **10,1** (1991), 107–117.

[126] A. Kufner, O. John and S. Fučík, *Function spaces*, Noordhoff, Leyden, Academia, Praha, 1977.

[127] A. Kufner, L. Maligranda and L.-E. Persson, *The Hardy inequality. About its history and some related results*, Vydavatelský servis, Pilsen, 2007.

[128] A. Kufner and L.-E. Persson, *Weighted inequalities of Hardy type*, World Sci., Singapore, 2003.

[129] K. Kuncová, *Lorentz spaces*, Bachelor Thesis, Charles University, Prague, 2009.

[130] K. Kuratowski, *Topology, Vol. 1*, Academic Press, New York-London; Państwowe Wydawnictwo Naukowe, Warsaw, 1966.

[131] S. Lai, Weighted norm inequalities for general operators on monotone functions, *Trans. Amer. Math. Soc.* **340** (1993), 811–836.

[132] J. Lang and L. Pick, The Hardy operator and the gap between L_∞ and BMO, *J. London Math. Soc.* **57** (1998), 196–208.

[133] J. Lindenstrauss and L. Tzafriri, *Classical Banach spaces I and II*, Springer, Berlin, 1977.

[134] L. A. Ljusternik and V. I. Sobolev, *Elements of functional analysis*, Hindustan Publishing Corp., Delhi; Halstadt Press, New York, 1974.

[135] G. G. Lorentz, On the theory of spaces Λ, *Pacific J. Math.* **1** (1951), 411–429.

[136] G. G. Lorentz, Relations between function spaces, *Proc. Amer. Math. Soc.* **12** (1961), 112–132.

[137] G. G. Lorentz and D. G. Wertheim, Representation of linear functionals on Köthe spaces, *Canad. J. Math.* **5** (1953), 568–575.

[138] G. Ya. Lozanovskij, On almost integral operators in KB-spaces, *Vestnik Leningrad. Univ.* **21** (1966), 35–44.

[139] J. Lukeš, *Topics in functional analysis (in Czech)*, Charles University Praha, 2003.

[140] J. Lukeš, L. Pick and D. Pokorný, On geometric properties of the spaces $L^{p(x)}$, *Rev. Mat. Complutense* **24, 1** (2011), 115–130.

[141] W. A. J. Luxemburg, *Banach Function Spaces*, Thesis, Delft, 1955.

[142] W. A. J. Luxemburg and A. C. Zaanen, *Riesz Spaces*, North Holland, Amsterdam, 1971.

[143] J. Malý and L. Pick, An elementary proof of sharp Sobolev embeddings, *Proc. Amer. Math. Soc.* **130** (2002), 555–563.

[144] J. Malý, D. Swanson and W. P. Ziemer, Fine behavior of functions whose gradients are in Orlicz space, *Studia Math.* **190** (2009), 33–71.

[145] J. Malý and W. P. Ziemer, *Fine regularity of solutions of elliptic partial differential equations*, AMS Mathematical Surveys and Monographs Vol. 51, Amer. Math. Soc., Providence, 1997.

[146] J. Martín and M. Milman, Symmetrization inequalities and Sobolev embeddings, *Proc. Amer. Math. Soc.* **134** (2006), 2335–2347.

[147] V. G. Maz'ya, A theorem on the multidimensional Schrödinger operator (Russian), *Izv. Akad. Nauk* **28** (1964), 1145-1172.

[148] V. G. Maz'ya, *Sobolev Spaces*, Springer, Berlin, 1985.

[149] V. G. Maz'ya, *Sobolev Spaces*, Springer, Berlin, 2011.

[150] E. J. McShane, Linear functionals on certain Banach spaces, *Proc. Amer. Math. Soc.* **11** (1950), 402–408.

[151] N. G. Meyers, Mean oscillation over cubes and Hölder continuity, *Proc. Amer. Math. Soc.* **15** (1964), 717–721.

[152] D. P. Milman, On some criteria for the regularity of spaces of the type (B), *Dokl. Akad. Nauk SSSR* **20** (1938), 243–246.

[153] M. Milman, *Extrapolation and optimal decompositions*, Lecture Notes in Maths. **1580**, Springer, Berlin, 1994.

[154] M. Milman and E. Pustylnik, On sharp higher order Sobolev embeddings, *Comm. Contemporary Math.* **6** (2004), 495–511.

[155] C. Miranda, *Partial differential equations of elliptic type*, Second revised edition. Translated from the Italian by Zane C. Motteler. Ergebnisse der Mathematik und ihrer Grenzgebiete, Band 2. Springer-Verlag, New York-Berlin 1970.

[156] B. S. Mitjagin, The homotopy structure of a linear group of a Banach space, *Uspehi Mat. Nauk* **25** (1970), 27–33.

[157] C. B. Morrey, On the solutions of quasi-linear elliptic partial differential equations, *Trans. Amer. Math. Soc.* **43** (1938), 126–166.

[158] C. B. Morrey, Second order elliptic equations in several variables and Hölder continuity, *Math. Z.* **72** (1959/1960), 146–164.

[159] B. Muckenhoupt, Hardy's inequality with weights, *Studia Math.* **44** (1972), 31–38.

[160] J. Musielak, *Orlicz Spaces and Modular Spaces*, Springer-Verlag, Berlin, 1983.

[161] M. A. Naimark, *Normed rings*, Noordhoff, Groningen, 1959.

[162] E. Nakai, On the restriction of functions of bounded mean oscillation to the lower dimensional space, *Arch. Math.* **43** (1984), 519–529.

[163] H. Nakano, *Modulared semi-ordered linear spaces*, Maruzen, Tokyo, 1950.

[164] I. P. Natanson, *Theory of functions of a real variable*, F. Ungar Publ. Co., New York, 1961.

[165] Yu. V. Netrusov, Embedding theorems for Lizorkin-Triebel spaces, *Notes of Scientific Seminars LOMI* **159** (1987), 103–112.

[166] C. J. Neugebauer, Weighted norm inequalities for averaging operators of monotone functions, *Publ Mat* **35** (1992), 429–447.

[167] J. S. Neves, *Lorentz–Karamata spaces, Bessel and Riesz potentials and embeddings*, University of Sussex, Preprint, 2000.

[168] P. Meyer–Nieberg, *Banach lattices*, Springer-Verlag, 1991.

[169] R. O'Neil, Convolution operators and $L(p,q)$ spaces, *Duke Math. J.* **30** (1963), 129–142.

[170] D. Opěla, *Spaces of functions with generalized bounded and vanishing mean oscillation*, Diploma Thesis, Charles University, Prague, 2001.

[171] B. Opic and A. Kufner, *Hardy-type inequalities*, Pitman Research Notes in Mathematics, Longman Sci&Tech. Harlow, 1990.

[172] B. Opic and L. Pick, On generalized Lorentz–Zygmund spaces, *Math. Ineq. Appl.* **2** (1999), 391–467.

[173] W. Orlicz, Über konjugierte Exponentenfolgen, *Studia Math.* **3** (1931), 200–212.

[174] R. E. A. C. Paley, A Remarkable Series of Orthogonal Functions (I), *Math. Ineq. Appl.* **2** (1999), 391–467.

[175] J. Peetre, Espaces d'interpolation et théorème de Soboleff, *Ann. Inst. Fourier* **16** (1966), 279–317.

[176] J. Peetre, On the theory of $\mathcal{L}_{p,\lambda}$ spaces, *J. Funct. Anal.* **4** (1969), 71–87.

[177] B. J. Pettis, A proof that every uniformly convex space is reflexive, *Duke Math. J.* **5** (1939), 249–253.

[178] L. Pick, *A remark on continuous imbeddings between Banach function spaces*, Colloq. Math. Soc. János Bólyai, 58. Approximation Theory, Kécskemét (Hungary), 1990, 571–581.

[179] L. Pick, *Optimal Sobolev embeddings*, Nonlinear Analysis, Function Spaces and Applications 6, Proceedings of the 6th International Spring School held in Prague, Czech Republic, May-June 1998, M. Krbec and A. Kufner (eds.), Olympia Press, Prague, 1999, 156–199.

[180] L. Pick, *Supremum operators and optimal Sobolev inequalities*, Function Spaces, Differential Operators and Nonlinear Analysis. Proceedings of the Spring School held in Syöte Centre, Pudasjärvi (Northern Finland), June 1999, V. Mustonen and J. Rákosník (eds.), Mathematical Institute AS CR, Prague, 2000, 207–219.

[181] L. Pick, *Optimal Sobolev embeddings – old and new*, Function Spaces, Interpolation Theory and Related Topics. Proceedings of the Conference held in Lund (Sweden), August 17-22, 2000, in Honour of Jaak Peetre on his 65th Birthday, A. Kufner, M. Cwikel, L.-E. Persson, G. Sparr, M. Engliš (eds.), W. de Gruyter, Berlin 2002, 359–368.

[182] S. I. Pokhozhaev, On eigenfunctions of the equation $\Delta u + \lambda f(u) = 0$, *Dokl. Akad. Nauk SSR* **165** (1965), 36–39.

[183] G. Pólya and G.Szegö, *Isoperimetric Inequalities in Mathematical Physics*, Princeton University Press, Princeton, 1951.

[184] E. Pustylnik, Optimal interpolation in spaces of Lorentz–Zygmund type, *J. d'Anal. Math.* **79** (1969), 113–157.

[185] I. K. Rana, *An introduction to measure and integration*, Graduate Studies in Mathematics, vol. 45, Second Edition, American Mathematical Society, Providence, 2002.

[186] M. M. Rao, and Z. D. Ren, *Theory of Orlicz spaces*, M. Dekker, New York, 1991.

[187] F. Riesz, Untersuchungen über Systeme integrierbarer Funktionen (German), *Math. Ann.* **69** (1910), 449–497.

[188] H. L. Royden, *Real Analysis*, MacMillan, New York, 2nd ed., 1968.

[189] W. Rudin, *Functional Analysis*, McGraw-Hill, New York, 1973.

[190] M. Růžička, *Electrorheological Fluids: Modeling and Mathematical Theory*, Lecture Notes in Mathematics Vol. 1748, Springer-Verlag, Berlin, 2000.

[191] J. Ryll, Schauder bases for the space of continuous functions on an n-dimensional cube, *Comment. Math. Prace Mat.* **17** (1973), 201–213.

[192] J. Ryll, Interpolating bases for spaces of differentiable functions, *Studia Math* **63** (1978), 125–144.

[193] D. V. Salekhov, On the norm of a linear functional in an Orlicz space and on a certain internal characteristic of an L_p space, *(Russian) Dokl. Akad. Nauk SSSR (N. S.)* **111** (1956), 948–950.

[194] E. Sawyer, Boundedness of classical operators on classical Lorentz spaces, *Studia Math.* **96** (1990), 145–158.

[195] J. Schauder, Eine Eigenschaft des Haarschen Orthogonalsystems, *Math. Z.* **28** (1928), 317–320.

[196] S. Schonefeld, Schauder bases in spaces of differentiable functions, *Bull. Amer. Math. Soc.* **75** (1969), 586–590.

[197] S. Schonefeld, Schauder bases in the Banach spaces $C^k(T^q)$, *Trans. Amer. Math. Soc.* **165** (1972), 309–318.

[198] I. Sharapudinov, On the topology of the space $L^{p(t)}([0;1])$, *Matem. Zametki* **26** (1978), 613–632.

[199] R. Sharpley, Counterexamples for classical operators in Lorentz–Zygmund spaces, *Studia Math.* **68** (1980), 141–158.

[200] G. Sinnamon, Spaces defined by the level function and their duals, *Studia Math.* **111** (1994), 19–52.

[201] G. Sinnamon, Embeddings of concave functions and duals of Lorentz spaces, *Publ. Math.* **46** (2002), 489–515.

[202] G. Sinnamon, Transferring monotonicity in weighted norm inequalities, *Collect. Math.* **54** (2003), 181–216.

[203] G. Sinnamon and V. D. Stepanov, The weighted Hardy inequality: new proofs and the case $p = 1$, *J. London Math. Soc.* **54** (1996), 89–101.

[204] L. Slavíková, *Relations between function spaces*, Bachelor Thesis, Charles University, Prague, 2010.

[205] L. Slavíková, *Functions with absolutely continuous norm in Marcinkiewicz spaces*, unpublished manuscript, Prague, 2011.

[206] L. Slavíková, Almost-compact embeddings, *Math. Nachr.* **285** (2012), 1500–1516.

[207] S. L. Sobolev, *Applications of Functional Analysis in Mathematical Physics*, Transl. of Mathem. Monographs, American Math. Soc., Providence, R. I. **7**, 1963.

[208] J. Soria, Lorentz spaces of weak-type, *Quart. J. Math. Oxford* **49** (1998), 93–103.

[209] S. Spanne, Some function spaces defined using the mean oscillation over cubes, *Ann. Scu. Norm. Sup. Pisa* **19** (1965), 593-608.

[210] G. Stampacchia, $L\mathscr{L}^{(p,\lambda)}$-spaces and interpolation, *Comm. Pure Appl. Math.* **17** (1964), 293-306.

[211] G. Stampacchia, The spaces $\mathscr{L}^{(p,\lambda)}$, $\mathscr{N}^{(p,\lambda)}$ and interpolation, *Ann. Scuola Norm. Sup. Pisa* **19** (1965), 443-462.

[212] E. M. Stein, *Singular Integrals and Differentiability Properties of Functions*, Princeton University Press, Princeton, 1970.

[213] E. M. Stein, *Harmonic Analysis*, Princeton University Press, Princeton, 1993.

[214] J. Steiner, *Gesammelte Werke*, Berlin, 1881–1882.

[215] V. D. Stepanov, Integral operators on the cone of monotone functions, *J. London Math. Soc.* **48** (1993), 465–487.

[216] V. D. Stepanov, The weighted Hardy's inequality for nonincreasing functions, *Trans. Amer. Math. Soc.* **338** (1993), 173–186.

[217] R. S. Strichartz, A note on Trudinger's extension of Sobolev's inequality, *Indiana Univ. Math. J.* **21** (1972), 841–842.

[218] G. Talenti, *An embedding theorem*, Essays of Math. Analysis in honour of E. De Giorgi, Birkhäuser, Boston, 1989.

[219] G. Talenti, *An inequality between u^* and $|grad\ u|^*$*, General Inequalities 6 (Oberwolfach 1990), Internat. Ser. Numer. Math., Vol. 103, Birkhäuser, Basel, 1992, 175–182.

[220] G. Talenti, *Inequalities in rearrangement-invariant function spaces*, Nonlinear Analysis, Function Spaces and Applications Vol. 5, Prometheus, Prague, 1995, pp. 177–230.

[221] L. Tartar, Imbedding theorems of Sobolev spaces into Lorentz spaces, *Boll. Un. Mat. Ital.* **8 1–B** (1998), 479–500.

[222] A. E. Taylor, *Introduction to functional analysis*, Wiley&sons, New York, 1967.

[223] H. Tietze, Über Funktionen, die auf einer abgeschlossen Menge stetig sind, *J. Reine Angew. Math.* **145** (1915), 9–14.

[224] M. F. Timan, Best approximation and modulus of smoothness of functions prescribed on the entire real axis (Russian), *Izv. Vysš. Učebn. Zaved. Matematika* **6(25)** (1961), 108–120.

[225] A. Torchinsky, Interpolation of operators and Orlicz classes, *Studia Math.* **59** (1976), 177–207.

[226] A. Torchinsky, *Real-variable methods in harmonic analysis*, Academic Press, New York, 1986.

[227] H. Triebel, Über die Existenz von Schauderbasen in Sobolev-Besov-Räumen. Isomorphiebeziehungen, *Studia Math.* **46** (1973), 83–100.

[228] H. Triebel, *Interpolation theory, function spaces, differential operators*, VEB Deutscher Verlag der Wissenschaften, Berlin, 1978.

[229] N. S. Trudinger, On imbeddings into Orlicz spaces and some applications, *J. Math. Mech.* **17** (1967), 473–483.

[230] I. Tsenov, Generalization of the problem of best approximation of a function in the space L^s, *Uch. Zap. Dagestan Gos. Univ.* **7** (1961), 25–37.

[231] P. L. Ul'yanov, The embedding of certain classes H_p^ω of functions (Russian), *Izv. Akad. Nauk SSSR Ser. Mat.* **32** (1968), 649–686.

[232] J. Vybíral, On sharp embeddings of Besov and Triebel-Lizorkin spaces in the subcritical case, *Proc. Amer. Math. Soc.* **138** (2010), 141–146.

[233] G. Weiss, A note on Orlicz spaces, *Portugal. Math.* **15** (1956), 35–47.

[234] D. Werner, Recent progress on the Daugavet property, *Irish Math. Soc. Bull.* **46** (2001), 77–97.

[235] H. Whitney, Differentiable functions defined in closed sets. I, *Trans. Amer. Math. Soc.* **36** (1934), 369–387.

[236] G. Wildenhain, Der Raum der H-stetig differenzierbaren Funktionen, *Math. Nachr.* **50** (1971), 217–228.

[237] G. Wildenhain, Über den Raum der hölderstetig differenzierbaren Funktionen, *Tagungsbericht zur ersten Tagung der WK Analysis (1970). Beiträge Anal.* **4** (1972), 35–38.

[238] P. Wojtaszczyk, Some remarks on the Daugavet equation, *Proc. Amer. Math. Soc.* **115** (1992), 1047–1052.

[239] S. Yano, Notes on Fourier analysis (XXIX): an extrapolation theorem, *J. Math. Soc. Japan* **3** (1951), 296–305.

[240] V. I. Yudovich, Some estimates connected with integral operators and with solutions of elliptic equations, *Soviet Math. Doklady* **2** (1961), 746–749.

[241] A. C. Zaanen, *Linear Analysis*, North-Holland, Amsterdam, 1960.

[242] E. Zeidler, *Beiträge zur Theorie und Praxis freier Randwertaufgaben*, Akademie-Verlag, Berlin, 1971.

[243] V. V. Zhikov, Averaging of functionals of the calculus of variations and elasticity theory, *Math. USSR Izv.* **29** (1987), 33–66.

[244] V. V. Zhikov, On passing to the limit in nonlinear variational problem, *Math. Sb.* **183** (1992), 47–84.

[245] W. P. Ziemer, *Weakly Differentiable Functions*, Springer, New York, 1989.

[246] T. Zolezzi, On weak convergence in L^∞, *Indiana Univ. Math. J.* **23** (1973/74), 765–766.

[247] A. Zygmund, *Trigonometric Series*, Cambridge University Press, Cambridge, 1957.

Index

$\gtrsim$, 204
$\lesssim$, 204
$\langle \cdot, \cdot \rangle$, 10
$\hookrightarrow$, 19
$\hookrightarrow\hookrightarrow$, 19
$\rightharpoonup$, 20
$\rightleftharpoons$, 19
$|\cdot|_{X \to Y}$, 17
BMO space, 188
$C^0(\Omega)$, 39
$C(\overline{\Omega})$, 40
$C(K)$, 55
$C_0^k(\Omega)$, 39
$C^k(\overline{\Omega})$, 40
$C^k(\Omega)$, 39
c_N, 179
$C_0^\infty(\Omega)$, 39
$C^\infty(\overline{\Omega})$, 40
$C^\infty(\Omega)$, 39
Δ_2-condition, 119
Dom(A), 16
E^Ψ-weak convergence, 162
exp X, 2
f^*, 238
f^{**}, 249
Id, 18
JN space, 188
(k, λ)-equicontinuous set, 54
$L^1 + L^\infty$, 294
$L^1 \cap L^\infty$, 294
Λ_φ, 269
$\mathcal{L}^\Phi(\Omega)$, 108
$L^p(a, b)$, 66
$L_C^{p,\lambda}$, 177
$L_M^{p,\lambda}$, 176
$L^p(\Omega)$, 66
$\mathcal{L}^p(\Omega)$, 62
$L^{p,q}$, 301
$L^\infty(\Omega)$, 79

$L^{p,\infty}(\Omega)$, 102
Λ_X, 270
$\mathcal{M}$, 203
$\mathcal{M}_0$, 203
M_φ, 264
$\mathfrak{M}_{N,k}$, 38
$\mathcal{M}_+$, 203
M_X, 270
$\mathcal{N}_\infty(f)$, 79
$\mathcal{N}_p(f)$, 62
Φ-mean continuity, 161
Φ-mean boundedness, 142
Φ-mean convergence, 138
p-mean continuity, 67
p-mean equicontinuous set, 94
Rng(A), 16
$(\mathcal{R}, \mu)$, 203
σ-algebra, 24
σ-finite set function, 25
X_a, 216
X_b, 221
$\overline{X}$, 257
X', 210
X'', 211
X^*, 20
X^{**}, 21

absolutely continuous embedding, 170
absolutely continuous measure, 33
absolutely continuous norm, 149, 216
abstract function, 16
addition, 1
algebra, 2
  of sets, 22
almost all, 26
almost everywhere, 26
almost everywhere convergence, 35
almost-compact embedding, 275
α-norm, 311
arithmetic measure, 25

Index

associate norm, 209
associate space, 210
atom, 25
average Hardy operator, 414

balanced set, 36
ball
    open, 6, 7
Banach function norm, 204
Banach function space, 204
Banach space, 12
Banach theorem, 17
Banach–Alaoglu theorem, 20
Banach–Steinhaus theorem, 17
Banach–Steinhaus theorem for weak convergence, 20
base, 3
basis
    Schauder, 14
        unconditional, 15
Bernstein polynomial, 47
Borel set, 24
bounded operator, 17
boundedness
    Φ-mean, 142
    norm, 142
broken logarithmic function, 313

Campanato space, 178
Cauchy sequence, 11
Cauchy–Schwarz inequality, 11
characteristic function, 26
Clarkson inequalities, 65
class
    B_p, 377
    $B_{1,\infty}$, 377
    Orlicz, 108
    reverse B_p, 377
    Zygmund, 315
classical Lorentz space, 375
    of type Γ, 376
    of type Λ, 375
closed set, 3, 7
closure, 4, 12
compact metric space, 15
compact operator, 17
complementary function, 115
complete measure space, 25

complete metric space, 12
completely atomic measure space, 25
completely continuous operator, 17
completeness, 12
condition
    Δ_2, 119
    (S), 41
    Hölder, 40
    Lipschitz, 40
conjugate function, 439
conjugate Lebesgue index, 62
constant
    embedding, 18
continuous linear functional, 19
continuous mapping, 4
continuous measure, 25
continuous operator, 16
convergence
    E^Ψ-weak, 162
    L^p-mean, 93
    Φ-mean, 138
    almost everywhere, 35
    in measure, 35
    in metric space, 11
    modular, 9, 138
    pointwise, 34
    strong, 11
    uniform, 5, 34, 43
        local, 34
        up to small sets, 34
    weak, 20
convergence in norm, 11
convergent integral, 28
convergent sequence, 11
convergent series, 11
countably additive set function, 22
countably subadditive set function, 22
counting measure, 25

Daugavet property, 455
Daugavet space, 455
dense set, 12
discrete measure space, 25
distribution function, 237
domain, 16
    of type $\mathcal{A}$, 176
dominating Young function, 125

Egorov theorem, 35
embedding, 18
   absolutely continuous, 170
   almost-compact, 275
embedding constant, 18
embedding operator, 18
equicontinuous set, 51
equimeasurable functions, 237
equivalence
   of Banach spaces, 268
equivalence of Young functions, 125
equivalent norms, 19
essentially faster growth, 126
extension of a function, 59

Fatou lemma, 28, 31, 206
Fatou property, 204
finite set function, 25
finitely additive set function, 22
friendly measure space, 278
Fubini theorem, 33
function, 16
   N, 114
   abstract, 16
   broken logarithmic, 313
   characteristic, 26
   complementary, 115
   conjugate, 439
   distribution, 237
   fundamental, 260
   Hölder continuous, 40
   Halperin level, 402
   integrable, 28
   Lipschitz, 40
   Lipschitz continuous, 40
   measurable, 28
   quasi-concave, 263
   set, 22
   σ-finite, 25
   countably additive, 22
   countably subadditive, 22
   finite, 25
   finitely additive, 22
   monotone, 22
   simple, 26, 204
   slowly varying, 373
   Steklov, 156
   Young

   dominating, 125
function norm, 204
functional
   continuous linear, 19
   linear, 19
   Minkowski, 8
functions
   equimeasurable, 237
function Young, 109
fundamental function, 260

generalized Hölder inequality, 66
generalized Lebesgue space, 438
generalized Lorentz–Zygmund space, 313
Gould space, 294
gradient, 413

Hölder condition, 40
Hölder continuous function, 40
Haar system, 97, 164
Hahn–Banach theorem, 20
Hahn–Saks theorem, 37
Halperin level function, 402
Hardy inequality, 88
Hardy lemma, 251
Hardy–Littlewood theorem, 242
Hardy–Littlewood–Pólya
   principle, 256
Hardy–Littlewood–Pólya
   relation, 251
Hilbert space, 12
Hölder inequality
   for Banach function spaces, 211
   for Lebesgue spaces, 64
   for r.i. norms, 254
   for r.i. spaces, 255
   for three functions, 64

ideal, 220
identity operator, 18
index
   Lebesgue conjugate, 62
inequalities
   Clarkson, 65
inequality
   Cauchy–Schwarz, 11
   Hardy, 88
   Hölder

Index

for Banach function spaces, 211
for Lebesgue spaces, 64
for r.i. norms, 254
for r.i. spaces, 255
for three functions, 64
generalized, 66
Jensen, 114
Minkowski, 65
  integral, 83
  reverse, 66
triangle, 6
Young, 62, 116
inner product, 10
inner product space, 10
inner regular measure, 26
integrable function, 28
integral
  convergent, 28
  of nonnegative function, 27
  of simple function, 27
integral nonexistent, 28
inverse operator, 16
isometric metric spaces, 18
isometrically isomorphic spaces, 18
isomorphic mapping, 18
isomorphic spaces, 18
isomorphism, 18

Jensen inequality, 114
John–Nirenberg theorem, 198

Landau resonance theorem, 211
lattice, 130
  rearrangement-invariant, 378
lattice property, 204
least concave majorant, 268
Lebesgue dominated convergence
theorem, 29, 30
Lebesgue measurable set, 30
Lebesgue measure, 30
Lebesgue measure space, 30
Lebesgue outer measure, 30
Lebesgue point, 177
Lebesgue set, 177
lemma
  Fatou, 28, 31
length of multiindex, 38
Levi monotone convergence
theorem, 30
limit, 11
linear functional, 19
linear hull, 133
linear operator, 16
linear subset, 13
Lipschitz condition, 40
Lipschitz continuous function, 40
Lipschitz function, 40
locally uniform convergence, 34
Lorentz endpoint space, 269, 270
Lorentz space, 301
Lorentz–Karamata space, 374
Lorentz–Luxemburg theorem, 212
Lorentz–Zygmund space, 372
Luxemburg norm, 8, 134
Luxemburg representation theorem, 257
Luzin theorem, 33

majorant
  least concave, 268
mapping, 16
  continuous, 4
  isomorphic, 18
Marcinkiewicz endpoint space, 264, 270
Marcinkiewicz space, 102, 173
mean continuity, 67, 161
measurable function, 28
measurable set, 24, 25
measurable space, 25
measure, 22
  absolutely continuous, 33
  arithmetic, 25
  continuous, 25
  counting, 25
  Lebesgue, 30
  outer, 23
    Lebesgue, 30
  regular, 26
    inner, 26
    outer, 25
  separable, 232
measure space, 25
  complete, 25
  completely atomic, 25
  discrete, 25
  friendly, 278
  nonatomic, 25

measure-preserving transformation, 313
metric, 6
metric space
　compact, 15
　complete, 12
　convergence in, 11
　relatively compact, 15
　separable, 12
m_φ, 265
Minkowski functional, 8
Minkowski inequality, 65
Minkowski integral inequality, 83
modular, 7
　continuous, 7
modular convergence, 9, 138
modular space, 7, 437
mollifier, 69, 156
monotone convergence theorem, 27
monotone norm, 130
monotone operator, 45
monotone set function, 22
Morrey space, 176
multiindex, 38
　length of, 38
multiplication, 1

neighborhood, 3
N-function, 114
nonatomic measure space, 25
nonexistent integral, 28
nonincreasing rearrangement, 238
nonnegative part of a function, 27
nonpositive part of a function, 27
norm, 6
　absolutely continuous, 149, 216
　　uniformly, 150
　associate, 209
　Banach function, 204
　Luxemburg, 8, 134
　monotone, 130
　of a linear functional, 20
　of operator, 17
　Orlicz, 126
　rearrangement-invariant, 253
norm boundedness, 142
norm-fundamental subspace, 215
normed linear space, 6

norms
　equivalent, 19
null set, 25

open ball, 6, 7
open set, 3, 6, 7
operator, 16
　bounded, 17
　compact, 17
　completely continuous, 17
　continuous, 16
　embedding, 18
　Hardy
　　average, 414
　identity, 18
　inverse, 16
　linear, 16
　monotone, 45
operator norm, 17
Orlicz class, 108
Orlicz norm, 126
Orlicz space, 126
outer measure, 23
outer regular measure, 25

point
　Lebesgue, 177
pointwise convergence, 34
polynomial
　Bernstein, 47
precompact, 15
principle
　Hardy–Littlewood–Pólya, 256
product, 2
　inner, 10
product of spaces, 14
property
　Daugavet, 455
　Fatou, 204
　lattice, 204
　Radon–Nikodým, 453
　Riesz–Fischer, 13
pseudonorm, 6

quasi-concave function, 263
quasinorm, 7

Radon–Nikodým property, 453
Radon–Nikodým theorem, 34

Index

range, 16
range of a measure, 244
rearrangement
    nonincreasing, 238
rearrangement-invariant lattice, 378
rearrangement-invariant norm, 253
rearrangement-invariant quasi-Banach
        space, 339
reflexive space, 21
regular measure, 26
relation
    Hardy–Littlewood–Pólya, 251
relatively compact metric space, 15
representation space, 257
resonant space, 243
reverse Minkowski inequality, 66
Riesz compactness theorem, 94
Riesz representation theorem, 75
Riesz theorem, 36
Riesz–Fischer property, 13
Riesz–Fischer theorem, 13

scalar, 1
Schauder basis, 14
second associate space, 211
semimodular, 7
semimodular space, 7
seminorm, 6
separable measure, 232
separable metric space, 12
separating system of seminorms, 36
sequence
    Cauchy, 11
    convergent, 11
sequence tending to empty set, 216
series
    convergent, 11
set
    (k, λ)-equicontinuous, 54
    balanced, 36
    Borel, 24
    closed, 3, 7
    dense, 12
    equicontinuous, 51
    Lebesgue, 177
    measurable, 24, 25
      Lebesgue, 30
    null, 25

    open, 3, 6, 7
    totally bounded, 15
    weakly bounded, 229
set function, 22
    σ-finite, 25
    countably additive, 22
    countably subadditive, 22
    finite, 25
    finitely additive, 22
    monotone, 22
Sharpley theorem, 372
simple function, 26, 204
slowly varying function, 373
space
    JN, 188
    BMO, 188
    associate, 210
      second, 211
    Banach, 12
    Banach function, 204
    Campanato, 178
    classical Lorentz, 375
      of type Γ, 376
      of type Λ, 375
    Daugavet, 455
    endpoint
      Lorentz, 269, 270
      Marcinkiewicz, 264, 270
    generalized Lorentz–Zygmund, 313
    Gould, 294
    Hilbert, 12
    inner product, 10
    Lebesgue
      generalized, 438
    Lorentz, 301
    Lorentz–Karamata, 374
    Lorentz–Zygmund, 372
    Marcinkiewicz, 102, 173
    measurable, 25
    measure, 25
      complete, 25
      friendly, 278
      Lebesgue, 30
    metric
      compact, 15
      complete, 12
      relatively compact, 15

separable, 12
modular, 7, 437
Morrey, 176
normed linear, 6
Orlicz, 126
rearrangement-invariant quasi-Banach, 339
reflexive, 21
representation, 257
resonant, 243
second associate, 211
semimodular, 7
strongly resonant, 243
topological, 3
uniformly convex, 105, 450
unitary, 10
vector, 1
weak L^∞, 412
weak Lebesgue, 102
spaces
  Banach
    equivalent, 268
    isometrically isomorphic, 18
    isomorphic, 18
  metric
    isometric, 18
  product of, 14
Steklov function, 156
strong convergence, 11
stronger topology, 4
strongly resonant space, 243
subset
  linear, 13
subspace, 13
  norm-fundamental, 215
support
  of a function, 26
system
  Haar, 97, 164
  of seminorms
    separating, 36

theorem
  Banach, 17
  Banach–Alaoglu, 20
  Banach–Steinhaus, 17
    for weak convergence, 20
  Egorov, 35

  Hahn–Banach, 20
  Hahn–Saks, 37
  Hardy–Littlewood, 242
  John–Nirenberg, 198
  Landau resonance, 211
  Lebesgue dominated convergence, 29, 30
  Levi monotone convergence, 30
  Lorentz–Luxemburg, 212
  Luxemburg representation, 257
  Luzin, 33
  monotone convergence, 27
  Radon–Nikodým, 34
  Riesz, 36
  Riesz compactness, 94
  Riesz representation, 75
  Riesz–Fischer, 13
  de la Vallée-Poussin, 151
  Vitali, 31
  Vitali–Hahn–Saks, 31
topological space, 3
topology, 3
  generated by a base, 3
  of pointwise convergence, 5
  of uniform convergence, 5
  stronger, 4
  weak, 229
  weaker, 4
totally bounded set, 15
triangle inequality, 6

unconditional Schauder basis, 15
uniform boundedness principle, 21
uniform convergence, 5, 34, 43
uniform up to small sets convergence, 34
uniformly absolutely continuous norm, 150
uniformly convex space, 105, 450
unitary space, 10

de la Vallée-Poussin theorem, 151
vector space, 1
Vitali theorem, 31
Vitali–Hahn–Saks theorem, 31

weak L^∞ space, 412
weak convergence, 20
weak Lebesgue space, 102

Index

weak topology, 229
weaker topology, 4
weakly bounded set, 229
weight, 85

Young function, 109
Young inequality, 62, 116

Zygmund class, 315